BUILDING CONSTRUCTION

The Sydney Opera House. Australian Information Service photograph.

BUILDING CONSTRUCTION
materials and types of construction

Fourth Edition

WHITNEY CLARK HUNTINGTON

ROBERT E. MICKADEIT
Allan Hancock College

JOHN WILEY & SONS

New York   London   Sydney   Toronto

Library of Congress Cataloging in Publication Data

Huntington, Whitney Clark, 1887-
  Building construction.

  Includes bibliographical references.
  1. Building. 2. Building materials. I. Mickadeit,
Robert E., 1923-   joint author. II. Title.

TH145.H8 1974   691   74-2474
ISBN 0-471-42215-0

Printed in the United States of America

10 9 8 7 6 5 4 3 2 1

PREFACE

This edition, like its predecessors, is a general survey covering the materials and methods of building construction. It is intended for students of architecture, civil engineering, building inspection, building technology, and for those active in the field who are reviewing for state registration examinations.

Several new developments in foundation construction are covered here. New material has been added on wood, concrete, steel, and plastering and articles have been included on concrete formwork and hardware. As with the third edition, an effort has been made to employ terminology which enjoys the widest usage; however, the reader is reminded that building terms are sometimes provincial. Certainly, neither all materials nor all methods of construction can be covered adequately in an elementary text of this size. A general contemporary updating has been made to an existing text that is authoritative, sound, informative and readable. Topics retained and expanded are those which the revision author has found to be the most useful during his employment in the field of construction. Surely, thoughtful readers will find some aspect which deserves broader and deeper exposure; however in the interests of editorial economy certain topics have been abbreviated. The reader will find ample material for further study in a list of recommended reading.

As in the third edition, numbered references have been included. Likewise, special acknowledgment is made to *Engineering News-Record* magazine, *Civil Engineering Magazine*, and especially to the publications of the American Concrete Institute, and the American Society for Testing and Materials. The student will find further reading in all of these publications informative and timely. Several manufacturers and manufacturers' associations have provided technical data and photographs. Appropriate acknowledgments are made where referred to in the text.

Once again we have included the timely and significant articles prepared by Milo S. Ketchum and W. S. Kinne, Jr. which appeared in the third edition.

Over the years of my employment in the field, hundreds of colleagues and several employers have unselfishly shared their experience and knowledge. It is nearly impossible and certainly impractical to credit individually all of those who have indirectly made their contributions to these pages. I gratefully thank them. Among those who have been particularly helpful in furnishing information for this edition are L. A. Fraikin of the Franki Foundation Company; C. W. Grimshaw of the Calweld division of Smith International, Inc.; C. W. Hoke of Mobile Drilling Company, Inc.; George F. Leyh of Concrete Reinforcing Steel Institute; A. G. MacKinnon of the Foundation Equipment Corp., and librarians Robert Johnson, James Russell, and Edward Wiseblood.

It was more than twenty-five years ago that I studied the second edition of this text in Professor Cowgill's classes at Virginia Polytechnic Institute; thus my debt to the late Whitney Clark Huntington is indeed longstanding and my gratitude sincere.

Lompoc, California   *Robert E. Mickadeit*

CONTENTS

1

INTRODUCTION

2

FOUNDATIONS

3

THE STRUCTURAL ELEMENTS

14
WINDOWS AND CURTAIN WALLS

15
DOORS AND DOOR HARDWARE

16
MISCELLANEOUS METALS

17
PLASTICS

18
INSULATING MATERIALS

BUILDING CONSTRUCTION

1

INTRODUCTION

1. BUILDING CODES

The purpose of building codes is to regulate the design and construction of buildings in a manner that safeguards the general public, although sometimes codes are viewed as inhibiting creativity.

Building codes usually include types of construction, function of the structure called *occupancy*, quality of materials, the imposed loads, allowable stresses, mechanical and electrical equipment, and other requirements related to buildings with special emphasis on fire safety.

Basically, the codes issued by various agencies are quite similar. There are, however, significant differences; therefore, the code adopted by a given municipality must be satisfied by buildings constructed within its jurisdiction. The requirements of various codes may be included in this text for instructional purposes, but for a specific application the governing code should be consulted.

Authority for Building Codes.

According to reference [2]:

"The building code derives its justification from the police power. This is the inherent power of government to protect the people against harmful acts of individuals insofar as matters of safety, health, morals or the like are concerned. It is the power forming the basis for State acts and municipal ordinances dealing with these matters and is of indefinite extent, although certain limitations concerning its use are to be found in the Federal and State constitutions and in court decisions. Fundamentally, under our system of government, the police power resides in the State and may be transmitted to local authorities through enabling acts authorizing the adoption of building requirements or may be conferred upon municipalities when a charter is granted."

If a code requirement is unnecessary to accomplish the protection referred to in the preceding quotation, judicial support is uncertain.

Model Building Codes.   Most municipalities have building codes. Jurisdictions that do not originate their own building codes often adopt, by reference, all or parts of certain established model codes.

Today there is a tendency in model codes toward the *performance code*. This type of code requires a specified result. The advantage of this approach, as opposed to the rigid specification type code, is that newer materials and methods of construction may be permitted.

Many states have established building codes which may control all building construction within the state; however, individual communities may adopt their own codes. An example is the State Building Construction Code adopted by the State of New York. As stated therein: "The municipalities of the state have the option to accept or not to accept the applicability" of this code.

Several regional agencies have formulated *model building codes*. These have been prepared to assist municipalities in preparing codes, or a municipality may adopt such a code, by reference, if authorized by the statutes of the governing state. The following are examples of model codes.

The International Conference of Building Officials, Whittier, California, publishes the *Uniform Building Code*. This model code has been widely adopted by various communities—nearly 2,000 municipalities in 44 states. This code, known as the *UBC*, is kept current by constant review, with a new edition published every three years. Additionally, this organization publishes a volume of building standards that defines various building tests and material qualities.

The Basic Building Code is published by Building Officials and Code Administrators International, Chicago, Illinois. This model code, a performance code, is continually updated by annual supplements and revisions, as required. BOCA publishes several other model codes such as *The Basic Fire Prevention Code* and *The Basic Housing Code*.

The Southern Standard Building Code is published by the Southern Building Code Congress, Birmingham, Alabama.

The National Building Code is recommended by the American Insurance Association, successor to the National Board of Fire Underwriters. This code contains appendices covering hurricanes and earthquakes. The American Insurance Association also recommends the *Fire Prevention Code* as a companion document.

Other Codes and Standards.   There are a number of organizations that advance the knowledge of materials and methods of construction and promote codes and standards. Some of these organizations are concerned with several materials, while others are involved with only one material and its applications. These organizations produce literature that is of interest to a technical audience—architects, engineers, students, and inspectors. A partial description of the activities and contributions of some of these organizations follows.

Among authoritative organizations is the American Society for Testing and Materials of Philadelphia. The work of this society includes the development of standards, test procedures, criteria, and specifications and definitions for a vast number of materials. ASTM standards cover most building materials, including soils and rock. ASTM standards are frequently cited in building specifications as the basis upon which quality decisions will be judged. Their standards are reviewed (if necessary, revised) and published annually.

The Construction Specifications Institute (CSI), Washington, D.C., promotes the standardized specifications that cover the field of building construction. Their format for construction specifications and materials classifications has been widely adopted in the industry. Its literature covers not only specifications but includes commentary on the uses of materials.

The American Concrete Institute (ACI) of Detroit, Michigan, is an organization that disseminates information for the improvement of design, construction, manufacture, use, and maintenance of concrete products and structures. Its interests encompass all aspects of concrete production. The ACI *Building Code Requirements for Reinforced Concrete* (ACI 318), which is revised annually, serves as a building industry standard.

The American Institute of Steel Construction, New York, advances the use of fabricated steel framed structures. It performs research and development and serves to further the use of steel by technical publications, seminars, awards, scholarships, and the promotion of quality and safety in shop and field. Its *Manual of Steel Construction* is used as the industry standard for the design of structural steel. Also, the *AISC Specification for the Design, Fabrication and Erection of Structural Steel for Buildings* is usually referenced by engineers in building specifications and is adopted by reference in building codes.

The American Iron and Steel Institute, New York, has established standards and produces technical information on steel and the industry. This includes information on uses, processes, and standard designations.

The Copper Development Association, Inc., New York, provides technical information regarding copper, bronze, and brass for building use.

The Aluminum Association, New York, provides standards and technical information that covers structural, mechanical, physical properties, uses, specifications, coatings, finishes, and building products of aluminum.

The National Fire Protection Association, Boston, Massachusetts, publishes the *Fire Protection Handbook*, which provides not only essential information on fire prevention and protection but details the reasons and gives background information. Additionally, NFPA publishes the standards known as the *National Fire Codes* in seven volumes averaging 900 pages each. Most of these standards are published separately in pamphlet form. Other publications of NFPA include an inspection manual and books covering particular occupancies.

Several regional lumber associations encourage regeneration of the forests, establish standards and sizes for lumber, and supply technical data to specifiers, builders, and users.

Among these associations are:

- The Western Wood Products Association, Portland, Oregon, is the largest of these associations. It is an association of softwood lumber manufacturers of the 12 Western states grading pines, firs, spruces, and cedars.
- West Coast Lumber Inspection Bureau, Portland, Oregon, grades and inspects Douglas fir, hemlock, cedar, and other softwood lumber of California and the regions of Washington and Oregon west of the Cascade Summit.
- Redwood Inspection Service, San Francisco, California, grades and inspects products of *Sequoia sempervirens*.
- Southern Pine Inspection Bureau, Pensacola, Florida, covers longleaf pine, slash pine, shortleaf pine, and loblolly pine produced in the southeastern states from Maryland to Texas.
- Northeastern Lumber Manufacturers Association, Inc., Glens Falls, New York, grades northern and eastern pines, cedars, spruces, balsam fir, and other softwoods of New England, New York, New Jersey, and Pennsylvania.
- National Hardwood Lumber Association, Chicago, Illinois, performs services for the hardwood products industry parallel to those performed by the softwood associations. Among the myriad hardwoods that are its concern are walnut, maple, oak, birch, ash, hickory, and cypress, a softwood.
- The American Institute of Timber Construction (AITC), of Engle-

wood, Colorado, tests and disseminates information on the structural aspects of timber construction, including laminated timber.
• The American Plywood Association of Tacoma, Washington, produces literature on plywood. This includes, but is not limited to, testing, grades, systems, specifications, physical qualities, and the structural and finished uses of plywood.

Various agencies of the United States government, especially the Bureau of Standards, Washington, D. C., and the Forest Products Laboratory, Madison, Wisconsin, resemble the above associations.

Extensive lists of material and construction standards and their issuing agencies are included in reference 3. Much information can be obtained, often at little or no cost, by writing these agencies. Publications by agencies of the federal government are obtained by writing the U. S. Government Printing Office, Washington, D. C.

Fire-Resistance Ratings. Building codes commonly classify buildings according to type of construction and according to *use* or *occupancy*. The most important factor in the classification according to type of construction is the resistance to fire exposure. In classifying buildings according to fire endurance, it is necessary to measure the performance of the various structural parts of a building under exposure conditions. It is necessary that the fire-resistive properties of materials and members be measured and specified according to a common standard expressed in terms which are applicable to a wide variety of materials, situations, and conditions of exposure.

The *"Standard Fire Test,"* known also as the *fire endurance test*, accomplishes these objectives. This test is designated by the Underwriters' Laboratories as UL 263 and by the American Society for Testing and Materials as ASTM E119. It consists of exposing samples of the material, a building member, or an assembly to a fire of specified intensity and duration. In some cases, a $2\frac{1}{2}$ in. fire hose stream is applied to a heated sample from a distance of 20 ft. Performance is defined as the period of resistance to standard exposure elapsing before the first critical point in behavior is observed, and it is expressed in hours. For example, a material is given a 2-hour rating if it withstands the test for a period of 2 hours.

Most of the common building materials and assemblies of materials have been tested, rated, and the results published by the Underwriters Laboratory. For this reason, it is not usual to make fire tests in connection with the design of individual buildings. An extensive report on such ratings is given in reference [4].

Fire endurance tests on a bearing wall assembly are required for a specified time in hours. Bearing walls are required to support their loads without penetration by flame or gases sufficient to ignite cotton wastes. The hose stream test (for assemblies which are to be rated for one hour or more) is conducted on the loaded assembly in a heated condition. The assembly, in a cooled state, must also support specified loads. In no event may heat that penetrates through the wall assembly produce a temperature rise of more than 250°F on the unexposed surface.

The fire endurance tests for nonbearing walls are similar to those conducted on bearing walls. Naturally, the nonbearing wall assembly is not required to carry imposed loads during the fire endurance test.

Floor and roof assemblies carry imposed loads during the fire endurance test. Acceptance as a fire rated assembly requires that the fire and temperature transmission requirements be similar to the forementioned bearing wall assembly. The hose test is not required.

The fire protective material that surrounds beams and columns may also be tested for endurance. The protective assembly must limit the rise in temperature in the steel member to not more than a specified amount.

Another fire test is the Test Method for Fire Hazard Classification of Building Materials, which is known as the Underwriters' Laboratories UL 723. This tests the *flame spread* ability of a material. The test determines "the comparative burning characteristics of the material under test by evaluating the flame spread over its surface, fuel contributed by its combustion, and the density of the smoke developed when exposed to a test fire." [22] The *flame spread classification* (FSC) of a material is graded on a scale from 0 to 100. As a basis, the classification of 0 is assigned to asbestos cement board and a classification of 100 is designated to select grade red oak flooring.

Flame spread ratings are usually specified for materials which are interior surface finishes, such as plastics, plywood, acoustic materials, and wall coverings.

Building Content and Fire Severity.   Burnout tests conducted in fire-resistive buildings indicate that the fire severity due to combustion of such materials as wood, paper, cotton, wool, silk, straw, grain, sugar, and similar organic materials may be considered to be as shown in Table 1-1. The values in this table enable fire severity to be visualized [11]. Fires in "fire-resistant" multi-floor buildings have resulted in loss of life and property. These fires, in buildings which are intended to be fire resistant, nevertheless contained furnishings which

TABLE 1-1
Relation of Amount of Combustibles and Fire Severity

w	hr	w	hr	w	hr	w	hr	w	hr
5	½	10	1	20	2	40	4½	60	7½
7½	¾	15	1½	30	3	50	6		

Key: w is average weight in lb. per square foot of floor area.
hr is fire severity in hours.

supported fire and resulted in smoke damage and death due to inhalation.

Fire Limits, Zones, or Districts.   The New York State Code includes the following definition: *"Fire limits.* Boundary line establishing an area in which there exists, or is likely to exist, a fire hazard requiring special fire protection."

Two classes of fire limits are included in this code.

"Fire limits A comprising the areas containing highly congested business, commercial and, or industrial occupancies, wherein the fire hazard is severe; and or, Fire limits B comprising the areas containing residential, business and, or commercial occupancies or in which such uses are developing, wherein the fire hazard is moderate.

"All of those areas not included in fire limits A or B are designated herein as outside the fire limits."

The areas within fire limits are often called *fire zones* or *fire districts.*

Building Classification

Building code requirements are based primarily on building occupancy or use, type of construction, and location.

Classification According to Occupancy.   The classification of buildings according to occupancy, as included in the National Building Code, is as follows.

Assembly occupancy means the occupancy or use of a building or structure or any portion thereof by a gathering of persons for civic, political, travel, religious, social, or recreational purposes.

Business occupancy means the occupancy or use of a building or structure or any portion thereof for the transaction of business, or the rendering or receiving of professional services.

Educational occupancy means the occupancy or use of a building or structure or any portion thereof by persons assembled for the purpose of learning or of receiving educational instruction.

High hazard occupancy means the occupancy or use of a building or structure or any portion thereof that involves highly combustible, highly flammable, or explosive material, or which has inherent characteristics that constitute a special fire hazard.

Industrial occupancy means the occupancy or use of a building or structure or any portion thereof for assembling, fabricating, finishing, manufacturing, packaging, or processing operations.

Institutional occupancy means the occupancy or use of a building or structure or any portion thereof by persons harbored or detained to receive medical, charitable, or other care or treatment, or by persons involuntarily detained.

Mercantile occupancy means the occupancy or use of a building or structure or any portion thereof for the display, selling or buying of goods, wares, or merchandise; except when classed as a high hazard occupancy.

Residential occupancy means the occupancy or use of a building or structure or any portion thereof by persons for whom sleeping accommodations are provided but who are not harbored or detained to receive medical, charitable, or other care or treatment, or are not involuntarily detained.

Storage occupancy means the occupancy or use of a building or structure or any portion thereof for the storage of goods, wares, merchandise, raw materials, agricultural or manufactured products, including parking garages, or the sheltering of livestock and other animals, except when classed as a high hazard.

Classification According to Type of Construction. Buildings are classified in building codes according to types of construction based on the fire resistance of their structural members or assemblies. Codes vary in the details of such classifications, but the objectives sought are similar. The *New York State Building Construction Code* will serve as an example.

According to this code, if the temperature required to ignite and support combustion of a material, or combination of materials, is below 1382°F., the material is designated as *combustible*, and if a higher temperature is required, it is designated as *noncombustible*.

The principal combustible materials used in buildings are wood, organic fiber boards, and plastics. The principal noncombustible materials are steel, aluminum, concrete, and masonry materials such as brick, stone, and structural clay tile, as well as plaster and glass.

According to the New York State Code, buildings are classified into five types as follows.

"*Type 1. Fire-Resistive Construction.* That type of construction in which the walls, partitions, columns, floors, and roof are noncombustible with sufficient fire resistance to withstand the effects of a fire and prevent its spread from story to story.

Type 2. Noncombustible Construction. That type of construction in which walls, partitions, columns, floors, and roof are noncombustible and have less fire resistance than required for *fire-resistive construction.*

Type 3. Heavy Timber Construction. That type of construction in which the exterior walls are of masonry or other noncombustible materials having an equivalent structural stability under fire conditions and a fire resistance rating of not less than 2 hours; in which interior structural members including columns, beams and girders are timber, in heavy solid or laminated masses, but with no sharp corners or projections or concealed or inaccessible spaces; in which floors and roofs are of heavy plank or laminated wood construction, or any other material providing equivalent fire-resistance and structural properties. Noncombustible structural members may be used in lieu of heavy timber, provided the fire resistance rating of such members is not less than 3/4 hour.

Type 4. Ordinary Construction. That type of construction in which the exterior walls are of masonry or other noncombustible materials having equivalent structural stability under fire conditions and a fire resistance rating of not less than 2 hours, the interior structural members being wholly or partly of wood of smaller dimensions than those required for *Heavy Timber Construction.*

Type 5. Wood Frame Construction. That type of construction in which walls, partitions, floors, and roof are wholly or partly of wood or other combustible material."

Each of the five types of construction which have been described, except *Type 3*, is divided into two subtypes which vary according to the degree of fire resistance required. The requirements for subtypes *a* are more severe than those for subtypes *b*.

The fire-resistance ratings required for the various types and subtypes are shown in Table 1-2.

From Table 1-2, it will be noted that the highest required fire rating is 4 hours. This rating is required in the higher type of *Fire-Resistive Construction* for exterior bearing walls, for party walls, and for interior firewalls, bearing walls, and partitions. It is also required for columns, beams, girders, and trusses supporting more than one floor. These are the most important structural elements in a building. It is also required for the party walls and fire walls of *Heavy Timber Construction.*

The lowest fire rating required is 3/4 hour. It is required for panel and curtain walls in *Fire-Resistive Construction* and in the higher type of *Noncombustible Construction*, and interior partitions and

TABLE 1-2

Minimum Fire-Resistance Requirements of Structural Elements (By types of construction; fire-resistance ratings in hours)

Structural Element	Construction Classification								
	Type 1 (fire-resistive)		Type 2 (non-combustible)		Type 3 (heavy timber)	Type 4 (ordinary)		Type 5 (wood frame)	
	1a	1b	2a	2b		4a	4b	5a	5b
Exterior									
Bearing walls	4	3	2	nc	2	2	2	3/4	c
Nonbearing walls	2	2	2	nc	2	2	2	3/4	c
Panel and curtain walls	3/4	3/4	3/4	nc					
Party walls	4	3	2	2	4	2	2	2	2
Interior									
Fire walls	4	3	2	2	4	2	2	2	2
Bearing walls or partitions	4	3	2	nc	2	3/4	c	3/4	c
Partitions enclosing stairways, hoistways, shafts, other vertical openings, and hallways									
on outside exposure	2	2	2	2	2	2	2	3/4	3/4
on inside exposure	1	1	3/4	3/4	3/4	3/4	3/4	3/4	3/4
Nonbearing walls and partitions separating spaces	1	1	3/4	3/4	3/4	3/4	3/4	3/4	3/4
Columns, beams, girders, and trusses (other than roof trusses)									
supporting more than 1 floor	4	3	2	nc	c	3/4	c	3/4	c
supporting 1 floor	3	2	3/4	nc	c	3/4	c	3/4	c
Floor construction, including beams	3	2	1	nc	c	3/4	c	3/4	c
Roof construction, including purlins, beams, and roof trusses	2	1	3/4	nc	c	3/4	c	3/4	c

Key: nc=noncombustible, c=combustible.
The code includes special requirements and exceptions which are not included in this table.

nonbearing walls for all types of construction except *Fire-Resistive*. It is also required for bearing walls and nonbearing walls for the higher type of *Wood Frame Construction* and for various other members in *Ordinary and Wood Frame Construction*. *Noncombustible Construction* without a fire rating is required for various members in the lower type of *Noncombustible Construction*. *Combustible Construction* is permitted in some parts of *Heavy Timber Construction* and in several parts of *Ordinary and Wood Frame Construction*.

The highest type of construction, designated as *Fire-Resistive* in Table 1-2, is often designated as *fireproof*. However, it is not feasible to construct a building which is really fireproof, and therefore that term is being replaced in building codes by the term *fire-resistive*. The term *mill construction*, which often is used instead of *Heavy Timber Construction*, had its origin in New England where construction using heavy timber was developed many years ago to decrease the fire hazard in textile mills. This type of construction is also called *slow-burning construction* for obvious reasons. These terms are occasionally used in building codes. The types of construction known as *Heavy Timber Construction, Ordinary Construction*, and *Frame Construction* are used in nearly all codes. A classification not given here but often found in codes includes buildings with unprotected steel structural members. It is considered to have a fire resistance lower than *Ordinary Construction* because the strength of steel decreases at high temperatures. Such a classification is designated as *unprotected metal, metal frame*, or *light incombustible frame*.

Classification According to Fire Hazard. All codes recognize the differences in fire hazard as determined by the kind of occupancy within each general occupancy class. For example, the New York State Building Construction Code includes the following classifications according to fire hazard.

Low hazard.   Business buildings.

Moderate hazard.   Mercantile buildings. Industrial and storage buildings in which combustible contents might cause fires of moderate intensity as defined in the code.

High hazard.   Industrial and storage buildings in which the combustible contents might cause fires to be unusually intense, as defined in the code, or where explosives, combustible gases, or flammable liquids are manufactured or stored.

Height and Area. The height and fire area of a single story are defined in each code. For example, the following definitions in the New York State Code are typical.

"Building Height. Vertical distance measured from the curb or grade level to a flat or mansard roof, or the average height of a pitched, gabled, hip, or gambrel roof, excluding bulkheads, penthouses and similar constructions enclosing equipment or stairs, providing that they are less than 12 feet in height and do not occupy more than 30 per cent of the area of the roof on which they are located.

Fire Area. The floor area of a story of a building within exterior walls, party walls, fire walls, or any combination thereof."

The permissible height and fire area of a building are determined by its occupancy, construction, and fire hazard classifications, the fire protection equipment installed in the building such as sprinkler systems, and the accessibility for fire protection equipment.

The permissible heights and fire areas included in the New York State Code for Low-Hazard buildings are illustrated in Table 1-3. This code also includes tables for other classes of occupancy and moderate and high hazards. Other codes have similar requirements stated in various ways.

The fire areas in this table are based on a frontage on one street or legal open space at least 50 ft. wide. If a fire area faces or abuts such streets or spaces on two sides, it may be 50% larger than the area shown in the table; on three sides, 75% larger; and on four sides, 100% larger, provided that such open areas are served by fire hydrants and are unobstructed and accessible at all times for fire-

TABLE 1-3

Permissible Height and Fire Areas, Low-Hazard Business, Industrial, and Storage Buildings

Maximum Height		Basic Fire Area in 1000 sq. ft.								
		Type 1 (fire-resistive)		Type 2 (Noncom-bustible)		Type 3 (heavy timber)	Type 4 (ordi-nary)		Type 5 (wood frame)	
st.	ft.	1a	1b	2a	2b		4a	4b	5a	5b
1	un	un	un	un	18	21	18	12	9	6
2	40	un	un	21	15	18	15	9	6	3
3	55	un	un	18	np	15	12	6	np	np
4	70	un	un	15	np	12	9	np	np	np
5	85	un	un	12	np	np	np	np	np	np
6	100	un	un	np	np	np	np	np	np	np
Over 6	Over 100	un	un	np	np	np	np	np	np	np

Key: st = stories, un = unlimited, np = not permitted.

fighting equipment. Specified increases in height and fire area are permitted if approved automatic sprinklers are installed.

Restrictions Based on Fire Districts. Most building codes prohibit *Wood Frame Construction* within fire districts with specified exceptions such as small private garages, greenhouses, and sheds. If buildings with such construction are located outside the fire limits, there are fire restrictions concerning the distance from property lines, and the fire area and the height, as illustrated in Table 1-3 for Type 5, and there are restrictions on kinds of roofing materials.

Other types of construction are permitted within the fire limits subject to certain requirements about the fire ratings of their various parts (illustrated in Table 1-2), the fire areas and heights (illustrated in Table 1-3), and other regulations mentioned in the preceding paragraph.

Other Regulations. Codes include requirements such as those for exits, stairways, elevator shafts, corridors, roofs, protection of openings in exterior walls, interior finish, flooring materials, doors and doorways, and building contents.

Many references are made elsewhere in this treatise to code requirements for various parts of buildings. The code requirements for the loads carried by buildings are considered in Article 2. Codes include allowable working stresses for the many materials included in a building. They are of great importance in structural design but do not fall within the scope of this book. The allowable bearing pressures for foundations, illustrated by Table 2-4, are included in codes.

Codes also include requirements for ventilation, sanitary features such as plumbing and wastes disposal, electric wiring, and mechanical equipment, which do not fall within the scope of this treatise.

2. BUILDING LOADS

Definitions. The loads to which buildings are subjected may be divided into four classes: dead loads, live loads, lateral loads, and other loads.

Dead loads include weights of all parts of the building such as walls, structural frame, permanent partitions, floors, roofs, stairways, and fixed equipment.

Live loads include the weights of the occupants, furniture, movable equipment, stored material, and snow that may accumulate on the roof.

Lateral loads include distributed horizontal loads which act above the ground level and are assumed to have effects equivalent to those produced by wind pressures. For some regions they include lateral loads which are assumed to have effects corresponding to those produced by earthquake shocks. Also included are horizontal loads on foundation walls caused by earth pressure and water pressure.

Dead Loads.   The magnitudes of dead loads to be used in design are determined by the weights of the building materials, the assemblies of materials, and the fixed equipment which are used in the construction of buildings.

Live Loads.   The magnitudes of the live loads on floors are determined by the types of occupancy of the various units into which a building is divided. There is a marked uniformity in building code requirements for the minimum live loads for which the floors of buildings are to be designed. These loads refer to normal conditions. Any special conditions to which a floor may be subjected must be considered. The floor loads recommended by the American Standards Association are given in Table 1-4 [8]. It will be noted that the floor

TABLE 1-4
Minimum Uniformly Distributed Live Loads, American Standards Association, 1955 [8]

Assembly halls, fixed seats	60	Residential. Dwellings	
Movable seats	100	First floor	40
Corridors, first floor	100	Second floor	30
Other floors, same as occu-		Habitable attics	30
pancy served except as		Uninhabitable attics	20
indicated		Residential. Hotels	
Dance halls	100	Guest rooms	40
Dining rooms and restaurants	100	Public rooms	100
Garages, passenger cars	100	Public corridors	100
Gymnasium floors and		Private corridors	40
balconies	100	Schools. Classrooms	40
Hospitals, operating rooms	60	Corridors	100
Private rooms and wards	40	Stairs, fire escapes, exitways	100
Libraries, Reading rooms	60	Stores. Retail	
Stackrooms	150	First floor	100
Manufacturing	125	Upper floors	75
Office buildings, offices	80	Stores. Wholesale	125
Lobbies	100	Warehouse, light storage	125
Residential. Multifamily		Heavy storage	250
Private apartments	40		
Public rooms	100		
Corridors	60		

Key: Pounds per square foot of floor area.

load for offices included in this table is 80 lb. per sq. ft. Many codes require only 50 lb. per sq. ft. for this occupancy. An additional load of 20 lb. per sq. ft. is often included for masonry partitions not included in the initial plans, if there is a possibility that such partitions may be installed at a later date. Many codes provide for a load of 2000 lb. concentrated on any 2½-ft. sq. area if such a load on this otherwise unloaded area produces stresses greater than the required uniform load. Other special provisions for live loads are included in various codes.

Codes include requirements for minimum live *roof loads*. The simplest requirement is a load of 20 lb. per sq. ft. of horizontal projection in addition to the wind load regardless of the slope of the roof. In most parts of the United States, provision must be made for a *snow load*. It may cause a greater load than the required minimum. Of course, the possible snow load varies widely throughout the country. According to the New York State Code, the required snow load on a flat roof varies from 20 to 60 lb. per sq. ft. according to the building's location on a snow map included in the code.

Another factor in the snow load is the slope of the roof. In the New York State Code, no load is required for slopes of 60 degrees or more with the horizontal. The required load varies for roof slopes between 0 and 60 degrees in a specified manner which is somewhat higher than would be given by a linear variation.

If a roof is also to be used as a floor, it must be designed for the live load it is to carry but for not less than the minimum live load required for floors, the magnitude depending upon the type of use of the roof.

Reductions in Live Floor Loads.   The probability that the required live floor load will not cover the entire floor area contributing to the load on the supporting members is recognized by all codes. For example, the New York State Code permits the following reductions.

"Uniformly distributed live loads on beams or girders supporting other than storage areas and motor vehicle parking areas, when such structural member supports 150 square feet or more of roof area or floor area per floor, may be reduced as follows:

When the dead load is not more than 25 lb. per sq. ft., the reduction shall not be more than 20 per cent;

When the dead load exceeds 25 lb. per sq. ft. and the live load does not exceed 100 lb. per sq. ft. the reduction shall be not more than the least of the following three criteria: 60 per cent or 0.08 per cent per sq. ft. times square feet of area supported; or 100 per cent times (dead load in lb. per sq. ft. plus live load in lb. per sq. ft.) divided by (4.33 times live load in lb. per sq. ft.).

TABLE 1-5

Percentage of Each Live Floor Load To Be Included

Roof	80	5th floor below roof	65
Floor immediately below roof	80	6th floor below roof	60
2nd floor below roof	80	7th floor below roof	55
3rd floor below roof	75	8th floor below roof	50
4th floor below roof	70	All other floors	50

For columns, girders supporting columns, bearing walls, and foundation walls supporting 150 square feet or more of roof area or floor area per floor other than storage areas and motor vehicle parking areas, the uniformly distributed live loads on these members shall not be less than the following percentages of total live loads on the following levels [Table 1-5]."

According to the Modern Standard Code: "For determining the total live loads carried by columns, the following reductions [in Table 1-6] shall be permitted, the reduction being based on the assumed live loads applied to the entire tributary floor area."

It will be noticed that these requirements are on different bases. In the first, the percentage of the total live load on *each* floor and the roof contributing to the column load is given. In the second, the percentage of reductions of the total live floor and roof load on *all* the areas contributing to a *specific* column is given. For a specific column and the same live floor and roof loads, the first requirement would result in larger column loads than the second.

Wind Pressures.   The wind pressures, or wind loads, on buildings vary with the geographical location and, for a given location, with the height above the ground surface. Winds are assumed to act in any

TABLE 1-6

Percentage Reductions in Total Live Floor Loads

Tributary Area	A	B	C	Tributary Area	A	B	C
Roof	0	0	0	4 floors and roof	15	30	30
1 floor	0	0	0	5 floors and roof	20	30	40
2 floors and roof	5	10	10	6 floors and roof	20	30	45
3 floors and roof	10	20	20	7 or more floors and roof	20	30	50

Key: A = Warehouse and Storage, B = Manufacturing, Stores, and Garage, C = All Other.

horizontal direction in spite of the direction of prevailing winds. There are positive pressures on the windward side of a building and negative pressures, or partial vacuums, on the leeward side. Design pressures are based on the total resultant pressure, which is equal to the sum of these pressures.

Extensive studies of wind pressures by the National Bureau of Standards, are reported in reference 7 at the end of this chapter and they form the basis for the Building Code Requirements of the American Standards Association as given in reference 8. The comments in this article are based on these references.

The recommended design pressures are computed from measured velocities of gust winds, and the rectangular shape of exposed building surfaces and many other factors that affect wind pressures are taken into consideration.

The minimum allowable resultant wind pressures for various parts of this country at a height 30 ft. above the ground surface are shown on the map in Figure 1.1.

The design wind pressures for various height zones above the ground surface and various minimum wind pressures for the height of 30 ft. above this surface are given in Table 1-7. For example, if the pressure at the height of 30 ft. as given on the map in Figure 1.1 is 30 lb. per sq. ft., the design wind pressure at a height of 100 ft. is 45 lb. per sq. ft. as shown in the table.

As stated in reference 7,

"The wind pressures and suctions in the immediate path of tornadoes are considered to be so great that construction strong enough to withstand them is not economically feasible. Therefore, in recommending design wind pressures, building codes do not take into con-

TABLE 1-7
Design Wind Pressures for Various Height Zones

Height Zone above Ground Level, ft.	Wind Pressure Map Areas, lb. per sq. ft.						
	20	25	30	35	40	45	50
	Design Wind Pressure, lb. per sq. ft.						
Less than 30	15	20	25	25	30	35	40
30 to 49	20	25	30	35	40	45	50
50 to 99	25	30	40	45	50	55	60
100 to 499	30	40	45	55	60	70	75
500 to 1199	35	45	55	60	70	80	90
1200 and over	40	50	60	70	80	90	100

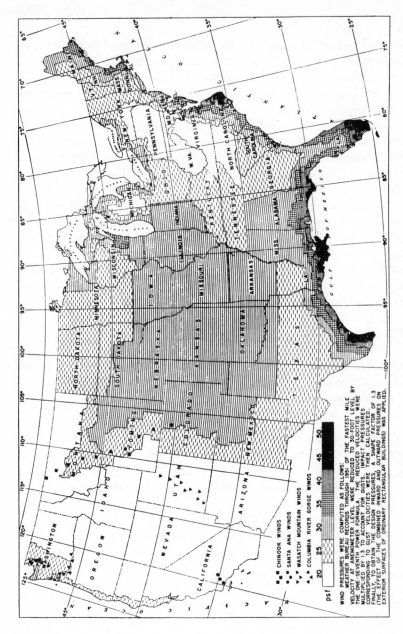

FIGURE 1.1   Minimum allowable resultant wind pressures.

sideration the violent forces that can be expected within the narrow path of a tornado."

Experience has shown, however, that buildings designed for wind pressures even less than those recommended have not collapsed but have survived without irreparable structural damage and relatively few, if any, casualties.

Wind pressures tend to overturn buildings as a whole and parts of buildings above any elevation. To avoid this, codes require that the resistance of the dead load of a building to overturning above any elevation be one and a half times the overturning effect of wind, both factors being measured by their moments about the leeward edges. For the condition which includes the effects of wind loads in the computed stresses in the resisting members, codes permit the allowable stresses to be increased one-third. However, the allowable stresses must not be exceeded for the condition that does not include wind load.

Building codes include factors for adjusting horizontal wind loads to sloping roofs and for the uplift pressures on roofs caused by wind loads.

Earth and Water Pressures. A portion of a foundation wall may be below the ground surface (Figure 1.2(a)). For this condition, the lateral earth pressures may be assumed to equal one-half of the corresponding water pressures.

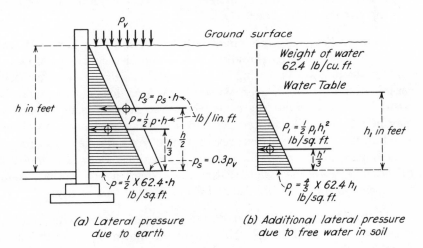

(a) Lateral pressure
due to earth

(b) Additional lateral pressure
due to free water in soil

FIGURE 1.2

If there is a uniform vertical pressure of p_v lb. per sq. ft. on the ground surface, the unit lateral pressure p_s acting on the height h may be assumed to equal one-third of p_v.

If the groundwater level is above the bottom of the wall (b), the additional pressures caused by the water may be assumed to equal four-fifths of the water pressures which would act on the submerged portion of the wall if no soil were present.

The reduction of one-fifth in the water pressure is assumed to offset the reduction in earth pressure on the submerged portion of the wall owing to the buoyant effect of the water on the submerged soil.

Foundation walls retaining expansive clay soils may be subjected to very large earth pressures because of their increase in volume as their moisture content increases, even if they are entirely above the groundwater level. If these are thin walls supporting low buildings, they may fail under the action of such pressures. It may not be feasible to construct a wall thick enough to resist these pressures. Under these conditions, a wall should be backfilled with a nonexpansive soil for a considerable distance from the wall.

Earth pressures against free-standing walls require special consideration, as explained in reference 9.

If the groundwater level is above the bottom of a basement floor and the ground water is in contact with this surface, the floor must be designed to resist an *uplift pressure* equal to the water pressure to which it is subjected. This is equal, in pounds per square foot, to 62.4 times the vertical distance, in feet, between the groundwater level and the bottom of the floor.

Earthquake Forces.   Only small portions of the U.S. have not witnessed at least minor earthquakes in recorded history, the most severe earthquake damage occurring in Alaska and California. The fact that seismic activity has not been evident in an area does not preclude the possibility of future destructive earthquakes, although the probability may be minimal. The procedures and investment necessary to design and build conforming structures that meet earthquake resistant criteria is little compared to the reassurances gained. Perhaps a thoughtful passage from the Brick Institute of America publication, *Reinforced Brick Masonry and Lateral Force Design* will clarify:

"All buildings, whether so designed or not, have some inherent resistance to lateral forces. This has been shown by the survival of some buildings of little or no designed resistance to severe earthquakes, tornadoes, and bombing attacks. The fact of such a survival does not guarantee future survivals, no more than does a soldier coming from a battle unwounded guarantee him to be bullet-proof." [23]

The probability of earthquakes of a destructive severity are determined from geological and historical data. The susceptibility to earthquakes of areas in the U.S. is indicated in Figure 1-3.

Earthquake loads, which are also called *seismic loads*, are treated as lateral loads that involve the entire structure. The seismic load is applicable to all parts of the structure, whereas other lateral loads (wind loads) are applicable to only portions of the structure. It is for this reason that earthquake resistant structures must be designed with connections capable of resisting seismic forces. For example, beams and columns *per se* seldom fail as a result of seismic stress, although the connections to them are more likely to fail. Thus, seismic forces require careful design and analysis of the entire structure, especially its connections. Seismic forces, often thought of as a lateral force, are unpredictable in their application; they may actually be a vertical acting force in severe earthquakes.

Listed below are several definitions of building components relating to earthquake resistant construction. These definitions, from the National Building Code, are similar to terms found in most building codes that have earthquake resistant provisions.

Space frame means a three dimensional structural system composed of interconnected members, other than shear or bearing walls, laterally supported so as to function as a complete self-contained unit with or without the aid of horizontal diaphragms or floor bracing systems.

Space frame—vertical load-carrying means a space frame designed to carry all vertical loads.

Space frame-moment resisting means a vertical load-carrying space frame in which the members and joints are capable of resisting design lateral forces by bending moments. Such moment resisting space frames may be enclosed by or adjoined by more rigid elements which would tend to prevent the space frame from resisting lateral forces, where it can be shown that the action or failure of the more rigid element will not impair the vertical or lateral load-resisting ability of the space frame.

Space frame-ductile moment resisting means a moment resisting space frame with the necessary ductility provided by a frame of structural steel conforming to ASTM A7, A36, A441 with moment-resisting connections or by a reinforced concrete frame. The provisions for more rigid elements used in combination with moment resisting space frames also apply to ductile moment resisting space frames.

Box system means a structural system without a complete vertical load-

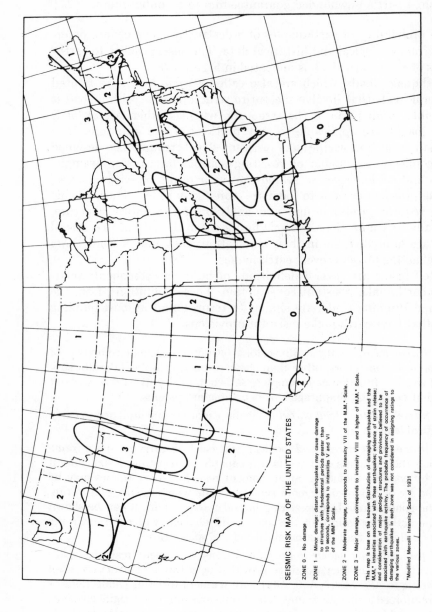

SEISMIC RISK MAP OF THE UNITED STATES

ZONE 0 – No damage

ZONE 1 – Minor damage: distant earthquakes may cause damage
to structures with fundamental periods greater than
1.0 seconds; corresponds to intensities V and VI
of the M.M.* Scale.

ZONE 2 – Moderate damage; corresponds to intensity VII of the M.M.* Scale.

ZONE 3 – Major damage; corresponds to intensity VIII and higher of M.M.* Scale.

This map is base on the known distribution of damaging earthquakes and the
M.M.* intensities associated with these earthquakes; evidence of strain release;
and consideration of major geologic structures and provinces believed to be
associated with earthquake activity. The probable frequency of occurrence of
damaging earthquakes in each zone was not considered in assigning ratings to
the various zones.

*Modified Mercalli Intensity Scale of 1931

FIGURE 1.3   Seismic risk map of the U.S. From *The Uniform Building Code*, Vol. 1, 1973. Courtesy
International Conference of Building Officials.

carrying space frame. In this system, the required lateral forces are resisted by shear walls as hereinafter defined.

Shear wall means a wall designed to resist lateral forces parallel to the wall. Braced frames subjected primarily to axial stresses shall be considered as shear walls for the purpose of this definition.

Examples of code requirements to resist earthquake shocks are included in the following:

* The *National Building Code* of the American Insurance Association.
* The *Uniform Building Code* of the International Conference of Building Officials, of which the Pacific Coast Building Officials Conference is a subsidiary.

Structural engineers well versed in earthquake-resistant design should be consulted on the design of major buildings to be constructed in regions subject to severe earthquake shocks [10].

3. BUILDING COSTS AND SITE WORK

Types of Estimates. Two general types of estimates may be prepared for the cost of a building. The first is a preliminary estimate, which may be required to determine approximate cost before detailed plans and specifications are prepared. Such an estimate may be used to inform the prospective builder, who knows the approximate amount of money he will have available for the project, of the approximate floor area that can be provided with that amount. Or he may know what floor area he requires and wish to know approximately what it will cost.

Such estimates are usually necessary before the architect is asked to incur the expense of preparing the complete plans and specifications.

The second type of estimate is prepared by the contracting organization to determine the amount of the *bid* or *proposal* on which he would enter into an agreement or contract with the owner to complete the building. This is called a *lump sum contract*. Under this form of agreement, bids are received from several contracting organizations.

Both the owner and the contractor may prefer a *cost plus* contract, one by which the contractor agrees to construct a building for the amount it costs him plus a percentage of this cost, or plus a fixed fee. Sometimes the contractor agrees to a *guaranteed maximum* cost to the owner. When a cost plus contract is used, the owner usually selects the contracting organization with which he wishes to enter into an agreement, and there is no competitive bidding. The owner usually requires, however, that he be provided with at least an approximate cost estimate.

Gross Floor Area Estimates.   A standard for computing the *gross floor area*, or architectural area as it is sometimes called, adopted by the American Institute of Architects is as follows:

"The *architectural area* of a building is the sum of the areas of the several floors of the building, including basements, mezzanine, and intermediate floored tiers and penthouses of headroom height measured from the exterior faces of exterior walls or from center line of walls separating buildings. Covered walkways, open roofed-over areas which are paved, porches and similar spaces shall have their architectural areas multiplied by an area factor of 0.5. The architectural area does not include such features as pipe trenches, exterior terraces or steps, chimneys, roof overhangs, etc."

The areas of pilasters and buttresses should also be excluded.

A preliminary cost estimate is made by multiplying the gross floor area in square feet by an appropriate total cost per square foot. This cost is selected on the basis of costs of similar buildings. Such costs may be based on the architect's experience, a contractor's suggestions, or published values.

Costs vary from year to year and have increased significantly in past and recent years as indicated by the *Engineering News-Record* Building Cost Index (Figure 1.4). In this index, the cost for the year 1913, just before World War I, is considered to be 100. In 1973, this

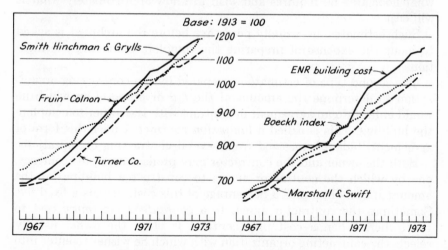

FIGURE 1.4   Building cost indexes.

index was more than 1,100 or eleven times the 1913 value. By using this index, given monthly in that publication, available square-foot costs for buildings already constructed can be brought up to date. Cost for buildings erected in a city other than that in which the contractor is to build can be adjusted to the locality by other cost indexes included regularly, for many cities, in issues of the *Engineering News-Record*.

Cubage Estimates.   A procedure similar to the one explained in the preceding paragraph is based on the cost per cubic foot, or *cubage*, of a building. A standard for computing the cubage or architectural volume adopted by the American Institute of Architects is as follows: "The *architectural volume* of a building is the product of the total areas defined above (architectural areas) and the height from the average depth of footings to finish floor, floor to floor, to the average height of the surface of the finished roof above, for the various parts of a building." In other words, it is the total volume of the building included between the horizontal surfaces at the average footing depth and roof height.

The bases of the selected cubage costs and the adjustments for date of construction correspond to those given in the preceding paragraph.

Other Approximate Methods.   Other even cruder methods are sometimes used in the early stages of a project. Some of these are based on the cost per unit of capacity, such as the cost of a school per pupil or classroom, a hospital per bed, or a hotel per room.

Detailed Estimates.   Final estimates on which a contractor's bid is based require complete plans, or working drawings, and specifications. They involve a *quantity survey* or *takeoff*, that is, a survey of the quantities of various materials and equipment items required in the building. The number of hours and kinds of labor involved in placing these items must be estimated, and finally the material and labor costs must be estimated to determine the *direct costs*.

To the direct costs must be added the *indirect costs* for construction equipment, supervision, purchasing, transportation, financing, administration, insurance, cleaning the building for occupation, site development, and numerous other items not included in the direct costs of the building. Finally, the anticipated profit must be added.

In addition to these costs, allowance must be made for the fees of the architects, engineers, and special consultants.

Approximate Costs.   The costs of buildings of the same type of construction and occupancy vary with the date of construction and the location, as has been explained. Other factors such as the labor

TABLE 1-8
Approximate Contract Costs per Square Foot of Gross Floor Area

Type	Cost	Type	Cost
Residences	$20	Schools	$30
Apartments	20	Hotels	35
Warehouses	16	Hospitals	50
Retail stores	25	Office Buildings	50

supply, weather conditions, and economic conditions affect building costs. Table 1-8 includes recent values for contract prices to give the reader some idea of the magnitude of building costs.

The qualities of materials used in building construction are controlled by standard specifications, most of which have been prepared and recommended by ASTM. Usually, these have been adopted by the various agencies which have prepared codes and by architects and engineers who are preparing specifications.

The *allowable* or *working stresses* used in the design of the various structural elements which enter into the construction of a building depend upon the properties of the materials of which they are made, the condition under which the elements are used, and many other factors. These stresses are specified in the codes.

Environmental Considerations. Environmental impact reports are now required in many municipalities and states. Environmental studies are now more catholic in application, whereas formerly they were thought of as involving only mechanical equipment of buildings —heating, ventilating, plumbing, and illumination.

The current concerns for environmental issues do increase construction problems and the cost of buildings. Environmental requirements for building construction at this time are unresolved as to what precisely constitutes environmental impact. A project in one locality may create a different environmental controversy than a nearby project.

Geological reports (in addition to soils reports) are now required in areas of high earthquake probability and of doubtful geologic stability. Prepared by engineering geologists, these reports should include, but not be limited to, information on rock core samples, the location of known faults, caverns, the bearing values of rocks, rock formations and their extent, and the classification and types of rocks.

Environmental aspects may be classified broadly as those which are of only temporary concern (the construction period) and those which have longlasting effects, including the ecological balance of the area. Some short-term concerns might be classified as public nuisances: noises of the building process, such as riveting and pile driving; dust, which may be controlled by water spraying or using a dust pallative.

The demolition of existing structures or the removal of rock overburden may require the use of explosives. This activity, closely regulated by governmental agencies, may require blasting mats and scheduling of blasting shots at certain times only.

The removal of unsightly construction debris, especially in demolition work, may be a problem. The rupture of gas lines and underground power lines can prove fatal. The rupture of water lines can cause flooding of the site and adjacent property.

Some long-term factors causing concern for the environment are: traffic congestion, parking availability, the utilization of ground surface for parking, landscaping, obstruction of views, unsightly and oversized signs, destruction of sites that have intrinsic, archaeologic or historic values, disruption of plant and animal life, water demands, sewage disposal, overhead utility lines, exposed pipe lines, malodorous emissions produced by a manufacturing process, and noise levels that are abusing.

The effects of the building operations and the building itself may be detrimental to the natural life of the area (the ecological balance) and may require studies and reports by experts.

Radiation hazards are a factor in certain structures. Flooding, causing floods, mud slides, and earth movement in such a manner that would abuse the public or a neighbor should be a matter for concern.

Other Site Conditions and Costs.   There are innumerable problems that often have some effect upon the cost of building construction. Access to the site by all-weather roads may not be available. Moreover, inadequate bridge capacities and clearances might require added expense. The availability of potable water, sanitary disposal, electrical power, and communications may require special planning. And certainly weather conditions of the site, if a long-term project is planned, should forewarn that there will be other expenses.

Many of these site work factors add to the cost of a structure before any construction progress is realized. Another such concern is *mobilization costs*. This includes the complete start up of the construction activity, which can be costly. Because these costs are not reimbursable directly to the builder, they must be borne as overhead expenses. There are attempts to overcharge on early actual construction work, usually excavation, in order to allay some of these mobilization costs.

REFERENCES AND RECOMMENDED READING

1. *Building Codes, Their Scope and Aims*, National Board of Fire Underwriters, 85 John Street, New York, New York.
2. *Preparation and Revision of Building Codes*, Building Materials and Structures Report 116, National Bureau of Standards, 1949.
3. *Selected Bibliography on Building Construction and Maintenance*, Building Materials and Structures Report 140, National Bureau of Standards, Second Edition, 1956.
4. *Fire Resistance Ratings*, National Board of Fire Underwriters, 85 John Street, New York, New York. Copies of this publication will be sent on request.
5. B. L. Wood, *Fire Protection through Modern Building Code*, American Iron and Steel Institute, 1945.
6. *Fire-Resistance Classifications of Building Constructions*, Building Materials and Structures Report 92, National Bureau of Standards, 1942.
7. *Wind Pressures in Various Areas of the United States*, Building Materials and Structures Report 152, National Bureau of Standards, 1959.
8. *American Standard Building Code Requirements for Minimum Design Loads in Buildings and Other Structures*, A58.1, American Standards Association, 1955.
9. Whitney C. Huntington, *Earth Pressures and Retaining Walls*, John Wiley and Sons, 1957.
10. *Recommendations, Earthquake Resistant Design of Buildings, Structures, and Tank Towers*, Pacific Fire Rating Bureau, 1950.
11. *Building Materials List*, Underwriters Laboratories, Inc., sponsored by National Board of Fire Underwriters, issued periodically.
12. *Uniform Building Code*, 1970 Edition, Volume I, International Conference of Building Officials, Whittier, California.
13. *National Building Code*, 1967 Edition, Recommended by the American Insurance Association, 85 John Street, New York, New York.
14. *1970 Standard Grading Rules for Western Lumber*, Western Wood Products Association, Portland, Oregon.
15. *No. 16 Standard Grading Rules for West Coast Lumber*, West Coast Lumber Inspection Bureau, Portland, Oregon.

16. *Standard Specifications for Grades of California Redwood Lumber*, Redwood Inspection Service, San Francisco, California.
17. *1970 Standard Grading Rules for Southern Pine Lumber*, Southern Pine Inspection Bureau, Pensacola, Florida.
18. *1970 Standard Grading Rules for Northeastern Lumber*, Northeastern Lumber Manufacturers Association Inc., Glens Falls, New York.
19. *Rules for the Measurement and Inspection of Hardwood and Cypress Lumber*, National Hardwood Lumber Association, Chicago, Illinois.
20. *Standard for Fire Tests of Building Construction and Materials UL 263*, Underwriters' Laboratories, Inc., Chicago, Illinois.
21. *1971 Annual Book of ASTM Standards, Part 14*, American Society for Testing and Materials, Philadelphia, Pennsylvania.
22. *Test Method for Fire Hazard Classification of Building Materials UL 723*, Underwriters' Laboratories, Chicago, Illinois.
23. Plummer and Blume, *Reinforced Brick Masonry and Lateral Force Design*, Structural Clay Products Institute, Washington, D.C., 1953.
24. *Uniform Building Code*, 1973 Edition, Volume I, International Conference of Building Officials, Whittier, California.

2

FOUNDATIONS

4. SOILS

Formation of Soils. The earth's surface is solid rock, called *bedrock*, either exposed as outcroppings or overlaid with water or unconsolidated material formed by the weathering of the solid rock. This unconsolidated material (mantle rock, soil, or earth) rests on the rock from which it is derived and gradually grades into that rock, or it erodes and is redeposited by wind, water, or glacial ice. *Residual soil* remains over the rock from which it was formed. Soil transported in water and redeposited at another site is called *alluvial soil* or *alluvium*. The sizes of the particles in alluvial soil depend upon the water velocity with grain size decreasing as water velocity decreases. Changes in velocity may occur gradually along the water course resulting in fairly uniform deposits over an extensive area, whereas rapid flow within short reaches may produce deposits with marked variations. As a result, alluvial soil deposits usually are *stratified*. *Loess* is a soil type formed of fine windblown material. Rock fragments and particles picked up by glaciers and redeposited by the melting glaciers form sizeable *moraines*. These deposits may be ridges crossing valleys, called *terminal moraines* and formed where the glacial movement had a balanced melting rate for an extended period. *Lateral moraines* are ridges paralleling the valley formed by debris deposited along the sides of melting glaciers. Debris deposited to a considerable depth over a wide area by a melting receding glacier forms *ground moraines*. Unless modified by subsequent water erosion, the glacial deposit of mixed size material forms an unstratified mass called *till*. Fine particles, ground off bedrock by abrasive glacial action, are known as *rock flour*.

Classification of Soils. Soils are formed by the disintegration of rocks. For engineering, the principal types of soil may be divided into two general classes according to the grain sizes. The *coarse-grained soils* are gravels and sands; *fine-grained soils* are silts and clays.

For comparative simplicity, the discussions in this article contrast the properties of sands and clays. Other soils and other terms applied to soils are considered briefly.

Soil deposits usually consist of mixtures of various classes of soil. They may contain organic materials such as *peat*, which is partially decomposed, highly compressed vegetable fibers, and *muck* (often found with peat), which is fine clay or mud.

There are various systems for classifying soils. A system of grain size classification of grain soils, adapted from that of the U.S. Department of Agriculture, indicates the following:

TABLE 2-1
Classification of Soils by Grain Size

| Gravel, 1/12 in. and over | Silt, 1/12,500 to 1/5000 in. |
| Sand, 1/500 to 1/5000 in. | Clay, less than 1/12,500 in. |

This classification should not be confused with the grain size separation between fine and coarse aggregates used as concrete aggregates.

This classification of soils according to grain size is convenient because grain size analyses are relatively simple. However, this means of classification is unsatisfactory because grain size is only one factor in determining the behavior of soils. Sand may be ground to a fineness leading to a classification as clay according to grain size; however, finely ground sand will not possess the other properties usually associated with clay.

Although the grains of sand may be granite, basalt, and other rocks, most sands and gravel grains are silicon dioxide and quartz. The grain shapes vary from angular to round. The table above shows that the grain size of clay is comparatively small. The grains are composed of hydrous aluminum silicate and various compounds such as quartz and iron oxide. Particle shape is flat or flake-like, similar to mica.

Silt has properties intermediate between fine sand and clay. The grain size of silts is not a true indication of their behavior. Some silts or rock flour resemble rounded sand grains in shape and behavior. They are usually silicon dioxide. Other silts resemble clay in their composition, particle shape, and behavior. Some soils intermediate between silts and clays in composition and behavior are called *silty clays*.

Structure of Soil.   The structure of a soil deposit bears a direct relation to its properties. Sand grains deposited in water gravitate to relatively dense deposits, whereas fine clay grains adhere to other clay grains without dense consolidation. These processes are illustrated in Figure 2.1. A clay deposit resembles a network of solid grains with the intervening water-filled spaces called *voids* or *pores*. Applied loads are transmitted through the deposit along the network. The transformed network results in increased water pressure in the pores termed *pore pressure*.

If the network is broken or *disturbed*, the grains become wholly or partially surrounded with water and the soil mass loses all or partial rigidity and is called disturbed, *remolded,* or *recompacted*. Samples of soil obtained intact are *undisturbed samples*.

Voids in Soils.   The voids may be filled with air, other gas, water, or a combination of these materials. The specific gravity of soil grains varies between 2.5 and 2.8, regardless of composition or size. Even though the specific gravity of most soil grains is fairly constant, a cubic foot of soil may vary from 70 lb. to 150 lb. This results from differences in proportions of voids to solids which are dependent upon the grain size gradation; moisture content; and the degree of *consolidation* or compression by overlying material and superimposed compaction, tamping, and vibrating loads.

In a volume occupied by equal spheres, the voids between spheres may vary from a minimum of one-fourth to a maximum of one-half the total volume. This variation in the void percentage is dependent

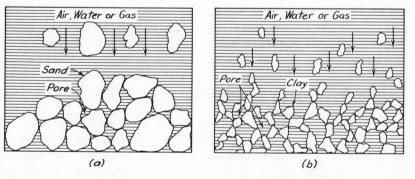

FIGURE 2.1   Formation of sand and clay deposits. (*a*) Sand. (*b*) Clay.

TABLE 2-2
Percentages of Voids and Void Ratios

	Percentage of Voids	Void Ratios
Sands	30 to 45	0.43 to 0.82
Clays	35 to 95	0.54 to 19.00

upon the arrangement of the spheres, not upon the sphere size. If the spheres are mixed sizes, the percentage of voids is reduced—the smaller spheres progressively filling in the voids between the larger spheres. The volume of voids in a soil sample to the total volume of the sample is called the *percentage of voids*. This ratio is expressed as a percentage. The ratio of the volume of voids to the volume of the solids is the *void ratio*. The percentages of voids, and therefore the void ratios of sand, are relatively small compared to those of clay.

From Table 2-2, it is obvious that, in a sample of sand, the volume of voids is less than the volume of solids, whereas in a sample of clay, the volume of voids may exceed the volume of solids manifold. The voids in sand above ground water will be filled mostly with air; however, if below the groundwater level, they will be filled with water. Except near the surface, the voids in most natural deposits of clays are filled with water irrespective of the groundwater level. Placed and compacted clay fills may contain considerable *trapped air*. Sometimes gases other than air occupy the voids of clay. Occasionally these are flammable and form explosive mixtures with air. Natural clays and silts do not contain measurable volumes of air if located below the zone of temporary desiccation.

Permeability.   A significant property of a soil is its *permeability*— that property of a soil that permits water to pass through its voids or pores. Because of its large grain and large pore size, sand offers little resistance to the passage of water. In contrast to sand, the grains of clay are microscopic and the pores minute, consequently clay has greater resistance to the passage of water. Sand, therefore, is very permeable and clay is very impermeable. Silts are intermediate in permeability. Because of compositional variations, wide latitudes exist in the permeabilities of different clayey soils—for example, the presence of a small amount of clay may greatly affect the permeability of sand. Some clays are a thousand times more permeable than others. As explained below, the permeability of a soil has a marked effect on the rate that a soil compresses or consolidates under foundation loads and also on the watertightness of the foundations.

Soil Water. Water exists in soil in four states: capillary, gravitational, hygroscopic, and adsorbed water. *Capillary water* is contained in minute pores small enough to cause capillary action. It is considered further below. Water that percolates through soil and that can be drained or pumped out is called *gravitational water*. Water that surrounds the individual grains with a thin film which cannot be removed by air drying is *hygroscopic* or *film water*. It can be removed by oven drying. *Adsorbed water* clings to the surface of the soil grains and cannot be removed by drying. The soil grains themselves do not absorb water but contain *chemically combined* water, not considered here. In a given sample of soil, the ratio of the water weight to the dry soil weight is the *moisture content*. This is expressed as a percentage.

The water in the voids or pores of soil is called *pore water*, and the pressure exerted by this water constitutes *pore pressure*. Pore pressure resists external pressures tending to compress or consolidate the soil. By partially holding the soil grains apart, pore pressure diminishes the frictional resistance between grains and therefore reduces the shearing strength. For this reason, tests of soil shearing strength that neglect the pore pressure are not valid. Because sand is very permeable, its pore pressure nearly equals the hydrostatic pressure resulting from the head in static water above the determined pressure point. Because clay is very impermeable, the pore pressure under a loaded area may indicate higher values than the surrounding soil. This inequality is gradually reduced as the pore water is forced from the pores in the loaded area and assimilated pressure equilibrium approached. This process may require centuries.

The density or weight per cubic foot of clayey soil in compacted fills is dependent on the soil moisture content. For every soil sample, there is a moisture content with a given compaction that will result in a maximum density. The *optimum moisture content* is the amount of moisture in a given soil sample that will produce the maximum density with a given compaction.

Capillarity. In a soil where pore water contacts air, a concave surface is formed called a *meniscus*. This adsorption is often called a *film*. The *surface tension* in the adsorbed water binds the grains, amply illustrated by damp sand. The soil voids also serve as capillary tubes, causing the water to rise above the groundwater level. This *capillary rise* is the height above groundwater level that water is elevated by capillary action. In soils, the amount of capillary rise varies with the diameter of the pores which is a function of the grain size. The grain size of sand is large with resultant large pore size, causing this type of soil to possess practically no capillarity; but in clays, the small grain size contributes to a large capillary rise. For this reason, the pores of subsurface clays are usually filled with water, though the groundwater level may be many feet below. The rate at which capillary water permeates a soil depends upon the size of the pores; it is very slow in clay. Capillary water cannot be drained out of soil by any system of drainage.

Swelling and Shrinking. The volume of clay soils tends to change as the moisture content changes, even though no change occurs in the external load. This *volume change* may be a *swelling* or a *shrinking*. The change results from stresses exerted by capillary action in the soil pores. The pores may be considered as bundles or networks of irregularly shaped capillary tubes permeating the soil. At each point where a pore is exposed to air, a meniscus forms on the water in the pore and over the end of the pore. As water evaporates across the surface, the volume of water in the pore decreases while the menisci remain at or near the ends of the pore. This combined action in all pores results in a pulling together, compressing, or shrinking of the soil. If menisci contact water from rains, floods, or other causes, the menisci are destroyed and swelling occurs. There is a limit to the amount of shrinkage that can take place. The *shrinkage limit* is the moisture content below which no volume decrease occurs and above which the volume increases.

The pore size of sands is so great that no volume change occurs because of this capillary effect; but in clays, this phenomenon is pronounced, as manifest by the deep cracks that form in clay deposits as they dry out and then dissipate after rains.

The causes of shrinkage in clay as it dries are illustrated by comparing the action to changes that take place in a water filled elastic tube (Figure 2.2). These changes are greatly exaggerated for clarity of presentation. A small diameter tube is shown in (a). This tube is filled with water (b). The meniscus is formed at each end, enlarged in (c). Surface tension at each meniscus exerts a pull T around the perimeter of the tube. This is a constant resultant force for each unit of length of the perimeter and exerts a corresponding component force C on the walls of the tube. This force shortens the tube. Through evaporation across the meniscus, the water decreases (d); correspondingly, the radius of curvature of the meniscus decreases

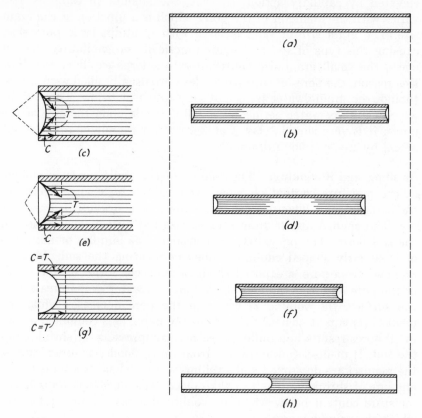

FIGURE 2.2 Shortening of tube due to capillary action (greatly exaggerated).

(e); thus, the angle decreases between T and the tube walls. Because the magnitude of T remains constant, C must increase correspondingly, and the tube is further shortened. This action continues until the menisci become tangent to the walls (g). The magnitude of C now equals T, its maximum value reached. Thus, the tube has attained its minimum length. Further decrease in water volume owing to evaporation causes the meniscus to recede (h). The tube now increases in length because only a portion of the tube is compressed by the meniscus activity. If all water evaporates or if the menisci are dissipated by flooding, the tube will return to its original length (a). Because of the complexity of the network of voids, the shrinking-swelling activity in soils is much more complicated. However, the basic causes of shrinking and swelling are the same as described.

Unlike the theoretical elastic tube, soil does not expand to its original dimensions when all water evaporates. In soil, menisci form through the entire mass where water and air are in the pores.

Shearing Strength. The *shearing strength* of soils depends upon several factors, namely, the frictional resistance between soil grains referred to as *internal friction,* the mutual molecular attraction of the soil grains called *true cohesion,* and *apparent cohesion* which is the binding of the soil mass by capillary action of the pore water. Molecular attraction is a negligible factor in the cohesive strength of sands because the distance between the centers of gravity of sand grains is comparatively large, the sand grains being large for a given volume. Conversely, clay has small, flat grains and relatively numerous particles in a given volume, resulting in a molecular attraction that is a factor in the cohesive strength. Additionally, the large grain size and correspondingly large pore sizes produce little capillary activity; thus, apparent cohesion is not a factor in the shearing strength of sand. However, clay has small particles and small pore size contributing to a large capillary action and therefore great apparent cohesion. Therefore, the shearing resistance of sand is due only to internal friction, while the shearing resistance of clay is due to both internal friction and cohesion. If clay is submerged, capillary action is destroyed, and as a result, apparent cohesion disappears. Submerging has little effect on internal friction in sand. There are many difficulties in determining the shearing resistance of clay, but this is beyond the present discussion.

Compressibility. The magnitude of foundation settlement depends to a large extent on the *compressibility* of the underlying soil. For the unit pressures under foundations, the unit stresses are relatively low considering the materials that make up the soil grains. The amount of internal deformation of the soil grains is negligible, leaving the decreased volume of the soil due to the decreased volume of the voids. For sands, there can be relatively little decrease in the voids unless their bulky grains are rearranged by a sudden shock or by vibration. For clays, with their scale-like grains and high percentages of voids, a large decrease in the volume of voids may result from foundation pressures. The compressibility of dense sand is therefore quite low, but for clays it is quite high. For some clays, such as those in Mexico City with void ratios as high as 14, the compressibility is very large. Because the compressibility of water is negligible, any settlement resulting from compression may be due to the squeezing of water from the turgid voids. Because of its permeability, sand offers little resistance to the passage of water, and the final degree of compression is reached almost as soon as the load is applied, even though the pores may be water filled. Conversely, clay is quite impermeable, with the pores of natural deposits normally water filled with compression taking place slowly and continually over an extended period. Air is forced from the voids of sand almost instantaneously by applied pressure while offering little resistance. In clay, entrapped air and pore water may lie dormant for a considerable time, compressing when pressure is applied; therefore, a certain amount of settlement can occur instantaneously if air is present in the voids. The process of squeezing water out of soil and thereby decreasing its volume is *consolidation*. Consolidation can take place only when the stresses are large enough to break the cohesive bond between soil particles.

Elasticity.   When a body is stressed, its dimensions change. A body that returns to its original form and dimensions when applied stresses are released is elastic, possessing the property of *elasticity*. Sands are not elastic. Clays have considerable elasticity when the stress values are small compared with the rupture values.

Plasticity.   The property of a material permitting it to undergo changes in shape or to flow when subjected to steady forces without noticeable change in volume is called *plasticity*. Sands are not plastic. Clays are plastic at high stresses.

Solubility.   Solubility is not a major factor in soil used to support building foundations. The most soluble material found in soil is gypsum. Limestone is moderately soluble.

Contrasting Properties of Sand and Clay.   Sand and clay are two soils with properties that are at either extreme, as shown in Table 2-3. Their properties are of special interest to the architect and engineer. Table 2-3 was prepared from a similar one by H. S. Gillette [13]. Sands, and especially clays, exhibit variations in compositions and properties while possessing certain aforementioned characteristics, but to varying degrees.

Uses of Soils.   Soil is used directly or processed as a building material. Naturally, it serves as the foundation support for most structures if rock beds are not readily accessible. Some foundations of major structures are carried to bedrock.

Clay is the principal material in the manufacture of fired brick, adobe brick, structural clay tile, vitrified clay pipe, terra-cotta,

TABLE 2-3
Contrasting Properties of Sand and Clay

Property	Sand	Clay
Grain size	Large and bulky	Minute and scaly
Pore size	Large and wide	Small and narrow
Void ratio	Relatively small	Usually high
Internal friction	Large	Very small*
Cohesion	Small	Usually large
Capillary effects	Very small	Very large
Permeability	High	Low
Compressibility	Low	High
Shrinkage	Very low	High
Elasticity	Low	High for low stresses
Plasticity	None	High for high stresses

* Internal friction of clay is large when no pore water is present. Pore water is usually present and holds the grains apart, causing the "apparent" internal friction to be low.

roofing tile, ceramic floor and wall tile, and porcelain plumbing fixtures. Clay deposits selected for a particular purpose are processed into the finished product primarily by heat.

Clay is an ingredient of portland cement, in which the clay is mixed with the proper amount of limestone, pulverized, burned to a clinker, and repulverized.

Sand is an essential ingredient in masonry mortars and plasters. Used thusly, it is mixed with a cementing material such as portland cement, lime, or gypsum plaster and water. Sand is the principal component of concrete. Sand and gravel are used as a free-draining base material under concrete floor and pavement slabs and as a cushion under brick floors and pavements to secure an even bearing. Sand is one of the principal constituents of window and structural glass.

Gravel is a soil grain-size term that applies to rock particles that range in diameter from 1/4 in. to 6 in. They are chiefly quartz but may be other types of rocks. The term *pebble* applies to gravels with diameters up to 3 in. The larger (rock) particles sized 8 in. up to diameters measured in feet are called *boulders*.

Gumbo is a dark-colored sticky clay. *Adobe* is a light-colored sandy calcareous clay that is used for sun-dried or kiln-fired masonry products. *Bentonite* is a highly expansive clay formed by the weathering of volcanic ash. It swells perceptibly up to 10 times its original volume when wet and is used in *slurry* or in panels to impede the passage of water.

Soil Testing. Soil tests are conducted in both field and laboratory. Laboratory testing usually verifies or determines some property of soils, such as moisture or permeability. Field testing, in conjunction with laboratory operations, usually determines or indicates the bearing capacity of the soil.

Field tests may consist of in-place density or compaction tests. There are several methods commonly used to ascertain the density of soils, namely, the *Sand-cone method,* designated as D 1556 by the American Society for Testing and Materials, and the *Rubber-balloon method* (ASTM D 2167).

In either test, a small hole is dug, the soil sample removed, and the moisture content of the sample calculated in the laboratory. The volume of the hole is measured by the amount of sand or by the volume of liquid that fills the hole. The in-place density is then calculated.

Another test performed for moisture-density-compaction relationship is ASTM D 698. This compaction test employs a 5.5 lb. hammer dropped from a 12 in. height on a confined sample. The moisture-density relationship is then calculated.

FIGURE 2.3 Sand-cone density apparatus. This device is used in performing ASTM D1556, which is the standard method of Test for Density of Soil in Place by the sand-cone method.

These tests are simple to perform and give plausible results of surface compaction. Their use in building foundations is confined to those foundations that support light loads, usually a slab on grade.

5. GENERAL DISCUSSION

The part of a building below the ground surface is called the *foundation* or *substructure*, and the portion above ground is called the *superstructure*. Herein, the part of a building bearing directly on the supporting soil or rock is called the foundation, whereas the supporting rock or soil is referred to as the *foundation material*. The function of the foundation is to transmit and distribute the building loads and other imposed loads to the foundation material. Walls below the ground surface resting on the foundations are *foundation walls*.

Codes use various terms to describe foundation design. The *bearing pressure* is the pressure intensity between the foundation bottom and the foundation bed. The *allowable bearing pressure* is the recommended design bearing pressure that will safely insure against excessive settlement or rupture of the foundation material. If the foundation material is soil, allowable bearing pressures are called *allowable soil pressures*. *Bearing capacity* is a general term referring to the ability of a foundation material to support loads safely. The *ultimate bearing capacity* is the maximum bearing pressure that a foundation material can sustain without rupture.

The difference between the ultimate bearing capacity and the allowable bearing pressure is a safety factor. These values are usually expressed in pounds per square foot. The ultimate bearing capacity divided by the allowable bearing capacity yields the *factor of safety*. This is a number usually from 2 to 3 as mentioned in Article 9.

Foundation Materials. Foundations may be supported on soil or *bedrock*, consisting of *bedding planes* exposed on the surface or underlying several hundred feet. Bedding planes may be horizontal, as originally formed, or inclined. Marked variations of rock classes and properties may occur within short distances or at different depths. Soils are also subject to wide variations in general type and in physical properties. A soil with a high bearing capacity may be underlaid by a soil with low bearing capacity. Layers of peat (or other highly compressive soil) may cause excessive settlement if not discovered and conditioned. Clay soils may cause extensive heaving due to swelling and contraction if exposed to alternate wetting and drying cycles.

When possible, expansible clays should be avoided as foundation materials.

Ground Water.   A large portion of our land surfaces are underlaid with *ground water* occupying the pores and other spaces in the soil and rock. The groundwater surface is called the *groundwater table* or simply the *water table.* Generally, the groundwater table is not horizontal but follows roughly the contour of the ground surface. The groundwater table elevation at a given point is termed the *groundwater level.* Reference is often made to the *permanent groundwater level,* but in most locations, no permanent groundwater level exists. The water table elevation fluctuates more or less with the amount of rainfall and with the depletion by pumping for water supply. The water table may be lowered markedly by the construction of sewers, drains, subways, and other underground works.

The location of the groundwater table is of particular concern in selecting the type of foundation, in foundation design, and in planning construction procedures.

Subsurface Exploration.   Because of uncertainties in underground conditions, adequate subsurface explorations should be made before the foundations are designed or the site purchased. Methods for making proper subsurface explorations are briefly described later.

Foundation Types.   Building loads and other imposed loads are transmitted to the earth in the following ways:

(1) By *spread foundations* bearing over a sufficient area preventing undue settlement or rupture. Foundations of this type consist of individual *footings* under walls or columns, as shown in Figure 2.4, or a single, heavy, reinforced concrete slab called a *mat* or *raft* which

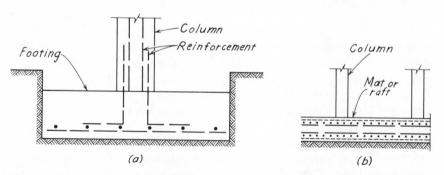

(a)   *(b)*

FIGURE 2.4   (a) Reinforced concrete column footing. (b) Portion of reinforced concrete mat or raft.

may cover the entire foundation area and support all building loads. Foundations of the latter type are often called *floating foundations*, although this term is commonly applied to all types of spread foundations.

(2) By *pile foundations*, which are slender vertical members of timber, concrete, steel, or a combination of these materials. The load is distributed into the soil by *friction piles* or transmitted directly to hardpan or rock by *end bearing piles*. Pile foundations require not less than three piles to a footing to insure stability. Groups of piles are called *clusters* and are joined at their tops by a *cap* which is designed to distribute the imposed loads to the piles. *Batter piles* are driven at a slope.

(3) By *pier foundations* in which concrete piers pass through soil of low bearing capacity until a satisfactory foundation bed is reached. This may be a firm soil or hardpan, requiring the pier to be *belled out* at the base to increase the bearing area. Piers resting on rock seldom require belled out bases. If soil characteristics preclude belling out, or if dimensions of the superstructure determine the dimensions of a pier, piers may be hollow. Hollow piers are frequently used for bridge piers, but rarely are they used for buildings.

Minimum Depths.   The foundation depths depend upon the building space requirements, the adjacent foundation depths, the imposed loads, the foundation material characteristics, and the climatic conditions. Except where bearing on solid rock, foundations exposed to

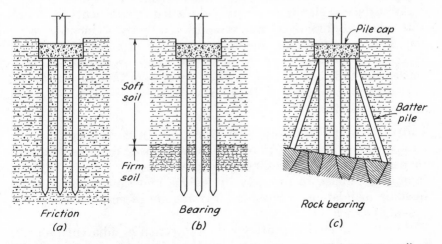

FIGURE 2.5   (a) Piles in soil. (b) Piles through soft soil into firm soil or hardpan. (c) Piles bearing on rock.

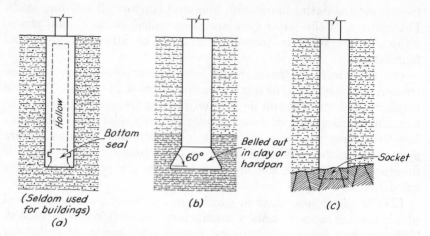

FIGURE 2.6   (a) Pier in firm soil. (b) Belled pier in firm soil. (c) Pier in rock.

freezing temperatures must be placed below *frost line* to prevent heaving. A minimum depth of 1½ ft. below the frost line is recommended. The frost line is the maximum depth that the ground freezes —varying with the locality. In colder regions, the frost line may be 6 ft. or more below the surface, whereas in Southern climes the ground never freezes. In most cities or localities, the depth of the excavation required to prevent frost heaving is well established by codes.

Subsurface explorations should be carried to sufficient depth and with adequate investigation to establish clearly the foundation material characteristics. This investigation should establish that no soil layers of low bearing capacity underlie to cause excessive foundation settlement and that no rock cavities, fissures, or faults exist. Layers of peat below pile tips have been known to cause excessive settlement of pile foundations.

Excavations Affecting Adjoining Property. In excavating the foundations of new buildings, provisions must be made to prevent damage to nearby land and buildings. Bank cave-ins may cause personal injury or damage to property by serious undermining of adjacent building foundations. Prevention may require extensive shoring, discussed in Article 14.

Because the responsibility for the protection of adjacent property varies, a careful investigation of local requirements should establish this responsibility prior to estimating or before initiating construction operations.

6.   SUBSURFACE EXPLORATION

Preferably before a building site is purchased, and certainly before the foundation plans are prepared, an investigation must be conducted to determine the character of the underlying material, including the depth to the groundwater table. Even though ground water may not affect building design, it may have adjunct effects resulting in costly construction difficulties. Major structures located on soils of little known properties require calculations and estimates of expected foundation settlement and investigation of swelling and shrinkage probabilities. Settlement predictions necessitate securing soil samples at various depths below the foundation levels. Soil samples should be taken in a manner that will preserve them in their natural, intact state. These are *undisturbed samples*. Various devices have been developed to secure such samples.

Loading tests are made to determine the soil bearing capacity. Before conducting such tests, their informative value should be carefully considered. As will be mentioned later, these tests may be of limited or of no value because the bearing capacity of some soils does not vary in direct proportion with the area; the unexposed soil underlying the test area may be of very different character from the soil contributing to the test results. In a loading test conducted over a small area (usually 1 sq. ft.), only the soil immediately underlying transmits sufficiently high stresses to affect the results significantly, whereas the soil characteristics for a considerable depth below an actual foundation contribute to foundation settlement. The larger the foundation, the greater the depth affected. This depth is usually about $1\frac{1}{2}$ times the width of the foundation.

Methods of Subsurface Exploration.   The usual methods that determine the character of the material underlying a building site are:

(1) Test pits   (5) Churn drilling
(2) Sounding rods   (6) Diamond drilling
(3) Soil auger borings   (7) Shot drilling
(4) Wash borings   (8) Geophysical methods

Test pits, sounding rods, soil auger borings, wash borings, and churn drilling are appropriate methods for soil explorations, whereas rock investigations are conducted by test pits, churn drilling, diamond drilling, and shot drilling. Where buildings are to be founded on bedrock, it is necessary to drill into bedrock far enough to insure the solid rock is thick enough to support the proposed loads. Wash borings and churn drillings are used to penetrate boulders in soil. Test pits can be used in rock, but the information necessary for examining rock can be secured more economically by diamond drilling, churn drilling, and shot drilling. Geophysical methods determine distinguishable subsurface conditions in the properties of rock and soil formations by analyzing their ability to transmit sound and electricity. This method is not used in building site investigations but more for geological and oil exploration. Each of these methods, except the geophysical method, will be explained briefly, with a concluding statement devoted to methods of obtaining soil samples. Surface soil tests were discussed briefly in Article 4.

Test Pits.   The most satisfactory method for securing reliable information concerning subsurface soil conditions is *test pits*, for this procedure permits soil examination in its natural undisturbed state. This method is relatively expensive because the pits must be large enough to work in. They are shored and lined by properly supported horizontal and vertical members to prevent caving. See Figure 2.7.

Sounding Rod.   The *sounding rod* is a steel rod or pipe about ¾″ in diameter, in about 5 ft. lengths, joined by standard couplings and provided with a pointed lower end. It is hand driven with a maul or drop weight, lubricated with water, and turned with a pipe wrench to reduce sticking. The number of blows required to drive the rod indicates the nature of the underlying soil. Driving should continue until the rod "refuses" further penetration. This may mean that rock has been struck, particularly if the rod "brings up" with a sharp ring. Because the effect is the same when hitting a large boulder or a rock ledge, other probings should be made nearby to determine if refusal is at about the same elevation or if the rod penetration had been stopped by a boulder. After driving has been completed, the rod is removed by a lever and chain. The sounding rod yields no soil sam-

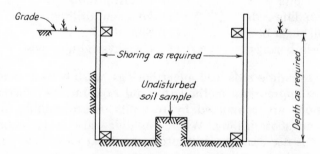

FIGURE 2.7   Test pit.

ples, but experienced operators can estimate the soil character by observing the manner by which the rod penetrates the soil. The results offer little value beyond indicating the presence of rock (within reach of the rod).

Soil Auger Boring.   Holes bored into soil with *soil augers* are rotated by hand or power driven. Augers are of various types, one type being similar to the wood auger. Diameters vary from 1½ to 12 in. The type selection depends upon the material to be penetrated and the power system employed. Augers are mounted on sections of pipe with new sections added as the boring progresses. If the penetrated material is damp sand or contains considerable clay, the hole may not cave; but if caving occurs, the auger is operated inside a metal casing which sinks or is driven as the boring progresses. If the soil adheres to the auger, indications of the nature of the soil can be obtained by examining this soil. Below the groundwater level in sand or in silt, it may be necessary to bail out the soil. Sometimes this method is employed above the groundwater level by adding water and mixing it with the soil to be removed from the hole. If boulders are present, they are drilled with a churn drill or shot with explosives.

Wash Boring.   One method for boring test holes into unconsolidated materials is *wash boring*. A bit is mounted on the lower end of a pipe called a *drill rod* through which water is forced. The drill rod is worked up and down or churned and rotated slowly. The bit strikes the soil or rock, gradually penetrating it. Various types of bits are used, selection depending upon the material to be penetrated, but all are provided with holes permitting water to pass from the drill rod through the bit and against the sides of the hole. For clearance, the width of the bit is greater than the diameter of the drill rod. Cuttings are washed by water rising to the surface in the annular space between the rod and sides of the hole. The process derives its name from this operation. In soil, the drill is removed and the casing is installed before caving occurs. The casing is driven at intervals as the drilling proceeds. The nature of the soil penetrated is often judged by examining the borings or cuttings surfaced by the wash water, but such information is questionable. Experienced operators may form some estimate of the soil characteristics by the "feel" of the equipment. A more satisfactory sampling method relies upon removing the bit and the drill rod and replacing the bit with an open-ended pipe or with special forms of samplers. The apparatus is again placed in the hole and the pipe is driven into the soil at the bottom. The pipe or sampler is brought to the surface; the sample removed and examined. Although not dry, the sample is called a *dry sample* to distinguish it

FIGURE 2.8   Mobile drill model B-30. Auger drill. Mobile Drilling Company, Inc.

from the samples brought up in wash water. Such a sampling procedure is, of course, not adaptable to rock. But this boring method is used for penetrating sand, gravel, clay, boulders, and solid rock.

Churn Drilling.   The *churn drilling* or *dry churn drilling* method is similar to the wash boring method, except the means of cuttings removal is different. The method is used in soil, boulders, and rock. In removing the cuttings, only enough water is used to fill the bottom of the hole. The cuttings mix with the water as the churning proceeds, and at intervals the drill stem is withdrawn; the cuttings and water are removed by a sand pump or bailer. The samples obtained from the bailer are unreliable indicators of the material penetrated. Reliable samples may be obtained in the same manner as "dry samples" in the wash boring method; or core samples of rock may be secured at intervals with diamond drills, shot drills, or saw-toothed drills, with the churn drill removed while obtaining cores.

Diamond Drilling.   The *diamond drill* is suitable only for rock drilling. The drill is a hollow, steel cylinder with embedded black diamonds outside and inside of the bottom edges, forming a bit. The diamonds are set to provide a small clearance between the bored hole and the outside diameter of the drill cylinder, and as a result, the core will be slightly smaller than the inside diameter of the cylinder. The drill is rotated by power driven equipment cutting into the rock by abrasion. Water forced down the drill rod to the bottom rises in the annular space between the drill rod and the sides of the hole. This water removes the cuttings and cools the bit. The bit is not attached directly to the drill rod, but a cylinder called a *core barrel* is intermediate, providing space for a core length up to 10 ft. The drilling apparatus with the sample in the core barrel is removed periodically. The *core lifter* automatically grips the core so that it is removed in the barrel. At its best, the core provides a plausible record of the material penetrated. Gaps usually exist in the record because core recovery is often incomplete. Diamond drilling is expensive but rapid. The core size is commonly 1⅛ in., with a 2 in. hole. Larger holes can be drilled, but the cost increases rapidly with the diameter of the hole. The diamond drill can drill in any direction.

Shot Drilling.   The *shot drill* obtains cores from rock. It is a hollow cylindrical bit rotated by an attached drill rod cutting a circular groove in the rock. The cutting is by chilled steel shot, which are fed into the hole and which find their way under the rotary bit. A core barrel is provided between the bit and the sides of the hole, as in the diamond drill. The cuttings are washed from the cut surface by pumped water surging through the drill rod and bit orifices. The water rises in the annular space between the drill rod and the sides of the hole. The drill rod has a considerably smaller diameter than the core barrel and bit; thus, the velocity of water flowing upward is decreased above the core barrel. For this reason, the cuttings accumulate in a space provided on top of the core barrel and are removed when each section of core is withdrawn.

The presence of cavities delays drilling because the shot disappears into the spaces, losing contact with the lower edge of the bit. After filling the spaces with cement grout, drilling is resumed with the bit cutting through the hardened grout. Holes are usually 4 or 5 in. in diameter. This is similar to a widely used method of taking in-place concrete core samples on structural slabs, such as highway pavements.

Soil Sampling.   As has been stated, the information obtained about the soil penetrated by some exploratory methods is usually inadequate for foundation design. To obtain information about the proper-

FIGURE 2.9 Mobile drill model B-61. Pacemaker drill flushing or jetting casing on a diamond coring project. Mobile Drilling Company, Inc.

ties of the soil at various depths, the bits attached to the drill rods are replaced at intervals by various types of samplers forced or driven into the bottom of the hole. The objective of the sampling operation is to obtain an undisturbed soil sample.

The simplest type of sampler, which is suitable only for cohesive soils such as clay, is a thin-walled seamless steel tube not less than 2 in. in diameter and from 2 to 3 ft. long; the tube is filled with soil by forcing it into the bottom of the hole. In the testing laboratory, the tube with its contents are cut into lengths of about 6 in. The soil is forced from these short lengths to obtain specimens for testing.

Another device is an open-ended cylinder referred to as a *sampling spoon* and especially designed for either sand or clay. The cylinder is driven into the ground and filled; when extracted, it provides a soil test specimen.

A *standard penetration test* records the number of blows required (by a 90 lb. weight falling 30 in.) on a 2 in. outside diameter soil sampler to penetrate the ground one ft.

7. TYPES OF SPREAD FOUNDATIONS

Spread footings distribute the building loads over a sufficient area of soil to secure adequate bearing capacity. They may be classified according to the manner in which they receive the loads.

Wall Footings. Wall footings supporting light loads may be *simple footings* constructed of plain concrete that do not project more than about 6 in. beyond the edges of the wall. The minimum footing depth is usually equal to twice the projection. Minimum code requirements state that light longitudinal reinforcement in simple concrete wall footings shall be imbedded to minimize shrinkage and objectionable temperature cracks and to bridge over soft foundation material.

The most common wall footing is the reinforced concrete slab footing of constant depth called a *T-footing*. The main reinforcement on a spread wall footing is perpendicular to the wall and near the slab bottom to keep projections from cracking off near the wall. The reinforcing bars tend to slip in the concrete because of high bond stresses. This necessitates a large surface area on the reinforcing bars which is obtained by using small, closely spaced bars.

Column Footings. Column footings are the most common type of *isolated* or *independent footings*. For light loads, the simple type may suffice, but most column footings are constant depth slab footings

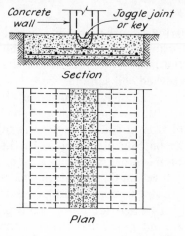

FIGURE 2.10 Wall footing types.

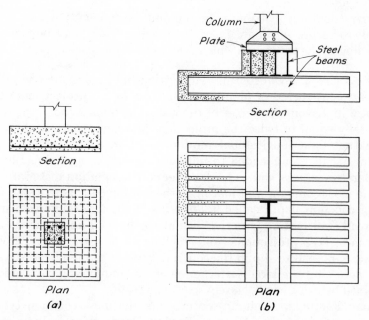

FIGURE 2.11 Column footings. (a) Isolated column. (b) Steel grillage footing.

with two-way reinforcing (Figure 2.11(a)). As in wall footings, small closely spaced bars provide greater bond strength. *Timber grillage* footings constructed in tiers, similar to the steel grillage footing in Figure 2.11(b), are permitted only for temporary buildings and for frame buildings with footings below the "permanent" groundwater level.

In *steel grillage footings*, Figure 2.11(b), the beams are held by spacers placed between them, a layer of concrete 6 or 8 in. thick is placed under the lower beams, and the entire footing is encased in concrete with a minimum edge thickness of 4 in.

Combined Footings. This type of footing frequently supports walls and columns near the property line. If such a column were centered on an isolated footing, the footing would project over the property line. If the column were near the edge of such a footing, the foundation pressures would not be symmetrically distributed, and the footing would tend to settle unevenly or to rotate. This is solved by combining the wall column footing and the nearest interior footing into a single footing, Figure 2.12(a). To avoid rotation, combined footings are proportioned so that the centroid (of the area which bears on the soil) is on the line of action of the resultant column loads. Combined footings are usually trapezoidal in plan. It is assumed that this produces uniform pressure distribution under the footing.

The footing for a wall column and two adjacent interior columns may be combined, particularly at the corners of the buildings. Com-

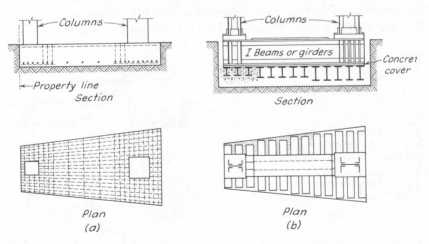

FIGURE 2.12 Combined footings. (a) Reinforced-concrete combined footing. (b) Grillage combined footing.

bined footings are usually reinforced concrete, although grillage footings (Figure 2.12 (*b*)) have been used. A combined footing may be rectangular in plan. If the interior column load is greater than the wall column load, a rectangular footing can be proportioned so that the centroid will be on a line of action of the column loads. In order to simplify the foundation construction, some building codes permit adjacent foundations to project into streets and alleys.

Cantilever Footings. These footings serve as *combined footings* by permitting a wall column load near the edge of a footing. The principle of the *cantilever* is illustrated in Figure 2.13(*a*). The load is supported near the end of a beam which has one support over the center of the wall footing and the other support at the adjacent interior column. Because the beam projects beyond its support, it is "cantilevered." Cantilever footings, (*b*), are usually reinforced concrete, but steel grillage footings have been used. The cantilever is not evident in actual footings because a fulcrum is not used, and in the reinforced concrete footing, the beam is merged into the slabs. However, it is evident that the beam prevents uneven settlement by holding the wall footing in a horizontal position. The beam connecting the footings is called a *strap beam*, and footings of this type are often called *connected footings*.

Continuous Footings. Continuous footings of reinforced concrete slabs extend continuously under three or more columns, Figure 2.14. They tend to reduce the *differential settlement* between columns. This action is more effective if the foundation wall is constructed as a reinforced concrete girder (Figure 2.14(*b*)). If the footing and the wall are

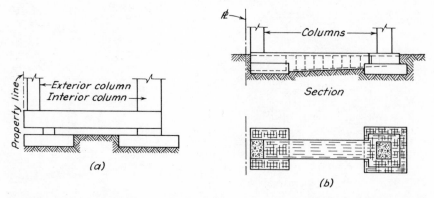

FIGURE 2.13 Cantilever footing. (*a*) Principle of cantilever footing. (*b*) Reinforced-concrete footing.

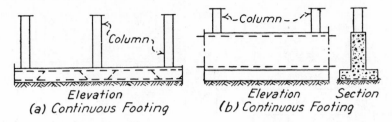

Elevation
(a) *Continuous Footing*

Elevation   *Section*
(b) *Continuous Footing*

FIGURE 2.14   Continuous footings.

poured separately, as is usually done, the lower reinforcing should be at the bottom of the wall; but if monolithic, this steel is more effective near the bottom of the footing. Transverse reinforcing must be near the bottom of the slab. Corresponding steel is required in the footing illustrated in Figure 2.14(*a*). Continuous foundations shown in (*b*) may support bearing walls as well as columns. This is desirable because it reduces differential settlement caused by variations in the soil and reduces cracking in the walls bearing on the foundation wall, as well as in the foundation wall itself. If there are windows in the foundation walls, the upper band of reinforcement should be placed below the windows, with light reinforcement over the openings, and reinforced diagonally at the corners of the openings. Cast-in-place concrete foundation walls reinforced in this manner are appropriate even for small structures such as residences. The additional cost is small and the results worth many times the cost. Two-way continuous footings may reduce differential settlement more effectively than one-way continuous footings. Foundations tied together in this manner resist earthquake stresses.

Raft or Mat Foundations.   These are usually reinforced concrete slabs from 4 to 8 ft. thick covering the entire foundation area, Figure 2.15 (*a*). These slabs or mats are reinforced with layers of closely spaced bars oriented at right angles to each other and placed 6 in. below the top and bottom surface of the mat. To avoid construction joints through lateral dimensions, the entire slab (*b*) may be poured in one operation. Another form of raft or mat is the reinforced concrete *inverted T-beam*, with a slab covering the entire foundation area (*c*). The beams are two-directional, intersect under the columns, and are poured integrally with the slab, forming a monolithic structure. Before the basement floor is placed, the space between these beams is filled as shown in the figure. If the slab is placed at the top of the beams and monolithic with the mat, the mat serves as the

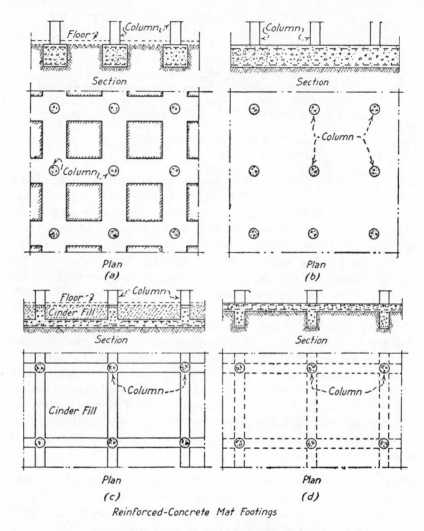

Reinforced-Concrete Mat Footings

FIGURE 2.15   Raft or mat foundations.

basement floor and saves excavation and filling, (d). This construction is suitable for a soil that will stand without caving so that the space occupied by the beams can be excavated and the whole mat poured without forms. Raft or mat foundations are used when the bearing capacity of the soil is so low that other footing cannot be used and where piles cannot be used advantageously or are not necessary. Foundations of this type are commonly called *floating foundations*.

Rigid Foundations. In recent years, there is a tendency to use the term floating foundations in a more restricted sense. It applies to foundations where the weight of the earth excavated will equal, or about equal, the building load. In such a case, the total vertical pressure on the foundation material is about the same after the building is completed as it was before the excavation was started, and thus settlement is held to a minimum. All settlement may not be eliminated because there may be an elastic rebound in the soil when the earth load is removed and a corresponding deformation when the building load is added. If a foundation is not rigid and is below the ground surface, the central portion of the building will probably settle more than the outer portion.

To minimize differential or uneven settlement, foundations must be constructed rigidly. Two methods may be needed to accomplish this. One method employs a rigid reinforced concrete box-like structure consisting of outside and cross walls in the basement (Figure 2.16(a)). The other method uses the basement floor and the first or second floor as chords and the columns as posts of rigid frames, (b). These frames are called *Vierendeel girders* or trusses. They are not trusses in the traditional sense because they are not composed of triangular frames. The joints must be designed to take bending stresses because the diagonal members have been omitted. In order to secure the required rigidity, girders may extend through two or more floors. This is conveniently done where a subbasement is required.

Reinforced concrete trusses with diagonals have been used instead of Vierendeel girders. Reinforced concrete interior walls restrict the use of basement space. Openings in these walls reduce the rigidity of the walls and introduce problems in design. Reinforced concrete trusses with diagonals also restrict the use of the space, but Vierendeel girders do not.

8.   SOIL PRESSURES UNDER SPREAD FOUNDATIONS

Spread foundations and settlement design computations are based on the pressure distribution between the *foundation bed* and the underlying soil. Investigators have measured these pressures and have developed methods for computation. The measured soil pressures, subject to the uncertainties in most pressure cell methods, correlate experiment and theory with reasonable satisfaction. The computed pressures, determined by analytical methods, assume that soils have certain properties which they possess only to a limited degree.

A widely used method, developed in 1885 by Boussinesq, is for a

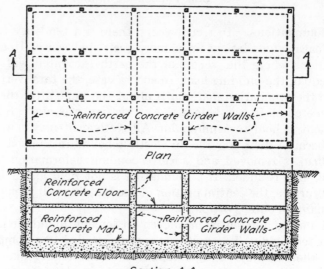

Plan

Section A-A

(a) MONOLITHIC FOUNDATION WITH GIRDER WALLS

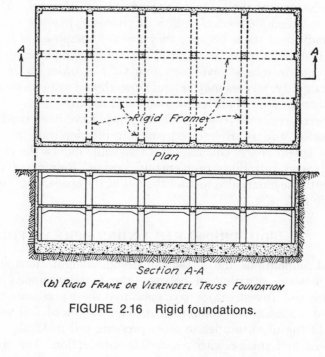

Plan

Section A-A

(b) RIGID FRAME OR VIERENDEEL TRUSS FOUNDATION

FIGURE 2.16 Rigid foundations.

homogeneous, isotropic, elastic solid with an applied normal surface load. Various proposals attempt to modify Boussinesq's solution for application to soils, but no approach is generally recognized. The pressures of interest in foundation computations are the unit vertical pressures on horizontal surfaces at various depths. The calculation of these pressures, using Boussinesq's method directly, is quite laborious; however, Newmark [2] developed a procedure that has greatly simplified the computations. Boussinesq's solution applies to soil pressures produced by superimposed loads—pressures of interest in settlement computations.

Contact Pressures.   The distribution of the normal pressure on the surface of contact between the bottom of a footing and the soil is worth noting. The analytical solution based on idealized properties and assuming the footing to be perfectly rigid gives the stress distribution in Figure 2.17 with infinite, unit, normal pressures at the edges. Actually, no material could be perfectly elastic for infinite stresses, and therefore, the pressures under the edges will have some large finite value.

The distributions for rigid plates on clay and sand have been determined experimentally, with the results shown in Figure 2.18(b). The contact pressures for clay were found to have a distribution similar to that given by the analytical solution. As the depths of the footing increased, the proportionate difference between the edge pressures and the center pressure decreased. Probably the pressure distribution in clay will gradually become more uniform as the clay consolidates under pressure [3]. The contact pressures for sand were found to be a maximum at the center and zero at the edges, where the footing

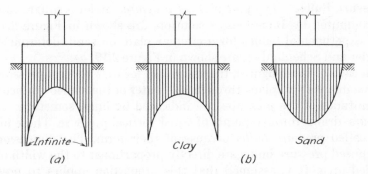

FIGURE 2.17   Contact pressures on surface. (a) Analytical by Boussinesq solution. (b) Clay and sand.

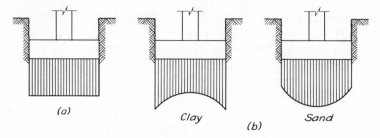

FIGURE 2.18   Contact pressures below surface. (a) Conventional assumption, on or below surface. (b) Experimental on clay and sand.

was on the surface. This would be expected because the sand grains under the edge of the footing have no lateral support and so can carry no vertical stress.

With clay, cohesion provides this lateral support. When the footing is below the surface, edge pressures are developed because the sand grains under the edge now have lateral support which is proportional to the depth below the surface. Lateral support increases, however, as the center of the footing is approached, and the pressures increase accordingly. The usual practice in the design of footings is to assume a uniform pressure distribution (Figure 2.18 (a)). The effect of other distributions should be considered, especially in combined and continuous footings and in raft foundations. The elastic deflection of the foundation on the distribution of foundation pressures should be considered.

The effect of the distribution of contact pressures on the unit vertical pressures under a footing does not extend deep enough to be of any importance in settlement computations.

Pressure Bulbs.   The *unit vertical pressures* under a square footing, as computed by Boussinesq's solution, are shown in Figure 2.19(a) and experimental values for a circular plate on sand, determined by Kogler and Scheidig [4], are shown in Figure 2.19(b).

In each of these figures, the values are for unit vertical pressures at points in a vertical plane through the center of the loaded surface. The magnitudes of the pressures are indicated by lines, sometimes called *isobars*, drawn through points of equal vertical pressure. These figures are called *pressure bulbs* because of their form. The dimensions of computed pressure bulbs are directly proportional to the width of the loaded area. It is assumed that this proportion applies to pressure bulbs in actual soils, but the variations in the soil do affect the dimensions of the bulb. The distribution of the contact pressures is different

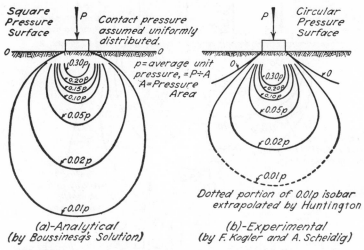

FIGURE 2.19   Bulb of pressure for an isolated spread footing.

for clay and sand, but the effect of this difference extends only below the immediate surface. The upper portions of the pressure bulbs for clay and sand are therefore different, but in other respects they are considered identical.

The isobar for zero stress is at the surface for Boussinesq's solution, (Figure 2.19); but for the experimental bulb, it extends downward and outward from the edge of the footing (Figure 2.19(b)). Note that the isobar does not close. It cannot pass under the footing, as do all other isobars, because nothing would be supporting the load along the surface of which this line is a trace.

Buildings are usually supported on several footings. The pressure bulbs for the individual footings gradually merge into a single bulb for the entire foundation (Figure 2.20).

The pressure bulb for a square mat supporting all the columns of a building is identical to that for a square footing (Figures 2.19 and 2.21).

Most of the settlement of a footing, a group of footings, or a mat is due to the unit vertical pressures included within the bulb for pressures of $0.2p$. For convenience, this will be referred to as the *significant bulb of pressure*. Pressures outside this bulb can be neglected.

Bell Diagrams. The distribution of unit vertical pressures along horizontal planes at various depths is illustrated in Figure 2.21. These

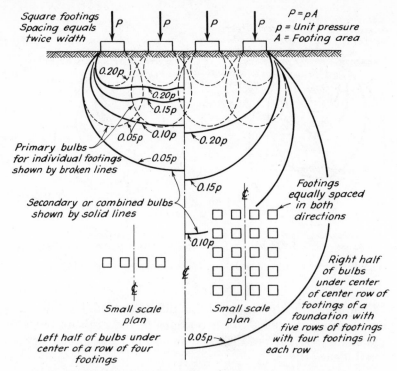

Analytical for groups of footings (by Boussinesq's solution)

FIGURE 2.20   Bulbs of pressure for a group of spread footings.

are *bell diagrams*. The "stress volume" included under each of these diagrams is equal to the load. The pressure bulb corresponding to the bell diagrams is shown. At each point where an isobar crosses the corresponding horizontal line, the ordinate of the bell diagram is equal to the pressure represented by the isobar. The distribution of unit vertical pressures along a vertical line through the center of a footing is shown in Figure 2.21. The abscissas in this figure are equal to the corresponding midordinates of the bell diagram. A pressure bulb is analogous to a contour map; and a bell diagram, to a profile.

9. PROPORTIONING THE BEARING AREAS OF SPREAD FOOTINGS

The bearing areas of spread footings must be proportioned to satisfy the following requirements:

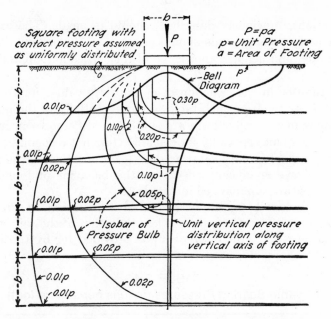

FIGURE 2.21   Relation between bell diagrams and bulb of pressure.

(1) There must be an adequate factor of safety against failure of the underlying soil, which ruptures when the shearing stresses produced by the footing loads exceed the shearing strength of the soil. The factor of safety commonly required for stresses caused by dead and live loads only is 3. If stresses produced by wind loads or earthquake shocks are included, the factor is reduced to 2; these loads are not considered to act simultaneously.

(2) The settlement of any footing from the compression of the underlying soil must not be more than use conditions permit.

(3) The relative, or differential, settlement between any adjacent footings must not be excessive. The differential settlement permitted depends on the consequences of such settlement. In the absence of complete information, differential settlement is often limited to ¾ in.

Code Values for Allowable Soil Bearing Pressures.   The most common procedure for selecting values for the *allowable pressures* on various soils is to consult the governing building code or values recommended by appropriate agencies. This information is sometimes supplemented by bearing capacity tests conducted in accordance with specified procedures, described later. Usually, appropriate subsurface investigations, previously described, should be made. The extent of these investigations depends upon occupancy, economic importance of the building, and general knowledge of difficulties anticipated at the site.

Uncertainties in the selection of appropriate values for the allowable soil pressures from such tables are due partially to the indefinite descriptions of the soils. They are subject to differences in interpretation. This is true especially for clays. Factors not recognized in code values or in the conventional procedure for proportioning the bearing areas may have significant effects on the behavior of foundations. Some of these are summarized in Article 10.

The allowable soil pressure for an underlying stratum, as determined by settlement, is in some instances less than that of the soil on which a foundation rests. This discrepancy is provided for by the require-

TABLE 2-4
Allowable Bearing Pressures (in tons per square foot)

	American Standards Association [5]	
Class	Material	Pressure
1. Massive crystalline bedrock such as granite, diroite, and trap rock in sound condition		100
2. Foliated rocks such as schist and slate, in sound condition		40
3. Sedimentary rocks such as hard shales, siltstones, sandstones, limestones; also thoroughly cemented conglomerates, in sound condition		15
4. Soft or broken bedrock of any kind except shale		
5. Exceptionally compacted or partially cemented gravels, sands, and hardpan		10
6. Gravel, sand-gravel mixtures, compact		6
7. Gravel, loose; coarse sand, compact		4
8. Coarse sand, loose; sand-gravel mixtures, loose; fine sand, compact		3
9. Fine sand, loose		2
10. Stiff clay and soft shales		4
11. Medium stiff clay		2.5
12. Medium soft clay and soft broken shales		1.5
13. Except in cases where, in the opinion of the building official, the bearing capacity is adequate for light frame structures, fill material, organic material, and silt shall be treated as being without any presumptive bearing capacity; the bearing value may be fixed by the building official on the basis of tests or other satisfactory evidence		

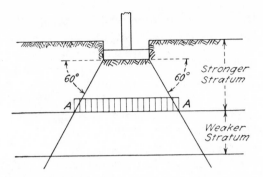

FIGURE 2.22   Assumed distribution of vertical pressures under a footing.

ment that the unit pressure due to the footing load computed on the top surface A-A of the weaker stratum (Figure 2.22) shall not exceed the allowable soil pressure for the material in that stratum. The load here is considered uniformly distributed over the area intercepted on the top surface of the weaker stratum by planes sloping from the edges of the foundation 60° with the horizontal, as shown in the figure. This is an arbitrary assumption and does not follow the form of the bell diagram in Figure 2.21. However, the procedure is considered sufficiently accurate for its intended purpose and is simple in its application.

In estimating settlement, the allowable pressures on underlying soil are considered as pressures that the building can add to the pressures already there. This is equivalent to neglecting the weight of the soil in determining the pressures on underlying soil for comparison with allowable values.

Bearing Capacity Tests.   Tests help select the allowable soil pressure. The tests are conducted in a pit bottom several feet below the ground surface. Test loads are usually applied in pits which have bottoms at the identical elevation as the bottoms of the proposed footings. Loads are applied to a test area at least one foot square on the bottom of the pit. There should be no other loads on the bottom of the pit.

In one procedure, a load equal to the proposed allowable soil pressure is applied and settlement readings taken at least every 24 hours. The test is continued until there is no more settlement during a 24-hour period. The load is then increased 50% and the settlements observed. The proposed allowable soil pressure is considered satisfactory if the initial settlement is not more than ¾ in. and if the additional settlement is not more than 60% of the initial settlement due to the load equal to the proposed soil pressure. The settlement resulting from the load equal to the proposed soil pressure is sometimes limited to ½ in. The test is repeated, if required, until an appropriate allowable soil bearing value can be selected.

A test area of 4 sq. ft. is sometimes required, especially for soils with the lower allowable soil pressures. Tests for several test areas are required for selecting allowable soil pressures when there are major differences in the proposed footing sizes.

Procedures for interpreting loading-test data are arbitrary and not entirely satisfactory. The results are affected by the relation between the test area and the area of the pit bottom. If the test area occupies a large portion of the pit bottom, the loaded soil receives considerable support from the soil surrounding the pit, but if only a small portion of the bottom is loaded, this support is minimal. Experienced judgment, guided by examination and tests of the underlying soil, must be used in the interpretation of test results.

This type of test may have little value in determining allowable soil pressure. The soil pressures under a footing that account for most of the settlement are those included within the pressure bulb equal to $0.2p$ (Figures 2.19(a) and 2.21). The pressure bulb extends to a depth approximately one and one-half times the width of the footing. This is the *significant bulb of pressure* (Figure 2.23). For a load test,

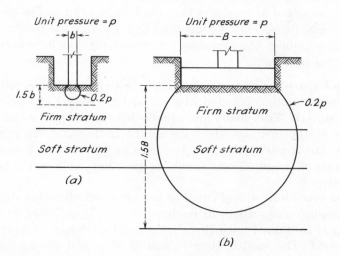

FIGURE 2.23   Comparison of significant bulb of pressure for test load and for actual footing. (a) Test load. (b) Actual footing.

this bulb extends only a short distance below the loaded surface, but for an actual footing (Figure 2.23 (b)), the bulb extends much deeper. If the actual bulb passes through an underlying soft stratum, much of the footing settlement might be due to the compression of the stratum. The settlement (due to the soil compression under a given unit pressure) varies directly with the dimensions of the loaded area, similar to pressure bulb increases.

Ultimate Bearing Capacity. The bearing capacity of a building foundation may be determined by the permissible settlement or by actual rupture of the soil. When a load is placed on soil, shearing stresses set up in the soil cause the grains to move in relation to each other. This fact is demonstrated by the photograph in Figure 2.24; a time exposure is taken as a rod was forced into sand contained in a box with the exposed side made of plate glass. A flat side of the rod was placed next to the glass. During the test, the path of each sand grain movement in contact with the glass is photographed by time exposure [6]. The exaggerated vertical movement of the rod is greater relatively than permissible footing settlement, but the photograph illustrates the tendency for the soil to displace laterally and upward. A similar phenomenon has been observed in clay [7]. The soil grains near the rod move outward and upward along curved paths, and

FIGURE 2.24   Displacement of sand by surface load.

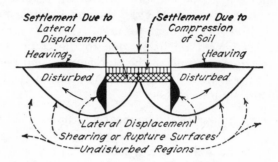

FIGURE 2.25   Causes of settlement of footing.

heaving occurs at the surface. Outside a fairly definite boundary, the soil is not displaced.

In the disturbed region, the shearing stresses along the curved paths exceed the shearing resistance of the soil, but in the undisturbed region the shearing resistance exceeds the shearing stresses. The displacements are illustrated in Figure 2.25. Analytical solutions have been devised for computing the maximum load that a soil with a known shearing strength will carry.

Failure of Earth Slopes.   Earth slopes of cohesive soil may fail by rupture with no load except the weight of the earth itself, as manifest by landslides. Buildings placed on or near a sloping earth surface may contribute to slide failures and are damaged or destroyed. Some earth slopes fail very slowly, and failure is foretold by excessive cracking in the buildings or by gradual settlement. In other earth slopes, the failure may be sudden because of the weakening of the soil by absorbed rainwater, the increased soil weight, and the force exerted by percolating water. The effect of earthquake shocks compounds the situation.

A common failure of earth slopes is diagrammed in Figure 2.26. A portion of the earth embankment designated as $ABCD$ tends to slide on the curved surface BC (a weakened shear plane) and to pile up as shown by the irregular lines. If a part or all of a building occupies the horizontal surface affected, the building will be damaged or will collapse. Experience has shown that earth slopes, except clean sand or gravel, fail on roughly curved surfaces. In computations of earth slope stability, the assumption is usually made that failure will occur on a smooth arc, called a *slip circle*. The portion of the slope that

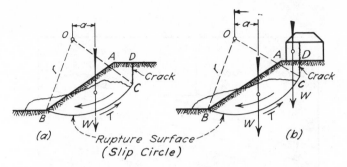

FIGURE 2.26 Failure of earth slopes.

slides is considered to rotate about 0, the center of the arc. The resistance to failure is offered by the shearing strength of the earth along the arc. If the moment $W \cdot a$ (in Figure 2.26), representing the forces tending to cause failure, exceeds $T \cdot r$, the moment of the resisting forces, failure results. If $W \cdot a$ is less than $T \cdot r$, the slope is stable, the factor of safety is measured by $T \cdot r \div W \cdot a$.

The arc BC, on which failure is most likely, must be located by trial, because the arc will give the least factor of safety in the computations. If a building is on the summit of a slope, or within the slip circle, (Figure 2.26(a)), the weight W' with the moment arm, b, must be included in the computations. Where forces other than these exist, they must be considered. A slope that is stable prior to building may fail because of the added building load. Buildings should not be on or near clay slopes unless measures are taken to prevent failure from the causes outlined in this paragraph. Instead of passing through a building (Figure 2.26(b)), the rupture surface may emerge a considerable distance behind the building, as in deep-seated failures. The sliding block of soil carries the building to destruction. Analytical methods for investigating the stability of earth slopes are recommended.

Bearing Capacity of Rock. Rock gradually grades into soil, and therefore no definite value can be given for the bearing capacity of rock. The bearing capacity of rock is given in Table 2-4. Sound bedrock usually will support any load transmitted by a concrete pier because the allowable unit compressive stress in the concrete is less than the allowable bearing stress in the rock. A given material is stronger in bearing when the load is applied to a small portion of its surface than when formed in a prism such as a pier. Failure in bearing could occur only by shear, in a manner similar to the failure of soil that carries a surface load (Figure 2.25). As stated in Article 4, some shales yield

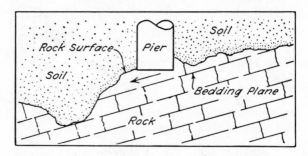

FIGURE 2.27   Tendency to slide along bedding plane.

by plastic flow when subjected to bearing pressures. Thus, shales should always be carefully investigated before subjected to heavy bearing pressures.

When rock is considered for foundation support, compression tests should be made to determine a safe bearing value. Allowable bearing pressures given in Table 2-4 are in general terms.

Other accountable factors in examining rock for proposed foundations are the direction of stratification or dip, as illustrated in Figure 2.27, structural defects, such as excessive fractures, caverns and solution channels in limestone, faults and fissures, and possible weak underlying strata. If a fault crosses the building site, the possibility of movement along the fault and possible resulting consequences should be considered.

As previously stated, a test hole drilled 10 ft. or more in the bottom of a rock excavation may yield a core that would reveal any objectionable structural defects or unsatisfactory underlying material.

10.   FOUNDATION SETTLEMENT

All foundations settle except those on sound rock. When the supporting soil is dense, coarse sand or gravel, the amount of settlement is relatively small, unless the loads are excessive and unless the settlement occurs as soon as the load is applied. In contrast to sand and gravel, settlement of clay may be large and may continue over many years. In extreme cases, the settlement of structures founded on clay is measured in feet.

Settlement may not be objectionable if the entire foundation settles uniformly, if the settlement is not excessive, and if the settlement is predictable within reasonable limits. Such settlement causes no structural damage. However, *differential settlement*, or different amounts of settlement in adjacent parts of a foundation, introduces internal stresses that may seriously weaken a building by throwing various parts of the building out of plumb. Differential settlement produces unsightly cracks, causes floors to be out of level and uneven, causes roofs to leak, interferes with door operation, and unbalances mechanical equipment.

In the design of spread foundations, it is usually assumed that, if the supporting soil is fairly uniform over the building site, settlement will be uniform when the unit pressures exerted are the same for all parts of the foundation. In computing pressures, only the weight of the building and imposed long-term static loads are considered to cause settlement. Additionally, the maximum unit pressure in any part of the foundation for any possible condition of loading is usually limited to some arbitrarily adopted allowable soil pressure—that this practice is unsatisfactory, particularly for cohesive soils such as clays, has become increasingly evident. Experience with actual foundations supported by experimental and theoretical investigations show that, within certain limits, the settlement for a given unit pressure on a given soil increases as the lineal dimension of that area increases. This practice has also been unsatisfactory because of the inadequate and faulty methods used for determining the allowable bearing pressures for soils.

Causes for Settlement. The possible causes of foundation settlement are:

(1) Reduction in the volume of the voids of the supporting soil and, therefore, in the volume of this soil, which is the *compressibility*. This is the principal and usually the only important cause of foundation settlement.

(2) Actual rupture of the soil.

(3) The lateral displacement of the soil is so small that it need not be considered unless the lateral support of the soil is reduced by adjoining excavations or unless shallow foundations bear on soft clay.

(4) The compression of the soil grains themselves is negligible because the unit stresses in the soil grains, in relation to the loads placed on foundations, are too small to be of any consequence.

As explained in Article 7, the most effective method for reducing the settlement of foundations supported on a deep bed of soft clay is to excavate the basement of the building to a depth sufficient enough to make the weight of the soil removed equal to the weight of the building and its permanent contents. Thus, increased soil pressure from the building load is avoided.

Settlement Predictions.   Settlement predictions for structures founded on clay are based on the consolidation phenomenon mentioned in Article 4. They are made by computing the unit vertical pressures, caused by the building loads, below the foundation and by making consolidation tests at various depths on representative samples of the soil that supports the building. The procedures followed in computing the unit vertical pressures were considered in Article 8. The consolidation test is made in a specially designed apparatus in which the soil sample is compressed and the time rate at which the sample deforms under a given unit pressure is observed, provision being made for water to escape at the top and bottom surfaces of the sample. Mathematical relationships have been developed which enable the settlements to be estimated from the unit foundation pressures and the consolidation characteristics of the samples, due consideration being given to the location of the porous strata. These samples in the test apparatus should be *undisturbed samples*. Settlement predictions require specialized knowledge and experience and are the province of an engineer soil specialist.

Lowering of Water Level.   The groundwater elevation may affect the settlement of foundations bearing on soil. The water level often lowers after a building has been completed. It may occur because sewer or drainage systems are installed or subways, depressed highways, and other underground works are constructed. Many years may elapse before this groundwater decline occurs.

If foundations are on sand or gravel with an underlying stratum of clay and the water level is in these permeable materials, the pressures on the underlying clay stratum may be increased by lowering of water level because of the reduced buoyant effect of water or the effective weight of submerged sand or gravel.

If the foundations are on clay, lowering the groundwater level will promote the consolidation of the clay. Also, the clay soil may be compressed further by the increased capillary tension in the soil, as explained in Article 4 and illustrated in Figure 2.2.

Effects on Adjoining Buildings.   Foundations of existing buildings supported on water-bearing sand or silt may be undermined by adjoining excavations because of pumping operations to control the water. Soft clay soils may squeeze into excavations for piers and walls. This condition is shown by the excess of the excavated material over the volume of the excavation. Excess of excavation, or *lost ground*, may result in a corresponding settlement of adjoining foundations.

Even though a building is constructed only to the limits of its own site, it causes vertical as well as lateral pressures in the soil of the adjoining sites. The increased vertical soil pressures under the foundations of an existing building due to the construction of a new building on an adjoining site may cause the existing building to settle.

A building may be damaged from upheaval of the soil resulting from the reduction of soil pressures when an adjoining building is removed, but soils that will produce this effect are rare.

Shrinking and Swelling. As explained previously, clay soils shrink and swell as the moisture content varies markedly during long periods of drought or rainy weather or with seasonal changes. These effects are most pronounced in the upper few feet of soil and therefore involve buildings on shallow foundations. Some clays are more susceptible to this action than others; those with high percentages of bentonite are especially objectionable. Shrinking and swelling tend to be more pronounced on exterior footings than on interior footings because of the greater protection given to the latter by the building. Extensive damage may occur only after a lapse of many years after construction.

Some exterior footings have settled considerably more than the interior footings because of shrinkage, and in others, the exterior footings have risen considerably above the interior footings because of soil expansion. The possibility of heaving and settlement should be considered when building in regions where this has taken place or where expansive clays exist.

The only apparent means of avoiding these difficulties is to carry the foundations to depths where they will not be affected.

Summary. Comments about settlement may be summarized as follows:

(1) For footings bearing on a given soil and causing equal soil pressures on the bearing area, the following relationships prevail for uniform deposits of sand, gravel, and clay foundation materials.

　(a) The settlements are approximately proportional to the widths of wall footings and to the linear dimensions of isolated footings which have the same ratios between their lateral dimensions.

　(b) The settlements of wall footings are greater than those of isolated footings.

(2) The settlement of footings founded on sand or gravel occurs during a

short period of time and then ceases, but for those founded on dense clays, the settlement may continue for many years.

(3) For uniform deposits of sand or clay, most of the settlement of a footing acting independently is due to the vertical soil pressures included in a depth equal to one and one-half times the width of the footing, as indicated by the pressure bulb for $0.20p$ in Figures 2.19 (a) and 2.21, which may be called the *significant bulb of pressure*.

(4) In uniform sand or clay deposits, most settlement of one row isolated footings is indicated by the significant bulb of pressure shown on the left side of Figure 2.20. The pressure bulb for several rows of isolated footings is shown on the right side of Figure 2.20.

(5) If the underlying deposits of soil are more compressible than the soil on which a footing is founded, the underlying deposits may contribute significantly to the settlement of the footing or to a foundation consisting of several isolated footings, especially if they lie within the significant bulbs of pressure.

(6) The settlement of shallow footings founded on some clays may significantly increase because of a diminution of the moisture content during periods of drought, or such footings may heave during periods of heavy rainfall.

(7) Foundations of unconsolidated sand or gravel may settle owing to a decrease in void spaces caused by excessive vibrations within a building or by operations adjacent to the building. Foundations on clay are not affected.

(8) Foundations on clay or soil underlaid with clay deposits may settle if the ground water is lowered.

11. PILES AND PILE FOUNDATIONS

Types of Pile Foundations. Pile foundations are similar to spread foundations in some respects. Spread foundations transmit loads directly to the soil, whereas pile foundations transmit loads first to the piles which, in turn, transmit the load to the soil or rock. The upper end of a pile is called the *head* or *butt*, and the lower end the *point*, the *tip*, or the *nose*. An *isolated* or *independent pile footing* is illustrated in Figure 2.28 (a). The member into which the butt ends of the piles are embedded is called the *pile cap*. A *cantilever pile footing* is illustrated in Figure 2.28 (b) and a *raft* or *mat* in Figure 2.28 (c). *Combined* and *continuous pile footings*, corresponding to these spread footings, are used extensively. Several types of pile are used: wood piles, precast concrete piles, cast-in-place concrete piles, structural steel piles, and steel pipe piles. Each of these types is discussed briefly in the following paragraphs.

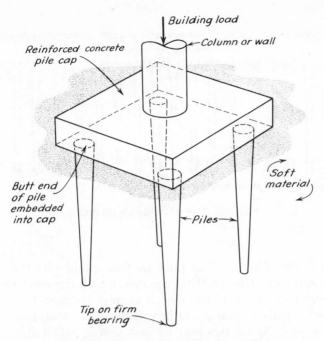

FIGURE 2.28(a)· Isolated pile foundation.

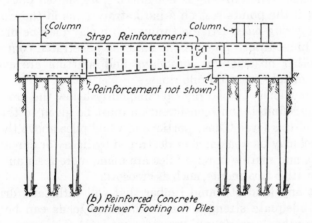

FIGURE 2.28(b)   Reinforced concrete cantilever footing on piles.

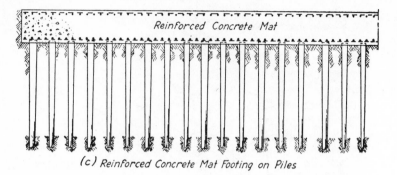

(c) Reinforced Concrete Mat Footing on Piles

FIGURE 2.28(c)   Pile foundations.

Wood or Timber Piles.   Wood piles are tree trunks with the branches trimmed and bark stripped. They are driven with the small end down, except under special conditions, such as in permafrost [31]. The tip may be cut square or pointed or provided with a metal point called a *shoe*, or a *pilot*. Metal tips may be cast iron or rolled steel. Pointed or blunt, metal tips may have *fins* that give more assurance for straight driving. A steel *pile ring* may protect the butt end of the pile from damage. When driving is not difficult, unpointed piles are used, especially if the points rest on a hard stratum, as in the *end-bearing* or *point-bearing piles.* Driving through firm clay may be made easier by using blunt-tipped piles. If the material penetrated contains boulders or other obstructions, *brooming* and splitting are avoided by installing metal shoes and pile rings.

Those portions of wood piles permanently below the ground water will last indefinitely, but consideration must be given to the possible lowering of this level. Those portions of wood piles above the ground-water level may be weakened or destroyed by decay or damage caused by insects and marine borers. Piles are made more resistant to decay by preservative treatments, such as creosote.

Almost any kind of sound timber that will withstand driving and that has adequate strength for the imposed loads can be used for piles below the "permanent" groundwater level. The most commonly used woods are the cedars and cypress (which are very decay resistant), Douglas fir (very long piles of excellent quality can be obtained from Douglas fir), southern pine, red oak, white oak, and spruce. Douglas fir, southern pine, and red oak respond better to treatment processes than the other woods. In general, heartwood is more resistant to decay than sapwood, but better treatment process results are secured with sapwood than with heartwood.

Wood piles are classified by ASTM D 25 according to use as *friction piles* or as *end bearing* piles. Further classification is made with respect to bark peeling, namely, *clean-peeled, rough-peeled,* or *unpeeled.*

Wood pile quality is based on the presence and severity of wood defects such as size, soundness and frequency of knots, holes, splits, checks, and the amount of grain twist.

Other requirements specify a relationship of butt diameter to tip diameter. The tip diameter is rarely less than 6 in. or the butt less than 8 in. Wood piles should be free from short or reversed bends and have a uniform taper, and a straight line extending from the center of the butt to the center of the tip should lie wholly within the pile. Limbs and knots should be trimmed flush with the surface of the pile [32].

The dimensions for wood piles are determined by the load to be carried and the nature of their use.

Precast and Pretensioned Concrete Piles. Piles constructed of concrete, both reinforced and pretensioned, are used extensively. Sizes vary from 9 in. to 54 in. in diameter, with lengths up to 300 ft. Tapered or of uniform section, piles are usually square or octagonal; but they may be cylindrical (hollow core or solid), triangular, or rectangular (tongue and groove sheet pile). The piles must be designed for erection stresses and the stresses produced by the subjected loads. In general, the bearing capacity is the resistance of the soil to the load that moves the piles through the soil, rather than by the strength of the pile itself.

The reinforced pile has longitudinal bars with hoops or spirals (Figure 2.29(a)). If the reinforcing is provided only because of the handling and driving stresses, the concrete coverage outside the reinforcing need only exceed 1 in. Reinforcing required by the service loads should be protected by at least 1½ in. of concrete; in extreme exposure conditions, 3 in. coverage is required. The top and point should be provided with additional lateral reinforcement to withstand the driving stresses. For piles driven in plastic clay soils, a blunt or flat point is preferable, whereas a long tapering point should be used in sand or gravel or for those that must penetrate hard strata. Where conditions are extremely severe, a metal point or shoe is desirable. If piles are jetted into position, as will be described later, a pipe is cast into the pile along its axis, with the pipe end contracted to a nozzle.

Precast piles are sometimes pretensioned in the same manner as columns. The principal advantage of prestressing piles is to avoid cracks that could lead to corrosion of the reinforcement by ground water.

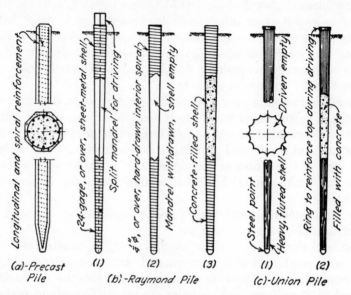

FIGURE 2.29 Precast concrete pile and cast-in-place shell piles. (a) Precast pile. (b) Raymond pile. (c) Union pile.

Precast concrete piles may be spliced by bars embedded in epoxy grout or by embedded steel welded together.

Pretensioned concrete piles are exceptionally well adapted for passing through water. Their durability, which assures less corrosion, ideally suits them for waterfront construction.

Cast-in-Place Concrete Piles. Concrete piles that are cast-in-place, or *in situ*, in the ground are of two general types: the shell or cased piles, and shell-less or uncased piles.

A *shell pile* is formed by driving a closed point sheet steel shell or casing into the ground and filling it with concrete. In the *Raymond pile*, shown in Figure 2.29(*b*), the tapered shell is made of relatively thin sheet steel with an 8 in. minimum diameter. Straight-sided shells are also used. This pile is formed by: (1) inserting a steel *mandrel* or core inside the shell and driving the mandrel and shell; (2) removing the mandrel and inspecting the shell interior; and (3) filling the shell with concrete. For piles up to 40 ft. in length, the shell is in 8-ft. sections with overlapping joints; for piles with lengths exceeding 100 ft., the shell consists of sections screw jointed. The *Union* or *Monotube pile* (Figure 2.29(*c*)), is formed by: (1) driving a heavy, fluted, sheet steel, tapered shell into the ground and inspecting the interior; and (2) filling it with concrete. For piles up to 40 ft., the shell is one piece;

and for piles to 100 ft. or more, sections are factory welded or field welded. The *MacArthur cased pile* is formed by: (1) driving a heavy steel casing with an inserted core; (2) removing the core and inspecting the casing; (3) inserting a corrugated steel sheet; (4) filling the shell with concrete; and (5) placing the core on top of the concrete and withdrawing the casing, leaving the concrete-filled shell in the ground.

A *shell-less* pile is formed by: (1) driving a shell with an inserted driving core; (2) withdrawing the core; and (3) filling the shell with concrete while withdrawing the shell. This type is the *Simplex pile* in Figure 2.30(*a*) and the *Pedestal pile* in Figure 2.30(*b*). As the casing is removed, the concrete in the Simplex pile is forced against the soil by the fluid pressure of the concrete and by impact during placing. Greater pressure may be exerted, if desired, by tamping the concrete.

The *Pedestal pile*, which may also be a pressure injected footing, (Figure 2.30(*b*)) is formed by: (1) driving a casing with a core into the ground; (2) removing the core and placing concrete in the bottom of the casing; (3) replacing the core and pulling the casing up $1\frac{1}{2}$ to 3 ft. while exerting pressure on concrete with the core; (4) ramming concrete to form a pedestal; and (5) removing the core, filling the casing with concrete, replacing the core to exert pressure on concrete, and pulling the casing to form the finished pile.

The *Franki* (Figure 2.30(*c*)) concrete foundation is a *pressure injected footing*. The completed footing features an enlarged compressed concrete bulbous tip which is forced into intimate contact with the surrounding densified soil. This unique footing forms by compacting a charge of 3 to 5 cu. ft. of no-slump concrete through a steel sleeve or "drive-tube" that guides a free-falling ram. The concrete plug is rammed into the earth, followed by the guide sleeve. At the desired depth, the penetration of the guide sleeve is stopped and the concrete is bulbed-out by compressing it against the bearing soil by further ramming. In the *uncased shaft*, the pipe is then withdrawn in stages of 12 to 24 in. increments. The ram is withdrawn while the pipe guide-sleeve is recharged with concrete—concrete which is rammed into close contact with the surrounding soil, compressing and forming a dense compact shaft. The piles are usually unreinforced; however, reinforcement can be added to meet design requirements.

Another method, used by the Franki Foundation Company, employs a *cased shaft* which is a steel casing that is inserted through the drive guide sleeve after the bulbous tip has been formed. This steel casing is embedded into the compressed pressure injected footing,

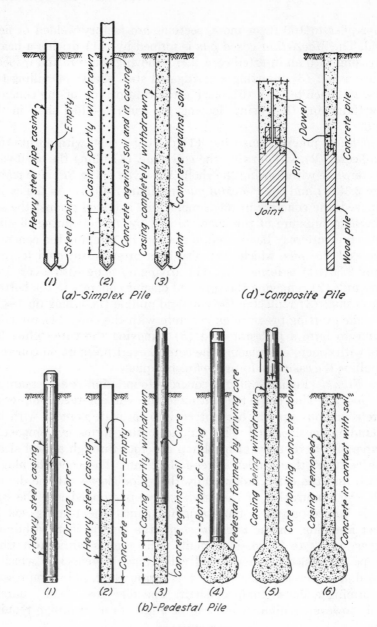

FIGURE 2.30

FIGURE 2.30   Cast-in-place shell-less piles, or pressure injected footing.

and then the guide sleeve is withdrawn. If reinforcement is required, it is placed in the casing, and then the casing is filled with concrete. This pressure injected footing has the distinct advantage of great resistance to uplift and is easy to form and construct.

These piles, which may be *battered* (sloped) or vertical, have been seated in soils ranging from fine sands to coarse gravels [33].

Composite Piles. Wood piles and cast-in-place concrete are combined to form *composite piles* (Figure 2.30(d)) to take advantage of the lower cost of wood piles for the portion below ground water and durability of concrete piles for the upper portion. If the Raymond pile is used, a wood pile is first driven nearly full length. The shell and mandrel are then placed on the wood pile and driven. The mandrel is removed and the shell filled with concrete, forming a concrete pile on top of a wood pile. The joint must be designed for driving and for excluding water and soil. The joint must withstand separation due to heaving soil when adjacent piles are driven. If the Mac-

Arthur cased pile is used, the wood pile is driven inside the casing and the concrete pile is formed in the usual manner.

Pipe Piles. Piles of this type consist of heavy steel pipe filled with concrete (Figure 2.31(a)). Pipes are driven with open ends or, for the smaller sizes, with the bottom end closed. The usual types of pile hammers are used, or hydraulic jacks are used if there is a load to jack against, as in underpinning.

The resistance to penetration is reduced by cleaning out open-end piles as the driving progresses. This is particularly true for piles driven through sand because the sand plugs the end of a pile and the resistance offered is the same as though the end were closed. Cleaning out is usually accomplished by pushing a 2½ in. pipe into the soil within the pile and forcing compressed air through at a pressure approaching 100 lb. per sq. in. This air pressure blows the soil out of the pile. Water may be mixed with the soil to facilitate its removal. Pipe piles are also cleaned with small orange-peel buckets and earth

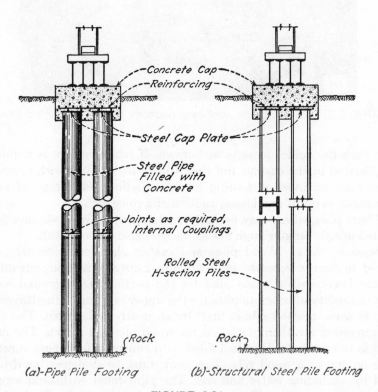

(a)-Pipe Pile Footing   (b)-Structural Steel Pile Footing

FIGURE 2.31

augers. Clay cores that enter open-end pipe piles have been removed by withdrawing the pipe and core, forcing the core out with a plunger, and redriving the pipe. The pipes may be placed in sections with joints holding the sections in position, assuring good bearing between the ends of the sections, and no projecting on the outside of the pipe. Concrete is placed by dumping it at the top of the pile, but the bottom portion of long piles should be placed with drop-bottom buckets to avoid segregation. Water should be removed from the pipe before the concreting, but if dewatering is impossible, a concrete seal is placed by a drop bucket. After sealing, the water can be pumped out and the remainder of the concrete placed in the dry. After the concrete is set, a steel bearing plate may be grouted over the upper end of the pile to secure good bearing on the steel and the concrete. In determining the bearing capacity of pipe piles, both the steel and concrete are considered as carrying stress with allowance for the different moduli of elasticity or stiffnesses of the two materials. The pipe walls are commonly $\frac{3}{8}$ in. thick but may be thicker. The diameters vary from a 10 in. minimum. In calculations, a deduction of $\frac{1}{16}$ in. is sometimes made in the thickness to allow for corrosion. Available information indicates this is unnecessary for piles driven in undisturbed soils [25].

Structural Steel Piles. Hot rolled-steel H-sections, called *H piles*, (Figure 2.31(*b*)) are increasingly used for bearing piles. This type of pile is particularly advantageous in driving to bearing on sound rock passing through soil where driving is difficult because of boulders or thin strata of hardpan or rock which are underlaid with weaker strata and therefore do not have adequate bearing capacity. Because of their small sectional area, steel bearing piles displace only a small volume of soil and are ineffective in compacting loose sands or increasing their bearing capacity. They are uneconomical as ordinary friction piles. Moreover, the full surface area is unavailable for frictional resistance because soil wedges between the flanges, causing the pile to act as a square pile. Available information indicates that corrosion is not a serious factor and that protection against corrosion is required only for structural steel piles driven in disturbed soils. Steel H piles range in depth from 8 to 14 in. and weigh up to 177 lbs. per linear ft.

Comparison of Types. The life of wood piles can be prolonged by impregnating them with creosote under pressure. Untreated, fire resistant wood piles should be used only below the permanent groundwater level. Concrete and steel piles do not have this limitation. A considerable reduction in excavation, masonry, and imposed loads may be realized by using concrete piles.

Where the piles project above ground surface, precast concrete piles are usually more suitable than cast-in-place piles, although the shell type can be reinforced longitudinally to give satisfactory service. Precast piles must be reinforced to withstand the flexural stresses which occur during handling and the stresses due to driving, but the stresses are usually compressive after the piles are in the ground and do not require reinforcing. Tensile stresses may be induced, however, by soil heaving brought on by driving adjacent piles, causing the rupture of shell-less, cast-in-place, under-reinforced piles. The rebound of the soil and the driving of adjacent piles may reduce the section or may otherwise damage the piles before the concrete has set. The driving of adjacent piles may collapse the shells of shell-type cast-in-place piles before the concrete is placed or may damage the piles themselves immediately after the concrete is placed. To avoid damage to cast-in-place piles by adjacent driving operations, all the shells for removable casings for a pile group are usually driven before any of the concrete is placed or no pile is filled within 5 ft. of a future pile location. The shell-type cast-in-place pile excludes water and earth from contact with the fresh concrete of the pile.

The required lengths of precast piles must be determined before they are ordered. They must be cut off in the field and excess length wasted if they cannot be driven to the required depth. Withdrawing the casing in forming shell-less cast-in-place piles or encased piles in which the shell is inserted in the casing before withdrawal releases the stresses set up on the soil during driving and makes any dynamic pile driving formula inapplicable.

To develop their full capacity, pipe piles and structural steel piles are usually founded on rock or hardpan and are usually used to support heavy loads. Structural steel piles can penetrate hard strata and gravel. Corrosion of pipe or sheet piling in structural piles is not a consideration.

The allowable load on individual piles varies from 15 tons to 150 tons, mainly depending upon the type of pile and the supporting material. The minimum value is for a wood-friction pile; the maximum value is for a large steel pipe pile bearing on rock. Larger loads can be carried by steel-encased concrete cylinders constructed in the same way as pipe piles, but cylinders with 24 in. or larger diameters are sometimes classed as *piers*.

Almost any type of pile can be secured or constructed in lengths as great as 140 ft. Structural steel H-bearing piles 200 ft. long were used to support an office building in New Orleans [40].

Further comparisons are made later in this chapter.

Pile Driving. Piles of various types and dimensions are driven by a pile driver operating in conjunction with a drop hammer, power hammer, or water jet.

When a drop hammer is used, the pile is lifted into vertical position by the pile driver. A heavy weight called a *drop hammer* is then dropped on the head of the pile, driving it into the ground, the weight guided in its fall by the *leads* of the pile driver. The hammer is raised by a steam engine, a gasoline engine, or an electric motor, and the process is repeated until the desired penetration is reached. This type of hammer is obsolete.

The *steam hammer* is used extensively. It rests on the head of the pile continually while its driving power, derived from the reciprocating parts of the hammer itself, strikes blows in such rapid succession that the pile is kept in almost continuous motion. Steam hammers are *single-acting* and *double-acting*. A single-acting steam hammer is a heavy ram which is raised 2 to 4 ft. by steam admitted under pressure to a cylinder located above the ram, which falls by gravity when the steam is exhausted. The steam pressure acts against the underside of a piston connected to the ram by a piston rod. The ram is guided in its fall, and the various parts are held together by a frame (Figure 2.32(*a*)). The hammer is positioned on top of the pile

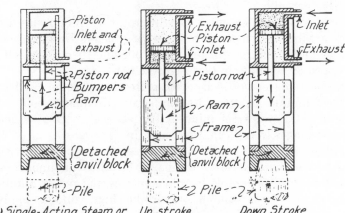

(a) Single-Acting Steam or Air Hammer on Up Stroke (b) Double-Acting Steam or Air Hammer

FIGURE 2.32 Steam or air hammers. (*a*) Single-acting steam or air hammer on up stroke. (*b*) Double-acting steam or air hammer.

by using a pile driver or by suspending it from the boom of a derrick or from a gin pole. A double-acting steam hammer is similar to a single-acting hammer. The ram is raised 4 to 20 in. by steam admitted under pressure to the cylinder on the lower side of the piston, but instead of falling by gravity alone, as in the single-acting hammer, the ram is forced down by steam under pressure admitted to the cylinder on the upper side of the piston synchronous to the exhaustion of the steam on the lower side. The principle of the double-acting hammer is shown in Figure 2.32(*b*).

Steam hammers can be operated with compressed air as *pneumatic hammers*. Some types are converted to operate under water. Steam hammers may be inverted for use in pulling sheet piles and sheeting. Power hammers drive piles more rapidly and with less damage than drop hammers.

One of the most noteworthy developments in pile driving has been the invention of the *vibratory pile driver-extractor*. In soils that previously caused driving difficulties—particularly sands and gravels —pipe, H piles, and sheet piles have been driven with dramatic ease. The theoretical principles supporting this vibratory hammer are not discussed here, although the mechanical device is briefly described.

This hammer, which features relatively silent operation, drives and extracts piles or caissons with amazing speed. An example of the capabilities of this hammer is illustrated by one project using 355 HP 14 x 177 lbs. by 55 ft. piles, where each pile was driven to *refusal* in about 7 minutes—by other methods this would require nearly 1½ hours for each pile.

Two motor-driven eccentric weights rotating in opposite directions impart a vibratory motion to a driving head. Undiminished, this vibratory motion is transmitted to the pile and then to the soil, which is displaced by the pile tip (see Figure 2.33).

The pile vibrates vertically with a frequency equal to the revolutions per minute of the rotating eccentric weights—or vibration ranges to 1800 cycles per minute. The vertical motion or *amplitude* of the pile ranges from ¼ in. to about 2 in.

In addition to the stated advantages, the vibratory driver does not distort or batter unprotected pile butts as an impact hammer would. It is easily operable in cold weather, and power requirements may be satisfied by available or portable high voltage supply [38].

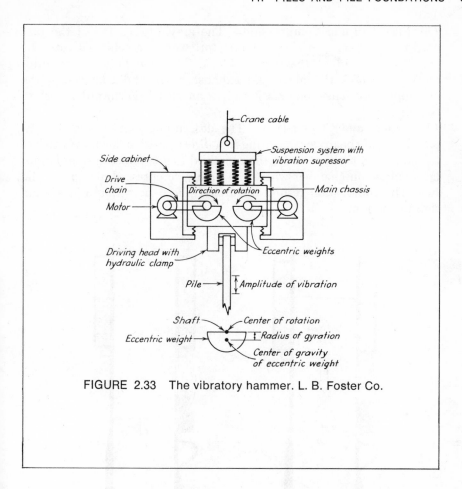

FIGURE 2.33   The vibratory hammer. L. B. Foster Co.

Because of their simplicity, efficiency, and capabilities, *diesel hammers* are becoming a wider and much preferred choice for many pile driving jobs. Water-cooled or air-cooled, they may be used for vertical or inclined driving. The energy ratings range from about 868 ft.-lbs. for the smallest hammer to 117,175 ft.-lbs. for the largest.

The extremely simple operating principle is described and illustrated (Figure 2.34) below [39]:

(1) **Free Fall and Compression.** The gravitational fall of the ram activates the cam on the fuel injector, which injects fuel into the cup-shaped head of the impact block. The free falling ram automatically closes the air intake-exhaust ports and compresses the entrapped air. This compression gives an initial downward thrust to the pile.

(2) **Impact and Combustion.** The descending ram strikes the impact block of the driver, resulting in a direct positive downward thrust to the pile. Simultaneously, the atomized fuel is forced into an annular combustion chamber where the highly compressed air ignites the fuel. The combustion results in further pile penetration and a reaction to the ram.

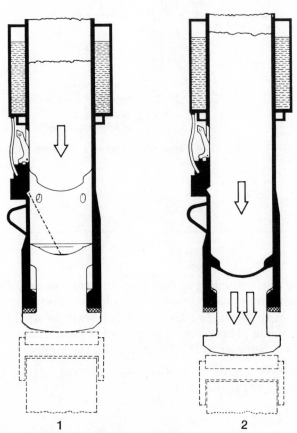

1   2

FIGURE 2.34   Diesel pile hammers. The Foundation Equipment Corp. and Delmag-Maschinenfabrik.

(3) **Power Stroke.** The reaction causes the ram to recoil upwards. Exhaust ports open, and burnt gases are expelled.

(4) As the ram elevates to its summit, it opens the air intake ports and activates the fuel system. The injector is refueled for the next cycle.

Working Principle of the DELMAG Diesel Pile Hammers

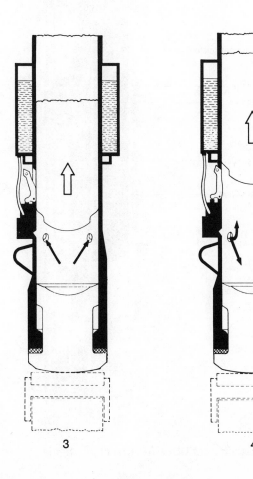

3   4

Cross of DELMAG–D22

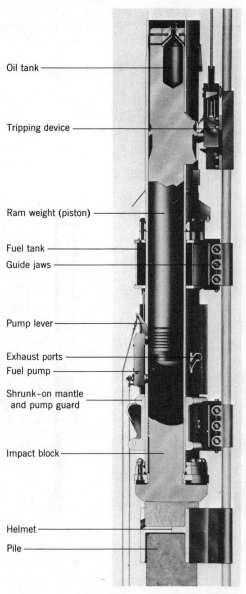

Oil tank

Tripping device

Ram weight (piston)

Fuel tank

Guide jaws

Pump lever

Exhaust ports

Fuel pump

Shrunk-on mantle
and pump guard

Impact block

Helmet

Pile

Cross-section of DELMAG Diesel pile hammer.

The *water jet* or *jetting* is used in driving piles. It consists of a pipe (placed at the side of a pile) with water under pressure washing the material away from the tip of the pile. The pipe is about $2\frac{1}{2}$ in. in diameter, with the lower end decreasing to form a $1\frac{1}{2}$ in. nozzle. The upper end of the pipe is connected to a force pump by a hose. The pile drops into the space formed by the water jet. Some water from the jet rises to the surface and lubricates the sides of the pile and decreases the friction of the surrounding earth, thus assisting materially in driving. A load placed on the pile assists in forcing it down, or else light blows are struck with a pile hammer. In addition to the jet that delivers water to the point of the pile, jets may be used to deliver water along the side of the pile to assist in decreasing the frictional resistance to correspond with that developed in driving without the jet. The final penetration is usually given with a pile hammer after the jet has been turned off.

Concrete piles often have the jet pipe cast in place along the axis of the pile. Piles driven with a water jet are not damaged in driving, and therefore this method is particularly suitable for precast concrete piles. The water jet may be used in many classes of material, but it operates most successfully in sand. Building codes restrict or ban the jetting except by special permission because soil may be disturbed around adjacent foundations, thus causing settlements.

The pipe for pipe piles used in underpinning is often forced into position by *hydraulic jacks* that act against the building. This operation requires less space than a pile driver and causes less disturbance.

12. SETTLEMENT OF PILE FOUNDATIONS

For convenience, it is usually stated that piles support foundations; however, piles are intermediate members which transmit foundation loads to the underlying soil or rock. Such piles are called *bearing piles* to distinguish them from *sheet piles* and other kinds of piles, such as *fender piles* and *guide piles*. Interlocking sheet piles are driven juxtaposed to form a continuous barrier or sheet, retaining earth or forming an effective watertight diaphragm. Fender piles and guide piles have special functions as indicated by their names.

The bearing capacity of the soil into which piles are driven may be fairly uniform throughout the length of the piles which transmit the load from the pile to the surrounding soil. The piles may pass through soft soil with little or no bearing capacity and then penetrate into firmer soil to which the load is transmitted by friction, or the piles may pass through soft soil to hardpan or rock to which the load is transmitted by the pile tips. The entire load on a pile foundation is considered to be transmitted through the piles with no load transmitted directly to the soil by the pile cap.

Design Requirements. Pile foundations must satisfy the following requirements:

(1) The pile caps or mats must be structurally adequate to transmit the building loads to the piles.

(2) Each pile must have adequate strength to carry its load and adequate bearing capacity to transmit its load to the supporting foundation material.

(3) The entire foundation must be proportioned so that excessive building settlement is prevented.

(4) The differential settlements between adjacent footings must not exceed permissible values. The consequences of these settlements are the same as they are for spread foundations, as explained previously.

(5) The shearing stresses in the supporting soil must not exceed an adequate factor of safety against rupture of the soil. A safety factor of 3 is adequate for dead load plus full live load, and a factor of 2 is adequate when wind loads or earthquake shock are included; the design is based on the most critical condition.

(6) To insure stability, a minimum of two rows of piles should support a continuous or wall pile footing, and a minimum of three piles should support an isolated or column footing.

Causes of Settlement. Some causes of friction pile foundations settlement are:

(1) An individual pile may settle by moving through the surrounding soil, as shown in Figure 2.35(a). An isolated pile normally settles because of the compression of the surrounding soil to which it transmits stress, as indicated by the pressure bulb (b). However, piles are not used individually.

(2) A pile footing and the surrounding soil will settle because the foundation load transmitted by the piles will compress the soil, as indicated by the pressure bulb (Figure 2.35(c)).

(3) A cluster of piles supporting a footing together with the enclosed soil may settle by yielding as a unit and moving downward with respect to the surrounding soil (Figure 2.35(d)).

(4) An entire pile foundation of several pile footings will settle because the soil underlying the structure as a whole compresses. This condition is normal for building foundations having a number of

isolated footings; but unexpected settlements of large magnitude may be caused by undetected layers of low bearing capacity soil, such as peat. To avoid this difficulty, adequate subsurface exploration of the soil should be conducted for a considerable depth below the pile tips. Foundation settlement is caused by the compression of the soil and results from the stresses transmitted to it by the piles, as indicated by the pressure bulb in Figure 2.35(e).

Other causes of settlement are the lateral movement of soil under a foundation, removal of lateral support or undermining, compacting of loose sand by vibrations, and damage to piles during driving.

Pile Action.   Unless resting on hardpan or rock, a pile is supported by *skin friction* along its length and *point resistance* at its tip, both exerted by the soil that surrounds the pile. A pile supported primarily by skin friction is called a *friction pile* or a *floating pile*, and one supported principally by point resistance is a *point-bearing* or *end-bearing pile*. The distribution of the load between skin friction and point resistance depends upon the type of soil penetrated and is discussed subsequently. Below the pile tip, the load is supported entirely by the underlying soil. A pile, therefore, increases the bearing capacity of a soil by distributing the load through a wider area and deeper into the soil. A footing on the ground surface, a friction pile, and a pile that transmits its load primarily by point resistance illustrate their pressure distributions in Figure 2.36.

Piles driven in deposits of loose sand compact the sand and thereby increase its bearing capacity. Piles driven for this purpose only are called *compaction piles*.

Pile Clusters or Groups.   Piles are not used singly but are arranged in groups or clusters centered under the loads. These loads are distributed to the individual piles by rigid *pile caps* of reinforced concrete similar to spread footings. A footing bearing on piles is called a *pile footing*.

It is sometimes assumed that the bearing capacity of a pile group is equal to the bearing capacity of one pile multiplied by the number of piles. This is true when the piles bear on rock or if the piles are spaced so far apart that the overlapping of pressure bulbs is of no consequence. Generally, bearing piles can be spaced more closely than friction piles. The spacing factors are the compressive strength of the piles and the bearing capacity of the supporting foundation material.

There is no entirely satisfactory criterion for the spacing of friction piles. It is sometimes considered that the perimeter of the pile group should equal at least the sum of the perimeters of the indi-

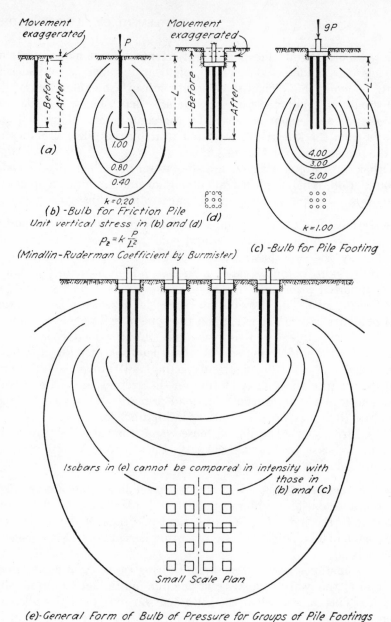

Movement exaggerated

Before
After

P

Movement exaggerated

Before
After

L

$9P$

L

1.00
0.80
0.40

$k = 0.20$

(a)

(b) -Bulb for Friction Pile
Unit vertical stress in (b) and (d)

(d)

$$p_z = k \frac{P}{L^2}$$

(Mindlin-Ruderman Coefficient by Burmister)

4.00
3.00
2.00

$k = 1.00$

(c) -Bulb for Pile Footing

Isobars in (e) cannot be compared in intensity with those in (b) and (c)

Small Scale Plan

(e)-General Form of Bulb of Pressure for Groups of Pile Footings

FIGURE 2.35   Bulbs of pressure and settlement of piles and pile footings.

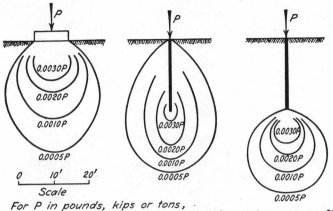

For P in pounds, kips or tons,
  vertical pressures are in pounds, kips or tons per sq. ft.

(a)-Surface Load   (b)-Friction Pile   (c)-Point-Bearing Pile
(Boussinesq's Solution)   (Mindlin-Ruderman Coefficients)

FIGURE 2.36   Comparison of bulbs of pressure.

vidual piles. This conclusion is ·based on the assumption that the shearing resistance per unit of surface area is the same for the piles as for the prism of earth included within the perimeter of the pile group (Figure 2.35(d)). The center-to-center spacing of piles is usually required to be not less than 2½ or 3 times the diameter of the round piles or the diagonal of rectangular piles, but the minimum of 2 ft. 6 in. is often used.

The distribution of vertical pressures on a horizontal plane at the points of the piles is somewhat as shown in Figure 2.37. In this figure, it is assumed that the piles act independently and that the total vertical pressure at any point on the horizontal plane is equal to the sum of the vertical pressures at that point produced by the individual piles. On this basis, the settlements of the center piles would obviously be greater than those of the outside piles. However, the rigid cap would cause all of the piles to settle nearly the same amount. This tends to increase the pressure under the outer piles and decrease it under the center pile, but the pressure under each center pile would still be greater than it would be under a single pile. Mat footings founded on piles as shown in Figure 2.28(c) tend to settle more at the center than at the end due to this effect.

A simple useful procedure assumes that the load from a group of piles is distributed outward and downward at an angle of 60° with the horizontal; the load extending from the perimeter of the group of piles at an elevation where the piles enter a stratum that has satis-

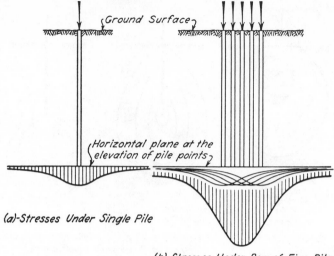

(a)-Stresses Under Single Pile

(b)-Stresses Under Row of Five Piles

FIGURE 2.37 Vertical pressure distribution at elevation of pile points.

factory bearing capacity. The soil pressures are considered uniformly distributed on the portions of horizontal planes included within the 60° limiting planes. This procedure is illustrated in Figure 2.38. Each pile is assumed to carry by point resistance a load equal to the allowable unit soil pressure at the point of the pile multiplied by the area of the point. The remainder of the load is assumed to be carried by skin friction and to be transmitted to the surrounding soil at a uniform rate along the pile. The actual soil pressures at the depths where the soil type changes are computed and compared with allowable soil pressures stated by the code. These allowable soil pressures must not be exceeded. Only the pressures due to the foundation load are included. Overlap of planes of adjacent footings must be considered.

Effect of Width of Foundation. Half pressure bulbs for a spread footing and a friction pile footing are compared in Figure 2.39(a) indicating that the piles lower the pressure bulb significantly and thereby increase the bearing capacity. Half pressure bulbs for a mat or raft foundation supported on the ground surface and on friction piles are shown in Figure 2.39(b). It is evident that the piles have little effect on the pressure bulb, except near the ground surface, and therefore are not effective in reducing the settlement or increasing the bearing capacity of the mat footing [9].

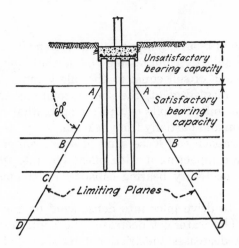

FIGURE 2.38 Assumed distribution of vertical pressures under pile footings.

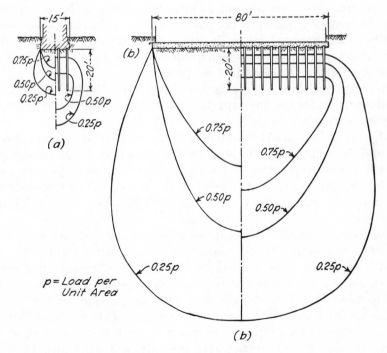

FIGURE 2.39 Relation of width of foundation to effectiveness of piles.

Effects of Type of Soil.   When a pile is driven, it occupies a space made available only by compressing the surrounding soil or by displacing the volume of soil equal to the pile volume. To compensate for this volume displacement, the soil surrounding the pile rises or *heaves*. In some soils, the pile occupies a space that results partially from compression and partially from heaving.

The bearing capacity of loose sand or other compressible fill may be improved by compacting it with piles. The pile spacing required to produce the necessary bearing capacity by compaction is determined by trial.

It is difficult to drive piles into dense sand or gravel without the aid of jetting. The water jet decreases the point resistance and, in some instances, decreases the frictional resistance along the length of the pile. The jet washes the sand grains away from the point and up along the pile surface to the top of the ground. If a jet is used, the final penetration of the pile should be secured by driving with the hammer only. Jets are not as effective in clay soils. If heaving must be reduced, possibly because it affects adjacent piles, holes may be bored or cored out with driven open-end pipe to receive the piles which are driven into final positions. The depth of the holes is usually less than the length of the piles.

The point resistance of a pile driven into clay does not contribute significantly to its bearing capacity, but it is a factor in the bearing capacity of the piles driven into dense sand.

When piles are driven through a layer of partially consolidated soil and into firm soil, the load on the piles may be gradually increased as the layer consolidates and settles. The increase is caused by the frictional drag on the piles as the settlement progresses and may result in increased settlement of the foundation or even failure of the piles.

The bearing capacity of a pile may increase or decrease after the driving is completed. If the pile is driven into coarse sand that can drain freely, there is little change after a period of rest. In fine sand and silt, which are less permeable than coarse sands, and in clay, the displacement of soil by the pile will build up pressure in the water in the voids of the soil surrounding the pile. This pressure resists the penetration and increases the point resistance. Since the soil is slowly permeable, however, water pressure will gradually be dissipated and the point resistance will decrease accordingly. Piles driven into clay displace the corresponding volume of clay, causing heaving and compressing the clay very little, but a small amount of water is squeezed out of the clay adjacent to the pile tip and rises around the pile. This lubricates the pile surface and reduces its frictional resistance. The vibration of the pile during driving makes the hole slightly larger than the pile and forms a space around the pile into which the water can flow. After driving, the clay surrounding the pile gradually reabsorbs the water and "sets" around the pile, thereby increasing the skin friction. Since most bearing capacity of piles driven in clay is due to skin friction, the increase in skin fricton after a period of rest increases the bearing capacity [10]. This is the reason that piles driven in clay "freeze up" after driving operations have been suspended for a short period. The resumption of driving will be more difficult and require breaking this frictional bond before further penetration is made.

Except in clean sands and gravels, the adhesion between the soil and the pile is usually greater than the shearing strength of the soil itself, so that the resistance along the sides of the pile is usually the shearing strength of the soil immediately surrounding the pile. This shearing resistance depends upon the internal friction and the cohesive strength of the soil. Because of the compacting effect of the pile during driving, the shearing strength of the soil close to the pile may be greater than that of the soil at some distance from the pile. The shearing area, of course, increases as the distance from the pile increases.

When clusters of piles are driven in clay soils, the driving of one pile may cause adjacent piles already in place to rise and become unseated. If this occurs, all of the piles so affected must be seated by partial redriving.

Bearing Capacity of Individual Piles.   The bearing capacity of individual piles is estimated by pile formulas or measured by applying loads to *test piles*. The formulas may be static or dynamic. These formulas calculate the resistance of a pile to loads tending to cause movement with respect to immediate surrounding soil and not to the bearing capacity of the surrounding soil. Pile settlement is movement of the pile through the soil or movement of the pile and the surrounding soil as a unit. The latter is caused by the compression of the soil under the action of the pressures induced in it by the load transmitted from the pile to the soil.

Static formulas estimate the bearing capacity by computing the shearing resistance along the surface of the pile and the bearing point resistance. The shearing resistance and point resistance are obtained experimentally. The extreme variation in soils of the same general class makes it difficult to determine suitable values for the pertinent properties in estimating shearing and bearing resistance.

Dynamic formulas compute the pile bearing capacity from its behavior during driving. The factors used in all formulas are the energy using in driving and the average penetration produced by the last few blows. Other factors may be the pile weight, its cross-sectional area, its length, and its modulus of elasticity. These factors compute the energy lost during the hammer impact on the pile.

There are a plethora of driving formulas available to serve some expressed or assumed soil condition. Any formula must be used with caution and measured judgment and only after a careful investigation of the prevailing conditions. The formula used most often is the *Engineering-News* formula, which has the advantage of simplicity. For an interesting and fuller discusion of this subject, the student is directed to read reference [31].

The frictional resistance and the point resistance of a pile in permeable materials, such as sand, gravel, and permeable fills, are nearly the same during driving as under static load. When a pile is driven into these materials, a theoretically sound, dynamic formula can be expected to give good results. The frictional and point resistances that prevail during driving, however, differ greatly from resistances that act when a pile is at rest if a pile is driven in relatively impermeable materials, such as fine-grained silts and soft clays. As explained in the previous section, skin friction in soils of this type is reduced during driving by water squeezed out of the soil, which lubricates the surface between the pile and the soil, and point resistance is increased by the pressure built up in the pore water as the point displaces the soil. After a period of rest, the water along the pile surface is absorbed by the soil and frictional resistance is restored, but point resistance decreases because the excess water pressure at the point is dissipated as water gradually moves from the region in which the pressure is built up by driving.

Dynamic pile driving formulas are applied to cast-in-place concrete piles. The formula may be logical when applied to shell piles where the shell is rigid enough to maintain the compression in the soil surrounding the pile. With shell-less piles, the conditions are changed so radically by removing the casing that such a formula is not a good indicator of bearing capacity.

The most satisfactory method of determining the bearing capacity of individual friction piles is by actually loading piles and observing their behavior during driving and under static load. These are called *test piles.* Enough test piles should be used to certify that the variations in the soil conditions in different parts of the site are determined. A common requirement is that the allowable load causes no settlement for 24 hours, and the total settlement shall not exceed 0.01 in. for each ton of test load. Another requirement is that the design load shall not be greater than 50% of the load, which causes a permanent settlement of ¼ in. in 48 hours. There are other modifications of this requirement. To be of value, the test loads should not be applied immediately after the pile is driven. Sufficient time should elapse enabling the soil to adjust itself to static conditions, thus avoiding the temporary effects of driving on frictional and point resistance. The bearing capacity of a pile may increase after a period of rest if driven in some soils; it may decrease if driven in others; or it may not change at all. In driving a test pile, the penetration per blow or the average penetration for several blows should be observed. When the penetration per blow of the other piles is equal to or less than the final penetration per blow of the test pile driven in the same material, those piles can be assumed to have a bearing capacity at least equal to that of the test pile. In using the test pile as an index of the bearing capabilities of other piles, the piles themselves, the kind of soil, the hammer, and the driving procedure must be the same as those associated with the test pile. This comparison of the final penetrations of piles with that of the test pile is not usually followed.

The procedure just described is satisfactory for determining the bearing capacity of individual piles as controlled by permissible settlement; but as has been explained, the bearing capacity of a group of friction piles is not always equal to the bearing capacity of one pile multiplied by the number of piles because of the overlapping of their pressure bulbs. Occasionally, groups of three or four piles are loaded with test loads, but the cost of applying loads to pile groups is so large that pile groups are not often tested. Any adjustment of the design load for the number of piles in a group is based on judgment, although building codes include formulas for this purpose.

Settlement Prediction. The procedure for predicting settlement of pile foundations supported by clay is similar to that for spread foundations outlined briefly in Article 10. The vertical pressures in the soil supporting the foundation are computed on the basis of Boussinesq's method (Article 8). Representative undisturbed samples of the soil are subjected to consolidation tests, which give data about the amount that they are compressed when subjected to given pressures. By mathematical procedures, the total compressive effect of all the soil under a building producing settlement is predicted from the vertical pressures and from the behavior of the samples. Piles complicate the situation in regions covered by the depth to the pile tips. Below that elevation, the conditions are identical with those under a spread foundation. It is necessary only to include the soil to a depth twice the entire foundation because at greater depths the soil pressures are so low that they have very little effect on the settlement, provided that there are no layers of peat or other soft soil at greater depths.

Formulas for the bearing capacity of piles can only give information about the loads that individual piles will carry without the pile moving through the soil that surrounds it. They give no data concerning the settlement. A loading test gives data concerning the settlement of an individual pile under a given load produced by any movement of the pile with reference to the soil and to a compression of the soil surrounding the pile during the relatively short period covered by the test. Because clay compresses very slowly, the total settlement of test piles driven in clay is not obtained from the tests. Furthermore, the overlapping pressure bulbs of the individual piles and pile groups form a composite pressure bulb for the entire foundation, as indicated in Figure 2.35; therefore, the settlement of an isolated test pile due to the soil compression is not an index of the settlement when that pile is under a foundation. The loads on adjacent piles contribute to settlement because of the overlapping pressure bulbs. The pressure bulbs indicate that deep-lying strata would have no effect on the settlement of isolated piles or on individual pile footings but contribute to the foundation as a whole. This condition may be particularly significant if deep-lying soil strata yield excessively. Friction piles are ineffective in reducing raft foundations settlement as illustrated in Figure 2.39. The same situation exists in buildings supported on several isolated, friction pile footings. There are cases in which settlement has greatly exceeded the settlement of an individual pile carrying the same load during a loading test. The common assumption that settlement will be uniform if all piles are loaded equally is but a dream.

Buildings founded on sand or gravel usually have little settlement, and friction piles are not often used for these buildings except when the sand is unconsolidated and vibrations from machinery may cause the sand and the building to settle. Any settlement will take place within a relatively short period. Present methods for computing settlements are based on the theory of consolidation and apply only to soils such as clay, in which settlement is due to the squeezing of water out of the soil.

13.   PIERS

Buildings are frequently supported on concrete piers extended to a material, such as dense sand or gravel, firm clay, hardpan, or bedrock, having the necessary bearing capacity. Such buildings, usually of steel or reinforced concrete, have foundations supporting columns rather than bearing walls. The column load is distributed over the pier by grillage beams or by rolled or cast-steel slabs frequently 6 in. or more in thickness. The top 2 or 3 ft. of each pier is usually reinforced with steel spirals or hoops. If the height does not exceed 12 times the diameter, or if the diameter is 6 ft. or more, heavy reinforcement is not required; but special reinforcement is sometimes used to reduce the pier size. Piers deeper than twelve diameters and less than 6 ft. in diameter should be reinforced with vertical steel spaced uniformly around the pier not less than 3 in. from the surface.

If a pier is to rest on bedrock, the uniform section shown in Figure 2.40(a) is used. If supporting material is clay or hardpan, the bearing area must be increased; therefore, the pier is *belled out* (Figure 2.40(b)). Firm clay or hardpan will stand without support while the bell is excavated. The side slopes of the enlarged section usually make an angle of about 60° with the horizontal. The sides are again made vertical for about one foot at the bottom. The bell should be designed as a spread footing, with reinforcing near the bottom to carry tensile stresses if they exceed allowable values for plain concrete footings.

Various excavation methods are used. In these methods, some device, such as sheeting, sheet piles, or caissons, holds back the earth and keeps out water. In one method, soil-laden water in the excavation prevents caving of clay until some form of casing is provided. The sheeting or sheet piles may be removed as the concreting progresses, but they are frequently left in place. Caissons are always left in place, becoming a part of the piers. Formerly, piers were constructed of brick and stone; now concrete is used exclusively. Timber and steel, used during the excavating process and left in place, will not deteriorate excessively in ground water.

The eccentricity of wall columns adjacent to the property line is counteracted by reinforced concrete or steel girders extending from the wall piers to the nearest interior columns. (Figure 2.40)

The methods used for excavating *wells* (*open pits*) for concrete

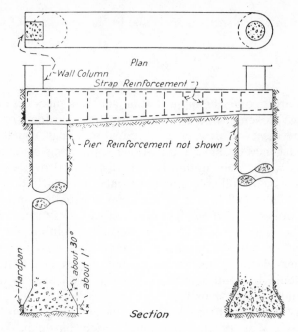

Plan

Wall Column

Strap Reinforcement

Pier Reinforcement not shown

Hardpan

about 30°

about 1'

Section

(a) *Reinforced Concrete Cantilever on Concrete Piers*

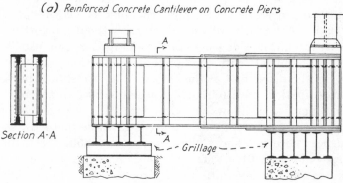

Section A·A

Grillage

(b) *Steel Cantilever Girder on Concrete Piers*

FIGURE 2.40   Cantilevers on concrete piers.

piers divide into two general classes, with several subdivisions, as follows:

(1) Open-Well Methods
   A. Simple excavation
   B. Vertical sheeting
   C. Vertical lagging
   D. Horizontal sheeting
   E. Sheet piling
   F. Steel cylinders
   G. Drilling

(2) Caisson Methods
    (a) Open caisson   (b) Pneumatic caisson

In the open-well methods, the excavation is performed under atmospheric conditions; the earth and ground water is retained in various ways. Caissons are used only where ground water is held back by an earth plug in the bottom of the caisson, enabling the use of open-air methods. In the pneumatic caisson method, water is held back by compressed air while the excavation is carried on by men working in compressed air.

Definitions. The construction used in open-well methods can be classified as cofferdams if ground water is present. A *cofferdam* is a structure built to exclude water from a given area enabling work to proceed under atmospheric conditions. This allows working *in the dry*. Some leakage usually occurs, but it is controlled by pumping. A cofferdam may be a temporary installation or part of a permanent structure. Sometimes the difference between a cofferdam and a caisson is difficult to distinguish. In general, a self-contained structure independent of the surrounding soil for support is a *caisson*; but if the structure requires lateral support, it is a cofferdam, as in the case of sheet piling and sheeting. In some localities, all foundation piers are referred to as caissons regardless of the meaning of the term. Foundation piers are sometimes called *subpiers* to distinguish them from piers above ground.

Excavations usually require support on exposed vertical faces to prevent caving. Thus, a diaphragm, ordinarily made of wood planks, is braced against the exposed face. Various terms designate the *diaphragm* or its components, some having identical or similar meanings. The most common term is *sheeting*, also called *sheathing*; it is usually vertical, but may be horizontal. Vertical sheathing is often called *lagging*, but this term implies that the members are parts of a curved diaphragm, such as the staves of a barrel or drum. Occasionally, lagging members are called *poling boards*. Horizontal sheeting members are often called *breast boards* because they are placed against that face or breast of an excavation as the work proceeds. They are also called *cribbing* if they form a rectangle.

The diaphragm retaining the earth is often constructed of vertical members designated *sheet piles* made of wood, steel, or occasionally of reinforced concrete. Sheet piling is distinguished from vertical sheeting because sheet piling is driven in advance of the excavation.

Wide-flange steel beams are often driven vertically into the ground spaced a few feet and along the edge of a proposed excavation. These are called *soldier beams* because they stand erect rather than in the usual horizontal position of beams. As the excavation progresses, horizontal sheeting, or breast boards, is supported on the beam flanges.

Vertical sheeting, sheet piles, and soldier beams are supported by horizontal members called *wales*, or *waling*, or *rangers* placed at horizontal intervals along the height of the soldier beam. The wales are in turn supported by one end of transverse struts or braces; the means of support for the other end depends upon prevailing conditions. Inclined *struts* or *braces* supporting wales are also called *rakes*, or *rakers*, and *spur bracing*. The meaning of these terms will be clarified.

Type of Soil.   The excavation method adopted depends upon several factors; a few are: (1) the nature of the soil, (2) the depth of the excavation, and (3) the type of foundation under adjacent buildings. The simplest material to excavate is firm clay, which is usually sufficiently water-tight to excavate below the groundwater level by open-well methods. Any water leakage or seepage is pumped out. Firm clay will stand unsupported without caving while excavation proceeds several feet and will give ample time to provide support. Some open-well caissons excavated in clay more than 100 feet deep have stood unsupported for several hours. Water-bearing seams of sand and gravel introduce complications that may necessitate a different procedure than would normally have been followed; however, it may be impossible to control and seal off these seams.

Hardpan presents no unusual difficulty in excavating if reasonably water-tight and will normally stand unsupported until the concrete is poured.

Saturated sand offers little resistance to the penetration of caissons, but special methods must be used to keep the surrounding soil from flowing into the excavation. Excavated soil in excess of the volume required by the pier may enter from under adjacent foundations, causing them to settle. This is called *lost ground*. Every effort should be made to avoid lost ground, although this may not be serious if the piers are isolated or are not located adjacent to water mains, sewers, or subways.

Layers of boulders are often on rock strata that support piers. Boulders may be difficult to remove and may interfere with lowering the caisson and establishing the necessary seal between the caisson and the bedrock.

Usually piers founded on rock rest on the rock surface or are embedded into the rock except for deep basements. Two or more excavation methods are frequently combined in a single well where different types of soil are uncovered during excavation.

The various methods used in excavating for piers will now be described.

Simple Excavation. Unsupported wells or pits may often be excavated in stiff clay with no provisions for excluding water. The excavating is by pick and shovel or with air spades. The excavated material is removed in buckets hoisted manually or mechanically.

Vertical Sheeting. In excavating wells or pits, the earth is held in place by sheeting, which is driven and braced as the excavation proceeds. It is impractical to drive sheeting more than 10 to 16 ft. long. Deeper excavations necessitate driving another set of sheeting a few inches inside the previous set until the required depth is reached, Figure 2.41(b). This method is little used for excavating deep wells, principally because the well area decreases as the depth increases.

The decreasing section featured by this method may be avoided by sloping the sheeting outward sufficiently to permit driving the next set without decreasing the well area. This method is illustrated in Figure 2.41(c). The outward inclination of the sheeting creates an objectionable condition at the corners which makes this method unsatisfactory in loose sand, for it is difficult to retain this material at the corners.

Vertical Lagging. Vertical sheeting or poling boards and lagging is placed in lengths of 4 to 5 ft. as the excavation proceeds in open wells. Wells excavated using lagging for retaining soil, such as clays, are frequently circular. The 4 or 5 ft. is excavated and a set of sheeting is placed with the boards retained by metal rings placed inside, forcing the lagging against the earth. The well must be accurately excavated

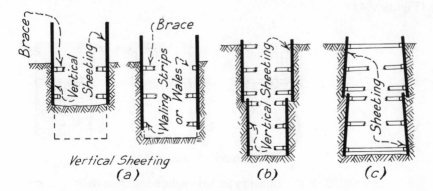

FIGURE 2.41   Vertical sheeting for open wells.

to avoid lost ground. After the first set is placed, another section is excavated, and another set of sheeting is placed. This process is repeated until the desired depth is reached. If difficult material is excavated, the sections may be as short as 18 in. Excavation may proceed manually.

The sheeting is 2-in. or 4-in. tongue and groove lumber with edges beveled to fit the curve, and with the metal rings varying in size from 3 in. by ¾ in. to 4 in. by 1 in. The rings are divided into semi-circles with bolted flanges at the ends and joined in pairs to complete the circle. In the wells for the Cleveland Union Terminal Building, clay squeezed into the excavation, threatening to collapse the lining. A disaster was prevented by inserting, where necessary, heavy wooden *drums* divided into two segments and forcing them against the lining by jack screws. These drums were removed as the wells were filled with concrete. (See Figure 2.53 later in this article.)

When using this method below groundwater level, pumping will be necessary because water will enter through seams in an otherwise impervious clay. Because it is mandatory to supply fresh air to the workmen, the exhaust of pneumatic spades may supply sufficient air for this purpose. Occasionally poisonous or explosive gases are encountered, and it is necessary to supply a large amount of air by blowers which force the air through lines extending to the bottom of the well. (Figure 2.42)

This method is designated the *Chicago method* or Chicago Board Method, for it originated in Chicago. The sheeting placed by this method is often called a caisson, but it is really a cofferdam.

An example of a deep foundation placed by this method was sixteen piers for the Cleveland Union Terminal Building, which was excavated 262 ft. below the curb and nearly 200 ft. into ground water. A combination of horizontal sheeting, steel sheet piles, and sheeting was used (Figure 2.42).

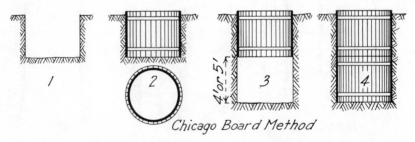

Chicago Board Method

FIGURE 2.42   Chicago Board method for open wells.

Horizontal Sheeting. Planks to retain earth may be placed horizontally; the wells are square or rectangular. The excavation need extend only a few inches below the last set of sheeting to provide room for the next set; thus, the method is suitable in soils likely to cave if a considerable depth is exposed. *Breast boards* or *curb planks* are planks, usually 2 × 8 or 2 × 10, placed on edge. In difficult material, the width may be reduced to 6 in. or even 4 in. Excavation is conducted in the same manner as the Chicago method. See Figure 2.43.

Sheet Piling. Instead of retaining the earth with sheeting, as in the above methods, sheet piling may be driven around the perimeter of a well in advance of the excavation (Figure 2.43). Excavation is conducted manually. If driving is easy, the sheet piling may be driven its entire length before excavating, but in all cases, the piles are kept in advance of the excavation. The piles are braced by horizontal and transverse frames as progress permits.

The simplest form of wood-sheet piling is wood planks driven side by side (Figure 2.43). This type of piling will retain earth but will not keep out water. A common form of wood-sheet piling is the *Wakefield piling* (Figure 2.44), which is three planks spiked to form a tongue and groove. Other forms of wood-sheet piling are shown in Figure 2.44. With the exception of the simple planks, all the forms are intended to keep out water and to hold back earth. Two forms of interlocking steel piling are shown in Figure 2.44. If wood piling is used, the well should preferably be square or rectangular (Figure 2.45), but if steel piling is used, a circular plan gives good results. Wakefield piling will withstand the impact of drop hammers, but steel

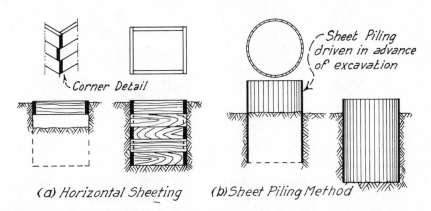

(a) Horizontal Sheeting   (b) Sheet Piling Method

FIGURE 2.43 Horizontal and vertical sheeting and sheet piling for open wells.

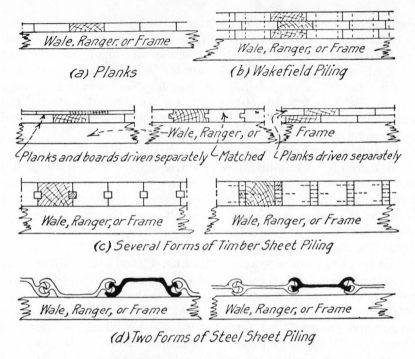

(a) Planks

(b) Wakefield Piling

Planks and boards driven separately   Matched   Planks driven separately

(c) Several Forms of Timber Sheet Piling

(d) Two Forms of Steel Sheet Piling

FIGURE 2.44   Types of sheet piling.

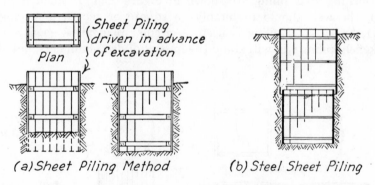

(a) Sheet Piling Method

(b) Steel Sheet Piling

FIGURE 2.45   Use of sheet piling.

sheet piling is usually driven with steam or air hammers or with resonant, vibratory, or hydraulic hammers.

For wells more than 25 ft. deep, two or more sections of piles may be driven offset as shown in Figure 2.45. The second section is a few inches inside the first section, after excavating to the bottom of the first section, and so on for other sections, allowing sufficient overlap to provide a water seal.

Wells to 60 ft. deep have been excavated by this method. Some forms of water-tight steel piling are used in this method for digging wells below ground water where water that enters the excavation neither exceeds pumping capabilities nor washes material under the piling into the well. Frequently, the piles are driven through porous water-bearing material until the ends of the piles are embedded in clay, which effectively seals the bottom.

Telescoping Steel Cylinders.   Telescoping steel cylinders from 5 to 8 ft. long may supplant vertical sheeting or sheet piling. These cylinders differ in diameter by 2 in. increments. The largest cylinder is sunk (Figure 2.46(a)) by excavating below the cutting edge and driving the cylinder down. The excavation is by hand. After the first cylinder is positioned, the second, which is telescoped inside, is sunk in the same manner and others in succession to the desired depth.

A bell is excavated if the soil permits (Figure 2.46 (b)). After the excavation is completed, the bell is filled with concrete (Figure 2.46(c)), and the cylinders are withdrawn as concreting proceeds until the pier is completed (Figure 2.46(d)). The excess concrete conforming to the increased section can be avoided by using a smaller cylindrical form. The annular spaces between the concrete and the earth walls are then filled with sand. Telescoping cylinders can be used in water-bearing soil that will not hold its shape, as required by the Chicago method, and where the cylinders are driven in advance of the excavating to avoid lost ground. A bell cannot be formed unless the final excavation is a suitable material such as clay. This is one form of *Gow pile* and might be classed as an open caisson.

Drilling Methods.   Various types of maneuverable equipment, truck

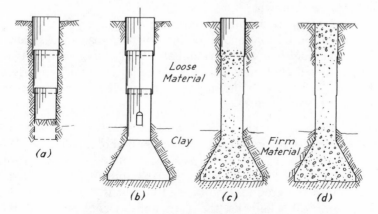

FIGURE 2.46   Telescoping steel cylinders.

or crawler mounted, are used in drilling wells for poured concrete foundation piers. The material penetrated is loosened or broken up and removed by drills, which are suspended from power cranes and guided by the cylindrical steel casings.

There are three types of drill used, namely, the auger and the bucket, both penetrating by rotation, and the grab-hammer or hammer-grab, which penetrates by vertical impact.

In some wells, partial protection is required; and in others, precautions are required to the full depth. Protection against caving is provided by vertical lagging or steel cylinder casings. Casings are $\frac{1}{4}$ in. to $\frac{3}{4}$ in. thick, as described later. If water-bearing formations are present, water is deterred from entering the well by steel cylinder casings, as will be explained.

Auger Drill. An auger drill (Figure 2.47(a)) is similar to a wood auger. It is mounted on the lower end of a power-driven solid or telescoping vertical shaft, called a *Kelly bar* (Figure 2.48). When drilling operations start, the drill tip is placed on the ground. As the shaft rotates, the drill bores a hole into the ground. The auger has two cutting edges and from three to five *helical turns* or *flights*. After the drill rotates sufficiently to fill the helical turns, the earth-laden auger is raised to the surface and spun rapidly to throw the earth from the auger by centrifugal force. This forms a ridge of loose earth around the hole. The earth is removed. Reaming devices attached to the drill are employed for excavating holes larger than the drill or for excavating bells at the bottom [13, 20]. Drills range in diameters to 20 ft. with capabilities of drilling to depths of 160 ft. [41].

The auger drill is suitable for excavating clay, sand, some hardpans, and other material that will remain on the helical turns while the auger is withdrawn.

Although piers are usually for large structures, small-diameter piers are sometimes used for small buildings such as residences. Piers may be desirable if suitable foundation material is at a considerable depth and especially if the soil is an expansive clay, the bottoms are often belled out. Wall bearing buildings are supported by *grade beams* on top of piers (Figure 2.64).

Bucket Drill. A *bucket drill* is a short steel cylinder or bucket mounted with its axis vertical on the lower end of a Kelly bar (Figure 2.48). The bottom of the bucket is so constructed that, when rotated in contact with the soil, it scoops up the soil, fills the bucket, and advances into the ground. One type of bucket is illustrated in Figure 2.47. When the bucket fills, the shaft and bucket are raised and rotated horizontally to clear the hole. The load is then dumped through

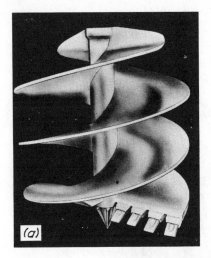

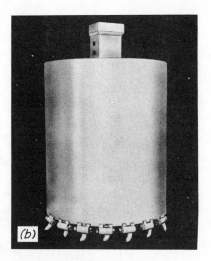

FIGURE 2.47   Earth driller. (a) Auger drill. (b) Core barrel. (c) Bailing bucket.

the movable bottom, and the bucket is lowered and refilled. Buckets have diameters up to 20 ft. and are capable of drilling to a depth of 200 ft. Wells with enlarged diameters and wells belled at the bottom can be drilled by attaching adjustable reamers or trimming arms to the bucket or shaft so that the trimmed earth falls into the bucket as illustrated in Figure 2.48 [13].

The bucket shown in Figure 2.47(b) is called a *core barrel*. It has

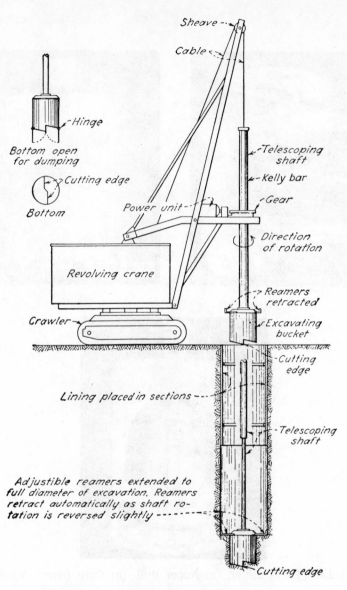

FIGURE 2.48 Well bored in clay with bucket drill.

Labels in figure:

Sheave

Cable

Hinge

Bottom open
for dumping

Cutting edge

Bottom

Power unit

Telescoping
shaft

Kelly bar

Gear

Direction
of rotation

Revolving crane

Crawler

Reamers
retracted

Excavating
bucket

Cutting
edge

Lining placed in sections

Telescoping
shaft

Adjustible reamers extended to
full diameter of excavation. Reamers
retract automatically as shaft rotation is reversed slightly

Cutting edge

replaceable teeth around its bottom edge. This type of bucket is suitable for excavating into solid or broken consolidated material in intermediate hardness.

The *bailing bucket* in Figure 2.47(c) is designed to remove water or mud that may contain small rocks. A valve at the bottom opens and admits the material while the bucket is lowered, closes as it is hoisted, and can be opened to discharge. It is not rotated.

Protection against Caving and Ground Water.   Caving of soil penetrated by auger and bucket drilling operations is prevented by installing vertical wood lagging when necessary, as in the Chicago method, Figure 2.42, or by inserting steel cylinder casings. These linings are placed in sections as the drilling progresses or after the excavation is completed, depending upon conditions. Steel casings are also employed where required to exclude ground water as the drilling progresses. When they are used, the bottom of the casing is kept far enough in advance of the excavation so that a plug of earth which excludes the water is formed. A steel casing can be driven in advance of the excavation until it penetrates an impervious formation to form a water seal which excludes ground water. Penetration into impervious formations is obtained by impact on top of the casing. Rock is sealed by cutting a circular seam while rotating a saw-toothedged drill casing during vertical thrust. The latter procedure is facilitated if the casing is somewhat smaller than the well. To avoid lost ground, it is necessary to fill the annular space solidly between the casing and the well. Wood wedges have been used for this purpose below the permanent groundwater level [13].

Lagging or steel casings may be left in place, or they may be removed for further use if economical. Some installations have characteristics of cofferdams and others of open caissons.

Hammer-Grab.   The *Benoto drill* is a grab-hammer or hammer-grab that is several feet long and weighs a ton or more. The hammer is similar to a drop hammer used in pile driving but has a grab at the lower end similar to an orange-peel or clam-shell bucket. The unit, alternately raised and dropped by a power driven cable hoist, is guided by a steel cylinder casing. When the hammer-grab is dropped, the jaws are open. They penetrate or chop into the material by impact. When the cable is hoisted, the jaws close, grab a load of soil, hoist and dump it. In addition to guiding the hammer-grab, the casing prevents caving and groundwater penetration. The tube or casing is extended by adding sections that are joined by special interlocking devices or by butt welding. In material difficult to penetrate, the jaws are locked open temporarily and the grab is dropped several times.

The lower end of the casing has a toothed cutting edge. It may be kept 3 or 4 ft. ahead of the grab, forming an earth plug which excludes water from the well in water-bearing soil; or if hard dry material is being penetrated, the grab is kept slightly ahead of the casing to facilitate the downward progress of the casing. After the well excavation has been completed, the bottom is prepared to receive the pier

FIGURE 2.49   Benoto hammer-grab.

and concrete. The casing is extracted progressively while the concrete is placed. Hydraulic rams withdraw the casing and exert longitudinal forces combined with partially reciprocating rotary motions.

The hammer-grab is suitable for excavating any type of material, including solid rock. In a large-diameter well containing water, a short vertical steel cylinder with an air lock on top, called a *diving bell*, can be lowered into a well to provide a working chamber where one or two men can work with pneumatic drills or place explosives. This arrangement can be used for the excavation of very hard ground or rock.

General Comments. The diameters and depths of piers depend upon the structural criteria, the depth to a formation capable of sustaining the loads, other geological conditions, and the drilling equipment available. There are no arbitrary limits on the diameters or depths of wells. When conditions are favorable, drilling is usually the most rapid method. In selecting a method for a specific project, relative costs and the time requirements are major considerations. Wells can be drilled in any type of material with appropriate equipment. A firm clay not requiring support against caving and containing no water-bearing seams is the ideal soil for drilling, but ideal conditions are rarely found. Auger and bucket drills are suitable only in soil and hardpans, but the hammer-grab can penetrate rock. Churn drills and special rotary drills excavate rock.

Open Caisson Method

The open caisson method consists of the following operations (Figure 2.50):

Constructing the caisson and preparing the site to receive it, *placing* the caisson over the site, *excavating* the soil on the interior of the caisson, *advancing* the caisson so that its cutting edge contacts or sinks below the bottom of the excavation and continuing this process to the foundation stratum, *sealing* the bottom of the caisson to exclude water and soil, *preparing* the excavated space to receive concrete, *examining* the foundation bed, and *placing* the concrete to form the pier.

Construction. Caissons are usually cylindrical, from 2 ft. or more in diameter, and are made of welded steel plates or of reinforced concrete. The thickness of the steel plate depends mainly upon the caisson, upon the material to be penetrated, and upon the method of advancing. Plate thickness varies from $\frac{1}{4}$ in. to about $\frac{3}{4}$ in. The cutting edges are often reinforced with steel bands placed on the in-

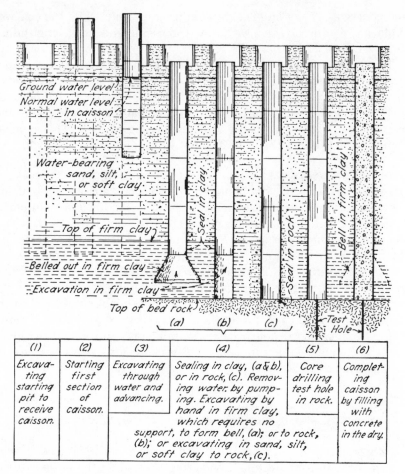

FIGURE 2.50   The open caisson method.

side. A steel caisson may be a full length single unit or its length increased by adding section units as the sinking progresses.

Placing.   A caisson is usually started in a sheeted *starting pit* above the groundwater level. To maintain the caisson in a vertical position, it may be propped against the sides of the pit, but it is often necessary to erect towers to maintain vertical alignment.

Excavating.   Excavation within a caisson depends upon the method used to advance it; therefore, these two phases must be considered together. Each operation, however, is listed separately. The soil is removed from the interior of the caisson by *manual* and *mechanical* methods. A pick or pneumatic spade is used to loosen the soil, which is then shoveled into buckets that are hoisted to the surface. Such a method is suitable above the groundwater level and is suitable below the water table only if the water flowing under the cutting edge and into the caisson can be controlled by pumping (Figure 2.51(*b*)). Mechanical methods employ the orange-peel and small clam-shell buckets which can be operated through water (Figure 2.51(*a*)). These are often *grab buckets* because they "grab" the soil and are self-filling.

Advancing.   Methods of advancing the caisson are as follows:

- By the weight of the caisson itself—always a factor in advancing the caisson; and in some soils, reinforced concrete caissons may weigh enough requiring no other applied force.
- By loads applied to the top of the caisson, such as concrete blocks, steel rails, pig iron, etc.
- By driving the caisson with pile hammers. This procedure is applicable only to steel tubular caissons reinforced at the top and thick enough to withstand driving. One hammer, acting on a beam across

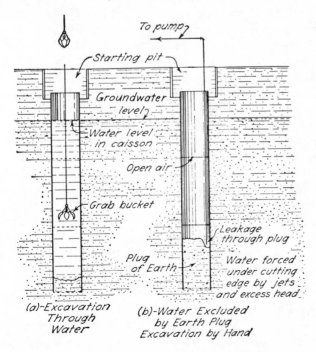

FIGURE 2.51   Methods of excavating in open caisson method.

the top of the caisson, may be sufficient; or two hammers acting simultaneously, but not necessarily synchronized, may be required.

• By reducing the resistance to penetration with water jets around the cutting edge to loosen and displace the soil. This is called *jetting* and is usually restricted by codes. Water jets supplement the caisson weight and are often used with loaded caissons or caissons driven by hammers.

• By drilling and blasting below the cutting edge to break up material that is difficult to penetrate, and by blasting near the cutting edge to jar the caisson loose momentarily and permitting it to fall a short distance. These procedures are occasionally used to supplement other methods.

• By driving the caisson with hydraulic jacks. This is possible only when there is a load to jack against, as there usually is in underpinning.

The usual procedure advances the caisson by placing loads on its top and excavating by hand to groundwater level. Below this level, water will usually enter the caisson under the cutting edge at such a rate that it cannot be controlled satisfactorily by pumping. The excessive inflow will wash large quantities of soil into the caisson. A serious objection to jetting and pumping is the possibility of undermining adjacent footings by the lost ground brought into the caisson by the incoming water, even when excavating through water. If the penetrated soil is fine sand, this action is quite pronounced, but it may cause no difficulty in some clays. Excavation is, therefore, continued by means of an orange-peel or clam-shell bucket. If the penetrated soil is clay with seams of permeable water-bearing soil, it may be possible to hand excavate the clay and to resort to grab buckets only while penetrating the permeable seams. Penetration may be facilitated by jetting if excavation is through water. In some soils, water jets make the addition of loads to the top of the caisson unnecessary.

If the caisson is driven with pile hammers or is forced down with hydraulic jacks, the cutting edge may be driven far in advance of the excavation. The soil plug in the bottom, as shown in Figure 2.51, reduces the inflow of water, allowing control by pumping and resulting in little lost ground, thus permitting manual excavation.

Sealing.   If the pier is founded on firm clay or hardpan, the caisson is *socketed* into the material to form a seal, as shown in Figure 2.50(b). Seepage is pumped out. Bells for the enlarged pier bases are excavated in the dry. If clay or hardpan overlies a rock foundation bed, the seal is established (Figure 2.50) and excavation is carried on through these materials to rock, using the procedure just given. If necessary, clay walls are supported by sheeting. Usually no support is required in hardpan.

If the porous water-bearing soil continues until rock is reached, a seal is formed between the bottom of the caisson and the rock.

Preparing the Bearing Surface. Where the excavation has been carried through a water-filled caisson, the rock is cleaned with a grab bucket or other device. A diver may make preparations and visual examination of the foundation bed. A *concrete seal* is then placed in the bottom of the caisson using a bottom-dump bucket. After this seal hardens, the caisson is pumped out and is ready for concreting.

When the caisson has been driven or forced with a hydraulic jack, an effort is made to socket the bottom of the caisson far enough into the rock to form a seal. If a plug of soil has been maintained to retard water inflow (permitting manual excavation), the plug cannot be removed until the seal is established. Cement grout pumped around the cutting edge may assist this operation. An uneven or sloping rock surface compounds the difficulty.

If the various methods for sealing between the bottom of a caisson and bedrock are unsuccessful, or if sealing through water is considered unsatisfactory because of the inability for preparing and examining the foundation bed, the open caisson can be converted into a pneumatic caisson. A top is placed on the caisson and air locks attached. Water is removed and investigating, preparing, and sealing the foundation bed is carried on under compressed air. The pneumatic caisson method, fully described later, may be necessary to employ before bedrock is reached if boulders or other obstructions are uncovered that cannot be passed in a simpler way.

Concreting. After the seal is established, water is pumped out and the caisson is filled with concrete using bottom-dump buckets.

Concreting Piers with Structural Steel Core. The load capacity of concrete piers in steel-plate open caissons can be increased by encasing a structural steel W or HP section (Figure 2.52(a)). Because of the inserted steel core, the bearing capacity of the rock stratum surface may be insufficient to carry the load transmitted by the pier. The required additional capacity is secured by drilling a hole into rock, the hole having a diameter equal to that of the outside of the shell. The bearing pile section is carried several feet below the surface of the stratum. The shell and the hole in the rock are then filled with concrete as shown (Figure 2.52(b)). By this means, the pier load is

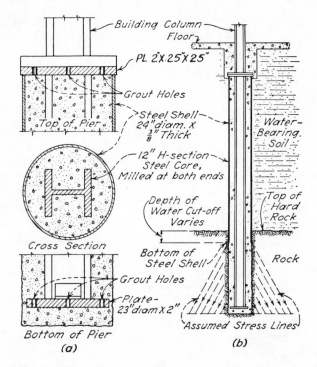

FIGURE 2.52   Concrete pier with structural steel core.

distributed over a much greater area at the bottom than if the pier were supported solely on the rock surface. The assumed stress distribution in the rock is shown in the figure. The stress distribution is determined by the allowable bond strength between the pier and the surrounding rock and by the assumed direction of the lines of stress.

In constructing this pier in water, the steel caisson shell is driven to rock. A hole in the rock is drilled deep enough to form a watertight seal after the caisson has been driven down into the rock. The caisson is then pumped out and the remainder of the rock excavation carried on in open air.

Pneumatic Caissons Method

The pneumatic caisson is used for constructing building and pier foundations through water-bearing material to bedrock. It consists of the following operations:

Construction and Placing.   The essential parts of the pneumatic caisson are the *working chamber*, the *shaft*, the *air locks*, and perhaps the cofferdam. The working chamber and cofferdam may be constructed of timber, steel, or reinforced concrete. The shaft is usually constructed of steel. See Figure 2.53(a) and (b).

The pneumatic caisson process is illustrated in Figure 2.53. The lower end of the caisson, including the working chamber, is first constructed and placed on the ground surface or in an excavated hole as shown in Figure 2.53(c).

Excavating and Advancing.   Excavation proceeds in the working chamber and the caisson advances by sinking of its own weight or because of an imposed load. As soon as the cutting edge reaches the ground water, water rises in the working chamber; therefore, the air locks are at the upper end of the shaft as shown in Figure 2.53(d), and sufficient air pressure is applied to the working chamber to force out the water. Men who have been acclimated to the increased air pressure can now work in the working chamber without interference of water. As the caisson sinks, the pressure in the working chamber must be increased to balance the increased water pressure.

The part of the caisson above the working chamber may be surrounded with a timber *cofferdam*, as shown in Figure 2.53(a), to retain the earth and water as the caisson sinks. Concrete is placed in the cofferdam to form the pier and as a weight to assist in forcing the caisson down. Removable forms may be used instead of the cofferdams, as shown in Figure 2.53(b). The level of the concrete is always kept above the top of the ground. Frequently, the entire pier is completed before the sinking starts.

The earth exposed in the working chamber is excavated by any convenient method and is hoisted to the surface in buckets passing through air locks. If the material is quite fluid, it may be blown to the surface through a pipe, the air in the chamber providing the necessary pressure. Soil that is blown out is heaped over the end of a stiff hose placed on the bottom of the working chamber and connected to the pipe leading to the surface. A valve is provided to close this pipe when not in operation.

Water Sealing and Concreting.   When the cutting edge of the caisson has reached bedrock, as shown in Figure 2.53(e), the surface of the rock is prepared to receive the pier. A layer of concrete heavy enough to resist the hydrostatic pressure is placed in the bottom of the working chamber. This seals the caisson to exclude water. The air pressure is released, the air locks and possibly the shaft are removed, and the concrete is carefully placed under atmospheric pressure in the remaining portion of the working chamber and the shaft, as shown in Figure

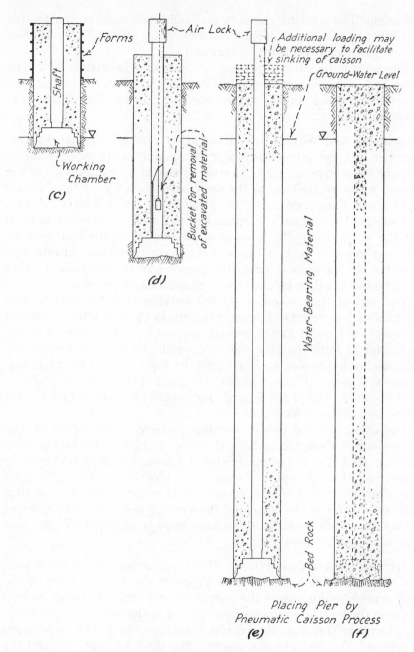

Placing Pier by
Pneumatic Caisson Process

FIGURE 2.53 Pneumatic caisson method.

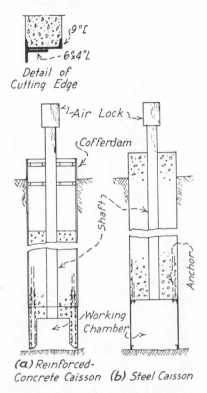

Detail of
Cutting Edge

(a) Reinforced-
Concrete Caisson   (b) Steel Caisson

2.53(f). Care must be exercised to avoid air pockets and to overcome shrinkage in the hardening concrete.

Remote Controlled Excavation. A system of excavation has been tested, permitting a 30 ft. diameter caisson to be sunk 165 ft. This system, invented by Mitsubishi Heavy Industries [34], controls television cameras and excavating equipment remotely. A remotely controlled *backhoe* excavates and moves the material to the center of the caisson while a *clamshell* on a cable hoist removes the material. The *clamshell* passes through a set of air locks. As excavation proceeds, the concrete caisson sinks of its own weight as a conventional caisson does (see Figure 2.54).

The system permits safe continual monitoring and final inspection during excavation and concreting. The hazards of working on a pneumatic caisson are eliminated, as are the depth limitations imposed by manual excavation.

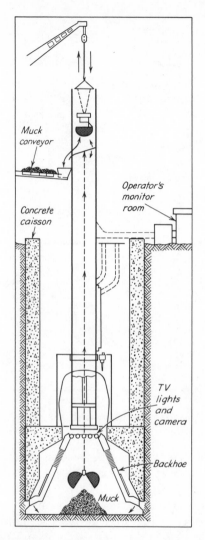

FIGURE 2.54   Self-sinking caisson unit. Reprinted with permission of *Engineering News-Record* [34].

Advantages and Disadvantages. The virtues of the pneumatic process are:

(a) The only excavation required is that of the volume of the pier.
(b) Adjacent foundations are not undermined as they might be in the open caisson method.

(c) An opportunity for proper examination and preparation of the foundation bed to receive the pier. This is not always possible in the usual form of the open caisson method.

The disadvantages are as follows:

(a) The depth below ground water is limited to about 110 ft. because men cannot work safely in the air pressure required by greater depths. The working period that men can endure without ill effects decreases as the required pressure increases.

(b) The labor cost is relatively high compared with other methods.

Combined Methods for Pier Construction

Usually it is unfeasible to excavate a well exclusively by one method. Two or more methods may be employed.

If a stratum of water-bearing material, such as water-bearing gravel or sand, overlies a thick stratum of clay which is underlaid with hardpan or rock, sheet piling may be driven through the top stratum of sand into the clay a sufficient depth to form a watertight seal. The clay may then be excavated by the Chicago method.

At the site of the Cleveland Union Terminal Building, this condition existed, but the ground water was a few feet below the level at which the well excavation started. The upper part of the well down to the groundwater level used horizontal sheeting (Figure 2.55). Below this and extending into the clay, steel-sheet piles were used; the remaining depth employed the Chicago method.

Another combination of methods, illustrated in Figure 2.56, meets the situation in which a thick layer of clay overlies a stratum of water-bearing gravel and boulders lying on a rock stratum on which a pier is to rest. The Chicago method is used in the clay. Before water-bearing gravel stratum is reached, the well is stepped out, as shown, to give sufficient room to drive steel-sheet piling through the gravel to bedrock. The bottom of the lagging is kept far enough above the top of the gravel stratum that the remaining clay acts as a seal to exclude water. The sheet piling is driven into the excavation and extends to bedrock. The piles are driven to form a seal between their lower ends and the rock; the enclosed earth, gravel, and boulders are cleaned out; the foundation bed is prepared; and the concrete is poured. Water-bearing strata between layers of clay can be passed with sheet piling following this procedure (Figure 2.56) or can be sealed off with steel cylinders with inside diameters large enough to clear the sheeting.

Because of their relatively low cost, open caissons are used to penetrate water-bearing stratum of sand, gravel, or silt; caissons can be

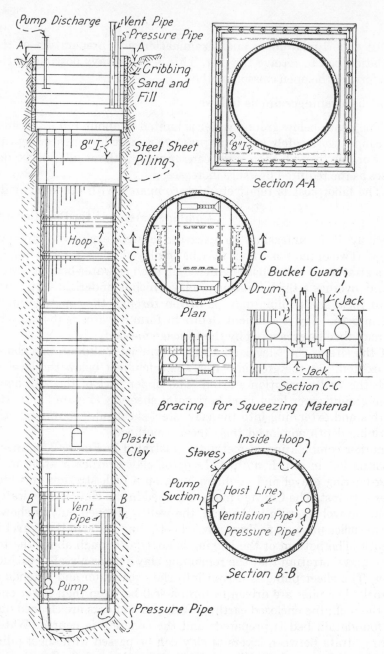

Pump Discharge

Vent Pipe
Pressure Pipe

A

A

Cribbing
Sand and
Fill

8" I

Steel Sheet
Piling

Hoop

C

C

Plan

Bucket Guard

Drum

Jack

Jack

Section C-C

Bracing for Squeezing Material

Plastic
Clay

Inside Hoop

Staves

Pump
Suction

Hoist Line

Ventilation Pipe

Pressure Pipe

Section B-B

B

B

Vent
Pipe

Pump

Pressure Pipe

Section A-A

8" I

FIGURE 2.55 Horizontal sheeting, steel-sheet piling, and lagging in a single well.

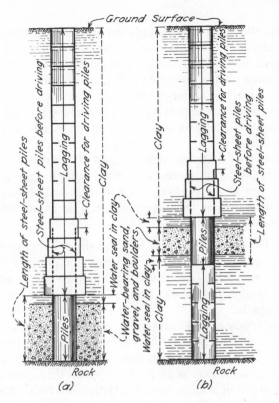

FIGURE 2.56   Combined vertical wood lagging and steel-sheet piling.

equipped with locks converting them into pneumatic caissons, a precautionary procedure adopted when encountering material which the open caisson will not penetrate, or it may be desirable to secure the advantages of the pneumatic caisson method in preparing the foundation bed and in filling the caisson with concrete.

14.   DEEP BASEMENT WALLS

Tall buildings may have basements that extend several floors below the ground level and even below the ground water. The methods employed in constructing basements are similar to construction methods explained for pier foundations. Procedures for bracing basement walls to resist lateral pressures while the permanent building frame is erected, however, are quite complex. After the frame is in place, it resists these pressures.

The construction methods include vertical and horizontal sheeting, sheet-pile cofferdams, and pneumatic caisson cofferdams that surround the area to be excavated. Open caissons are rarely used for buildings because of the impending danger of undermining adjacent foundations and other underground works by excessive lost ground. They may be suitable, however, for undeveloped areas or in some built-up areas if special precautions are taken to avoid settlement of adjacent buildings.

A combination of methods may be used. The excavation above the ground water may be protected by a single diaphragm of sheeting or sheet piling; the lower portion, by a double-wall steel-sheet piling cofferdam within which the basement wall is constructed or by a pneumatic caisson cofferdam.

Methods of Bracing.   For relatively shallow excavations, vertical sheeting or sheet piling may be suported temporarily by *rakes* or *rakers*. The upper end of each *raker* bears against a *wale* and the lower end against a timber *foot block*, Figure 2.57, all struts bearing against a single foot block.

Before the bracing, the interior is excavated to the level below which another construction method is employed. Portions to be excavated, called *berms*, are undisturbed until the wales and inclined *struts* can be placed successively to support the sheeting or sheet piling. The basement wall is then constructed within the confines of the sheeting or sheet piling, the bracing altered as required to serve as a temporary bracing. The bracing must be wedged tightly to minimize movement. Actually, *berms* are horizontal surfaces that form breaks in earth slopes. The term, however, is often used as indicated above.

The bracing method just described is not usually used for deeper excavations. For deep excavations, the bracing spans the entire well, extending to the opposite pit wall. Thus, the pit walls give mutual support to prevent caving (Figure 2.57). This is called *cross-lot bracing*. Normally, it extends in both directions across the excavation. Temporary bracing must be arranged to clear the members of the permament structure. The excavation is conducted as shown by the dashed lines in the figure, leaving temporary beams for lateral support while the bracing is installed progressively downward. To reduce the cost, members of the permanent structure may serve as temporary bracing. In both cases, the supporting members must be stressed with wedges or jacks to minimize lateral movement.

Soldier Beams and Breast Boards.   A common retaining method for

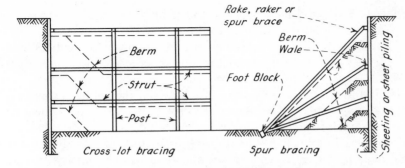

FIGURE 2.57   Lateral bracing for sheeting or sheet piling.

basement excavation uses wide-flange *soldier beams* driven full length before excavating. Horizontal timber *breast boards* or sheeting inserted between them are supported on the beam, Figure 2.58.

The soldier beams are spaced a few feet apart. The breast boards are timber planks, from 2 to 4 in. thick and long enough to provide end bearings on the flanges of the beams. Breast boards, placed as the excavation progresses, are supported against the inner flanges, (*a*)

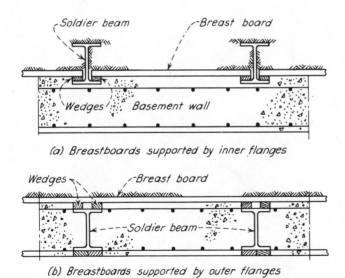

(a) Breastboards supported by inner flanges

(b) Breastboards supported by outer flanges

FIGURE 2.58   Soldier beam and breastboard construction. (*a*) Breast boards supported by inner flanges. (*b*) Breast boards supported by outer flanges.

and (*b*). Wedges driven between the flanges and the breast boards force them against the trimmed earth.

The soldier beams are supported laterally by wales at vertical intervals. The wales are supported by spur or cross-lot bracing (Figure 2.57).

In Figure 2.58(*b*), the wall is cast between the soldier beams, and for that shown in (*a*), in front of the beams to which it is anchored. The wall is reinforced to resist lateral pressures.

Where rock is available for tieback anchorage, soldier beams may be braced by *tiebacks*. One end is anchored in rock outside the area to be excavated, and the other end attached to a wale beam (Figure 2.59). First a hole is drilled through the soil and into the rock surface. After the soldier beams are driven and excavation and breast boards placed to a depth just below the top line of wales, a pipe which serves as a sleeve is driven at a 45° angle from the wale to the rock formation surface. A *socket* is then drilled into the rock formation by a drill

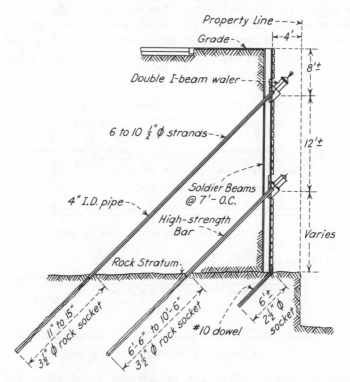

FIGURE 2.59 Tiebacks supporting soldier beams. *Engineering News-Record,* April 18, 1961.

operating through the pipe. The pipe sleeve and socket are cleaned out with an air or water jet. The lower end of the *wire tendon tieback* is grouted in the socket with high-early-strength cement grout. The tendon is prestressed by jacks and anchored to the wale. This procedure is repeated for each tieback located at each soldier beam. The lower ends of the soldier beams are anchored, as shown. The apparent advantage of this procedure is that unobstructed working space is available. It was originated by Spencer, White, and Prentis, Inc. [16].

Double-Wall Steel-Sheet-Pile Cofferdam. As illustrated in Figure 2.60, a deep basement and foundations are constructed at a site where explorations indicate underlaid filled material, an unconsolidated permeable water-bearing formation, a relatively impervious rock formation with low bearing capacity, and finally, sound bedrock on which the foundation is supported. The basement extends into the impervious rock with low bearing capacity.

The fill is supported by a single row of steel-sheet piles with *spur bracing* installed as the enclosed area is excavated, as shown in (*a*). Two rows of interlocking steel-sheet piles are driven through the permeable water-bearing formation until their bottoms form a water seal in the relatively impervious rock. Steel diaphragms driven at intervals between the two rows of piles form cells from which the material then is excavated. As the excavation proceeds, I-beam wales and braces are installed to resist the lateral pressure of the earth and water. In this manner, a double-wall steel-sheet-pile cofferdam is constructed completely around the area. Finally, a trench requiring no lateral support is excavated through the relatively impervious rock to bedrock.

All the preparations required for constructing the outside wall, as shown in (*b*), are now completed. First, the trench is filled with concrete, using forms on the exposed face in the basement. Above this point, the wall forms are constructed in sections in the cofferdam and concreted, the original wales and cross braces removed progressively and replaced with wales and shorter braces bearing against the completed concrete and the exposed row of sheet piles. The keyed horizontal construction joints between successive wall sections are provided with water stops.

While the construction of the outside walls is underway, the interior portion of the basement is excavated, leaving berms of unexcavated material to provide lateral support. These berms and the top of the general excavation are lowered progressively to permit the installation of the cross-lot bracing, as illustrated in Figures 2.57 and 2.60(*c*):

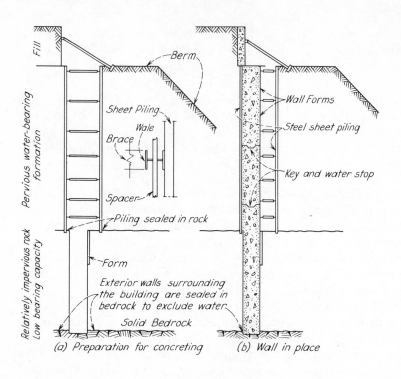

(a) Preparation for concreting   (b) Wall in place

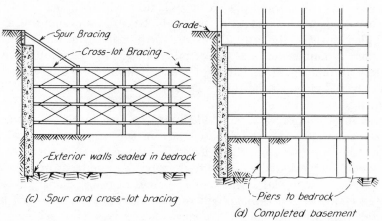

(c) Spur and cross-lot bracing

(d) Completed basement

FIGURE 2.60   Double-wall steel-sheet-pile cofferdam. (*a*) Preparation for concreting. (*b*) Wall in place. (*c*) Spur and cross-lot bracing. (*d*) Completed basement.

134

the spur bracing is supported by the cross-lot bracing. Spur bracing can be avoided by extending the top of the outer row of sheeting to the ground surface, if conditions permit.

After placing the cross-lot bracing, the piers supporting the interior columns can be constructed, and finally the basement floor, as shown in (d), together with the necessary waterproofing and under-drainage, can be installed as described on page 142. The remainder of the building is then constructed (d).

The above discussion is based on references 14, 15, and 18, actual projects using these procedures.

Open Caisson Cofferdams.   Monolithic open caisson cofferdams have been used in excavating for deep basements. One example is illustrated by the half-section in Figure 2.61. The caisson, which is cylindrical with an outside diameter of 187 ft., was sunk with its cutting edge 91 ft. below ground surface and sealed in compact clay. The diameter at the cutting edge is about 8 in. larger than that of the upper portion of the caisson. The resulting 4 in. annular space which surrounds the caisson was injected full of bentonite slurry, which acted as a lubricant to facilitate sinking and as waterproofing for the excavation. The portions of the structure were constructed progressively as the soil under the cutting edge was removed and the caisson sank. These comments are based on reference [19]. The architect was Fritz Jenny.

Pneumatic Caisson Cofferdams.   The exterior walls of deep basements may be constructed by the pneumatic caisson method using a cofferdam of rectangular reinforced concrete caissons (Figure 2.62). Each caisson is sunk to rock, sealed, air locks removed, and the work-

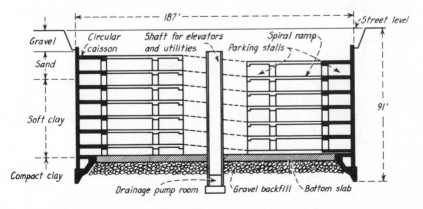

FIGURE 2.61   Open caisson cofferdam.

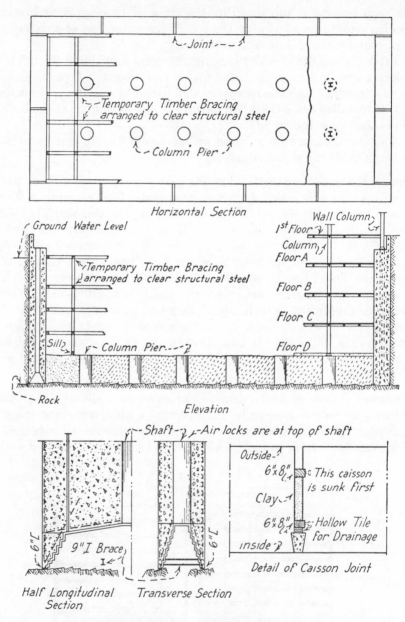

Horizontal Section

Elevation

Half Longitudinal Section

Transverse Section

Detail of Caisson Joint

FIGURE 2.62 Caisson cofferdam.

ing chamber and shafts filled with concrete. Excavation and bracing can then proceed.

Well Point Method.   A *well point* is a pipe with a pointed lower end and with a screen or filter along the lower 3 or 4 ft. of its length. The well point method for dewatering in water-bearing material surrounds the area to be excavated by one or more rows of closely spaced well points placed along each side. The water table is lowered by pumping the water through the well points. The excavation within the area can then be in the dry. Some well points are placed by driving with a maul, whereas others are positioned by jetting. The well points are spaced about 3 ft. apart; the top is connected to a header pipe connected to a pump. The water level may be lowered as much as 20 or 30 feet. Additional lowering may be accomplished by arranging well points in two or more vertical stages. This method is used chiefly in trench excavation and in dewatering an entire building site and is also applicable to pier excavation. To keep banks from caving, sheeting may be necessary, but it need not be watertight.

Grouting Soil and Rock Formations.   Certain soils can be stabilized and made more impervious. Structural defects in rock formations, such as cracks, joints, fissures, and cavities which weaken them or cause them to be permeable, can be filled by pressure injected grout. The grouts used may be chemical solutions or mixtures of water and portland cement, portland cement and finely ground sand, or clay.

Various methods are used for chemical grouting. In general, each makes use of two solutions which react chemically when mixed and gel or solidify. They may be mixed in the formation by injecting one and then the other. This is called a *two-shot method*. In the *one-shot method*, the two chemicals are mixed immediately before or during injection and gel at a predetermined time after injection. Injections by specially designed pumps force solutions into the formation through a driven pipe, a cased or uncased drill hole, depending upon the character of the formation. Techniques confine the grout to desired locations. After the grouting, the formations treated become more stable and relatively impermeable.

One chemical grouting method uses a mixture of sodium silicate and calcium chloride. Another utilizes a mixture of acrylamide and methylenebisacrylamide and is called AM-9 Chemical Grout [22]. Chemical grouting may be effective in gravel and sand formations but not in silt or silty sand. It has been used successfully in some hardpans.

Portland cement, portland cement and fine sand, and clay grouting are extensively used to make rock foundations for dams more impermeable or more stable, but these methods are rarely, if ever, used in constructing foundations for buildings. The grouts formed by mixing these materials with water are pumped into holes drilled in rock foundations.

Injection procedures are expensive and involve uncertainties because of unknown subsurface conditions, and the results are sometimes disappointing. The operations are complex and require skill and experience.

Bentonite Slurry. In some areas, a groundwater table cannot be lowered because any groundwater drawdown from dewatering would cause settlement to the surrounding buildings. This prevails as a potential hazard where surrounding buildings have foundations resting in the soil rather than extending to bedrock. Dewatering by conventional well points is usually possible; but if the groundwater table is lowered, the soils lose their buoyancy. Soil consolidation follows with possible disastrous results to structures adjacent to the excavated site.

If a building foundation is to be built on such a site, piles or caissons might be an obvious solution. If, however, a basement is to be excavated, a *cut-off wall* must be built to maintain the existing groundwater level. This protects the surrounding soils while excluding the ground water from the excavation site so that operations may proceed in the dry. Conventional underpinning methods to adjacent structures may be prohibitive or impossible.

A method devised to permit working in the dry is the *slurry-trench,* also called the *diaphragm wall* or *slurry-wall.* A trench about three feet wide is excavated in sections about 20 ft. long to bedrock or to a depth below the intended basement or foundation. This trench is then filled with a mixture of water and bentonite. This slurry prevents the collapse of the trench walls. A prefabricated, welded cage of deformed reinforcing bars is lowered into the slurry. Tremie concrete is then placed in the trench, displacing the bentonite slurry, which is pumped out and stored in tanks for future use.

Because a mechanical bond between the concrete and the rebars is deterred by the residual bentonite slurry, it is necessary to use deformed bars that are welded and made into a cage. Trenching and constructing the diaphragm wall along the entire perimeter of the building site proceeds in increments. After the wall sections have been made watertight, the site is dewatered. The excavation of the foundations or basement walls then progresses in the dry while maintaining the integrity of the groundwater level and soil conditions outside of the excavation.

This is basically the procedure; however, since its initial use in Milan, Italy, in 1961, there have been improvements in technique and procedural variation.

One of the greatest difficulties has been to effect a tight seal between adjacent slurry panels to prevent the tremie concrete from entering the adjacent panels. This seal has been accomplished by the cored-in large steel soldier beams spaced about 20 feet between slurry panels. This separates and forms the ends of the panels. The piles are concreted after they are placed.

A notable example of the method was the excavation of the World Trade Center in New York, in conjunction with the wall tieback system, using four rows of tiebacks. The concrete wall was anchored into bedrock as the site excavation proceeded. The completed wall was used as the perimeter basement wall. A full and interesting account of this procedure can be read in the April, 1969 and December, 1967 issues of *Civil Engineering* [35], [65].

15.   BASEMENT WALLS, DRAINAGE, AND DAMPPROOFING

Basement floors above ground water must be protected against dampness and infiltrating rainwater that penetrates permeable backfills. This protection is called *dampproofing*. Basements below the groundwater level must be waterproofed to resist water infiltration and hydrostatic pressure is equal to the unit weight of water (62.4 lb./cu. ft.) multiplied by the depth (in feet) of the point below the groundwater level. Unless alleviated, the pressure is exerted on the underside of basement floors.

Dampproofing. Water infiltrates basement walls through cracks, defective joints, and through small continuous passages in permeable concrete or masonry units. Moisture from backfills may also penetrate walls by capillary action through continuous channels causing damp interior walls.

Dampproofing is accomplished by any of the various procedures listed below. Some may be omitted where walls are not adversely affected by dampness, or where an arid climate or impermeable backfill exists. These methods are illustrated in Figure 2.63.

(1) Drainage of surface water away from the building by sloping the ground surface and by conducting the rainwater from downspouts that discharge on the ground surface into paved channels.

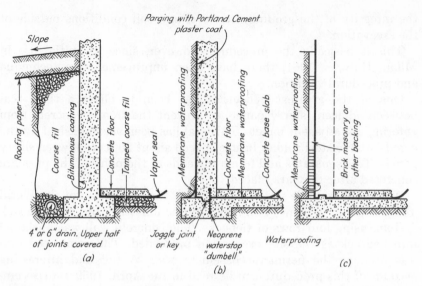

FIGURE 2.63 Dampproofing and waterproofing basements. (a) Damp-proofing. (b) Outside of wall accessible. (c) Outside of wall inaccessible.

(2) Applying a dampproof coating to the exterior surfaces below the ground level. This coating may be one of several materials:

(a) Bituminous materials or silicones applied in liquid form by brushing or spraying or in plastic form by troweling.
(b) One or two coats of portland cement mortar, including a water repellent that is troweled or pneumatically placed, giving a smooth finish.

The bituminous coating can be applied directly to smooth concrete surfaces, but if the surface is rough or if the coating is intended for a masonry wall, a smooth mortar coating should be applied first. The mortar coating may suffice without requiring the bituminous application.

The coatings will seal a wall against moisture penetration by capillary action from a damp backfill or from infiltration through minor defects, but coatings are, of course, ineffective in sealing cracks that develop after the coatings are applied.

(3) Providing granular fills under floors and installing vapor seals. Concrete floors laid directly on fine-grained soils tend to draw moisture from the soil by capillary action. Such a floor should always be on a tamped granular base from 4 to 6 in. thick made of gravel,

crushed rock, or cinders. As an additional precaution, a layer of heavy roll roofing or a 4 to 6 mil polyethylene film is laid on a sand bed over the granular base. Care should be taken to avoid puncturing the vapor barrier during concrete placement. The sheeting is usually lapped but taped and sealed when free water is present. See Figure 2.63(a).

(4) Installing a backfill drainage system to keep rainwater and wet soil away from basement walls. Included in this system are the *footing drains* which discharge into a storm sewer, or when feasible, into a dry well. A *dry well* is a pit excavated in permeable soil, or a pit with a bottom in such soil, and filled with gravel or crushed rock. Sometimes it is a lined empty pit with a cover and a soil bottom.

The footing drains are 4- or 6-in. vitrified clay drain tile perforated or plain laid with open joints. The upper half of open joints is covered with strips of roll roofing to exclude soil. The drain tiles are surrounded with gravel or crushed rock. A *blanket drain* of this material, at least 12 in. thick, is placed against the wall extending upward to within a foot of the ground surface. The remainder of the backfill is ordinary tamped soil—the top soil supporting vegetation and preventing free access of surface water into the blanket drain. The latter objective is accomplished more effectively by placing a strip or roll roofing over the top of the blanket drain and extending it into the ordinary soil; see Figure 2.63(a). This accomplishes the further objective of preventing the ordinary soil from working into the drain.

Other factors deserve consideration. In residence construction, the basement may occupy only a portion of the area under the first floor, thereby leaving an exposed ground surface. Fine-grained soil may serve as a "wick" in drawing moisture into the basement from the underlying soil by capillary action. This can be reduced or prevented by covering the exposed wall with a heavy polyethylene film held in place, especially at the joints, by earth. Crawl spaces should be treated in the same manner.

Dampproofing coatings are sometimes placed on the inside surface of exterior walls. Such coatings of either bituminous materials or cement mortar may eventually lose their bond with the wall and fail. If used, special care should be taken to insure good bonding.

Waterproofing Walls. A basement in the ground water must be designed to exclude water and to resist the earth and hydrostatic pressures. Prevailing conditions determine procedures.

An effective method for waterproofing basements provides a continuous membrane around the outside walls below the groundwater level and on the underside of the basement floor (Figure 2.63(b)). This membrane is built up of three to five or more layers or plies of bituminous saturated felt, or cotton fabric, or a combination of these materials. It is sealed and held to the wall surface with coal-tar pitch hot mopped similar to constructing a built-up roofing surface. Care in backfilling must be taken to prevent rupturing the membranes.

Parging.   A surface prepared for membrane application should be smooth and dry. A unit masonry surface will require a coat of smooth troweled portland cement plaster ⅜ to ½ in. thick. The surface is hot pitch mopped before the first ply is applied, and the exposed outer surface of the in-place ply is mopped. This is called *parging*. See Figure 2.63(b).

A wall membrane should be protected from puncture caused by rock backfill and by rupture caused by backfill settlement. A coat of portland cement mortar, a layer of asphalt-impregnated fiber board, or preferably a single wythe of brick masonry will protect the membrane.

Waterproofing Floors.   Before placing a floor membrane, a tamped base of gravel or crushed rock 4 in. or more thick must be provided. A sand bed or a cement mortar or concrete leveling bed at least 1 in. thick is placed on this base to receive the membrane. The membrane is placed, and then the concrete basement floor slab is poured. The membrane must be continuous under all columns, interior walls, and exterior walls, as shown in the figures. The membrane must also be continuous across the joint between the floor and exterior wall membranes.

Construction joints in walls not protected by membranes must be sealed with noncorrosive metal or neoprene waterstops which permit movement without rupture; see Figure 2.63.

The wall must be designed to resist the lateral earth and water pressures, and the floor designed to resist any hydrostatic uplift pressures.

Obviously, the basement cannot be excavated nor these procedures accomplished if the ground water cannot be pumped out and lowered below the basement floor level.

If the exterior surface of the basement wall is accessible, the membrane and its protection are applied to this surface (Figure 2.63(b)). If inaccessible, the procedure is reversed as shown in Figure 2.63(c).

According to the modified procedure, the protective wythe of brick masonry is placed, rigidly supported against outward movement and plastered with a coat of portland cement mortar. The membrane is then applied to this surface. Finally, the wall is constructed with the protective layer serving as one side of the form if the wall is of concrete. If it is a masonry wall, any spaces between the outer surface of the wall and the masonry must be filled tightly with mortar or grout.

The quantity of water pumped from a basement excavation depends upon the groundwater level and the soil permeability. If the basement walls enter clay stratum acting as a water seal, the quantity may be relatively small. To discharge this ground water, the built-up membrane under the basement floor can be replaced with a vapor seal and by providing a tile drainage system in the granular base. The water discharges into a pit called a *sump*, and the collected water is elevated by an automatic electric *sump pump* discharging intermittently into a storm sewer. This system alleviates uplift pressure on the underside of the basement floor.

Bentonite Panels.   One method of dampproofing takes advantage of the expansive qualities of bentonite clay. Kraft paper bags, made into thin 4-ft. square panels, are filled with a bentonite compound. Secured against the basement wall and back filled, the kraft paper deteriorates, exposing the bentonite compound. This material, when contacted by water, swells to about ten times its original volume, forming a positive moisture barrier.

Loose bentonite compounds placed under slabs are covered with a temporary sheet preventing premature wetting. The bentonite swells when wet, forming an impenetrable barrier that dampproofs the basement slab. This application should be supplemented by installation of area drainage that will deter a build-up of water that would impose a condition of hydrostatic uplift [61].

16. SLABS ON GRADE

Slabs on grade are frequently used for light buildings such as residences and those requiring only shallow foundations. The usual methods of constructing slabs on grade are the slab with *grade beams* and the slab with thickened edges, called a *thickened slab*. Thickened slabs are constructed in a manner that enables them to serve as a foundation—a foundation integral with the slab. Slabs on grade are also constructed with grade beams supported on piers, piles, or pedestal types of footings. In either case, the slab floats on the prepared sub-base.

Foundation Bed Preparation.   Using a bulldozer, grader, or scarifier, the top 8 to 12 in. ground surface is first *skinned* of all organic material, and the top soil removed and stockpiled for possible future use. Any new soil required for fill or sub-base is then imported and spread, watered, and compacted to the required density with the use of sheepsfoot rollers and rubber tired compactors. Testing is conducted as compaction proceeds. Tests are usually performed by using the sand cone density method or the Modified Proctor compaction test, ASTM D-1557 [60]. A density of 90% is normally specified.

The limits of compaction are the entire area of the building slab plus a 5 to 10 ft. perimeter border. This includes the area that will be occupied by the foundations, which are to be excavated later.

After a sub-base has been compacted to the specified density, well-graded rock or gravel is spread over the entire sub-base, including perimeters. This rock base is usually 4 to 6 in. thick and is compacted to a density of about 95% compaction. A sand bed spread over the rock serves to cushion and protect the waterproof film from puncture.

With the foundation bed compacted, excavation proceeds, using backhoes and trenchers, or bucket drills. The excavation for footings, piers, pedestals, and trenches for grade beams or thickened slabs are cut through the compacted rock base and into the compacted sub-base.

Grade Beams.   After excavation, the grade beams are formed where required on each side, reinforced, and the concrete poured. The grade beams may form the entire perimeter of the building and support the exterior walls and the imposed loads. The walls may be of masonry, or framed of steel or wood.

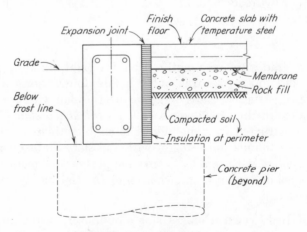

FIGURE 2.64   Grade beams.

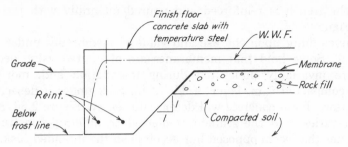

FIGURE 2.65   Thickened slabs.

Rigid insulation, such as asphaltic impregnated fiber boards or plastic foam types, may be placed on the interior side of the grade beam.

Grade beam depths are determined by the structural requirements designed to resist the imposed loads. The bottom of grade beams should extend below the frost line (see Figure 2.64).

Thickened Slabs.   Thickened slabs are poured integrally with the floor slabs. The interior side of the perimeter trench footing is sloped or battered 1 to 1. Only the exterior side of the trench may require forming. The slab is then reinforced and poured, Figure 2.65. The thickened portion extends below the frost line.

These economically constructed slabs are used only on light construction and carry only moderate loads.

Reinforcement.   Because these slabs support minimal loads, very little reinforcing is required. In most installations, electric Welded Wire Fabric (W.W.F.) is specified; the most common type is a fabric with 10 gage rods in both directions, forming 6 in. square holes, designated 6 × 6 —10/10 W.W.F.; see Figure 2.65.

Interior Walls.   The interior walls of slabs on grade are supported by thickening the slab. A shallow trench is excavated with sloping

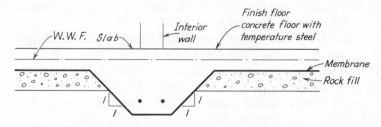

FIGURE 2.66   Interior walls.

sides; the trench is reinforced and poured integrally with the floor slabs, Figure 2.66.

Opinions differ upon the use and value of membranes under slabs on grade. Some hold that membranes are of limited value because they are invariably punctured during installation and, moreover, that a compacted 6 in. layer of 1½–2 in. rock under a dense, well graded, low slump concrete will deter water entry, Figure 2.66. Again opinion varies, and the student is alerted to differing construction techniques that seem opposed but accomplish the intended task.

REFERENCES AND RECOMMENDED READING

1. Ralph B. Peck, Walter E. Hanson and Thomas H. Thornburn, *Foundation Engineering*, John Wiley and Sons, 1953.
2. N. M. Newmark, *Influence Charts for Computation of Stresses in Elastic Foundations*, Engineering Experiment Station, Series 338, University of Illinois, 1942.
3. A. Casagrande, "The Structure of Clay and Its Importance in Foundation Engineering," *Journal of Boston Society of Civil Engineers*, April 1932.
4. F. Kogler and A. Scheidig, "Druckverteilung im Baugrunde" (Pressure Distribution in Building Soil), *Bautechnik*, 1927, Nos. 29 and 31; 1928, Nos. 15 and 17; 1929, Nos. 18 and 52.
5. *Building Code Requirements for Excavations and Foundations*, American Standards Association, A56.1-1952.
6. M. L. Enger, *Transactions American Society of Civil Engineers*, No. 85, 1921, p. 1581.
7. W. S. Housel, "A Penetration Method for Measuring Soil Resistance," *Proceedings of American Society for Testing Materials*, Vol. 35, 1935.
8. A. E. Cummings, *Lectures on Foundation Engineering*, Circular Series 60, University of Illinois, Engineering Experiment Station, 1949.
9. Karl Terzaghi and Ralph B. Peck, *Soil Mechanics in Engineering Practice*, John Wiley and Sons, 1948.
10. Charles Terzaghi, "Science of Foundations," *Transactions of the American Society of Civil Engineers*, Vol. 93, 1929.
11. H. A. Christine, "Boring Machine Digs Wells for Concrete Piers," *Engineering News-Record*, July 28, 1932.
12. "Rotary Drills Speed Caisson Shaft Construction in Wet and Dry Soils," *Construction Methods*, November 1932.

13. "Big Augers Go Deep So Big Buildings Can Go High," *Engineering News-Record*, May 25, 1961, p. 32.
14. Robert C. Johnson, "Deep Foundations for Pittsburgh Skyscrapers," *Engineering News-Record*, December 7, 1950, p. 39.
15. Robert C. Johnson and Nicholas W. Koziakin, "Foundation' Design and Methods Cut Skyscraper Cost," *Engineering News-Record*, July 24, 1958, p. 34.
16. "Tiebacks Remove Clutter in Excavation," *Engineering News-Record*, June 8, 1961, p. 34.
17. Edward E. White, "Deep Foundations in Soft Chicago Clay," *Engineering News-Record*, November, 1958, p. 36.
18. "Chemicals Seal Foundation for New York Building," *Civil Engineering*, October 1957, p. 47.
19. "Caissons Dig Out a Seven-Story Basement," *Engineering News-Record*, July 6, 1961, p. 42.
20. Ralph B. Peck and Sidney Berman, "Recent Practice for Foundations of High Buildings in Chicago," *Symposium on the Design of High Buildings*, Golden Jubilee Congress, University of Hong Kong Press, 1961.
21. *Construction Methods*, August 1957.
22. R. H. Carol, *Soils and Soil Engineering*, Prentice-Hall, 1960.
23. Chester W. Campbell, ,"Chemicals Seal Foundation of New York Building," *Civil Engineering*, October 1957, p. 47.
24. "3 Uses of Chemical Grout Show Versatility," *Engineering News-Record*, May 31, 1962, p. 68.
25. "Pile Foundations," Fourth Edition, American Iron and Steel Institute.
26. Charles Terzaghi, "Physical Differences Between Sand and Clay," *Engineering News-Record*, December 3, 1925, p. 912.
27. R. E. Grim, "The Clay Minerals in Soils and Their Significance," *Proceedings of Purdue Soil Mechanics Conference*, 1940.
28. R. R. Proctor, "Fundamental Principles of Soil Compaction," *Engineering News-Record*, Vol. III, 1933.
29. H. S. Guillette, *Elementary Soil Fundamentals*, University of Oklahoma Press, 1936.
30. Joseph M. Trefethen, *Geology for Engineers*, D. Van Nostrand Company, 1959.
31. Robert D. Chellis, *Pile Foundations*, Second Edition, McGraw-Hill, 1961.
32. *ASTM Annual Book of Standards, Part 16*, American Society for Testing and Materials, Philadelphia, Pennsylvania, 1971.

33. Catalog: *Franki Foundation Specialists*, Franki Foundation Company, New York, New York.
34. *Engineering News-Record*, November 5, 1970, p. 30.
35. Martin S. Kapp, "Slurry-Trench Construction for Basement Wall of World Trade Center," *Civil Engineering*, Official monthly publication of the American Society of Civil Engineers, April 1969, p. 36.
36. *Wood Handbook*, Agriculture Handbook No. 72, Forest Service Laboratory, U.S. Division of Agriculture, 1955.
37. Catalog: *Taywood Piling Service*, Taylor Woodrow Construction Limited, Southall, Middlesex, England.
38. Catalog: *Foster Vibro/Driver Extractors*, L. B. Foster Co., Pittsburgh, Pennsylvania, 1970.
39. Catalog: *Delmag Diesel Pile Hammers*, The Foundation Equipment Corporation, Newscomerstown, Ohio.
40. Walter E. Lessey, *High Capacity Long Steel Piles*, American Iron and Steel Institute.
41. Catalog: *Calweld Drilling Tools*, Calweld Company, Santa Fe Springs, California, 1970.
42. Donald W. Taylor, *The Fundamentals of Soil Mechanics*, John Wiley and Sons, 1948.
43. Clarence W. Dunham, *Foundations of Structures*, McGraw-Hill, 1962.
44. T. William Lambe and Robert V. Whitman, *Soil Mechanics*, John Wiley and Sons, 1969.
45. Robert D. Krebs and Richard D. Walker, *Highway Materials*, McGraw-Hill, 1971.
46. Ronald F. Scott and Jack J. Schoustra, *Soil Mechanics and Engineering*, McGraw-Hill, 1968.
47. Karl Terzaghi and Ralph B. Peck, *Soil Mechanics in Engineering Practice*, Second Edition, John Wiley and Sons, 1967.
48. Whitney C. Huntington, *Earth Pressures and Retaining Walls*, John Wiley and Sons, 1957.
49. Harmer E. Davis, George Earl Troxell, and Clement T. Weskocel, *The Testing and Inspection of Engineering Materials*, Third Edition, McGraw-Hill, 1955.
50. Sidney M. Johnson and Thomas C. Kavanagh, *The Design of Foundations for Buildings*, McGraw-Hill Book Company, 1968.
51. Elwyn E. Seelye, *Foundations Design and Practice*, John Wiley and Sons, 1956.
52. T. William Lambe, *Soil Testing for Engineers*, John Wiley and Sons, 1961.

53. N. J. Tomlinson, *Foundation Design and Construction*, Second Edition, John Wiley and Sons, 1969.
54. Joseph E. Bowles, *Foundation Analysis and Design*, McGraw-Hill, 1968.
55. G. A. Leonards, *Foundation Engineering*, McGraw-Hill, 1962.
56. Karl Terzaghi, *Theoretical Soil Mechanics*, John Wiley and Sons, 1966.
57. A. Brinton Carson, *Foundation Construction*, McGraw-Hill, 1965.
58. A. Brinton Carson, *General Excavation Methods*, McGraw-Hill, 1961.
59. George B. Sowers and George F. Sowers, *Introductory Soil Mechanics and Foundations*, Macmillan, 1970.
60. *ASTM Annual Book of Standards, Part 11*, American Society for Testing and Materials, Philadelphia, Pennsylvania, 1971.
61. *Sweet's Catalog*, 1971 Edition, F. W. Dodge Corp., New York, New York.
62. Catalogs: Ben C. Gerwick Co., San Francisco, California.
63. Catalogs: Spencer, White and Prentis, New York, New York.
64. Catalogs: Associated Pipe and Fitting Corp., Clifton, New Jersey.
65. *Civil Engineering*, December 1967.

3

THE STRUCTURAL ELEMENTS

The structural elements of buildings may be composed of various combinations and forms of walls, columns, beams, girders, trusses, arches, rigid frames, cables, and domes.

Walls are constructed of all the traditional materials of construction, that is, masonry, wood, and concrete. Additionally, walls are made of manufactured materials in various combinations.

Columns and beams are constructed of wood, steel, masonry, or reinforced concrete; however, stone columns are used for ornamental purposes.

Trusses are usually constructed of steel, although wood is used extensively and reinforced concrete occasionally. Reinforced concrete is not considered an appropriate material for truss construction, although there are examples of its successful use.

Rigid frames, which are a form of arch, are constructed of wood, reinforced concrete, and steel. Arches, vaults, and domes of long span are frequently constructed of wood, steel, reinforced concrete, and occasionally of aluminum. Masonry arches are constructed of brick or stone. Steel twisted-wire strand cables are sometimes used to support long-span roofs and cantilever projections of roofs.

The assembling of the various structural elements so that each may perform its function is known as *framing*. One classification of buildings is on the basis of the function of the walls. If the walls support the dead, live, and other loads (in addition to keeping out the weather), the building is classed as *wall-bearing construction*; but if the loads, including the weight of the walls, are carried by the structural frame of columns, beams, and girders, the building is classed as *skeleton construction*. This term is usually applied to office and similar structures but not to industrial buildings.

The classification of buildings according to type of construction, as included in building codes, should be reviewed, namely, frame, heavy timber, ordinary, etc.

Steel construction consisting of steel beams, girders, columns, and trusses supporting floors and roofs of light joist construction, of heavy timber construction, or of fire-resistive construction is extensively used on all classes of buildings except dwellings. The exterior walls may be bearing walls for the lower buildings, but for buildings of more than three or four stories, skeleton construction is usually used. In a more expensive class of buildings, fire-resistive construction is used throughout. For buildings up to about 60 stories high, steel and reinforced concrete are competitive; but for higher buildings, steel construction is without rival because faster erection is possible. Sometimes the structural frame and floor slabs have been constructed at the rate of a story a day. Loft buildings have been completed in a year.

Reinforced concrete construction may be used for nearly all classes of buildings. Private dwellings are rarely of this type, although apartment houses, hotels, office buildings, schools, warehouses, and industrial buildings are often built of reinforced concrete. In buildings with a steel framework, the floors and roofs are usually of reinforced concrete. For tall buildings, steel construction has the advantage of smaller columns for the lower floors. Even for the lower buildings, skeleton construction may be used because of faster erection time. When bearing walls are used, each floor must await the bearing wall to carry it; but with skeleton construction, the structural frame and floors may be constructed rapidly.

17. WALLS

Wall construction is classified according to functions, positions, and methods of construction. These classifications do not always meet clearly defined demarcations because some walls meet more than one classification.

Walls enclose buildings, exclude weather, functionally subdivide floor areas, and serve as one of the oldest architectural components —the bearing wall. As *bearing walls*, they carry vertical loads of the building; as *nonbearing* walls, they support only their own weight; as *shear walls*, they resist the lateral loads caused by seismic forces or wind parallel to the axis of the wall. These loads may be transmitted to a *foundation wall* which is built below ground level, below curb level, or below the floor level immediately above the ground level.

Curtain walls, sometimes called *panel walls* or *enclosure walls*, enclose the building and are supported and anchored to the structural frame. A special type of curtain wall is the *sandwich panel wall*. It consists of two thin slabs, usually reinforced concrete, separated by a core of rigid insulating material such as lightweight concrete, cellular glass, or plastic foam. The slabs are connected by metal ties. Although sometimes made of masonry, this type of wall is more frequently made of lighter materials, such as glass and metals.

Similar to the curtain wall is the *solar screen* made of perforated material that must be supported by a structural frame. The solar screen functions as a visual barrier for privacy, as a sunshade, and mercifully, as a screen to hide poor architecture. It should be used with caution on the lower floors because it serves as a wonderful ladder for playful children and adventurous second-story men.

Some walls, called *area separation walls*, divide the floor area of a building into separate parts for fire protection, for different uses, or for restricted occupancy. A *fire wall*, or more properly a *fire resistive wall*, is made of fire resistive materials and is given a fire rating. It subdivides the building by a continuous barrier from the foundation through the roof, restricting the spread of fire. Such a wall that extends through the roof is classed as a *parapet wall*—that portion above the roof is the *parapet*. An *area separation* wall is similar in function to a fire wall, but it is not necessarily continuous from the foundation through the roof and is called a *partition*.

A number of walls are identified by their method of construction. Frequently, *faced walls*, or *veneered walls*, are built with a facing material that is adhered or anchored to the backing. This facing, also called a *veneer*, is usually of material more durable and attractive than the backing to which it is anchored. Although bonded to the backing, it does not contribute structurally to the wall. These walls are sometimes called *composite walls*.

18.   COMPRESSION MEMBERS

Columns support beams and trusses and resist compressive stresses. They are constructed of wood, masonry, steel, or reinforced concrete. *Piers*, a prism of concrete or masonry, serve the same function as columns. They may be a relatively narrow protrusion as an integral part of the wall, or they are free standing as an isolated pier. Sometimes piers are called *engaged piers*, especially when serving as pilasters. Whereas *pilasters* serve to buttress a column, they are also an integral part of the wall with a projection from the wall that is usually about one-half of the wall thickness. A buttress, or an engaged pier, is a vertical projection built into the face of the wall, increasing lateral stability while supporting a beam or truss. It is similar to a pilaster but has greater projection.

Forces that tend to shorten or compress a member are called *compressive* forces, and the stresses set up in a member by these forces are called *compressive stresses.*

The vertical members of a structural frame, called *columns*, transfer floor and roof loads to the foundations. Such loads cause stresses in the columns which are chiefly compressive; however, eccentric loads, rigidity of joints, wind loads, earthquake shocks, and lateral loads subject columns to bending stresses which may be of considerable magnitude. *Bending* or *flexural stresses* are set up in a member when it bends.

Classes of Columns.   Columns are divided into three general classes according to the ratio of the longitudinal dimension to the least radius of gyration.

If the length of a column is relatively small when compared with its lateral dimension (Figure 3.1(*a*)), the column does not tend to bend to any extent when carrying a load; and if the load is applied so that its resultant load is on the axis of the column, the stresses are assumed to be uniformly distributed over each cross section of the column. If the length of a column is great when compared with its lateral dimension, the column will tend to fail by bending or buckling (Figure 3.1(*b*)) when carrying a load, the magnitude of the direct compressive stress being small. A third class includes columns intermediate in ratio of length to lateral dimension to the classes just mentioned; such columns tend to fail by a combination of direct stress and bending or buckling. Reinforced concrete columns are usually of the first class if there are not lateral loads. Timber and steel columns may be in either the first or third class. Columns of the second class are not used to any extent.

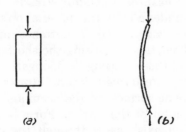

(a)   (b)

FIGURE 3.1   Classes of columns.

Other Terms Used. Columns are often called *posts*, especially when made of timber. Truss members carrying compressive stresses are called *struts*, but their action is the same as that of columns. In general, members that carry compressive stresses are called columns, posts, struts or *props*.

The light, closely spaced, vertical compressive members used in walls and partitions in wood frame and light steel construction are called *studs*. Stone or brick columns are sometimes called *pillars*, but this is not a technical term. The term pier has about the same meaning as pillar and is more commonly used.

19.   BEAMS AND GIRDERS

A *beam* is a horizontal member supported at one or more points along its length and carrying loads acting perpendicular to its length. The reactions at the supports are parallel to the direction of the loads (Figure 3.2).

If the line of action of the loads is not perpendicular to the axis of the beam, these loads are resolved into components acting perpendicular to the length of the beam and components acting parallel to the length of the beam (*b*). In supporting the transverse components, the beam performs its primary function; while carrying the components parallel to its length, it acts as a column.

A beam may be curved or bent (*c*) if the supports are so arranged that the reactions at the supports will be vertical for vertical loads. This may be accomplished by placing one end on rollers (*c*) or, to a certain degree, by using plates which permit sliding. If the ends are so arranged that horizontal movement is restricted as the structure deforms, the reaction will no longer be vertical, and the structure will be an arch (*d*).

Flexural Stresses.   There are two general classes of stress set up in a beam by the loads: *flexural* or *bending* stresses and *shearing* stresses. As a beam deflects under a load, the material on the upper side of the beam shown in (*a*) is compressed or shortened, and that on the lower side is elongated. The stresses causing the shortening are called *compressive* stresses, and those causing the elongation are *tensile* stresses. Taken together, they are termed *flexural stresses*. These stresses are greatest at the extreme surfaces of the beam and are zero at the *neutral axis* near the center of the section. For a beam composed of one material, the neutral axis passes through the center of gravity or *centroid* of the cross section. The variation of stress is illustrated by

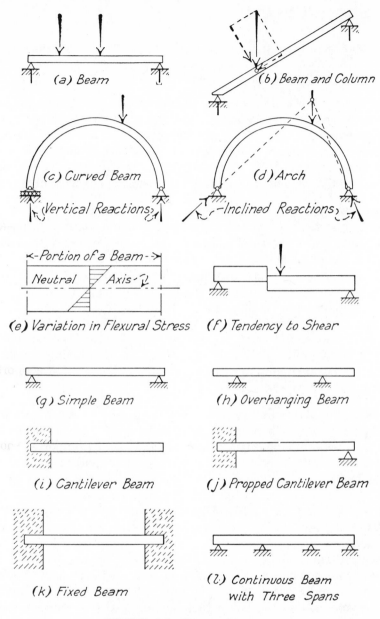

(a) Beam

(b) Beam and Column

(c) Curved Beam

(Vertical Reactions)

(d) Arch

-Inclined Reactions

<-Portion of a Beam->

Neutral   Axis

(e) Variation in Flexural Stress

(f) Tendency to Shear

(g) Simple Beam

(h) Overhanging Beam

(i) Cantilever Beam

(j) Propped Cantilever Beam

(k) Fixed Beam

(l) Continuous Beam
with Three Spans

FIGURE 3.2   Types of beams.

the triangles in (e), the intensity of the stress usually varying directly as the distance from the neutral axis, as indicated by the lengths of the horizontal lines in the triangles.

Shearing Stresses. It is evident that the forces acting on a horizontal beam tend to cut it along vertical cross sections, as shown in (f). The stresses set up in the beam by this action are called *shearing* stresses. Another stress is *horizontal shear*, which acts across the transverse section of the beam through its neutral axis. This form of shear is quite evident in wood beams of short spans with high loads.

Classification According to Support. Beams are divided into classes according to their method of support. A *simple beam* is supported at two points near its ends. (Figure 3.2(g)). A special case of the simple beam is the *overhanging beam*, which is supported at two points but projects beyond or overhangs one or both supports (h). Similar in some respects to the overhang beam is the *cantilever beam*, supported at one end only but rigidly held in position at the other end (i). The *propped cantilever beam* is supported at two points and rigidly held in position at one of them (j), whereas the *fixed beam* is supported at two points and rigidly held in position at those points as shown in (k). A beam supported at three or more points as shown in (l) is referred to as a *continuous beam*.

Classification According to Use. Beams are also classified according to their function. When any distinction is made between *beams* and *girders*, the beam is the smaller member and may be supported by the girder. A *lintel* is a beam with a short span supporting masonry or other loads over an opening (see Figure 4.7(i)). *Rafters* are closely spaced beams supporting the roof and running parallel to the slope of the roof (see Figure 5.10). Although rafters for flat roofs are called *joists*, joists are usually closely spaced beams supporting a floor or ceiling.

A horizontal member spanning between two opposing rafters, and usually, but not necessarily, located at some distance above the wall plates is a *collar beam*. It does not generally act as a beam, but ties the rafters together to prevent their spreading at the bearing plate. If the rafters are adequately anchored to the bearing plate, the collar beam functions as a strut, thus reducing the deflection of the rafters.

A *purlin* supports the roof sheathing while spanning between the principal roof supports, such as the roof trusses or arches. *Girts*, placed horizontally on the sides of a building, function similar to purlins. They are attached to the columns and support the sheathing and siding of buildings. They are used a great deal on mill buildings.

Framed openings in floors for stair wells or duct openings require members with special names. The beams that frame the side of the opening parallel to the floor joists are called the *trimmers*, while those that frame the other side and frame into the trimmers are called *headers*. *Tail beams* are the members that were cut out to make the opening. They frame into the headers.

Materials.   Beams are constructed of wood, steel, or reinforced concrete, as described in the following articles. Stone is occasionally used for lintels; but its low structural strength usually demands that it be supported by a steel lintel, which is frequently concealed.

20.   TRUSSES

A *truss* is a framed structure consisting of a group of triangles arranged in a single plane so that loads applied at the points of intersection of the members will cause only *direct stress* (tension or compression) in the members. Loads applied between these points cause flexural stresses. The framework shown in Figure 3.3 will illustrate the essential features of a truss, although a truss of this type is of no practical use.

In a truss, the ends of a framework must be supported so that the reactions at the supports are vertical for vertical loads. This result is accomplished by arranging one end so that horizontal movement may take place by sliding or rolling on the bearing plate (*b*) when loads are applied, or when changes of length occur due to temperature changes.

If the ends are so arranged that horizontal movement is restricted, the reactions will be inclined and the framework will act as an arch (*c*). It is not customary to provide sliding or rolling bearings for trusses whose span does not exceed 40 or 50 ft.

Parts of the Truss.   The points of intersection of the truss members are called *joints*, or sometimes *panel points*. The upper members form the *upper* or *top chord*; and the *lower* members, the *lower* or *bottom chord*. The *web members* connect the joints on the upper chord to those on the lower (*a*) by *gusset plates*. Web members carrying compressive stresses are *struts*, and those carrying tensile stresses are *ties*. The terms *end post, vertical post, hip vertical,* and *panel* apply to special forms of trusses and will be defined later. The distance from center to center of the supports is called the *span*.

Materials.   Trusses may be built wholly of wood, of wood and steel rods combined, or of rolled-steel sections. Concrete is used to a limited extent, but it is not usually suitable material for trusses.

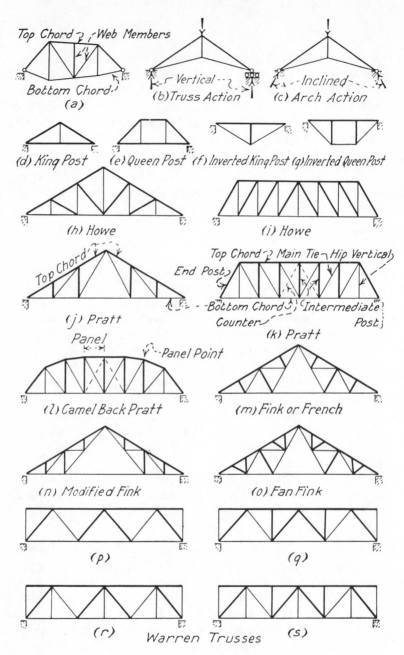

FIGURE 3.3 Types of trusses.

Types of Trusses.   Since a truss is composed of a group of triangles, it is possible to arrange innumerable types; but certain types have proved more satisfactory than others, and each has its special uses. The various types of trusses used in building construction are illustrated by line diagrams in Figures 3.3 and 3.4. The members indicated by heavy lines normally carry compressive stresses, and those indicated by light lines normally carry tensile stresses for vertical loads. The types shown with parallel chords may have gently sloping top chords that permit roof drainage without changing the truss type. The number of subdivisions or panels will depend upon the length of span and the type of construction.

The more common forms of trusses will be discussed in the following paragraphs.

The *king-post* truss (Figure 3.3(*d*)), the *queen-post* truss (*e*), and the *inverted king-post* and *queen-post* trusses (*f*) and (*g*) are all used for short spans in wood construction. The members indicated by heavy lines are made of wood, and those indicated by light lines are usually steel rods. The inverted king-post and queen-post trusses are often called *trussed beams* and are described in Article 31. The lower chords of the trusses in (*d*) and (*e*) carry tensile stresses but are usually made of wood and therefore are indicated by heavy lines.

The *Howe* truss may be constructed with inclined top chords (*h*) or with parallel chords (*i*). Howe truss members that are always constructed with wood are indicated by heavy lines. Those indicated by the light lines may be steel rods. The truss with sloping chords is usually used for pitched roofs, whereas the truss with parallel chords may be used to support flat roofs or floors. Howe trusses may be divided into any number of panels to suit any span or purlin spacing.

The *Pratt* truss, of wood or steel, may be constructed with inclined top chords (*j*), with parallel chords (*k*), or with broken upper chord (*l*), forming a camel-back Pratt truss. The truss with sloping chords is used for supporting sloping roofs, and the type with parallel chords is used for supporting flat roofs or floors. Pratt trusses may be divided into any number of panels to suit any span or purlin spacing.

The *Fink* truss is always constructed with inclined chords (*m*), and all the members are made of steel sections or of wood. Fink trusses are very widely used in supporting sloping roofs. They may be divided into any number of panels to suit any span or purlin spacing. A modified form of Fink truss is shown in (*n*), and a *fan Fink* or *fan* truss is shown in (*o*).

The *Warren* truss is always constructed with parallel or nearly parallel chords (*p*). Vertical members may be provided to reduce the distance between joints on the upper chord (q), on the lower chord (r), or on both chords (s). The Warren truss is very widely used for supporting floors or flat roofs.

The *bowstring* truss (Figure 3.4(a)) is usually constructed of wood.

A truss is said to be *cambered* when the bottom chord is raised at the center, as in the cambered Fink truss shown in (b). Camber improves the appearance of a truss and prevents sagging and the illusion of sagging.

Various special forms of wood roof trusses are shown in (c) to (h). The trusses shown in (c) and (d) have curved upper chords; the various members are built up of 1- or 2-in. lumber. A curved trussed beam or tied arch is shown in (e). This beam is built up of light lumber and steel rods. A *scissors* truss is shown in (g), a *hammer-beam* truss shown in (h), and a *Towne lattice* truss in (f). In the hammer-beam truss shown in (h), the parts of the truss marked a-b-c act as brackets to reduce the span to d-d. These brackets must be securely fastened to the wall along the vertical member b-c, and the wall must be capable of withstanding the outward thrust produced at the point c. The structural action is complicated.

Other special forms of roof trusses are shown in (i) to (n). *Monitors* are placed on top of roof trusses (i) and (j) to give better light and ventilation. The vertical face of a monitor, called the *clerestory*, is provided either with glass in sash, which will open to provide light and ventilation, or with *louvres* for ventilation only.

The *sawtooth* truss shown in (k) and (l) is used to provide light and ventilation; the steeper face of the roof is covered with glass so that a part of the sash will open. This face is usually turned toward the north to secure a uniform light. The type shown in (k) is constructed of steel, and that in (l) is constructed of timber and steel rods.

Another form of saw-tooth truss is shown in (m). Their vertical face is provided with top-hung sash. This type is always constructed of steel sections.

A *camel-back Pratt* truss with a heavy cambered lower chord is shown in (n).

Steel or timber roof trusses secured to columns give lateral rigidity (o) and (p). Such combinations of trusses and columns are called *transverse bents*. In the transverse bent shown in (o), the braces between the truss and columns are called *knee braces*. They are provided to give transverse or lateral rigidity.

A type of truss which is rarely used is the *Vierendeel* truss (Figure 3.5). It does not satisfy the definition of a truss but is given that

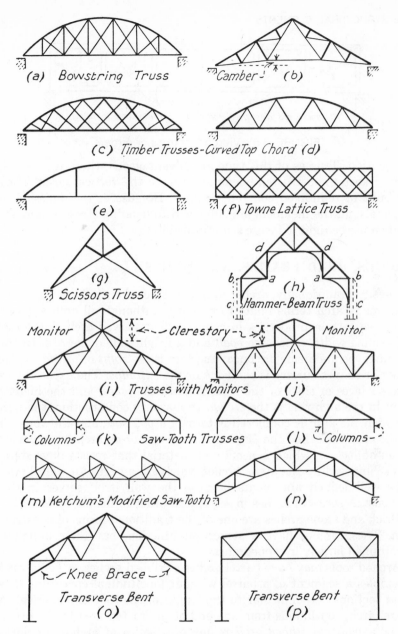

(a) Bowstring Truss

Camber (b)

(c) Timber Trusses-Curved Top Chord (d)

(e)

(f) Towne Lattice Truss

(g) Scissors Truss

(h) Hammer-Beam Truss

Monitor ←--Clerestory--→ Monitor

(i) Trusses with Monitors (j)

Columns (k) Saw-Tooth Trusses (l) Columns

(m) Ketchum's Modified Saw-Tooth (n)

Knee Brace

Transverse Bent (o)

Transverse Bent (p)

FIGURE 3.4 Types of trusses and transverse bents.

161

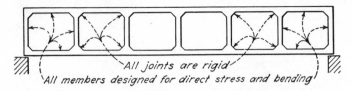

FIGURE 3.5　Vierendeel truss or girder.

designation. Trusses of this type are advantageous where it is necessary to keep unobstructed openings between the vertical posts. They are built of reinforced concrete or steel. Because of the omission of diagonals, all the members are subjected to bending stresses, and the joints must be rigid to insure structural stability.

21.　ARCHES, RIGID FRAMES, AND DOMES

Arches.　Traditionally, arches were constructed entirely of masonry; thus, many arch terms relate to masonry, although all arches generally have certain properties in common. The curved arch ring forms an opening width called the *span* and a height called the *rise*. In the opening, the underside or concave surface is the *soffit* or *intrados*. The upper or convex surface is the *back* or the *extrados*. The exposed vertical surfaces or sides of the arch ring are the *faces*, and the highest point in the arch ring is the *crown*. In masonry construction, the unit of which the arch is composed is termed a *voussoir*, and the voussoir at the crown is called the *keystone*. Each end of an arch is supported by an *abutment*, which is a mass of material that resists any lateral thrust imposed by the arch. Arches which are a part of a wall do not have well-defined abutments because the wall itself serves in this capacity. A series of arches in sequential order forms an *arcade*.

Brick and stone arches are one of the traditional forms of construction. These arches have a structural function, but their design is determined largely by appearance.

Arched roofs may have barrel arches or ribbed arches. A *barrel arch* resembles a segment of a barrel without longitudinal curvature. The cross sections perpendicular to the length of the roof are identical. A barrel arch, continuous from one end of the roofed area to the other, forms a *vault*. A *ribbed arch* is one of a series of arches providing structural support for a roof deck which spans between arches. The deck is continuous from one end of the roofed area to the other.

There are two general types of arch ribs (Figure 3.6). The *solid rib* (*a*) is subjected primarily to compressive stresses but also to flexural and shearing stresses. The *trussed* or *framed rib* (*b*) is made up of members arranged in triangles, as in trusses, each of which is subjected to either tensile or compressive stresses. Arch ribs are also called *arch rings*. Some types of framed arches are not arranged in the form of ribs.

Arched roofs consisting of two or more intersecting arched units are called *groined roofs, vaults,* or arches. The lines of intersection of the arched units of the structural members in these locations are called *groins*.

Arched roofs are used chiefly for the long spans required for such structures as armories, exhibition halls, field houses, gymnasiums, and assembly buildings. Barrel arches are usually constructed of reinforced concrete; solid arch ribs of reinforced concrete, masonry, steel, or wood; and trussed arched ribs of steel or wood.

Arches are constructed with their ends securely fixed to their foundations or *abutments* so that no rotation can take place at these points. *Hinges* may be installed to permit rotation and expansion. In the *three-hinged arch*, there is a hinge at each abutment and one at the crown; however, the ends of the arch ring are securely fixed at the abutments (Figure 3.7(*c*)); and in the no-hinged or *fixed* arch, no hinges are provided, leaving the ends of the arch rigidly anchored to the abutments (*d*). Steel and wood arches are usually three-hinged or two-hinged. Reinforced concrete arches may be of the three-hinged, two-hinged, or fixed type. The one-hinged arch is rarely used. Actual hinges may be provided for long-span arches, but often arrangements which offer negligible resistance to rotation are used.

The horizontal component of the thrust at each end of an arch may be transmitted through the foundation to the supporting soil if it has

(a) Solid rib arch   (b) Trussed or framed rib arch

FIGURE 3.6   Arch ribs.

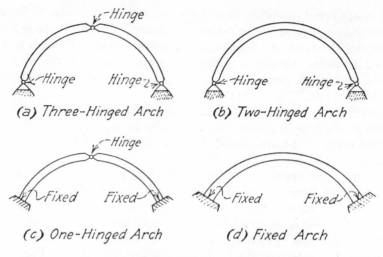

FIGURE 3.7   Types of arches.

adequate capacity to carry this component as well as the vertical component. Otherwise a steel tie placed below the floor level and encased in concrete forms a *tie beam* (Figure 3.8(a)).

If desired, the enclosed area is restricted by vertical side walls, shown by dashed lines in the figure, the ends of the arches left exposed or enclosed, thus increasing the building space. The ends of the arch may be supported on rigid abutments (b), but occasionally their construction is similar to that of the framed arch.

Curve of Arch Ring.   Curves and combinations of curves are used in forming arches. When strength and economy of material are principal factors, as they are in arch bridges, the curve of the arch ring is determined by the span and rise of the arch and the loading characteristics. For the ordinary arches over openings in walls, appearance determines choice in design rather than strength and economy of material. The failure of arches is caused by the spreading of the abutments or of the material that serves as abutments. Spreading occurs when arched window or door openings are placed too close to the wall corners or when the abutments at the ends of the arcade are insufficient.

Types of Arches.   The curve of the intrados is often a portion of the arc of a circle or a combination of the arcs of various circles with different radii and centers. Four types that have one center, shown in Figure 3.9, are the *semicircular*; the *segmental*, which includes less

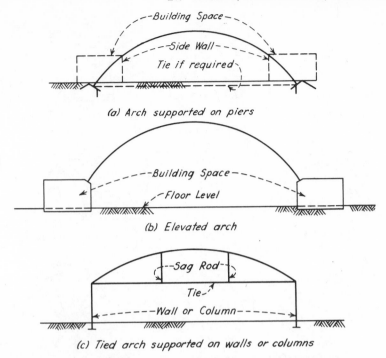

(a) Arch supported on piers

(b) Elevated arch

(c) Tied arch supported on walls or columns

FIGURE 3.8 Supports for ends of arches. (a) Arch supported on piers.
(b) Elevated arch. (c) Tied arch supported on walls or columns.

than a semicircle; and the *stilted*, which consists of a semicircular arch ring with straight vertical sections on each side.

Two-centered arches are shown in (*b*). The three types—blunt, equilateral, and lancet—differ only in the relation between the radius and the spacing of the centers. In the *blunt* or *drop arch*, the centers are within the arch. In the *equilateral* or *Gothic arch*, the radius of the intrados equals the span, and the centers are therefore on the springing lines. The centers of the *acute* or *lancet arch* are outside the arch.

There are two types of three-centered arch (*c*). In the first, one center is used for the arc of the central portion of the arch and two centers for the arcs at the ends of the arch ring. In the second, one center serves for the two arcs at the ends of the arch ring, and two centers are required for the central portion of the arch ring.

The four-centered or *Tudor arch* (*d*) is similar to the second type of three-centered arch, but the centers for the lower section of the

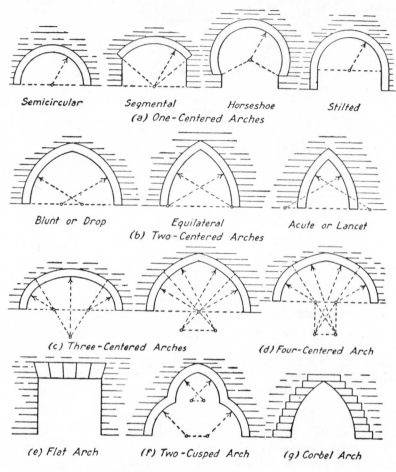

FIGURE 3.9   Types of arches.

arch ring do not coincide as in the three-centered arch. The vertical alignment of centers shown in the figure is not essential to this type.

The *flat arch* (*e*) may be supported by arch action, but it is usually carried on a concealed lintel.

The two-cusped arch is illustrated in (*f*). Many forms of cusped arch have been used for effect. Structurally, they are inefficient.

The elliptic arch is similar in shape to the three-centered arch shown in (*c*); however, the intrados is parabolic with its axis vertical.

The *corbel arch* (*g*), although not truly an arch, is called an arch because of its shape. There is no real arch action; each course is *corbeled* or cantilevered out over the course below until the two sides meet. This, the oldest form of arch, is not used in modern building construction. A *stepped arch* is one whose back is stepped to correspond with courses of masonry units in the wall above the openings.

Rigid Frames.   The structural action of rigid frames resembles that of arches. However, arches are so proportioned that the stresses in the principal members are primarily compressive, whereas rigid frames are proportioned to provide adequate ceiling height over the entire area, even though the structure is supported by foundations located just below the ground surface and not elevated, as the arch shown in Figure 3.8(b). Rigid frames are subjected to large bending and shearing stresses because of their form, illustrated by the several types in Figure 3.10. Rigid frames are often called arches, because their structural action is similar to that of an arch. Several types of

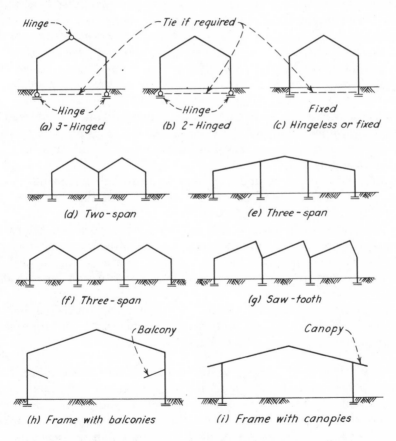

FIGURE 3.10   Types of rigid frames.

single-story rigid frames are illustrated in Figure 3.10. Multi-story rigid frames are also constructed.

The vertical members of rigid frames are called the *legs*. The top members are sometimes called *rafters*. The portion of the frame adjacent to the junction of a rafter and a leg is called a *knee*, or the *haunch*. If two rafters of adjoining frames meet on a leg, the adjacent portions of these members form a Y. The center of the top member of a symmetrical frame, or a single break in the direction of the top member, is called the *crown*.

Three single-span rigid frames are illustrated in Figure 3.10 (*a*), (*b*), and (*c*), the only differences being the types of connections at the bottoms of the legs and at the crown. The frame in (*a*) is called a *three-hinged frame* because of the hinges at the bottoms of the legs and at the crown; the frame in (*b*) is called a *two-hinged frame* because there are hinges only at the bottoms of the legs; and the frame in (*c*) is called a *hingeless, no-hinged,* or *fixed frame* because of the absence of hinges. Actually, no real hinges are provided at any point but, at points where it is assumed in the stress computations that hinges are present, the detail or arrangement provides a negligible resistance to rotation. When it is assumed that bottoms of the legs are fixed, the connection of the leg to the foundation and the foundation itself are so designed that there will be negligible rotation of the legs at these points. The usual assumption is that the connections at these points are hinged.

Steel ties (*a*), (*b*), and (*c*) may be provided under the floor if necessary to carry the horizontal components of the thrusts of the legs.

Balconies may be cantilevered out from the vertical legs to provide for spectators (*h*), or running tracks may be provided in gymnasiums in this manner. The top members may be cantilevered outward beyond the legs (*i*) to provide canopies for protection over loading platforms or other areas.

Rigid frames are used extensively in gymnasiums, field houses, assembly halls, churches, and industrial structures requiring large unobstructed floor areas and ceiling heights.

Domes. Domes (Figure 3.11) are frequently used as roofs over large circular floor areas for assembly halls, gymnasiums, field houses, and other buildings. They are constructed of self-supporting reinforced concrete shells as described in Article 50, or the structural members may be various types of wood, steel, and aluminum sections arranged in a great variety of patterns.

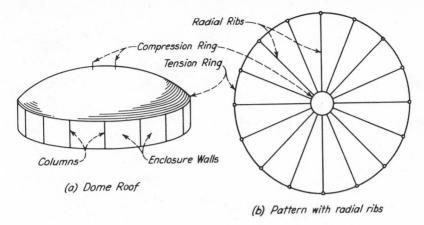

FIGURE 3.11   Circular dome roof.

A dome exerts outward thrusts continuously around its perimeter. These are resisted by a tension ring, as shown in the figure. The dome and tension ring are usually supported on columns spaced around the perimeter and braced to provide lateral stability for the structure. Bearing walls are also used for support. Purlins are commonly provided to span the space between ribs and support the roof deck.

Cables.   The roofs of buildings may be supported by steel cables. Each cable consists of a *strand* of several wires. The wires in a strand are arranged in a helical form. The cables used in supporting roofs are made up of wires produced by cold-drawing steel rods through steel ties. The process of cold-drawing produces wires with very high tensile strengths.

REFERENCES AND RECOMMENDED READING

1. *The National Building Code,* 1967 Edition, Recommended by the American Insurance Association, New York, New York.
2. *Uniform Building Code,* 1970 Edition, Vol. I, International Conference of Building Officials, Whittier, California.

4

MASONRY CONSTRUCTION

22. GENERAL MASONRY TERMS

Masonry construction is one of the traditional methods of building. Masonry structures from the Egyptian through classical and medieval architecture to our present efforts are marked with a quality that is found only in masonry.

Masonry is essentially an assembly of building block units that has a distinctive architectural character and impact of its own—its massiveness, variety of patterns, rich texture, warmth of color, aging patina, and enviable agelessness. Masonry still provides an opportunity for individual craftsmanship in a mechanistic time.

Let us investigate some of the contemporary technical aspects of this historic system.

Prerequisite to the study of masonry construction, certain masonry terms common to all masonry construction must be defined. *Masonry* is an assembly or combination of small building units made of clay, shale, concrete, glass, gypsum, or stone that are set in mortar [1]. Masonry consists entirely or partially of hollow or solid units laid contiguously in mortar. *Mortar* is a plastic mixture of cementacious materials with water and inert fine aggregates. *Grout* is mortar that is of pouring consistency; *dry grout* is used for dry packing, such as setting steel column base plates.

Mortars. Masonry units are bonded with mortar. It adheres to the units, joining them into a cohesive unified mass or bonding them to supporting structural members. Mortar strength is the critical factor in determining the design strength of masonry walls. Proportions are vital in obtaining desired strengths as indicated by Table 4-1.

Portland cements for mortars are usually Types I, II, or III (ASTM C 150). Air entraining cements or admixtures are permitted but not recommended where high bond strengths are required. Spe-

TABLE 4-1
Mortar Strengths and Mortar Proportions*

Mortar Type	Average Compressive Strength at 28 Days, lbs. per sq. in.	Parts by Volume of Portland Cement or Portland Blast Furnace Slag Cement	Parts by Volume of Masonry Cement	Parts by Volume of Hydrated Lime or Lime Putty	Aggregate, Measured in a Damp, Loose Condition
M	2500	1	1	...	Not less than 2¼ and not more than 3 times the sum of the volumes of the cements and lime used.
		½	...	¼	
S	1800	1	1	...	
		1	...	over ¼ to ½	
N	750		1	...	
		1	...	over ½ to 1¼	
O	350		1	...	
		1	...	over 1¼ to 2½	
K	75	1	...	over 2½ to 4	

* From 1971 Annual Book of ASTM Standards, Part 12, ASTM C 270, Table I and Table II [22].

cially manufactured *mortar cement* which is classified by Federal Specifications as Type I is used in masonry above grade and not subject to frost. Type II masonry cement is more widely used for general masonry construction. If used, mortar cement should be the product of only one manufacturer because the properties vary among manufacturers and may produce undesirable results. Any unrevealed additives, whose bonding and durabilities may be questionable, should be avoided; and an architect or engineer would be exercising sound professional judgment in rejecting any material whose contents or results are unsubstantiated. *Nonstaining cements* should be used for marble, limestone, and terra-cotta masonry.

Mortar sands should meet ASTM screening standards, C 136, be nonstaining, and be free from organic material. Sand is tested for organic impurities by ASTM C 40, which is performed by placing a 3% solution of sodium hydroxide into a bottle containing a sample of sand. A discoloration appearing after 24 hours indicates the presence of organic matter. A standard color chart indicates the suitability of the sand for mortar use [27].

Lime putty is added to improve *workability* and *water retentivity*. Hydrated lime, slaked not less than 24 hours before using, should be protected from the elements.

Water sufficient for a workable mix should be of the same quality required for concrete work, namely, be clean, free from acids, sugars, and other impurities.

Mortar colors are produced by adding paste or powdered coloring agents to the mix. They are usually red, umber, or black. Other colors are available, but architects favor the earthen colors, which are more reliable. Colors should be finely ground and have high purity.

After all ingredients have been placed in a mechanical mixer, mortars should be mixed at least 3 minutes. Mortars that have lost water through evaporation may be *retempered* (water added) to maintain workability; however, mortars older than 2½ hours should be discarded.

Types of Masonry Joints. The exposed mortar joints are finished in one of several ways. Joint finishing, called *tooling*, is formed with a *jointer tool* at the time the masonry unit is laid and not afterwards, as in *pointing* stone work.

The *struck* joint, shown in Figure 4.1(*a*), is the most commonly used finished joint. Suitable only for interior walls, it is likely to produce leaky exterior walls. It is used when masonry walls are to be plastered.

Joint illustrations

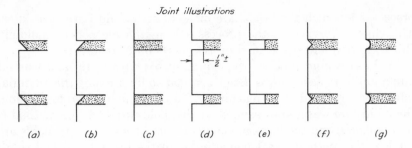

| (a) | (b) | (c) | (d) | (e) | (f) | (g) |

FIGURE 4.1 Joint illustrations.

The *weather* joint (*b*), similar to the struck joint, slopes to shed water more effectively. More difficult to form than the struck, weathered joints are rarely used, although they are more watertight.

In the *flush* joint (*c*), the mortar is cut off flush with the face of the brick. A joint common for unexposed interiors, it is also used for face brick.

The *raked* joint (*d*) is formed by raking out the mortar to the depth of about ½ in. The *stripped* joint (*e*), similar to the raked joint, is sometimes formed by placing wood strips in the joints as the brick are laid. These strips, insuring a uniform joint, are removed as soon as the mortar has set. Raked and stripped joints, more suitable for the interior, are used on exterior walls to produce a desired shadow. They are relatively expensive and not too watertight. The *Vee* joint (*f*) and *concave* (*g*) are made with a special tool which is run along the joint. These joints in exterior walls, including the vertical joints and the interior joints, should be entirely filled with mortar to reduce moisture penetration. The outside face of the joints should be made smooth and dense by exerting considerable pressure on the jointer tool.

23. BRICK MASONRY

Brick masonry consists of brick walls, columns, arches, floors, and beams. Mortars used between the brick forms the joints and holds the bricks together to act as a unit.

Kiln fired brick are made of clay or shale. They are used in all classes of construction, from private dwellings to the finest public building. Concrete or cement brick and sand-lime brick have limited use and are not burned. Adobe brick made of fired or sun-dried clay is now rarely used, except by the avant garde.

Brick Manufacture. Clay and shale brick are made by three different processes in which the bricks are molded, dried, and burned in kilns. The chief difference in the processes is in the method of molding. Brick are frequently specified by the method of manufacture.

In the *soft-mud process*, the clay is mixed with water and worked into a uniform plastic mass. Brick are shaped by pressing this material into molds by hand or by machinery. To keep the brick from sticking, the molds are watered or sanded. If the molds are wet, the method is *slop molding*, and the brick are called *water-struck brick*. If molds are sanded, the method is *sand molding*, and the brick are called *sand-struck brick.*.

In the *stiff-mud process*, only enough water is mixed with the clay to produce a mixture which may be forced through a die, forming a ribbon with a cross section equal to the flat side, or to the end of a brick. The brick, cut from this ribbon by tightly stretched wires, are *wire-cut brick*. If the beds are cut, the brick are called *side-cut*; but if the ends are cut, they are called *end-cut*.

By the *dry-pressed process*, clay of dry consistency is pressed into gang molds by plungers that exert a heavy pressure. This process produces the most accurately formed brick.

Brick pressed in oversize molds, dried, and repressed to the correct size are called *repressed brick*. Such brick are accurately formed and strong. Stiff-mud brick are frequently repressed.

Concrete or *cement* brick are usually made by compressing a moistened mixture of portland cement and sand into gang molds. A dry mixture permits the molds to be removed immediately. The brick are then *steam-cured* or lightly sprayed with water. This brick should be called *concrete brick* rather than cement brick.

Although concrete brick is a dead cement color, different fin'shes are made by using aggregates or colored cements. Concrete brick are used where suitable clay is scarce.

Sand-lime brick are a mixture of sand and hydrated lime compressed in molds and steam-cured.

Brick Sizes.   Any attempt to remember the plethora of brick sizes will surely lead to some confusion. The most cursory examination of the masonry literature convinces one that it is futile to memorize every brick size on the market.

Efforts have been made to standardize actual brick sizes and to eliminate the oddball sizes that plague designers. In some cases, there is more than one size assigned the same brick name. Brick sizes vary with locality or among manufacturers; moreover, brick made in the same size molds will differ because different clays vary in the amount of shrinkage in burning. Most brick manufacturers concentrate on producing certain sizes that are more marketable.

The smallest nominal brick measures 4 in. wide, 2 in. high, and 8 in. long; it is called a *baby Roman.*

Modular bricks are intended to meet even coursing dimensions. Mortar joint thickness, which varies from ¼ in. to ½ in., in combination with the brick is designed to conform to even coursing dimensions. There are about five *modular sizes* that in combination with joints meet standard course heights.

Brick Shapes.   The six surfaces of a brick are the *face*, the *side*, the *end*, the *cull*, and two *beds* (Figure 4.2).

Brick are cut into various shapes; the more common are shown in Figure 4.2. They are known by the following names: *half* or *bat*, *three-quarter*, *closer*, *king closer*, *queen closer*, and *split*. They are used where the full brick will not fit. A *soap* is a brick that has been split to a dimension less than 2 in.

Special shapes are used for moldings or finishing an opening. They are not usually used for ordinary brickwork but are available in face brick and in the glazed and enameled clay brick described later. Other useful shapes are *bull-nose* and the *double bull-nose*, the *external octagon*, the *internal octagon*, the *cove header*, and the *cove stretcher*.

TABLE OF MODULAR BRICK

Common Name	Height	Width	Length	Modular Coursing
Economy	4	4	8	
Utility	4	4	12	
6 Jumbo	4	6	12	1C = 4 in.
8 Jumbo	4	8	12	
Baby Roman	2	4	8	
Roman	2	4	12	2C = 4 in.
Modular	2⅔	4	8	
Norman	2⅔	4	12	3C = 8 in.
SCR	2⅔	6	12	
Double	5⅓	4	8	
Triple	5⅓	4	12	3C = 16 in.
Engineer	3⅕	4	8	
Norwegian	3⅕	4	12	5C = 16 in.
6 Norwegian	3⅕	6	12	

*Adapted from Ramsey and Sleeper, *Architectural Graphic Standards*, J. Wiley & Sons, 6 Edition, p. 148.

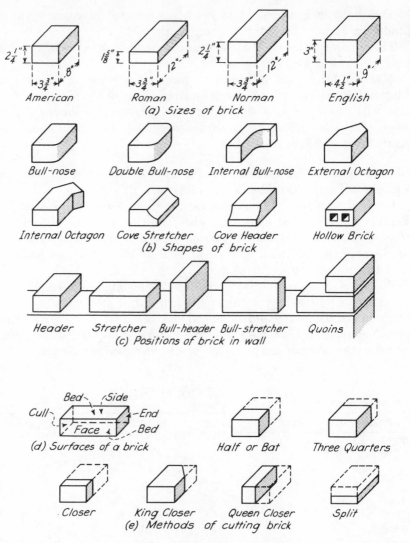

FIGURE 4.2 Brick positions and types.

Hollow brick used as face brick or backing brick reduce the weight of a wall and gain air space.

Cored brick have vertical cylindrical *cores* through the bed that reduce the weight. A brick may have a rectangular 3/8 in. deep recessed bed, called a *frog*. Neither a core nor a frog may come within 3/4 in. of any edge. Deeper frogs are used providing a section through the net area of the frog is at least 75% of the gross area.

Hardness of Clay Brick. The hardness and durability of clay and shale brick depends largely upon the degree of burning. Burned brick are sorted according to their hardness. The brick that immediately surround the fire are usually *overburns* and are badly warped and discolored. These, known as *arch* or *clinker brick*, are suitable for foundations, or similar places, for they are very durable though unattractive. Correctly burned brick are known as *hard* or *well-burned* brick and are for general use. Those remote from the fire are underburned and are known as *soft brick*. They are weak and not resistant to weather; therefore, they are suitable only for backing brick, where moisture is minimal and where strength is not a major factor. The terms *salmon*, *pale*, and *light* are often applied to soft red brick.

Color of Clay Brick. *Common* or *building brick*, used where appearance is not a factor, are usually made of clay which burns red, but sometimes common brick are white- or cream-colored. Very attractive walls are built of selected common brick, using well-tooled joints and the appropriate bond.

Face brick are used on exposed faces where appearance is a factor. By selecting mixing clays and introducing certain oxides, a great variety of colored face brick are produced—most commonly shades of red, brown, and gray.

Surface Finish of Clay Brick. The exposed face of brick may be smooth or roughened by wire cutting or by *combing*, as in *tapestry* brick.

Salt-glazed brick are smooth-faced brick of special composition. A glaze forms on the face exposed to the furnace gases when common salt is thrown into the fires near the end of the burning. Salt glazes are gray, brown, and green. Salt-glazed brick are impervious, smooth, and easily cleaned. They are suitable for exteriors as well as interiors with wainscots or the entire surface glazed. Glazed and facing hollow clay tile are made for use with corresponding brick.

Vapors, which are produced by the action of heat on the salt, react chemically when they contact the brick or tile surfaces. This reaction forms a glaze.

The face of smooth unburned clay brick of special composition are sprayed with a coating and burned, resulting in an *enameled* or *ceramic glazed* face. The common glazes are white, green, and brown finished in bright, medium, or dull. Enameled brick possesses a more impervious, smooth, and easily cleaned surface than salt-glazed brick. Enameled brick are particularly desirable for swimming pools, commercial kitchens, hospitals, and other locations where their properties are required. Those called *fire-flash* or *fire-mark* brick have acquired a surface marking by exposure to the fires of a kiln.

Face Brick and Backing Brick.   Brick are divided into *face brick* and *backing brick*, according to their position in the wall. Brick used as face brick should have a higher quality, greater durability, and better appearance than backing brick. Backing brick, often produced in the same kiln as the face brick, are of inferior quality due to underburning or overburning. *Back-up* is the material included between the facing and the back.

Brick are also divided into face brick and common brick. By this classification, *face brick* are made especially for facing purposes by selecting the clays to produce the desired color or for special surface treatment, whereas *common brick*, sometimes referred to as *building brick*, are made from natural clay without a special surface treatment. Selected common brick with attractive fire-marks are sometimes used as face brick.

Brick Quality.   Brick quality is determined by its strength, durability, and appearance. ASTM Specifications C 216 distinguishes between three types of face brick made from clay or shale. These three types are *FBS*, *FBX*, and *FBA*; the type distinctions are based on color range, variation in size, texture, architectural effects, and mechanical perfection. Further classification of facing brick provides for two grades—*SW* and *NW*—which is an index of durability and weathering resistance (ASTM 12).

Based on durability to exposure, building brick is divided into three grades: *SW* (*Severe Weathering*), *MW* (*Moderate Weathering*), and *NW* (*Negligible Weathering*). This standard further provides for physical properties, appearance, strength, size, frogging, and visual inspection of building brick.

Figure 4.3 below is a map of the U.S. showing weathering indices of brick. The ASTM further states that:

"Figure 4.3 indicates general areas of the United States in which brick masonry is subject to severe, moderate, and negligible weathering. The severe weathering region has a weathering index greater than 500. The moderate weathering region has a weathering index of 100 to 500. The negligible weathering region has a weathering index of less than 100.

"The use of Grade MW brick in a wall area above grade is structurally adequate in the severe weathering region, but Grade SW

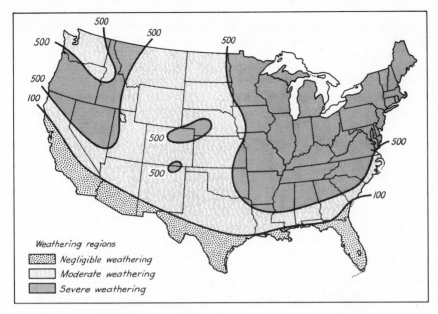

FIGURE 4.3 Weathering indices for masonry.

would provide a higher and more uniform degree of resistance to frost action. The degree of durability called for by Grade SW is not necessary for use in wall areas above grade in the moderate weathering region. Grade MW brick performs satisfactorily in wall areas above grade in the no-weathering region, where the average compressive strength of the units is at least 200 psi. Grade SW brick should be used in any region when the units are in contact with the ground horizontal surfaces, or in any position where they are likely to be permeated with water." [22].

Headers, Stretchers. Bricks are placed in a variety of positions in a wall. If they are laid on the *bed* with the *end* or *cull* exposed, they are called *headers*; but if they are laid with the *face* (long side) exposed, they are called *stretchers*. Half-brick are used as *false headers* giving the appearance of headers but not projecting into the backing. Brick placed on the side with the end exposed are called *bull headers* or *rowlocks* and are used for sills or for belt courses. Occasionally, they are laid on the side with the bed exposed forming *bull stretchers* or *flatters*. Belt courses and flat arches formed of brick set on end with the narrow side exposed are called *soldiers*; with the bed exposed, they are *sailors*. *Quoins* are brick placed at corners with one end and one face exposed. These positions and types of brick were illustrated in Figure 4.2(*c*).

Bonds.   The arrangement of brick tying units together in various ways is called the *bond*. A *course* is a bond of continuous horizontal layer forming part of a wall. See Figure 4.4. Courses are measured

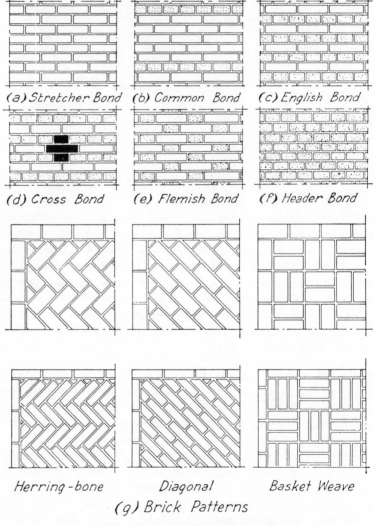

(a) Stretcher Bond   (b) Common Bond   (c) English Bond

(d) Cross Bond   (e) Flemish Bond   (f) Header Bond

Herring-bone   Diagonal   Basket Weave

(g) Brick Patterns

FIGURE 4.4   Brick bonds and patterns.

vertically. A *wythe* is a tie or a course of masonry units measured horizontally through the wall. See Figure 4.5.

In *running* or *stretcher* bond, the face brick are all stretchers as shown in Figure 4.4(*a*). The face brick are tied to the backing by metal or reinforcing, or by using clipped or secret bond as described, or by setting a queen closer face brick at every sixth course so that a continuous row of headers may project halfway into the face course and into the backing.

In *common* or *American* bond, every sixth course of stretcher bond is usually made a header course, resulting in a much stronger wall than that secured with metal ties.

English bond consists of alternate courses of headers and stretchers. The vertical joints in the header courses aligning or *breaking* over each other, and the vertical joints in the stretcher courses are also in line while bisecting every other brick in the header course.

English cross bond, *Dutch* bond, or *cross* bond is similar to English bond, but the alternate courses of stretchers do not align. Built up of interlocking crosses, this wall consists of two headers and a stretcher, forming a cross. For illustrative purposes, the cross is emphasized.

In *Flemish* bond, each course has alternate headers and stretchers; the alternate headers of each course are centered over the stretcher.

A *header* bond is composed entirely of headers laid to *break joints*. An entire wall would not usually be laid in header bond, but certain areas of a wall may be in header bond for decorative effect (Figure 4.4(*f*)). Some of the headers may be false headers.

In *clipped* or *secret* bond, face brick are laid in running bond, but the inside corners of the brick in every sixth course are clipped to permit a tie with the backing by headers laid diagonally. This bond offers little resistance to the separation of the face and backing, so it is desirable to add metal wall ties at every sixth course. These ties should be set on each brick of the course midway between the diagonal header courses. This construction gives a stronger wall if using wall ties only, but it is not as satisfactory as the other bonds.

Face brick, laid with the vertical as well as the horizontal joints continuous, is called *stack* bond. Brick may be laid with exposed faces or exposed headers. The objective is to secure an effect. Horizontal and vertical reinforcement is provided by including steel bars in the joints.

Patterns. Patterns are used primarily in brick floors, but can be laid in walls and fireplaces. A number of patterns may be created by arranging headers and stretchers in various ways. Emphasized headers may differ slightly in color from the stretchers. Other patterns are created by arranging face brick diagonally or vertically. The most common patterns are the *herringbone*, the *diagonal*, and the *basket-weave* shown in (*g*). Patterns should be bonded to the backing by using noncorrosive metal ties.

Skintled Brickwork.   Skintled brickwork is the setting of face brick (as running bond) out of line with the face of the wall. The corners project or recess from $\frac{1}{8}$ in. to $\frac{1}{4}$ in. or more. The mortar which protrudes may be allowed to remain, forming a *squeezed joint*. This type of brickwork leaks and collects dirt.

Bricklaying.   Brick masonry joints as thin as $\frac{1}{8}$ in. are formed for enameled or glazed brick presenting an easily cleaned surface, while joints as thick as $\frac{3}{4}$ in. meet aesthetic requirements, although the most common thickness is $\frac{1}{4}$ to $\frac{1}{2}$ in., meeting modular requirements.

The joints are made with a *trowel* or *jointer tool* by the following:

(1) Enough mortar is spread to form the horizontal joint for three or four bricks.

(2) The mortar that projects over the edge is cut off to keep it from running down over the face of the wall.

(3) The brick is bedded by tapping with the trowel handle.

(4) The mortar forced over the edge by the bedding process is cut off, the end to the brick is *buttered* to form the next vertical joint. This process forms a *rough-cut* joint.

(5) The exposed surface of the joints is tooled in one of the various kinds of joints shown in Figure 4.1.

In spreading mortar, a shallow longitudinal furrow is formed in the mortar with the trowel. Such a furrow is necessary to position the brick accurately. If the furrow is made too deep, the joint will not be completely filled with mortar as is required for watertight construction [1].

Push or *shove* joints are formed by placing a brick on a heavy bed of mortar and pushing or shoving it into position against a brick already placed in the same course so that the vertical joint between the two is entirely filled with mortar. Walls constructed in this manner are stronger and more watertight than walls constructed in the usual way.

Buttered joints are formed by buttering mortar on the bottom around the four edges and on the vertical end that will contact the last brick laid. The brick is then placed and tapped into place with the handle of the trowel. Narrow joints in enameled or glazed brickwork are commonly formed this way. At one time, face brick were set with buttered joints ⅛ in. wide, but this practice has been practically discontinued. The mortar is around the edges of the brick for buttered joints but not under the central bed of the brick.

The brickwork for all party walls, fire walls, and bearing walls should be laid solid with all joints filled with mortar.

Bond Required.   Every sixth course on each face of a wall should be a a header course; however, walls laid in Flemish bond or English bond have full brick headers at every fourth course bonded into the backing. The remaining headers may be half-brick, called false headers. Running bond should be bonded by using clipped bond combined with metal wall ties, as described, or by using split stretchers.

In walls more than 12 in. thick, the inner joints of header courses should be covered with another header course which breaks joints with the course below.

Methods of bonding brick walls of various thicknesses and wythes are shown in Figure 4.5.

Although face brick should be laid at the same time as the backing, backing brick is sometimes laid ahead by apprentice masons, while the journeyman and master masons lay the face brick.

At intermittent work stoppage, such as at the end of the day, masonry should be *stepped down*. *Toothing* is accomplished by building the wall to full height and ending the work with a portion of the brick projecting at alternate courses. Toothing is usually prohibited, except as noted below.

Anchorage at Wall Intersections.   All intersecting walls should anchor and bond to each other. When such walls are not built at the same time, the perpendicular joint may be (with the architect's per-

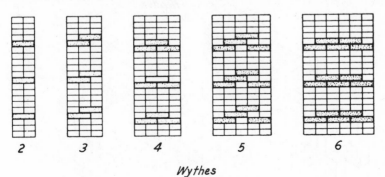

2   3   4   5   6

Wythes

FIGURE 4.5   Methods of bonding brick walls.

mission) toothed with 8 in. maximum offsets. Metal anchors 1½ by ½ in. by 2 ft. long with bent-up ends or cross pins extending 12 in. on each side of the joint. They should not be spaced more than 4 ft. apart vertically.

Wetting Brick.   Except in freezing weather, clay and shale brick should be wet down prior to setting. The brick are wet down to minimize absorption of water from the mortar. Loss of water will deter proper setting and lessen adhesion between brick and mortar; additionally, the accumulated jobsite dust is washed away by the wetting.

Hollow Brick Walls.   Brick walls may be made hollow; the open space prevents rainwater from passing through. Water penetrating the outer wythe is conducted out again by *flashing*. At the lowest course of face brick, an oiled, ¼ in. cotton rope about 12 in. long is sometimes laid across the mortar bed at about six-foot intervals. After the mortar has set, the ropes are pulled out, leaving a weep hole for the moisture to drain. There are three kinds of hollow wall: the *rolock* or *rowlock wall,* consisting of bull stretchers and bull headers, as shown in Figure 4.6(a); the *masonry-bonded hollow wall* (*b*), in which the face brick wythe is separated from the backing by a 2 to 3 in. air space that is bridged by headers that bond the face brick to the backing; and the *cavity wall* (*c*) which bonds the facing with metal ties.

The hollow wall with headers is not as effectively bonded as is the solid wall with the same number of wythes. The headers in the rolock and masonry-bonded hollow walls may allow moisture penetration.

Metal Wall Ties.   Metal ties, if properly made, do not permit mois-

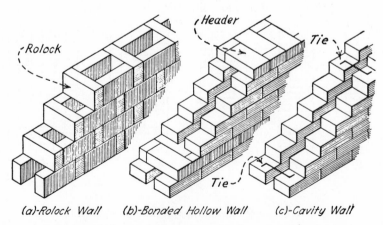

(a)-Rolock Wall   (b)-Bonded Hollow Wall   (c)-Cavity Wall

FIGURE 4.6   Hollow walls of brick masonry.

ture penetration. Metal masonry ties should be corrosive resistant and crimped or twisted. This forms a *drip* so that moisture does not bridge across the tie. They are effective in separating two wythes, but a load applied to the inner wythes cannot be transmitted through the ties to the outer wythe.

Other advantages of hollow walls are improved heat insulation, elimination of furring, and, as with the rolock wall, a saving in material. Hollow walls of brick are not widely used in this country, but the cavity wall is extensively used in Europe. They appear to have possibilities that deserve serious consideration because of their successful use abroad.

Stucco on Brick.   Brick walls are sometimes plastered with stucco. The brick should be rough, hard-burned brick set in portland-cement mortar with struck joints not less than ⅜ in. thick. This joint provides a better bond surface between the stucco and brick. The brick should be brushed free of all dust, dirt, or loose particles and wetted to prevent water absorption from the stucco.

Face Brick with Backing.   Face brick, backed with structural clay tile or concrete block, is either bonded to the backing with a row of headers every 16 in., or attached to the backing with embedded metal wall ties spaced not more than 1 ft. apart vertically nor more than 2 ft. horizontally. If metal ties are employed, face brick veneer cannot be considered in determining the required wall thickness; but if brick headers are used, the face brick contributes to the required wall thickness.

Structural clay tile backing is sometimes placed when panel or curtain walls are supported by a steel or reinforced concrete frame. They increase the resistance to heat losses or reduce the cost when tile backing is more economical than lightweight concrete block.

Hollow brick in Jumbo brick size are used on the inside face of exterior walls; the air cells in the brick increase the resistance to heat loss or gain.

Brick Walls Faced with Stone.   Brick is laid as backing for walls faced with stone masonry. Stonework veneer, not less than 4 in. thick, should be reasonably uniform in thickness, although not necessarily of the same thickness. Each stone should be bonded into the backing with noncorrosive metallic anchors.

Brick Arches.   Brick arches are used to span openings in brick walls. Various forms of arches were discussed in Chapter 3.

Brick arches are constructed of one or more wythes; they are soldier coursed with the end exposed (Figure 4.7(a)), or a rowlock arch

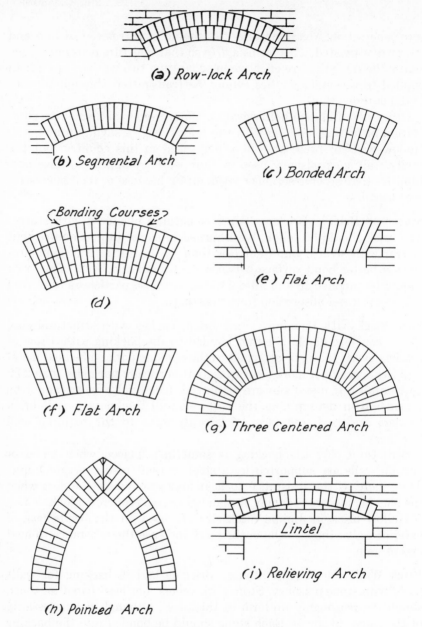

(a) Row-lock Arch

(b) Segmental Arch

(c) Bonded Arch

(d)

Bonding Courses

(e) Flat Arch

(f) Flat Arch

(g) Three Centered Arch

(h) Pointed Arch

Lintel

(i) Relieving Arch

FIGURE 4.7 Brick arches.

is formed with one or more wythes of brick with the narrow side exposed (*b*), with the courses forming the arch ring bonded (*c*), or with bonding courses at intervals (*d*).

The arch is formed by making wedge-shaped mortar joints or by beveling the brick to fit and forming joints of uniform thickness. Arches of shaped brick are termed *gaged arches* (Figure 4.7(*c*), (*d*), and (*g*)). Shaping is accomplished by laying the arch out on the floor and cutting and rubbing the brick to the proper shape before setting. Arches of other types than the rowlock are usually gaged.

Two forms of *flat* or *jack arches* are shown in (*e*) and (*f*). The flat arch is often supported on a concealed lintel; but if so, it is not a true arch.

Brick arches, often constructed over lintels in the backing of brick or stone walls (*i*) take the load from the lintels and are called *relieving* or *discharging arches*.

Reinforced Brick Masonry. Unreinforced brick masonry has considerable compressive strength but very little tensile or flexural strength. Embedded steel reinforcing bars in some of the horizontal bed joints provide horizontal tensile resistance. In addition, vertical reinforcing bars grouted in the vertical joints between wythes increase compressive strength, illustrated in Figure 4.8(*a*), (*b*), and (*c*) [6]. By omitting the center wythe of brick at vertical intervals in a three-wythe wall, a grout-surrounded reinforcing bar is placed as shown in (*d*); thus, larger bars provide greater resistance. The head and bed joints in the outside wythes are made in the usual manner; care must be taken to keep the joints between wythes free from loose droppings. This type of construction is also called *grouted masonry*.

High Lift Grouting System. Because of its ability to increase resistance to shock and blast, reinforcement is used in brick masonry subject to blasts and earthquake shocks. Reinforcement is also desirable for parapet walls subject to temperature extremes which tend to induce tensile stresses resulting in cracks from longitudinal contraction.

Reinforced brick columns are constructed by pouring concrete columns in vertical openings formed within brick masonry walls, piers, or columns. The vertical reinforcement should have an area of not less than ½ nor more than 4% of the gross column area and should consist of not less than four ½-in. bars. Lateral reinforcement or ties or spirals should be provided.

Many other forms of reinforced brick masonry have been developed [8, 11].

The following requirements are included in the *Building Code Re-*

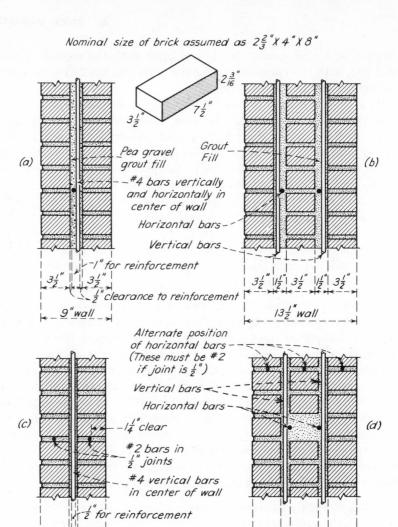

FIGURE 4.8 Reinforced brick masonry walls. Structural Clay Products Institute [6].

188

quirements for Reinforced Masonry approved by the American Standards Association as American Standard A41.2-1960, National Bureau of Standards Handbook 74, 1960.

"All masonry units used in grouted masonry shall be laid plumb in full head and bed joints and all interior joints, cores, or spaces that are designed to receive grout shall be solidly filled. The grouted longitudinal joints shall not be less than ¾ in. wide. Mortar fins shall not protrude into spaces to be filled with grout. When the least clear dimension of the longitudinal vertical joint or core is less than 2 in., the maximum height of the grout pour shall be limited to 12 in. When the least clear dimension of the longitudinal vertical joint or core is 2 in. or more, the maximum height of the grout pour shall not exceed 48 times the least clear dimension of the longitudinal vertical joint or core for pea gravel grout nor more than 64 for mortar grout but not to exceed 12 ft. Grout shall be agitated or puddled during placement to insure complete filling of the grout space. When grouting is stopped for 1 hr. or longer the grout pour shall be stopped 1½ in. below the top of a masonry unit. Masonry bonders (headers) shall not be used, but metal wall ties may be used to prevent spreading of the wythes and to maintain vertical alignment of the wall.

In reinforced grouted masonry, vertical reinforcement shall be accurately placed and held rigidly in position before the work is started. Horizontal reinforcement may be placed as the work progresses. . . . The thickness of grout or mortar between masonry units and reinforcement shall not be less than ¼ in. except ¼ in. bars may be laid in ½ in. horizontal mortar joints and No. 6 gage or smaller wires may be laid in ⅜ in. horizontal joints. Vertical joints containing both horizontal and vertical reinforcement shall be not less than ½ in. larger than of the diameters of the horizontal and vertical reinforcement contained therein."

Efflorescence. The white soluble salt deposit that frequently appears on masonry wall surfaces is called *efflorescence*. It consists of calcium and magnesium sulfates in the masonry unit or in the mortars, leached out by water that penetrates the masonry surface.

Efflorescence, which is formed by the salt only in the presence of water, may be minimized or prevented by selecting materials free of the salts and by impeding water intrusion into the wall. The latter is accomplished by observing a number of good construction practices, namely, by using water-repellent mortar and solidly filled joints, by capping walls with tight jointed copings sloped to drip free of the parapet wall onto the roof instead of toward the face of the wall, by effective flashing and calking, by providing a waterproof layer on top of foundation walls, and by protecting the walls during construction from rain and melting snow.

Cleaning.   At the conclusion of masonry work, all surfaces of face brick should be thoroughly cleaned with a 5–10%, solution of *muriatic acid* (HCl) in water. A stiff wire brush is used to remove spots and stains and efflorescence. After cleaning, the surface is flushed with water to remove all traces of acid. This treatment is not desirable for soluble limestones or for polished surfaces on other stones.

24.   ROCK AND STONE

General Discussion.   The term *rock* is used by geologists to include the solid and unconsolidated material forming the earth's crust, the former designated as bed rock and the latter as *mantle rock*. In engineering, rock refers only to the solid or *bed rock*. The unconsolidated material is soil or earth. In agriculture, the term *soil* applies only to the few inches of surface or *top soil* supporting vegetation; the underlying material is *subsoil*.

The terms rock and *stone* are often used synonymously, but generally rock relates to geologic formations, whereas smaller or quarried pieces of rock are called stone. However, this distinction is not always made.

Rock formations are of interest to the structural engineer and architect because rock formations are used to support the foundations of many structures. Stone is of interest because of its use in stone masonry, concrete aggregate, and as an ingredient in the manufacture of building materials.

Rocks are divided into three classes according to the method of formation. These classes are igneous, sedimentary, and metamorphic. Rocks are also divided into classes according to their chemical composition. The most abundant of these classes are argillaceous, siliceous, and calcareous. *Argillaceous* rocks are primarily alumina (Al_2O_3), the chief component of clay. *Siliceous* rocks are primarily silicon dioxide (SiO_2), the principal ingredient of quartz sand. *Calcareous* rocks are primarily calcium carbonate or lime ($CaCO_3$).

The earth's crust is composed chiefly of eight chemical elements. Nearly one-half is oxygen and more than one-quarter silicon. These elements in combination with silicon, aluminum, iron, calcium, sodium, potassium, and magnesium, form the minerals mentioned below.

Nearly all rocks are of one or more *minerals* and have definite chemical compounds usually with crystalline structures, but some consist of natural glass or volcanic dust. The most common rock-making minerals are *quartz* (silicon dioxide), the *feldspars* (potassium, sodium, or calcium aluminum silicates), the *micas* (complex hydrous silicates of aluminum with potassium, magnesium, and iron), *hornblende* (primarily calcium magnesium silicate), *kaolinite* (hydrous aluminum silicate), *calcite* (calcium carbonate), and *dolomite* (magnesium carbonate).

Igneous Rocks. Igneous rocks are formed from the solidification of molten rock. If solidification occurs below the earth's surface, plutonic or intrusive rock is formed; but if solidification occurs on the surface, the rock is *volcanic* or *extrusive*. *Pyroclastic* rocks are formed when molten rock is erupted violently into the air. The term *lava* is also applied to solidified extrusive rock. Molten rock forming igneous rocks is composed of a hot solution of feldspar, quartz, mica, water vapor, and carbon dioxide gases. The solidification of igneous rocks is due to a decrease in temperature and pressure. Igneous rock texture is determined greatly by the rate of cooling and by the volatile substances present. The texture may be coarse-grained if the cooling is very slow and fine-grained if the cooling is more rapid. If cooling is very rapid, *glass* is formed. The volatile substances facilitate crystallization. The entrapped gases in the solidifying mass form extrusive sheets with a *vesicular* or porous structure.

The most common igneous rocks are granite, felsite, basalt, and obsidian. *Granite* is usually a strong, durable, nonporous, and practically insoluble rock desirable as a foundation and building material. *Felsite* is a light-colored, fine-grained volcanic rock usually occurring as dykes or lava sheets. It is usually less porous than basalt. *Basalt* is dark-colored rock occurring chiefly in lava sheets or dykes. It is likely to be porous, cavernous, and badly fractured. Basalt is practically insoluble; the caverns are caused by methods of formation and not due to the solvent action of water. *Obsidian* is a volcanic rock glass formed by rapid cooling of molten rock. *Pyroclastic rocks* include: *volcanic ash*, which are fine, glassy particles deposited at considerable distance from the volcano; *lapilli*, which are formed from the gravel-like particles deposited closer to the volcano; and *braccias* and *tuffs*, which are formed by consolidation of the coarser particles falling near the volcano. The volcanic material comparing in fineness to sand is known as *puzzolana*. *Trap* or *trap rock* is a commercial term confined to certain fine-grained, dense, durable igneous rocks, such as basalt. The difficulty in quarrying and finishing limits their suitability for building stone.

Granite.   Most domestic granite is quarried in the states from Maine to Georgia, and in Minnesota and Wisconsin. Vermont produces the most granite. Granites are the hardest and most difficult to work but can be finished with highly polished surfaces. They are in a variety of colors, including white, gray, pink, red, and green. Granite is used for curtain wall panels, steps, platforms, sills, trimstone, and as thin veneer over other masonry. It is also used in base courses and locations requiring an extremely durable stone.

Sedimentary Rocks.   Sedimentary rocks are formed from the disintegrated products derived from igneous rocks or from other sedimentary rocks. This disintegration is brought about by *weathering factors*: *temperature changes*, which produce cracking because of unequal coefficients of expansion of the minerals; *temperature differentials*, which result in disintegration because of unequal temperatures in different parts of a rock mass; *alternate freezing and thawing*, which produce a repeated disruptive action caused by the expansion of ice in the rock pores; *abrasion*, which results from moving glacial ice, running water, and wind, with the effectiveness accelerated by solid particles; and *chemical action* by atmospheric gases, by rain water that has absorbed such gases, or by ground water carrying chemicals in solution.

These products of rock disintegration may remain in place or be deposited as sediments by running water, by wind, or by glacial ice. The deposits may remain unconsolidated, as soil, or be solidified through pressure exerted by overlying material in consort with cementing materials included in the deposits or supplied subsequently by infiltration. Another formational factor in the geologic evolution of rocks is seashells, which are formed from calcium carbonate originally derived from rocks and carried in solution in sea water. These shells, which accumulate on the bottom as the organisms die, are ground by the shifting currents and become solidified by pressure and the cementing action of the calcium carbonate itself. Millennia are required for this continuous cycle of rock distintegration, transportation, deposition, and solidification to reform rock.

Characteristics of sedimentary rocks are their layers, or *strata*. The process of deposition is rarely uniform and continuous. Variations occur in water and wind velocity and, therefore, in the size and composition of the material carried in suspension or in solution. These variations have resulted in stratified layers. The dividing surfaces between these strata are called *bedding planes* or *beds*. Because of this division into strata, many sedimentary rocks are also termed *stratified rocks*, although some limestones show so little stratification that they are called *freestones*, their structure being uniform. The *beds* between strata may have *weakened shear planes*, particularly if they contain clay which becomes very slippery when wet. Movement may take place easily along these surfaces, sometimes called *gliding planes*. Sedimentary rocks are formed chiefly of the minerals quartz, kaolinite, calcite, and dolomite. The more common sedimentary rocks are sandstone, conglomerates, limestone, and shale.

Sandstone. Sandstones are formed by the consolidation of sand beds deposited by water carrying sand in suspension. The consolidation is due to pressure exerted by overlying material and to a cementing material, namely clay, calcium carbonate, iron oxide, or silica. The character of the cementing material has a pronounced effect on the properties of sandstone; silica being the strongest and most durable adhesive. Pure sandstone is silicon dioxide. *Conglomerates* are similar to sandstones but made of cemented gravel instead of cemented sand. *Puddingstone* is a conglomerate in which the pebbles are well-rounded, and in *breccia* they are angular. Sandstones of pure silicon dioxide are white, whereas iron oxide impurities produce various shades of yellow, brown, and red.

Sandstones are found in nearly every state, with Ohio producing the most sandstone. Sold chiefly in the Middle West, the stones from Ohio are blue, gray, and buff, with good working qualities and durability. Sandstones are used for the same applications as limestones. *Bluestone* and *brownstone* are sandstones named for their color. Bluestone is very strong and durable and splits readily into thin slabs. It is used for sidewalks, steps, flagging, and sills. [9]

Limestones. Limestones are sedimentary rocks formed chiefly from the accumulation of shells on ancient sea bottoms which may now be many miles inland. Some limestones show fossils, but others show no trace of their origin because of the fine grinding imposed on the deposited shells. The cementing material is calcium carbonate, but magnesia, silica, alumina, and iron oxide are present in varying amounts.

Limestones are widely distributed with Indiana leading all states in production. Indiana, Bedford, or French Lick limestone, an oolitic limestone, is exported extensively. This stone varies from buff to gray; it is durable and can be sawn, planed, lathed, and carved easily, effectively, and economically. Low working costs permit Indiana limestone to compete with local stones in most parts of the country despite transportation costs. Limestone is used on interiors or exteriors for masonry veneer over other masonry, trimstone, steps, sills, coping, flagstones, and floor tiles.

Travertine is a limestone formed by the chemical precipitation of calcium carbonate from hot ground water. Characteristic of this stone are the small, irregularly shaped cavities. Travertine is used as floor tile and for limited ornamentation. Although this stone is imported from Italy, domestic travertine quarries are developing. *Oolitic limestone* contains many very small, rounded particles called *oolites*, which are concentric layers of calcium carbonate deposited around a nucleus. Most limestone is gray, but it may be buff or brown.

Shales are formed by the compacting or by the compacting and cementing of clays, muds, and silts. They may grade into sandstones if a large amount of siliceous material is present or grade into limestone if formed from silts containing an abundance of calcareous material. Shales have a finely stratified, impermeable structure. Those formed by compaction without cementation *slack* and disintegrate when acted upon by water after partial or complete drying. Such shales are unsatisfactory for supporting heavily loaded foundations because the shale gradually flows, a phenomenon known as *plastic flow*. Shales with grains cemented and compacted by pressure are not subject to plastic flow and do not disintegrate as do the shales which have been compacted only. For these reasons, nonplastic flowing shales are more suitable for foundation support. Shales do not possess the requisite durability for building stones.

Metamorphic Rocks.   *Metamorphic rocks* are either igneous or sedimentary rocks whose physical or chemical characteristics have been altered by pressure from the earth movements, temperature changes caused by intrusions of molten rock, or vapors or liquids permeating the rocks. The changes are in mineral composition, texture, and structure and include cementation by siliceous matter. The changes in mineral composition depend upon the chemical composition. Because all kinds of rocks are subjected to metamorphic action, there is a wide variation in the mineral composition of metamorphic rocks. Such rocks are highly crystalline in structure because of their origin. They are usually *foliated* or laminated similar to the stratified structure of sedimentary rocks and are known as *schists*. The common metamorphic rocks are *gneiss*, a laminated rock with a mineral composition similar to granite; *schist*, a laminated crystalline silicate rock that splits easily; *slate*, a fine-grained argillaceous rock that splits easily into slabs; *quartzite*, a hard durable crystalline quartz rock derived from sandstone; and *marble*, a crystalline rock that can be polished.

Marble is derived from limestone and is therefore chiefly calcium carbonate. There is no definite demarcation between many of the metamorphic rocks. For example, schist may grade into gneiss, or into slate.

Marble.   Marble deposits are rather widely distributed; about 90% is quarried in Tennessee, Vermont, and Georgia. Pure marble is white, but other substances and impurities produce a selection of beautiful colors and variations characteristic of marble. The color patterns, together with the ability to take a high polish, make marble a valuable medium for decorative uses. Marble is used as a curtain wall panel or as a veneer for monumental buildings. In general, it is used for exterior and interior decorative purposes, including wainscoting, panels, mantels, hearths, floor tile, and stairs.

Rock Structure.   The methods of formation of the various rocks have been considered. After rocks have been formed, their structures may be altered by movements of the earth's crust, resulting in bent and fractured formations, as shown in Figure 4.9 (*a*). The bends formed are called *folds*. Fractures are called *joints* if there is no displacement along the fracture, or *faults* if there is movement along the fracture, as shown in (*b*). The rock adjacent to a fault may be fragmented, as shown in (*c*), and cemented to form *fault breccia*, or the rock may become pulverized and referred to as *gouge*. Sedimentary

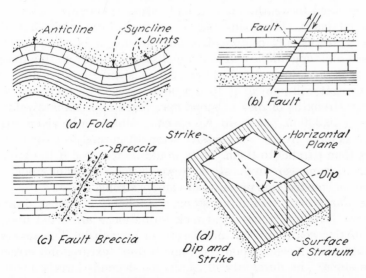

FIGURE 4.9   Rock structure.

rocks are usually deposited in horizontial layers or *strata*, but because of the folding activity, the beds may be sloped. The vertical angle that a bed makes with a horizontal plane is called a *dip*. The direction of a line parallel to the bed plane, perpendicular to the dip, with a horizontal plane is called the *strike* (*d*).

Because limestone is slightly soluble in water containing carbon dioxide, there is always the possibility that a limestone formation will be *cavernous* or contain *solution channels* caused by ground water. Such defects are particularly objectionable because they may induce failure of building foundations. If defects of this type are discovered before a structure is built, the cavities can usually be filled by *pressure grouting* or by cleaning out the openings and filling them with concrete.

Occasionally, large areas will be encountered in which rock formations have been broken up and crushed to great depths, though little surface evidence is apparent. Such areas are called *shear zones*.

Physical Properties. The physical properties of stone requisite to support foundations or for use in masonry structures are strength, durability, permeability, fire resistance, workability, solubility, color, and appearance. Some of these have already been considered.

Strength is mandatory for rocks serving as foundation support or as stone masonry. Sound bedrock is usually strong enough to support any imposed load transmitted by a concrete pier. Durable stone exposed to weathering is usually strong enough to support the subjected loads. The loads placed on rock formations and stone masonry are usually compressive, but flexural or bending loads are imposed on stone lintels and individual stones when resisting unequal settlement and improper bedding. Stone should not be used in a manner that will subject it to tensile stresses. The maximum compressive strengths of rock or stone may reach approximate values in excess of 50,000 lb. per sq. in. No specific values can be given. Stratified rocks have greater compressive strength normal to their bedding planes than parallel to them. Tests would be required if the strength of a given stone were in question. However, most rocks or stones considered for construction purposes are amply strong.

The *durability* of building stone is probably its most desirable property. Building stones (and rock formations) are acted upon by most of the weathering factors described previously. The resistance to weathering depends upon the composition, texture, and structure of the stone. The durability of sandstone depends largely upon the kind of cementing material that holds the grains together, silica being the most durable, cement and clay usually the least. Limestone and marble are slightly soluble in water containing carbon dioxide, but the dissolving rate is usually so slow that this action is negligible in building stones. However, this factor may receive consideration if a polished marble surface is maintained.

The best method for determining durability is by examining the rock *outcropping* or the parts of a quarry where the stone has been exposed for a long period. Similarly, examination of the same rock in older existing structures is a good index of the rock's weathering properties. Other properties that may indirectly indicate durability are its weight and its compressive strength; high values for these properties usually, but not always, being favorable indications. Accelerated freezing and thawing laboratory test cycles are also valuable. Blocks of stone quarried may contain a considerable amount of water called *quarry sap*. Such stones are frequently broken by freezing; but if allowed to season, they will probably not break.

The *porosity* is the proportion of pore volume to the total volume and is often an indication of a rock's resistance to frost. It is not the amount of pore space available, but their uniform continuity measured by the *permeability*—the ability to permit water to pass through its pores. A stone may be very porous and still be quite impermeable because of lack of continuity in the pores. For this reason, it may not be seriously affected by alternate freezing and thawing. Moreover, an open free-draining texture is more resistant to frost action than a fine texture that retains capillary water.

The *fire resistance* of a stone is a structural consideration. Few building stones will withstand very high temperatures. This fact is particularly true when the heated stone is subjected to a stream of high pressure water. Because of low resistance to fire, the use of stone interior piers, caps, and bond stones is prohibited by some building codes, but this practice is probably too severe. Building stones may be arranged in order of their resistivity as follows: (1) fine-grained sandstone with silica binder, (2) fine-grained granite and oolitic limestone, (3) ordinary limestone, (4) coarse-grained granite, (5) marble. Limestone fails by calcination at a relatively low temperature.

Workability is an economic factor in the selection of building stones. Marbles and granites are durable, strong, and attractive but are expensive for building purposes because of the labor required to shape them. Some stones unsuitable for ashlar may be satisfactory for rubble or squared-stone masonry requiring relatively little labor in shaping. Stones soft enough to work readily, such as shales, are frequently not durable. Ornamental work, such as moldings and carvings, requires a stone with an even grain free from seams and other defects. Stones easily worked in any direction and free from stratification are called *freestones*. Stones are more easily worked when they are *green* than after they have seasoned and the quarry sap has drained and evaporated.

Abrasive resistance is desirable in stair treads, door saddles, and floor finishes.

The *color* of a stone may be a determining factor in its selection for a given building.

25.   STONE MASONRY CONSTRUCTION

Stone masonry and cut-stone facings are used in exterior and interior walls where appearance is the deciding factor, and *trimstone* is set in exterior brick masonry walls. Stone is used on interiors for wainscoting, mantels, hearths, floor tile, steps, and stairways.

Many stones satisfactory for interiors are unsuitable for exterior climatic conditions, and some stones satisfactory for ordinary building purposes have low resistance to abrasion and are unsuitable for steps, door sills, and floors. Stones that are durable, strong, and attractive may be unsuitable because of excessive labor costs. Some stones not suitable for accurately shaped and carefully finished first-class masonry are satisfactory for cruder stone masonry. Stones soft enough to work readily frequently lack durability. Ornamental moldings and carvings require *freestones* with even grain, free from seams and other defects. Delicate carvings require a stone of considerable strength to withstand damage. Appearances may demand stone of a certain color.

Quarrying.   Quarrying separates rough blocks of stone from rock formations. In small quarries, hand tools may be used exclusively; but in larger quarries, special machinery is used. Because the use of explosives causes excessive waste, their use is confined chiefly to the removal of overburden to expose the solid stone.

The nature of the rock formation may determine the quarrying methods. The contact surfaces of adjacent layers or *beds* offer planes along which separation is easily accomplished. The beds may be so close that the stone is useful for *flagging* only, or the beds may be so far apart or so indistinct that they are of little assistance in quarrying; but the stone may be of greater value in spite of this fact, for the size of the stones is then not limited by the beds.

The unstratified rocks such as granite do not lie in separate layers but have a massive structure. Surfaces of separation have to be made by artificial means. Such rocks may split more easily in one direction than in another.

Both stratified and unstratified rock formations may be divided by *seams* running in any direction. Seams may be an advantage or a disadvantage because of the limitations of size and shape of the pieces. These seams may be very conspicuous and offer a distinct surface of separation; they may not be discovered until considerable work has been done on a stone; or they may even cause failure after a stone has been in a structure. *Streaks* may occur in stone without necessarily reducing its strength.

If a formation is badly broken by beds and seams, the rock may be removed with crowbars, picks, wedges; but the *dimension stones* are difficult to secure under these conditions except in small sizes.

Building stone is removed in rough blocks, which are later cut into the desired size. Rough blocks may be separated from the rock formation along a line by drilling a row of closely spaced holes and splitting the rock between the holes by *plug and feathers*. The plug is a steel wedge, and the feathers are semi-circular wedges that fit the hole; they

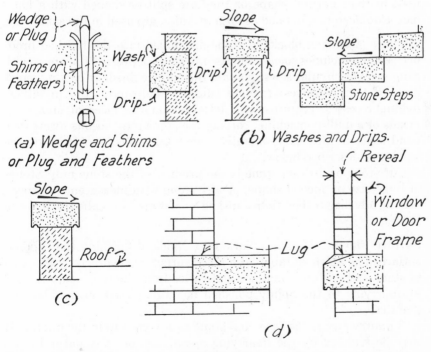

(a) Wedge and Shims or Plug and Feathers

(b) Washes and Drips.

(c)

(d)

FIGURE 4.10

are flat where the steel wedge is driven, as shown in Figure 4.10(a). Plug and feathers are placed in each hole, and each plug is gradually driven. By continuing this operation, sufficient force splits the rock. In stratified stone, the blocks are split along a cleavage plane parallel to the bed, which gives a natural surface of separation. In unstratified rocks, plug and feathers may be used along two planes at right angles to each other. Plug and feathers are also known as *wedge and shims*. The splitting is sometimes accomplished by wooden plugs driven in the holes and soaked with water, which causes the plugs to expand and split the rock.

Channeling machines, widely used for quarrying, cut narrow vertical channels along the face of the block, as deep as 10 or 15 ft. The blocks are separated from the quarry ledge at a bedding plane, or split along a horizontal plane by drilling holes and driving wedges.

The plug-and-feathers method, suitable on any kind of stone, is used in granite quarrying. The channeling method, not suitable for granite, is commonly used for quarrying sandstone, although the plug-and-feathers method is more suitable.

Fieldstones found loose in nature are used for building. They are used in their natural shape or they are split or shaped with a hammer. *Cobblestones*, defined as *large pebbles*, are used in the same way.

Milling.   Quarried blocks of stone are milled into the finished product. *Milling* includes *sawing* blocks into slabs of desired thickness with gang saws or circular saws; *planing* the surface finish or cutting moldings on their surfaces; *turning* columns, balusters, etc., in lathes; milling recesses, patterns, and lettering on the faces of stones by means of a milling machine; *carving* the stone into various forms with hand tools or with hand-operated pneumatic tools; and finishing the surface as described below.

Cut stone or dressed stone is the product of the stone mill. Stones of large size or special shape, or any stone with measurements specified in advance, other than finished cut stone, are called *dimension stones*.

Surface Finish.   There are various methods of finishing the exposed surface. The finish suitable for a given surface is governed by the kind of stone and its functional use and varies from the rough face formed in quarrying to the highly polished face often used on marbles and granites.

A *quarry face* is the face on a stone as it comes from the quarry. It may be *scabbed* by the quarrying operations, or by a natural seam called a *seam face*. Quarries producing seam-face stone are traversed in many directions by natural seams forming relatively small blocks of stone of irregular shape and size. Seam faces are often highly colored by deposits from mineral-laden waters which have penetrated the seams. A *split face* is formed by splitting a rock.

A *sawed finish* has visible saw marks. This may be similar to a *shot ground* or *shot rubbed* finish, which has arched markings left by drilled shot used in cutting. A *smooth finish* is produced by planers without handwork other than the removal of objectionable tool marks.

The *4 cut, 6 cut*, and *8 cut* are corrugated surfaces identified by the number of corrugations per inch.

A *thermal finish*, produced by heating a granite surface, causes certain crystals to expand and pop out and resembles a rough split face.

Rubbed and *honed finishes* are produced by machine grinding or hand rubbing a sawed or pointed surface. Small surfaces and moldings are usually hand finished. A coarse *rubbed finish* shows small scratches, but a *honed finish* gives a smooth dead surface practically free from scratches.

A *polished finish* is obtained by polishing honed surfaces. Granite and marble will hold a polish, but most other stones will not.

The margin or border of a stone may have one finish and the remaining area another. Figure 4.11(j) shows a tooled margin with the remaining surface *bush-hammered*. Stones finished with dual finishes are called *drafted stones*.

Sandblasting is used to cut lettering and designs into granite. In this operation, the polished surface is coated with a molten rubberlike substance called *dope*, which forms an elastic covering. The design is cut into this covering, exposing the stone surface. The exposed stone is then sandblasted, while the elastic coating protects the remaining stone.

Selecting Surface Finish.   Appropriate finish depends upon the type of masonry, the kind of stone, its position and function in the building, the appearance desired, the atmospheric conditions, and the funds available.

The finish selection giving the desired results at least cost is usually chosen. Satisfactory results are often produced at low cost with quarry, seam, or split-face rubble masonry.

All the types of finish may be used on granite and marble; and except for polishing, they may be used on sandstone and limestone. Limestone that can be polished is usually classed as marble. The hammered finishes are suitable only for the harder sandstones and limestones, for on the softer stones the ridges will not stand up but

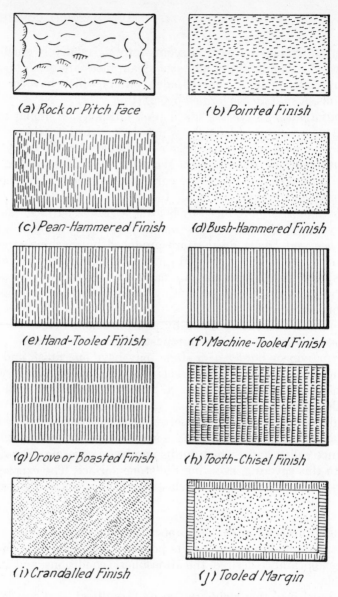

(a) Rock or Pitch Face

(b) Pointed Finish

(c) Pean-Hammered Finish

(d) Bush-Hammered Finish

(e) Hand-Tooled Finish

(f) Machine-Tooled Finish

(g) Drove or Boasted Finish

(h) Tooth-Chisel Finish

(i) Crandalled Finish

(j) Tooled Margin

FIGURE 4.11   Stone finishes.

will break off leaving a bruised face. These finishes, often called *hard-stone finishes*, are unsuitable for soft stone. The tooled finishes, similar to the hammered finishes, are more suitable for soft stones.

In general, rubble masonry is quarry face or *pitched face*; and ashlar or cut-stone masonry is a pointed or hammered face if granite and a sawed, smooth, or rubbed face if limestone or sandstone. Marble is usually rubbed or honed for exteriors and either honed or polished for interiors. A quarry face or pitch face may be used for ashlar.

Because of the ease in cleaning, polished surfaces are often the base courses for parts that might be soiled and for lower stories exposed to a smoky atmosphere. The fine finishes keep clean longer.

The sawed finish is the most economical finish for limestone and sandstone, except the harder grades.

The finer finishes are more suitable for interiors than the coarser finishes; but on the exterior the finer finishes fail to show if used above the first floor.

Washes and Drips.   The exposed top surfaces of cornices, copings, belt courses, and sills, steps, platforms, and other stones should be provided with sloping surfaces called *washes* (Figure 4.10(*b*)).

Projecting cornices, belt courses, and sills are provided with a groove or channel called a *drip* on the under surface near the outer edge. The drip causes water to drip from the lower edge of the stone rather than shedding along the wall. Drips should be at least ½ in. wide by ¼ in. deep. (*b*).

Stonework will usually streak or soil where washes pitch toward the face of the stonework. Thus, copings should pitch toward the roof and water drained off projecting stones.

Where other work is built on stones provided with a wash, it is usually necessary to cut *raised seats* and *lugs* at the jamb, forming level beds for the work built on them (*d*).

Classification of Stone Masonry.   Stone masonry is classed by the refinement in shaping the face stone arrangement and the surface finish of the stones.

There are no absolute classification standards; but in general, the crudest masonry, constructed of stones with little or no shaping, is called *rubble*, and the highest form, constructed of shaped stones, is *ashlar*. Between these two classes are degrees in shaping and various stone arrangements. The most common classification divides masonry into *rubble*, *squared-stone* masonry, and *ashlar*, according to the care in shaping the stones, and into *range*, *broken range*, and *random*, according to the arrangement of the stones in the wall. The latter classification does not apply to rubble. Rubble is classed as *coursed* or *uncoursed*. Ashlar is also called *cut stone*.

Lines of demarcation are undefined between ashlar, squared-stone masonry, and rubble. When stratified stone is used, the horizontal joints of rubble may be as narrow and as uniform as those of squared-stone masonry; the distinction between the two classes is the vertical joints. If only the sharp corners are knocked off loose rock, it is rubble; but if the stone is shaped to give a uniform vertical joint, it is squared-stone masonry. If the end joints are not vertical but are uniform in thickness, the class would be identical to squared-stone masonry but could not logically be placed in that class because of the shape of the stone.

Sometimes the joints are as thin as ashlar joints, but the end joints are not vertical. Such masonry should probably be classed as ashlar. *Polygonal masonry* is made of irregular stones without parallel surfaces; these stones are shaped to fit the spaces they are to occupy, and the joints are neither vertical nor horizontal. Stones in polygonal masonry are sometimes accurately cut with joints as thin and uniform as ashlar. Often the stones are only roughly shaped, and the joints are nonuniform. This type of masonry is often called *mosaic rubble* or *cobweb rubble*. It is structurally questionable because it is difficult to reinforce.

Arrangement of Courses.   In range masonry, the stones are laid in uniform courses throughout the length. All courses, however, are not the same thickness (Figure 4.12(*a*)).

In broken-range masonry, the stones are coursed, but for short distances only (*b*).

In random masonry, no attempt is made to form courses (*c*). The terms *range*, *broken-range*, and *random* are usually applied only to ashlar and squared-stone masonry; but where rubble masonry is constructed of stratified stones, the upper and lower surfaces may be parallel, resulting in random masonry, although rubble masonry is generally divided into coursed and uncoursed rubble.

In coursed rubble, the masonry is leveled at specified heights (*d*) or is laid in fairly regular courses marked *m-m*, *n-n*, *o-o* in the figure. Uncoursed rubble masonry is not leveled as in coursed rubble (*e*).

Backing.   In rubble masonry, the face and backing are usually rubble, the better stones being selected for the face; however, concrete backing may be used.

The backing for squared-stone masonry and ashlar masonry may be rubble masonry, brick, structural clay tile, or concrete block.

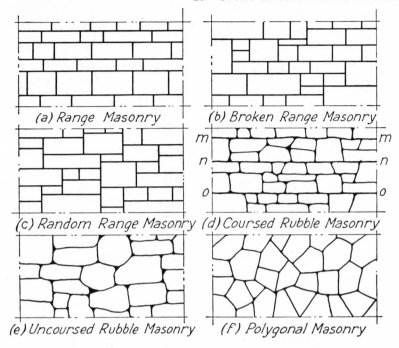

(a) Range Masonry   (b) Broken Range Masonry

(c) Random Range Masonry   (d) Coursed Rubble Masonry

(e) Uncoursed Rubble Masonry   (f) Polygonal Masonry

FIGURE 4.12   Classes of stone masonry.

Rubble masonry is unsuitable for backing thin panel walls. If rough stone blocks are shaped at the building site, the stone shards are used in the rubble backing. Ashlar or squared-stone masonry is not used for backing.

Concrete backing should not be placed against limestone facing or against brickwork in contact with limestone without providing a waterproof layer between stone and concrete. If a waterproof layer is not provided, the stone may be discolored. Certain stones may not require waterproof backing, but it should not be omitted without a thorough investigation. Ashlar is often used as a veneer over concrete walls or other surfaces.

Facing stones used over brick or structural clay tile backing should intermesh with the backing, creating horizontal through-wall joints at intervals desired for bonding.

For illustrations of wall sections with stone facing and structural clay tile backing, see Figure 4.18.

Setting.   Placing stone in position is called *setting*. Stones lifted by derricks are held with *grab hooks*, as shown in Figure 4.13(a), or with

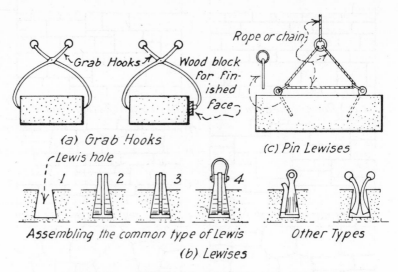

FIGURE 4.13   Methods of lifting stone blocks.

lewises (*b*), or with *pin lewises* in inclined holes (*c*). If lewises are used, lewis holes must be provided 3 in. to 4 in. deep. The most common type of lewis is assembled in the hole, as shown in (*b*).

Stratified stone should be dressed to set with the natural quarry bed horizontal because, when so oriented, it has greater strength and greater resistance to weathering. Water enters more freely, and the stone weathers and scales off badly and more rapidly when the quarry bed is vertical.

The setting of Indiana limestone on its natural quarry bed requires little concern. The great majority of such ashlar is sawed with the grain parallel to the face of the wall, and monolithic columns are produced with the grain running vertical. However, most limestones are somewhat weaker when loaded parallel to the bedding than when loaded perpendicular to the bedding.

Door and window sill ends built into the masonry should be bedded only at the ends, the space between left entirely free from mortar except for the pointing mortar which is applied later. If this practice is not followed, the sills are quite certain to break when the ends are loaded. This type of sill is called a *lug sill*. Window sills called *slip sills* are made slightly shorter than the intended opening and are independent of the masonry, except the mortar bed. Exterior steps and sills are pitched to the front, permitting drainage.

Nonstaining mortar should be used in setting limestone facing and its backing. The application of waterproof cement mortar to the back of the facing masonry or the face of the backing material is called *parging, pargeting,* or *back plastering.* Parging increases the resistance of the wall to moisture penetration and prevents efflorescence. The use of nonstaining cement mortar, however, deters staining. It is preferred that the backing masonry be free from soluble salts.

The practice of *back painting* the sides or backs of facing stones with bituminous waterproofing to prevent efflorescence is objectionable.

Bondstones and Anchors. Longitudinal stone masonry bond is secured by breaking joints as in brick masonry, although in broken-range and random masonry a vertical joint may be three stones high.

Bond between the face and backing is secured by headers or bondstones by bond courses, by metal anchors, or by a combination of these materials.

The bond or anchorage required between the face and backing depends on whether the wall is a bearing wall or a nonbearing wall and if the required wall thickness includes the facing.

Headers extending entirely through the wall are shown in Figure 4.14(a).

Bondstones projecting into the backing an amount equal to the thickness of the other facing stones are shown in (b), with one bed of every stone in contact with a bondstone. The arrangement shown in (c) makes every other course a bond course. Only 10% of the stones in a bond course are required to be bondstones; and in broken-range and random masonry, the bondstones are distributed at random throughout the wall, with each stone providing the tie for a given surface area. *Noncorrosive metal anchors,* usually stainless steel, tie the facing to the backing (d). Bond courses and anchors are usually used together. In (e), the anchors are supplemented by a bond course bearing on the spandrel beam of each story. The combination of bond courses and anchors is shown in (f). Every stone not a bondstone may be anchored to the backing, as shown in (g). The method of anchoring a cut-stone cornice is illustrated in (h). Special anchors should be provided for cornice and belt-course stones that have insufficient bearing on the wall. These anchors should be hooked into the stone at least 2 in. and spaced about 2 ft., with at least two anchors to a stone.

Further discussion of the bond requirements are below.

Anchorage. Anchors are usually placed in the horizontal joints and hooked into the tops of the stones, but at times it is desirable to hook the anchors into the side of the vertical joints near the top and bot-

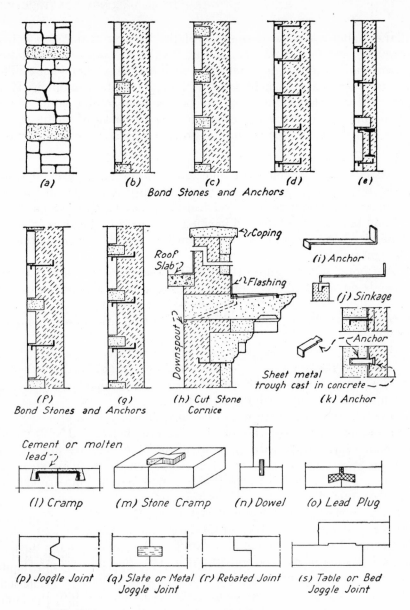

FIGURE 4.14 Bonding and anchoring stone masonry.

tom of the stone, especially when structural clay tile backing is used. A typical form of anchor is shown in (i).

Anchors should not be placed in the mortar joint, but a *sinkage* or depression of ample width and of a depth slightly greater than the thickness of the anchor should be provided at the back of anchor holes (j). There are two reasons for providing sinkages. With the $\frac{1}{4}$-in. joints usually used in ashlar masonry, there would be difficulty in placing the anchors in the joints; the stones would tend to rest on the anchors rather than to secure uniform bearing on the mortar joint. This condition would tend to cause the stones to crack. The sinkages do not fit the anchor accurately but are usually larger than necessary and are crudely formed. Anchor holes should be kept about 2 in. from the outer surface of the stonework so that the anchor may be adequately protected. The anchors should be completely embedded in mortar or lead.

Rustproof anchors should be made of stainless steel or noncorrosive metal or steel that is galvanized after cutting and shaping. Rust may stain the stonework, split the stone by expansion while rusting, or finally destroy the anchor. Coating anchors with paint will not supplant galvanizing, but it is an additional precaution.

A *dovetail anchor* convenient for anchoring a stone facing to a concrete wall is shown in (k). Before the wall is poured, sheet-metal troughs are tacked vertically at proper intervals to the side of the form for the outer face. When the forms are stripped, the trough is exposed in the face of the wall. It is beveled on the sides to receive an anchor with a dovetailed end. The anchors are adjusted up and down to fit into the joints. This anchor is more convenient and effective than most types of anchors, which are cast in concrete walls and are bent out into the mortar joints of the facing.

Lewis Anchors, Cramps, Dowels. Anchors other than those described are sometimes required. When suspending the soffits of openings from steel work, *lewis anchors*, similar to the ordinary lewis, are used. Flat *iron clamps* or *cramps* keep coping stones and stair rails from falling apart. Cramps vary from $1\frac{1}{4}$ by $\frac{1}{4}$ in. for light work to $1\frac{1}{2}$ by $\frac{1}{2}$ in. for heavier pieces. These are turned down to $1\frac{1}{2}$ or 2 in. at the ends and vary in length from 6 to 12 in. They may be set in sinkages in the tops of the stones, or they may be set under the stones with their ends turned upward into the stones. Cramps should be heavily galvanized after being bent to shape. On first class work, they may be protected by pouring molten lead around them (Figure 4.14(l)), the holes being larger at the bottom than at the top in order to hold the lead and anchor in place. Cramps or keys of stone may be used in place of steel cramps (*m*). The lead plug shown in (*o*) serves the same purpose but is rarely used. In forming the lead plug, molten lead is poured into the vertical channel and fills the dovetailed holes, which are sloped so that the lead can easily fill the holes. Brass or bronze dowels made of solid rods or of pipe are ordinarily used to hold the ends of balusters, window mullions, and similar pieces (*n*). The ends of inclined copings may be doweled into the kneelers.

Joggled, Tabled, and Rebated Joints.   The types of joints rarely used are the *joggled joint* shown in Figure 4.14(*p*) and the *tabled joint* or *bed joggle joint* shown in (*s*), designed to prevent movement along a joint. These joints are very expensive to form. A *slat* or *metal joggle joint* (*q*) may be formed more economically. The joint is not usually continuous. A *rebated joint* shown in (*r*) is sometimes used for coping stones placed on a slope.

Faced and Veneered Walls.   Stone walls may be classed as faced walls and veneered walls according to the provision made for bonding the face to the backing. If the facing and backing are securely bonded and acting as a unit, the entire wall thickness may be considered in strength calculations and in satisfying requirements for minimum thickness. Such walls are called *faced walls*. If the facing is not attached and bonded to the backing to the extent that it forms an integral part of the wall, the wall is called a *veneered wall*, and only the backing may be considered in strength calculations and in satisfying the requirements for minimum thickness.

Joints and Pointing.   The mortar layers between stones are called *joints*. Horizontal joints are *bed joints* or simply *beds*; vertical joints are known as *head joints* or *builds*; and joints between wythes are called *collar joints*.

In rubble masonry, the joints are neither uniform in thickness nor constant in direction; they simply are the spaces between stones of irregular shapes. Large spaces in the backing may have small pieces of stone called *spalls* embedded in the mortar.

In squared-stone masonry, the joints are horizontal and vertical and are uniform in thickness, the stones having been roughly dressed to shape. In general, the joints in squared-stone masonry are ½ to 1 in. thick.

In ashlar or cut-stone masonry, the stones are accurately dressed to shape so that the joints do not exceed ½ in. in thickness. A very common thickness of joint for the ashlar facing of buildings is ¼ in.

Joints ⅛ in. thick are sometimes used for interior stonework but are not desirable for exterior work.

The mortar in the horizontal and vertical joints of ashlar masonry is kept back from the face in setting the stone or is raked out to a depth of about ¾ in. In this space, a special mortar is placed to make a tighter and more attractive joint. This process is known as *pointing*. The various types of joints formed by pointing are shown in Figure 4.15. Pointing is done after the mortar in the joint has set and usually after all the stone has been placed and the wall has received its full load.

Squared-stone masonry may be pointed in the same manner as ashlar, or the joint may be finished at the time the stone is set.

The joints in rubble masonry are usually finished when the stones are set, and no pointing is done.

Often the joints of squared-stone masonry or rubble are made flush with the surface of the stones, and after the mortar has set, a narrow bead of colored mortar is run on the wide joints to give the effect of narrow joints. See Figure 4.15(*b*). The wide joint is frequently made the same color as the stone, and the narrow joint is made a contrasting color.

The joints are often emphasized by forming *rusticated* or *rebated* joints (*c*). This joint is frequently used in the stonework on the lower stories, giving a massive appearance.

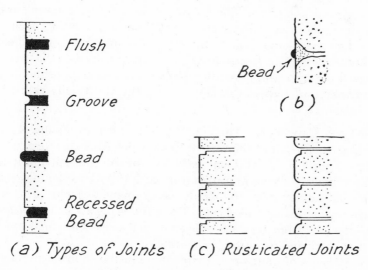

(a) Types of Joints   (c) Rusticated Joints

FIGURE 4.15   Joints for stone masonry.

Trimstone.   Cut stone is frequently used as a trim around window and door openings and for belt courses, copings, and cornices in walls constructed of brick or rubble masonry. Stone used in this manner is called *trimstone*.

Cast Stone.   The use of cast-concrete units, commonly known as *cast stone*, to replace cut stone has been growing in recent years. Cast stone consists of molded blocks of concrete with special surface treatment. They may be formed in any of the shapes obtained by cutting the natural stone and may have surface finishes which resemble the rubbed finish often used on limestone and other stones, or any of the tooled finishes. Special aggregates may be used next to the face or for the entire block so that when the cement surface film is removed by etching with acid or tooling, the face will resemble granite, marble, and other natural stones; or the aggregate may be chosen simply to produce an attractive finish without attempting to imitate any natural stone.

One of the manufacturing problems of cast stone is the prevention of crazing. *Crazing* consists of small web-like cracks which often form on the surface of cast stone.

26.   HOLLOW UNIT MASONRY WALLS

Classes of Hollow Units.   Hollow units made of burned clay or shale, concrete, gypsum, and glass are extensively used in the construction of walls and partitions. The terms tile and block represent a single unit or can be used collectively to represent a number of such units, as in the case of brick. The term *terra-cotta*, meaning "burned earth," is sometimes applied to structural clay tile, but this term is preferably reserved for ornamental building units of burned clay, usually classed as *architectural terra-cotta*. Hollow clay tile are usually called *structural clay tile*.

Structural Clay Tile.   These units, illustrated in Figure 4.16, are manufactured by the stiff mud process, which consists of extruding or forcing a plastic clay through specially formed dies, cutting it to the desired dimensions, and burning it in kilns to various degrees of hardness, depending upon the grade being manufactured. The hollow spaces in the tile are called *cells*. The outer walls are called *shells*, and the inner partitions between cells are called *webs*. The shells are ¾ in. or more thick and the webs ½ in. or more.

Many types of units other than those illustrated are manufactured. Special shapes are for corners and jambs so that the ends of the cells

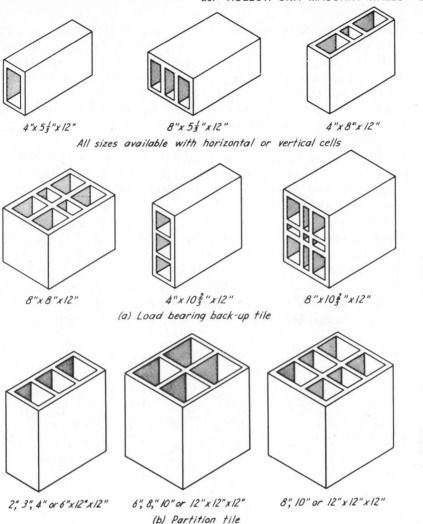

4"x 5⅓"x 12"   8"x 5⅓"x 12"   4"x 8"x 12"

All sizes available with horizontal or vertical cells

8"x 8"x 12"   4"x 10⅔"x 12"   8"x 10⅔"x 12"

(a) Load bearing back-up tile

2", 3", 4" or 6"x 12"x 12"   6", 8," 10" or 12"x 12"x 12"   8," 10" or 12"x 12"x 12"

(b) Partition tile

FIGURE 4.16   Structural clay tile.

will not be exposed if the cells are laid horizontally. Sills and headers receive the ends of brick headers so that cutting will not be required.

Specifications for structural clay tile include requirements for absorption and, for load-bearing units, compressive strength.

Classes of Structural Clay Tile.   Structural clay tile are divided into two classes (Figure 4.17) according to the orientation of the axes of the cells when the units are placed in masonry. In *side-construction* *tile*, the cells are placed horizontally; and in *end-construction tile*, vertically. The compressive strength, based on the gross area of a horizontal section included within the outer surfaces of a tile, is about 45% higher for end-construction tile than for side-construction tile. Structural clay tile are divided into two classes according to loading condition: *load-bearing* (ASTM C 34) (Figure 4.16(*a*) and (*b*) and *nonload-bearing* (ASTM C 56). Nonload-bearing tile are usually called *partition tile* because of their use in *nonbearing* partitions.

There are two grades of structural *load-bearing* tile. *LBX Grade* is used where exposure to weather is a factor, and *LB* is suitable in protected locations (ASTM C 34). Nonload-bearing tile has only one grade, *NB* (ASTM C 56).

Structural clay tile are divided into two broad groups according to their function in a multiple unit wall: *backing* or *back-up* tile, which backs up various facing units such as brick and stone or serves as a base for stucco; and *facing tile* (ASTM C 212), which has a finished surface on the one or two exposed sides. Facing tile may have *solid* shells, *double* shells, or *cored* shells (Figure 4.17). Facing tile grading is based on mechanical perfection, absorption rate, stain resistance,

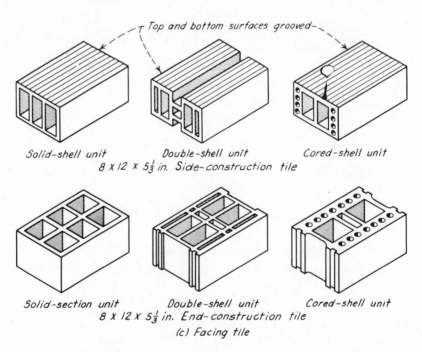

FIGURE 4.17   Structural clay facing tile.

ease of cleaning, color range, dimensional tolerances, surface texture, and suitability for exposure to weather. *Type FTX* is a better grade than *Type FTS*. Face tile is further classified as *standard,* which is for general use, and as *special duty,* which is used for areas requiring superior strength, resistance to impact, and better moisture impedance (ASTM C 212).

Sizes. Structural clay tile usually has a nominal length of 12 in., although some types are 16 in. The actual lengths are shy $\frac{1}{2}$, $\frac{3}{8}$, or $\frac{1}{4}$ in. The nominal dimensions allow for the mortar joints while still maintaining a 4 in. module.

Structural clay tile sizes are identified by first stating the width, then the dimension perpendicular to the cells, and finally the dimension that is parallel to the long axis of the cells. The side-construction tile shown in Figure 4.17 is called out as 8 by $5\frac{1}{3}$ by 12, whereas the tile of the same dimensions in end-construction tile is identified as 8 by 12 by $5\frac{1}{3}$—a switch in the last two dimensions.

Surface Texture and Color. The surface finish selected for structural clay is determined by the use to be made of the tile. The following finishes are available: natural or *smooth, scored, combed* or *roughened* (by wire cutting or brushing), *salt glaze,* and *ceramic* glaze.

The natural or smooth finish is specified for exposed surfaces that

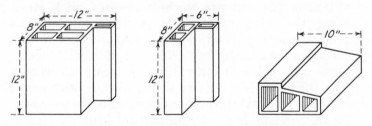

(a) Jamb and sill tile for end-construction

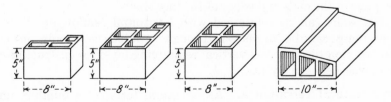

(b) Jamb, corner and sill tile for side-construction

FIGURE 4.18 Jamb, sill, and corner structural clay tile.

are to be untreated or painted. The scored finish provides increased bond for plaster or stucco finishes, although investigations have shown that the unglazed, smooth finish provides an adequate plaster bond. The roughened finishes achieve a desired texture, but they may be used to receive plaster. The salt and ceramic glazes provide a desired appearance as well as low water absorption and easy cleaning.

Tile with natural or smooth and roughened finishes are produced in an assortment of colors, depending upon the composition of the clay and the degree of burning. Such colors include cream, gray, buff, brown, and various shades of red, purple, and black comparable to the colors of face brick.

A salt glaze is produced on tile made of light-burning clays by the process described on page 177. The usual colors of salt glaze tile are cream and buff.

Ceramic glazes or *enamels* are formed on surfaces of facing tile by spraying them with mineral glazing materials before burning. They are so constituted that they will fuse during burning and form a glass-like coating on the tile surface. They may form single or multicolor shades. Some of the single-color shades are white, light gray, ivory, yellow, coral, tan, blue, and green. Multicolor glazes include mottles of white, gray, cream, and green with a trim shade of black.

Wall Construction. Structural clay tile backing units are laid with facings of ceramic tile, concrete block, terra-cotta, brick, stone, and structural clay tile facing units (Figure 4.19). The backing units may be either end-construction or side-construction.

This illustrates brick facing for 8-, 10-, and 12-in. walls, but similar arrangements are used with other types of facing. In most of the illustrations, the facing is bonded to the backing by headers or bonders. In (e), however, metal ties anchor the facing to the backing.

The tile are normally laid with the vertical joints staggered as in the stretcher bond of brick masonry (Figure 4.4), but they are altered to provide for bonders as required. Joints should have full mortar coverage of the ends and edges of the face shells.

Tile walls can be reinforced with horizontal reinforcing. Vertical reinforcing can be in-placed through grouted cells only in end-construction tile, whereas vertical reinforcing for side-construction tile must pass between wythes.

Architectural Terra-Cotta. Architectural terra-cotta building units are hard-burned clay. They are glazed or unglazed, plain or ornamental, machine-extruded or hand-molded, and generally larger than brick or facing tile.

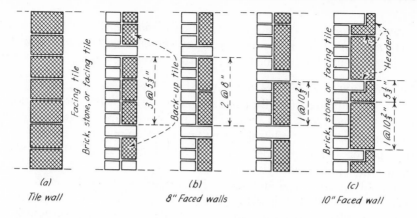

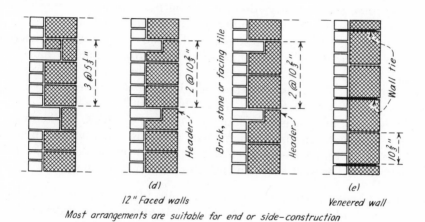

FIGURE 4.19 Structural clay tile in wall construction. Adapted from Structural Clay Products Institute.

The units are hollow or ribbed and made in a great variety of surface finishes, textures, colors, and designs. Their function is decorative.

Ceramic veneer is architectural terra-cotta characterized by large face dimensions and thin sections. The thinner sections are held in place by adhesion to mortar between the terra-cotta and the backing; and the thicker sections, by grout and wire anchors connected to the backing.

Selection of terra-cotta for outside installation must consider a product that will not spall or crack and one whose exposed surface will not craze or flake off under weather conditions.

Concrete Block.   Concrete block, which is identified as a *hollow concrete masonry unit*, is made of portland cement, aggregate, and water sufficient to produce a moist mixture that resembles damp, loose sand.

Manufacture.   Block are formed in machines with steel molds; the mix is consolidated by vibrating and compacting. This provides accurately formed block and permits the moist mixture. Thus, the molds can be removed immediately after impact. Commercial production is by a power-driven automated machine with gang molds that produce thousands of units each day.

The block are *cured* and *dried*. The tendency of the block to crumble, shrink, and crack is deterred by curing. During the curing operation, the blocks are kept moist so that setting and hardening is gradual. During the drying operation, the moisture content is reduced to average air humidity, preventing the moisture of the block from being reduced by the surrounding air at the time of laying.

Curing and drying are accomplished in the *open air* with protection from the weather, by blowing *heated air* through stacked block, by high-temperature *steam curing* in sealed kilns, or by high-pressure steam curing in a pressure vessel called an *autoclave* that is filled with *saturated steam* at a temperature of about 370°F.

The open air procedure requires from 2 to 4 weeks; the heated-air process is more rapid, although steam curing is accomplished in about 12 hours. Units stored, either at the plant or on the job site, should be protected against absorbing moisture from rain or snow.

Block Classification.   Concrete block falls into two general classifications as *load-bearing block* (ASTM C 90) and *nonload-bearing* block.

Classification of load-bearing units is divided into two grades, N and S, and two types, I and II. Grades are specified according to intended use; those blocks that qualify for the more severe exposures are designated *Grade N*. Block requiring protection from the weather are termed *Grade S*.

Block Weight.   Load-bearing block are classified by weight as *normal weight*, *medium weight*, and *lightweight*. Block weights are determined by the type of aggregate used. Ordinary aggregate such as gravel, crushed stone, and air-cooled slag produce heavier block. They weigh approximately 40 to 50 lbs. for a standard 8 by 8 by 16 stretcher block.

Lightweight aggregates are coal cinders, expanded shale, clay or slag, and natural lightweight materials, such as volcanic cinders, pumice, Haydite, and vermiculite. They are used extensively for lightweight block, which weighs approximately 50% of the block that is produced by heavyweight aggregates.

Sizes and Shapes.   The unit with the nominal dimensions of 8 in. width, 8 in. height, and 16 in. length has corresponding actual dimensions of 7⅝ by 7⅝ by 15⅝ in. to allow for ⅜ in. joints. This conforms to the *modular coordination of design* which is based on the 4 in. module. The same dimensional relationship is maintained on all other units. Other nominal widths are 3, 4, 8, 10, and 12 in., and other heights are 3½ and 5 in. The usual length is 15⅝ in., although *half blocks* are also manufactured.

Several special shapes are molded for specific locations. Single and double square, bullnose corner, sash or jambs, and header block receive the headers and serve as end block. Other special shapes are U-shaped *lintel* or *beam block*, *sill block* for openings, *pilaster block*, and *column block* for structural purposes.

Specifications for concrete block include requirements for moisture content, absorption, and compressive strength.

Surface Finish.   The exposed surfaces are usually plain. The surface finish or texture may be fine, medium, or coarse, depending upon the grading of the aggregates.

Lightweight concrete block are glazed with an impervious surface ⅛ to ⅜ in. thick. This glaze is formed by mixing selected mineral aggregates, pigments, and a thermosetting plastic binder. The plastic binder is formed and then hardened at elevated temperatures using the high-pressure curing process. Block of the usual sizes are manufactured in nearly 50 colors. The surface of the shell face conforms to the requirements to form coves, returns, caps, and the other shapes necessary for a complete glazed masonry surface. This type of block has the distinct advantages of a glazed moisture-resistant surface in a single wythe unit, thus eliminating applied finishes with their attendant ties.

Some have surfaces with a molded face shell of various geometric designs.

Patterned Block.   Recent developments have produced *patterned block* that have face shells with a raised or recessed geometric design. These blocks can be used in various orientations that give more variety and emphasis when used in conjunction with the flat surface shell faces. Striking shadow effects are achieved with these blocks.

Screen Block.   Pierced grill block, called *screen block*, are frequently used as solar screens. These blocks, usually 12 in. square, may have designs similar to the patterned face shells of the solid block. These block require independent support by a structural frame.

Wall Construction.   Concrete block walls and partitions are normally one block thick and laid with broken or staggered vertical joints. Concrete block are used as backing for exterior walls faced with brick or stone in the manner illustrated for structural clay tile backing (Figure 4.19). Special units are available for corners, jambs, and other purposes, as has been explained. Half-block units may also be obtained.

Cavity walls are constructed of two wythes of blocks, each at least 4 in. thick, separated by an air space of from 2 to 3 in. and tied together, as required, with noncorrosive metal ties.

The mortar in the bed joint usually covers only the shells; but if strength is an important factor, the webs are also covered. The mortar covers only the shells in forming the head or vertical joints. The normal joint thickness is $\frac{3}{8}$ in. The preferred joints are the Vee and concave joints, illustrated in Figure 4.1, which are made by exerting pressure while running a tool along full or flush partially set mortar joints. They are more watertight than other types of joints. Even then, exposed exterior wall surfaces should be treated as described on pages 141 and 142.

Reinforcement.   Horizontal steel reinforcement may consist of an assembly of two or more parallel *side rods* welded to *cross ties* or *cross rods*. The cross rods may run perpendicular to the side rods, forming a ladder assembly, or the cross rods may weave back and forth between the side rods, forming a truss-like shape. These zinc coated or galvanized reinforcement assemblies are made of #8, #9, or 3/16 in. rods or various combinations of these sizes, the size being determined by structural considerations.

Galvanized woven wire mesh, similar in pattern to poultry wire, with two longitudinal rods, is also made. Regular reinforcement bars —usually #4 bars—set in grout-filled bond beam block are frequently used. Continuous bond beams are usually about 4 ft. centers.

Vertical reinforcement invariably consists of reinforcement bars encased in the middle of the grout-filled cells. *Rebar* sizing is always dependent upon the structural requirements but usually consists of #5 bars spaced on about 24 in. centers. Bars usually extend out of the foundations at an optimum height of about 60 in., and the block are threaded over the reinforcement and set in place. Reinforcement continues to the full wall height with splices made as required, with laps measuring not less than 24 bar diameters.

Such reinforcement alleviates the stresses caused by block shrinkage or temperature change. The visible cracking on the surface that would otherwise result is thereby deterred. This action is similar to that of temperature and shrinkage reinforcement in reinforced concrete slabs. Reinforcement is also effective in reducing cracking caused by differential settlement or deflection over openings in the wall. By controlling cracking, increased resistance to water penetration is achieved. Lateral reinforcement ties the wythes of cavity walls together and laps or otherwise carries intersecting walls together around corners and anchors. In regions subject to seismic forces, lateral reinforcement is required to resist these shocks.

The cells of all blocks are usually filled with a grout mix.

Surface Treatments.   In regions subject to driving rains, exterior, above-grade concrete block walls are quite likely to leak because of water which penetrates the joints or the block. Such leakage can be prevented by the application of portland cement stucco or by coating with portland cement paint. Such paints are mixed with water and are available in an endless variety of colors.

The outside face of concrete block foundation walls should be parged.

Interior wall surfaces require no treatment except for decorative purposes. Neither exterior nor interior wall surfaces of glazed block, of course, require surface treatment, but extreme care should be used in laying exterior block to secure watertight joints.

Gypsum Tile or Block.   Gypsum tile or block (ASTM C 52) are primarily of calcined gypsum to which water is added, resulting in the chemical reaction and setting as explained in Article 22. They may or may not include aggregates. If combustible aggregates, such as wood fiber, are included, their weight must not exceed 15% of the weight of the dry tile.

Some of the forms of gypsum tile are shown in Figure 4.20. Rectangular in shape, they may be solid or cored. Special shapes are also on the market.

Gypsum tile are installed as interior nonbearing partitions and as fire protection of structural steel members. Their use is not recommended for walls subjected to continual dampness. They are set only in gypsum mortar. Gypsum tile does not form a suitable base for portland cement or lime plaster unless self-furring metal lath is used. Gypsum tiles provide an excellent base for gypsum plaster, especially if they are *scored*.

Glass Block.   Glass block are hollow, translucent, masonry units

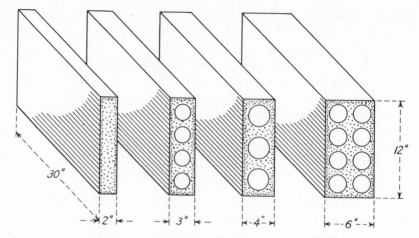

FIGURE 4.20 Gypsum partition tile or block.

formed by fusing, at high temperature, two sections that have been cast separately. A manufacturing process seals the block and produces a partial vacuum in an interior. The exposed faces are patterned or shaped. The interior or exterior surfaces of each face may be smooth, ribbed horizontally or vertically, or covered with a ceramic coat. If patterned on the interior, the exposed surfaces are smooth, which permits easy cleaning while assuring privacy and noise reduction. Their primary function is to transmit diffused light through walls. Suitable patterns direct the light to the ceilings from which it is reflected downward. They are more resistant to heat and sound transmission and condensation than ordinary glass.

Special block, more resistant to heat transmission, are more effective in brightness control—special properties are achieved by dividing the interior cavity with fibrous glass screens during manufacture. In some cases, the screen is a blue-green and imparts this color to the block [20].

Glass block are set in exterior walls and interior partitions but carry no load other than their own weight. Glass block are laid in stacked bond only.

Glass block were quite popular following their introduction in the 1930's. Their popularity waned, but renewed interest in this form is increasing; there are several innovative and tasteful variations of glass block units.

Sculptured glass modules measuring 12 by 12 by 4 in. wide have a shaped geometric design in the two exposed faces. Another type of unit features a design in gray ceramic that frames recessed antiqued clear glass while still another type is coated on the edges and has raised glass patterns on the block face.

Glass block are manufactured in numerous colors, as well as clear transparent glass.

REFERENCES AND RECOMMENDED READING

1. *American Standard Building Code Requirements for Masonry*, U. S. Department of Commerce, National Bureau of Standards, Miscellaneous Publication 211, July 15, 1954, American Standards Association Standard, A.41-1953.
2. Cyrus C. Fishburn, *Water Permeability of Walls Built of Masonry Units*, Report BMS 82, National Bureau of Standards, April 15, 1942.
3. Cyrus C. Fishburn and Douglas E. Parsons, *Tests of Cement-Water Paints and Other Waterproofings for Unit Masonry Walls*, Report BMS 95, National Bureau of Standards, March 15, 1943.
4. *Standard Specifications for Building Brick (Solid Masonry Units from Clay or Shale)*, ASTM Designation C62-58.
5. *Standard Specifications for Facing Brick (Solid Masonry Units Made from Clay or Shale)*, ASTM Designation C216-57.
6. *Technical Notes on Brick and Tile Construction*, Vol. 5, No. 1, January, 1954, Structural Clay Products Institute.
7. *Technical Notes on Brick and Tile Construction*, Vol. 11, No. 2, February 1960, Structural Clay Products Institute.
8. Plummer and Blume, *Reinforced Brick Masonry and Lateral Force Design*, Structural Clay Products Institute, 1953.
9. Oliver Bowles, *The Stone Industries*, McGraw-Hill, 1934.
10. *Technical Notes on Brick and Tile Construction*, Vol. 2, No. 3, March 1951; No. 6, June 1951; and No. 11, November 1951; Structural Clay Products Institute.
11. *Technical Notes on Brick and Tile Construction*, No. 17, November-December 1962, Part I of IV.
12. *Concrete Masonry Handbook*, Portland Cement Association.
13. *Building Code Requirements for Reinforced Concrete*, American Concrete Institute (ACI 318).
14. *Concrete for Industrial Buildings*, Portland Cement Association.
15. *Design and Control of Concrete Mixtures*, Portland Cement Association.
16. *National Building Code*, National Board of Fire Underwriters, 1955.

17. *Building Code Requirements for Reinforced Masonry*, National Bureau of Standards, American Standard A41.2-1960.

18. *Reinforced Brick Masonry*, Technical Notes on Brick and Tile Construction, Vol. 5, Nos. I to IV, January to April 1954, Structural Clay Products Institute.

19. *Mortars for Clay Masonry*, Technical Notes on Brick and Tile Construction, No. 8, August 1961, Structural Clay Products Institute.

20. H. F. Kingsbury, *Glass Block, Windows and Glass in the Exterior of Buildings*, Publication 478, 1957, Building Research Institute, National Research Council.

21. Charles G. Ramsey, and Harold R. Sleeper, *Architectural Graphic Standards*, Sixth Edition, John Wiley and Sons, 1970.

22. *ASTM 1971 Annual Book of Standards, Part 12*, American Society for Testing and Materials, Philadelphia, Pennsylvania.

23. *Sweet's Catalog*, 1971 Edition, F. W. Dodge Corp., New York, New York.

24. *Masonry Design Manual*, The Masonry Industry Advertising Committee, San Francisco, California, 1969.

25. Joseph M. Trefethen, *Geology for Engineers*, Van Nostrand, 1959.

26. *Handbook on Reinforced Grouted Brick Masonry Construction*, Brick Institute of California, Los Angeles, California, 1961.

27. *ASTM 1971 Annual Book of Standards, Part 10*, American Society for Testing and Materials, Philadelphia, Pennsylvania.

5

WOOD CONSTRUCTION

27. WOOD MATERIALS

This chapter is concerned with building construction primarily of wood. Various terms such as lumber, wood, wooden, frame, and timber are applied to parts of a building constructed of wood. The basic material is always wood, just as steel and concrete are basic materials. In general, the term timber designates heavy wood construction. There is a tendency to use the term wood in place of the term timber; therefore, it is difficult and probably not worthwhile to be entirely consistent in the use of these terms.

The terms wood, lumber, and timber, often used interchangeably, are terms that do have distinct meanings. *Wood* is the hard fibrous substance forming the trunk, the major stems, and the branches of trees. *Lumber* is the product of the saw or planing mill, not processed beyond sawing, resawing or lengthwise planing, crosscutting to length, and working. *Timber* is lumber 5 in. or larger [8].

The term *millwork* applies to the wood building materials manufactured in planing mills and millwork plants and includes doors, window frames, shutters, porch work, interior trim, stairways, mantels, panel work, and moldings but not flooring, ceiling, and siding.

Classification of Trees.   Timber is furnished by two classes of trees: the *needle-leaved\conifers* such as pines, firs, and spruces, which are classed as *softwood*, and the *broad-leaved* trees classed as *hardwoods* —for example, maple, oak, and sycamore. Basswood, which is classified as a hardwood, is softer in texture than most softwoods; some of the softwoods are as hard as the hardest hardwoods.

Softwood is invariably used in structural applications that include the framing, sheathing, and blocking of buildings. Seldom is hardwood used for these applications. Hardwood is used mostly in flooring, wall and ceiling paneling, decorative moldings, and for furniture. Softwood is used for these purposes too; however, hardwood usually offers better choice of colors, quality, and workmanship.

Growth and Structure.   Trees grow by adding a layer of wood to all parts of the tree each year. This layer shows in the cross section as a new *annual ring* surrounding the old wood. The cross section of a tree consists of an exterior bark over annual growth rings surrounding a *pith* at the center that varies in diameter from $\frac{1}{20}$ in. to nearly $\frac{1}{4}$ in.

The annular rings near the outside form the *sapwood* and are lighter than those near the center, which form the *heartwood*.

Plain- and Quarter-Sawed Lumber.   Boards sawed tangent to the annual rings are called *plain-sawed, slash-grained,* or *flat-sawed,* whereas those sawed in a perpendicular direction are called *quarter-sawed* or *edge-grain.*

If the annual rings are neither tangent nor perpendicular to the sides, they are classed as quarter-sawed if the grain makes an angle greater than 45° with the side of the board. If this angle is less than 45°, the boards are classed as plain-sawed. Quarter-sawed lumber shrinks and warps less than plain-sawed lumber. The exposed grains of quarter-sawed wood are different from those of wood that is plain-sawed. In some woods, such as straight grained woods—the spruces, Port Orford Cedar, and vertical grained redwood—more attractive effects are secured in quarter-sawed boards. In other woods, such as knotty pines, the plain-sawed boards with attractive knots and burls may be desirable. Because quarter-sawed lumber yield is less, it is more expensive than plain-sawed lumber.

Units of Measure.   The *board foot* is the unit of measure for lumber. It is the quantity of lumber contained in a piece of rough green lumber 1 in. thick, 12 in. wide, and 1 ft. long, or its equivalent in thicker, wider, narrower, or longer lumber. Lumber less than than 1 in. thick is considered 1 in. thick.

In referring to commercial lots of lumber, this method of measurement is called *board measure* and is abbreviated *b.m.* The common unit is 1000 board feet, designated as *M.* For example, a lot of lumber that contains 25,000 board feet is designated as 25 M.b.m.

Moldings are measured by the lineal foot and shingles by the number of pieces of a specified length.

Size Standards.   Lumber usually is designated by its *rough, green,* or *nominal size,* not its *actual size.* Rough lumber is larger than *sur-*

faced lumber by the amount necessary for surfacing. Because shrinkage and dressing reduce softwood yard lumber, sizes designated as 1 in. thick are reduced to $^{25}/_{32}$ in. thick after surfacing; and 2-in. material is planed to $1^9/_{16}$ in. thick and seasoned to $1\frac{1}{2}$ in. thick. Corresponding allowances varying from $\frac{3}{8}$ in. to $\frac{3}{4}$ in. are made in the widths; the amount depends upon the width and the species of the material. A smaller allowance is made for a 4 in. nominal width, while those with a nominal width of 12 in. have larger allowances.

Hardwood lumber is cut oversize so that it will be full size when dry. When it is surfaced on one side (*S1S*), the thickness is $\frac{1}{8}$ in. less than the rough size; when surfaced on two sides (*S2S*), it is $\frac{3}{16}$ in. less than rough size.

The standard lengths for softwood lumber vary by 2-ft. increments from 4 ft. up; the most common lengths are 12, 14, and 16 ft. A few odd lengths as 9, 11, 15, and 17 ft. are considered standard for specified sizes.

The standard lengths of hardwood lumber vary by increments of 1 ft. from 4 to 16 ft., but grading rules permit not more than 15% of the odd lengths.

Seasoning. Wood that is prepared for construction requires more preparation than simply shaking out the squirrels. It requires a cured product. Timber that is seasoned by natural drying by exposure to the air is called *Air Dried* (AD). This drying may be accelerated by subjecting the timber to high temperatures in a kiln, a process called *kiln drying*. The lumber is referred to as *Kiln Dried* (KD). The principal effects of seasoning are reduced moisture content; decreased shrinking, warping, and checking; and an increase in strength and resistance to decay.

Proper seasoning of lumber requires that lumber be stacked in a manner that will allow the free circulation of air upward through the stack. Improperly stacked lumber will prolong drying time. This could result in development of decay in the seasoning lumber. Air drying requires from 2 to 6 months for properly seasoned lumber—time depending upon the location, natural humidity, and the species and size of the lumber.

Kiln drying, which is accelerated drying, is more controlled and satisfactory seasoning. Excessively high temperatures will result in an increase in splits, checks, and warp of the lumber. This is the result of the internal stresses that develop in the wood. These stresses are caused by the differences in the moisture content of the dry surface and the moist interior. On the other hand, if drying is too slow, there is the likely danger of fungus and stain.

Chemicals may serve as an aid in seasoning. Spread over the surface of the lumber, urea salts extract moisture.

Moisture Content.   The Moisture Content (MC) of wood is a variable that affects several important mechanical and physical properties of wood. The moisture includes water or sap. The moisture is contained as *free water* in the wood cell cavities and intercellular interstices and as *absorbed water* in the capillaries of the fibers of the walls and ray cells. In seasoning, the free water moisture is first removed, thereby reducing the moisture content to a state referred to as the *fiber saturation point*. Below this point, a further reduction in the moisture content results in shrinkage of the wood. In most species, the fiber saturation point is reached at about 30% moisture content. Thus, shrinkage does not take place until there is a loss of absorbed water.

The shrinkage of lumber is directly related to the moisture content. In general, hardwoods shrink more than softwoods, and the heavier woods shrink more than the lighter woods, although there are notable exceptions.

A reduction of the moisture content by seasoning increases resistance to fungi attack and improves structural ability. The *moisture content*, which is expressed in terms of a percentage, is the weight of the moisture in a given representative sample of wood divided by the oven dry weight of the sample. As recommended by ASTM D 2016, the testing for moisture content is by four methods—oven drying method, electric moisture meter method, distillation method, and hydrometric method. The required moisture content of lumber is specified by the various grading rules for lumber.

Lumber Characteristics.   After lumber has been sawed, planed, and shaped in the mill, it is visually graded—its quality and intended use are established based upon lumber characteristics. In order that grading quality be reasonably uniform, terms have been defined that identify both natural characteristics and manufactured imperfections. These listed characteristics, frequently referred to as "defects," are most frequently cited in grading literature and in construction specifications.

Natural characteristics are those defects or deviations in the wood grain that have resulted from natural causes. They are usually the more frequent, although their severity or extensiveness in the finished product may not be serious. These defects may be cut out or, if left, the wood may be assigned a suitable usage that will not be impaired by the defect. Conversely, some of these characteristics, such as knots or burl, may be desired in paneling or furniture.

Below are listed some of the more frequent natural characteristics as defined by PS 20-70 [14]:

Check—Lengthwise grain separation, usually occurring through the growth rings as a result of seasoning.

Cross break—Separation of the wood across the width.

Gum pocket—Opening between growth rings which usually contains or has contained resin or bark, or both.

Gum seam—Check or shake filled with gum.

Gum spot—Accumulation of gumlike substance occurring as a small patch. May occur in conjunction with a bird-peck or other injury to the growing wood.

Gum streak—Well-defined accumulation of gum in more or less regular streak. Classified in the same manner as pitch streaks.

Peck—Channeled or pitted areas or pockets as sometimes found in cedar and cypress.

Pitch—Accumulation of resin in the wood cells in a more or less irregular patch.

Pitch pocket—Opening between growth rings which usually contains or has contained resin or bark, or both.

Pitch seam—Shake or check filled with pitch.

Pitch streak—Well-defined accumulation of pitch in a more or less regular streak.

Shake—Lengthwise grain separation between or through the growth rings; may be further classified as ring shake or pith shake.

Split—Lengthwise separation of the wood extending from one surface through the piece to the opposite surface or to an adjoining surface.

Stain—Discoloration on or in lumber other than its natural color.

Wane—This is bark or lack of wood from any cause on the edge or corner of a piece.

Warp—Any variation from a true or plane surface, includes bow, crook, cup, or any combination thereof.

  Bow—Deviation flatwise from a straight line from end to end of a piece, measured at the point of greatest distance from the straight line.

  Crook—Deviation edgewise from a straight line from end to end of a piece, measured at the point of greatest distance from the straight line; and classified as slight, small, medium, and large. Based on a piece 4 in. wide and 16 ft. long, the distance for each degree of crook shall be:

slight crook, 1 in.; *small crook,* 1½ in.; *medium crook,* 3 in.; and *large crook,* over 3 in. For wider pieces it shall be ⅛ in. less for each additional 2 in. of width. Shorter or longer pieces may have the same curvature.

Knots.   Knots are perhaps the most frequently cited defect in grading rules. Their frequency, size, shape, location, and occurrence are important factors in determining lumber grades. This is especially true for lumber that is intended for structural applications. Knots may be desirable for aesthetic purposes as in paneling. PS 20-70 defines knots as follows:

Knot—Branch or limb embedded in the tree and cut through in the process of lumber manufacture; classified according to size, quality, and occurrence. To determine the size of a knot, average the maximum length and maximum width, unless otherwise specified.

Pin knot—Not over ½ in. in diameter.

Small knot—Over ½ in., but not over ¾ in. in diameter.

Medium knot—Over ¾ in., but not over 1½ in. in diameter.

Large knot—Over 1½ in. in diameter.

Knot Quality:

Decayed knot—Softer than the surrounding wood and containing advanced decay.

Encased knot—Its rings of annual growth are not intergrown with those of the surrounding wood.

Hollow knot—Apparently sound except that it contains a hole over ¼ in. in diameter.

Intergrown knot—A knot partially or completely intergrown on one or two faces with the growth rings of the surrounding wood.

Loose knot—Not held tightly in place by growth or position, and cannot be relied upon to remain in place.

Fixed knot—Will hold its place in a dry piece under ordinary conditions; can be moved under pressure, though not easily pushed out.

Sound knot—Solid across its face, as hard as the surrounding wood, shows no indication of decay, and may vary in color from the natural color of the wood to reddish brown or black.

Star-checked knot—Having radial checks.

Tight knot—So fixed by growth or position as to retain its place.

Firm knot—Solid across its face, but containing incipient decay.

Water-tight knot—Its rings of annual growth are completely intergrown with those of the surrounding wood on one surface of the piece, and it is sound on that surface.

Knot Occurrence:

Branch knots—Two or more divergent knots sawed lengthwise and tapering toward the pith at a common point.

Corner knot—Located at the intersection of adjacent faces.

Knot cluster—Two or more knots grouped together, the fibers of the wood being deflected around the entire unit. A group of single knots is not a knot cluster.

Single knot—Occurs by itself, the fibers of the wood being deflected around it.

Spike knot—A knot sawed in a lengthwise direction.

Decay. Timber decay is caused by *fungi*, which feed on the cell walls. To develop, these fungi require warmth, air, and moisture. At low temperatures, the fungi lie dormant; however, they are killed by elevated temperatures. Timber that remains under water will last indefinitely because the air that fungi require is absent—pieces of timber that have been totally immersed in water have lasted for centuries. If moisture is not present, wood exposed to air will not decay. In *dry rot*, moisture is present. The development of dry rot fungi may be promoted by sealing a partially seasoned surface with paint or by embedding the timber in masonry so that the moisture present in the timber cannot escape.

Wood in the early stages of decay is usually characterized by slight discoloration that is called *incipient decay*. Although this type of decay has not usually progressed to the point of impairing soundness or strength, architects frequently specify that wood with incipient decay be rejected.

Decay is defined by PS 20-70 as follows:

Decay—Disintegration of wood substance due to action of wood-destroying fungi. Also known as dote and rot.

Advanced (or typical) decay—Older stage of decay in which disintegration is readily recognized because the wood has become punky, soft, spongy, stringy, shaky, pitted, or crumbly. Decided discoloration or bleaching of the wood.

Pocket rot—Typical decay which appears in the form of a hole, pocket, or area of soft rot, usually surrounded by apparently sound wood.

Water soak or *stain*—Water-soaked area in heartwood, usually interpreted as the incipient stage of certain wood rots.

Insect Damage. Various boring insects attack timber and cause considerable damage. Among these are *bark beetles*, which damage wood by tunneling under the bark; *ambrosia beetles; roundhead bee-*

tles, which get into freshly cut timber; *powderpost beetles*, which attack freshly cut and seasoned hardwood by burrowing holes about $\frac{1}{16}$ in. in diameter through the wood; and *termites* or *white ants*, which are the most destructive of all and will be given further consideration. Most insect damage in wood construction is caused by termites, which are light-colored, resemble ants in appearance, live in colonies (like ants), and thus are commonly called white ants.

Two types of termites, the *subterranean* and the *drywood*, are found in the U.S. The subterranean termites live in the soil but leave it in order to attack trees and wood structures. To live, they must have access to moisture from the ground. Therefore, they build shelter tubes from the ground to the areas where they are active. They bore holes along the grain while devouring interior wood, leaving only an intact exterior shell. A member appearing to be in prime condition may actually be on the verge of failure. The most effective control prevents the passage of the termites from the soil to the wood. By using concrete foundations reinforced to prevent cracks, or by using cement mortar to fill all the joints solidly in brick and stone masonry, termites cannot work their way through the joints. *Termite shields* of noncorrosive metal placed between the top of the foundations walls and the wood plates or sills (Figure 5.1) should extend 2 in. over the wall and bend down 45°. This projection prevents the termites from building shelter tubes from the foundation walls to the wood plates or sills. All scraps of lumber should be removed from the building site. Wood treated with preservative by a process that penetrates the interior will resist termites; however, termites attack timber from the interior, making surface coatings and sprays ineffective. Subterranean termites are prevalent in nearly all parts of the U.S.

Drywood termites attack wood directly and do not maintain contact with the ground. They are confined to the warmer climes of the U.S. and are not such a serious menace as the subterranean termites. All structures in this country constructed wholly or partially of wood,

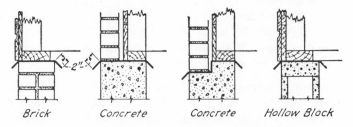

Brick   Concrete   Concrete   Hollow Block

FIGURE 5.1   Termite shields.

except those of temporary character, should be protected against termites.

Marine Borers.   Wood in salt water, such as timber piles for marine structures, may be attacked by *marine wood borers*, such as the *teredo* or *shipworm* and the *limnoria* or *wood louse*. Various methods of protecting wood against these borers have been devised, including impregnating the wood with creosote.

Manufactured Characteristics.   Manufactured characteristics are blemishes that may be imposed on the wood at some stage in its processing into salable lumber. The general classification of these blemishes, known as *mismanufacture*, may occur at any time between initial cutting and final shipment. PS 20-70 identifies several of the more common mismanufacture characteristics as:

Chipped grain—Area where the surface is chipped or broken out in very short particles below the line of cut. Not classed as torn grain and, as usually found, is not considered unless in excess of 25% of the surface involved.

Hit-and-miss—Series of surfaced areas with skips not over $\frac{1}{16}$ in. deep between them.

Hit-or-miss—To skip or surface a piece for a part or the whole of its length, provided it is nowhere more than $\frac{1}{16}$ in. scant.

Loosened grain—Small porton of the wood loosened but not displaced.

Machine burn—Darkening or charring due to overheating by machine knives.

Machine gouge—Groove due to the machine cutting below the desired line of cut.

Mismatched lumber—Worked lumber that does not fit tightly at all points of contact between adjoining pieces or in which the surfaces of adjoining pieces are not in the same plane.

Raised grain—Roughened condition of the surface of dressed lumber in which the hard summerwood is raised above the softer springwood but not torn loose from it.

Skip—Area on a piece that failed to surface.

Torn grain—Part of the wood torn out in dressing.

Variation in sawing—Deviation from the line of cut.

Wood Preservation.   Proper seasoning of timber is the simplest way to prevent decay but, when timber is used where moisture is present, seasoning naturally loses its effectiveness. Under these conditions, the best method for checking decay is to introduce substances into the timber that will poison the fungi. The substances commonly used for this purpose are called wood preservatives.

Wood preservatives are chemicals that are impregnated or applied to wood for the purpose of thwarting the activity of various decay-producing fungi and destructive insects. The preservatives, applied by the methods mentioned below, are usually induced by some vehicle that leaves the preservative impregnated in the wood. These wood preservatives are creosote and creosote solutions, oil-borne solutions, water-borne solutions, and gas-borne solutions.

Creosote is applied directly and is also applied as a creosote-coal tar solution and a creosote-petroleum solution. Creosote is widely used as a preservative for timber piles, railroad ties, and other rough wood construction intended for direct contact with the soil.

Creosote may be *brushed* or *sprayed*, but these methods are ineffective because the coating is thin and does not penetrate or fill all the cracks. The advantage of this method is the low cost. Another method, *dipping* the timber in a tank of hot creosote, is more effective than brushing or spraying, but only a slight penetration into the wood is realized.

In the *open-tank process*, the timber is first placed into a tank of hot creosote and then into a tank of cooler creosote. In the first tank, the air contained in the timber expands, and some air is forced out. When the timber is placed in the second tank, the entrapped air contracts, drawing the creosote into the wood; thus, a deeper penetration is obtained than by dipping into hot creosote only.

Various processes secure a deeper penetration of the preservative into the timber by pressure. These processes require extensive equipment and are expensive, but their use is justified where the most effective treatment is required, as in piling. These processes use cylinders as large as 8 ft. in diameter and 150 ft. long. The material is loaded on cars and run into a cylinder; the ends are then sealed.

In the *full cell process*, the first step is to pump air from the cylinder, thus drawing the air and moisture from the timber. Creosote, introduced into the tank, is forced into the wood cells by pressure. The treatment is completed by drawing off the creosote and removing the timber after the excess creosote drips off. This process leaves the cells, to the depth of penetration, full of creosote.

In the *empty cell process*, the creosote is first forced into the timber by pressure and then the creosote, which is in the cells, is removed by creating a vacuum. Only the creosote in the cell walls remains. The empty cell process is therefore less expensive than the full cell process.

When zinc chloride is used as a preservative, the full cell process is adopted.

Pentachlorophenol may be either gas-borne or oil-borne. By using a gas-borne solution, the wood is left in a condition which permits finish application or a surface that may be glued. Pentachlorophenol impregnated with an oil-borne solution leaves an oil film on the wood. This may restrict the usage of the wood.

Water-borne salt preservatives, which are frequently identified in specifications by their initials, are Acid Copper Chromate (ACC), Ammoniacal Copper Arsenite (ACA), Chromated Copper Arsenate (CCA), Chromated Zinc Chloride (CZC), and Fluor Chrome Arsenate Phenol (FCAP). These preservatives are generally used on wood that is intended for protected areas. The preservative is likely to leach out of the wood if used in locations exposed to moisture. Some water-borne solutions will result in discolorations to the wood, which may limit their use to wood that is covered or unsuitable for finished appearance [9].

Grading. It is appropriate to forewarn the student that grading may be a slippery field. An understanding of the art of lumber grading requires study, experience, and that rarest of qualities, common sense. The confusion that is certain to greet the tyro is in part due to the multiple definitions of some terms—i.e., dimension and boards. "Grade" as it applies to lumber may relate to manufacture, usage, sizes, and qualities.

Wood, being a product of nature, is varied in its properties—strength, appearance, size, and physical characteristics such as grain pattern. In order to establish reasonable uniformity for commercial use, lumber must be graded. The multiple uses of lumber, as in timber structures, require lumber of known specific qualities.

The process of evaluating the quality and use of wood that has been sawn, planed, and seasoned is called *grading*. The assigned designation is the *grade*. This *grade marked* lumber is usually stamped with a *grade mark* symbol that may designate quality, species, use, strength, and grading authority, or symbols denoting other wood properties.

Grading is a visual process based upon established *lumber grading rules*. These rules, usually established by a lumber inspection bureau, are generally applicable to only a single species or to several similar species.

Grading attempts to assign to lumber some defined value, the grade, based upon criteria that evaluate characteristics both natural and manufactured. The absence or the frequency of defects, the degree of severity, location, and the intended construction use are factors that determine grade.

Some of the most common defects that are visually graded are knots, stain, warp, splits, bark, and undesirable grain configurations.

Property variations may exist between sapwood and heartwood. These variations might have marked implications on the color (if color is a factor), the structural ability, and the resistance to decay. The cuts of wood determine the exposed grain pattern of grain configuration, which in turn affect the wearing quality and the natural beauty of the finished product.

Softwood Grading. Guidelines for softwood lumber have been established by the National Bureau of Standards. Voluntary Product Standard 20-70 [14] provides voluntary guidelines for standardized grading, inspections, and sizes of softwood lumber. PS 20-70 classifies softwood lumber according to *size*, *use*, and *manufacturing*.

Three size classifications, based upon nominal sizes, are identified as boards, dimensions, and timbers. *Boards*, according to size, are less than 2 in. in thickness and 1 in. or more in width. See Figure 5.2. Boards that are less than 6 in. in width are called *strips*. Lumber that is 2 in. but less than 5 in. thick, and 2 or more in. wide is called *dimension lumber*. Lumber that is 5 in. or more in width or thickness is *timber*.

Use classification is divided into structural lumber, factory-and-shop lumber, and yard lumber. *Structural lumber* includes dimension lumber and timbers that are used where stress graded material is required. *Factory-and-shop lumber* is for industrial and remanufacturing, or in other words, for wood products—boxes, etc. *Yard lumber* is for general building construction purposes. It is further classified as *select* and *common*, which are appearance grades. Common is broken down into four grades of descending quality.

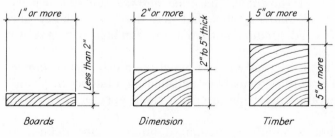

FIGURE 5.2

Three classifications based upon the extent of manufacture are rough lumber, dressed (surfaced) lumber, and worked lumber. *Rough lumber* is that which has rough saw marks visible on the lumber surfaces. Lumber that has been run through a planing machine and brought to a smooth finish is called *dressed lumber* or *surfaced lumber*. This material is machined to a uniform size. If surfaced on one side, it is identified as S1S; surfaced on two sides is S2S; and S4S is surfaced four sides. Another surfaced lumber is S1S1E, which is surfaced one side and one edge. Other examples of lumber abbreviations are in Table 5-3. *Worked lumber* is first dressed and then is matched, shiplapped, or patterned. Lumber that has been tongued and grooved is referred to as *matched lumber*. *Shiplapped* is material that has been rabetted on one or both edges. *Patterned* lumber may be shiplapped or matched and shaped to a determined pattern. See Figure 5.17.

The National Grading Rule.   The National Grading Rule for Dimension Lumber, which was developed by a committee established by PS 20-70, clarifies, simplifies, and unifies softwood dimension lumber sizes and grades. The National Grading Rule classifies dimension lumber into two width categories and five use categories.

The two width categories are dimension up to 4 in., and dimension that is 6 in. and wider. These broad classifications are further classified into the use categories and subclassified into grades as shown in Table 5-1.

Softwood Standard Sizes.   The sizes of dimension lumber, as standardized by the PS 20-70, are based upon an established moisture content of the lumber. If the moisture content of the lumber exceeds 19%, it is classified as *green* or *green lumber*. Lumber which is at 19% moisture content or less may be classified as *S-Dry*, or *dry lumber*. Lumber which has been air dried or kiln dried to a moisture content of 15% may be designated *MC 15*; however, if the lumber has been kiln dried to 15% moisture content, then the lumber may be stamped *KD*. Table 5-2 is from Table 3, PS 20-70.

Softwood Grading Rules.   Following the guidelines of the National Grading Rules, several associations or agencies have established grading and inspection services that conform to these standards. The grading rules usually apply to several species, although they may apply to only one species.

The Southern Pine Inspection Bureau classifies, grades, and inspects Southern Pines—namely slash pine, shortleaf pine, longleaf

TABLE 5-1

Width Category	Use Category	Grade
	Structural Light Framing	Select Structural No. 1 No. 2 No. 3
Dimension up to 4 in. wide	Light Framing	Construction Standard Utility
	Stud	Stud
Dimension 6 in. and wider	Structural Joists & Planks	Select Structural No. 1 No. 2 No. 3
	Appearance Framing	Appearance A

pine, loblolly pine, and several others that are of lesser commercial importance.

Two grading associations primarily serve the grading and inspection of the western lumbers—that is, the firs, spruces, and pines. The Western Wood Products Association (WWPA) and the West Coast Lumber Inspection Bureau (WCLIB) serve the lumber industry for Douglas fir, Sitka spruce, Western Hemlock, Western red cedar, and several other pines, spruces, firs, and cedars.

In certain species, special aspects or properties of the lumber may require grading terms that are peculiar only to that species. For example, redwood, which is graded by the Redwood Inspection Service (RIS), is graded in a manner that makes a marked distinction between lumber that is all heart redwood and that which contains sapwood.

Some trees are distinguished by their rate of growth. The rate of growth determines the spacing of their annual rings in the wood. If the rate of growth is relatively slow, the annual rings are usually closely spaced. This results in a lumber that is termed *dense*. Dense lumber is more likely to have better structural qualities than wood

TABLE 5-2

Nominal and Minimum-Dressed Sizes of Boards, Dimension, and Timbers (The thicknesses apply to all widths and all widths to all thicknesses)

	Thicknesses			Face Widths		
		Minimum Dressed			Minimum Dressed	
Item	Nominal	Dry	Green	Nominal	Dry	Green
		Inches	Inches		Inches	Inches
				2	1½	1 9/16
				3	2½	2 9/16
				4	3½	3 9/16
				5	4½	4⅝
	1	¾	25/32	6	5½	5⅝
				7	6½	6⅝
Boards	1¼	1	1 1/32	8	7¼	7½
				9	8¼	8½
	1½	1¼	1 9/32	10	9¼	9½
				11	10¼	10½
				12	11¼	11½
				14	13¼	13½
				16	15¼	15½
				2	1½	1 9/16
				3	2½	2 9/16
				4	3½	3 9/16
	2	1½	1 9/16	5	4½	4⅝
Dimension	2½	2	2 1/16	6	5½	5⅝
	3	2½	2 9/16	8	7¼	7½
	3½	3	3 1/16	10	9¼	9½
				12	11¼	11½
				14	13¼	13½
				16	15¼	15½
				2	1½	1 9/16
				3	2½	2 9/16
				4	3½	3 9/16
				5	4½	4⅝
Dimension	4	3½	3 9/16	6	5½	5⅝
	4½	4	4 1/16	8	7¼	7½
				10	9¼	9½
				12	11¼	11½
				14		13½
				16		15½
Timbers	5 & Thicker		½ Off	5 & Wider		½ Off

which is not so dense. Thus, a special grading usually denotes lumber that qualifies for a dense classification. Other terms that relate to density of the lumber are *close grain, medium grain, open grain,* and *coarse grain.*

Hardwood Grading.   Hardwood is graded and inspected by several associations, most of which are concerned with two or three species and their applications. However, the National Hardwood Lumber Association has extensive rules that apply to a great variety of hardwoods, especially domestic hardwoods, that are commercially important. Comments below are based upon these rules [26].

Standard hardwood grades are based upon the number and size of cuttings that can be obtained from a board or, restated, the amount of clear usable lumber in a piece. Except in certain cases, standard grade is determined from the poorer side of the boards. In certain grades, such as No. 1 Common Face, grade determination is made from the better face of the board. There are eight standard grades of hardwood and several combination grades. The *standard grades* are Firsts, Seconds, Selects, Sound Wormy, No. 1 Common, No. 2 Common, No. 3A Common, and No. 3B Common. The standard grades are frequently formed into *combination grades*, such as Firsts and Seconds (FAS), which contains a specified amount of Firsts. Another is Selects and No. 1, which may combine as one grade except in mahogany, walnut, and cherry. A combination grade made up of No. 3A Common and No. 3B Common is known as No. 3. Other combination grades are No. 1 and Better, No. 2 and Better, and No. 3B and Better. The term "Better" means that no material of inferior grade is permitted below the stated standard grade.

Sound Wormy, a standard grade, may have defects such as worm holes, bird pecks, stain, or sound knots ¾ in. or less in diameter. Sound Wormy and the more restrictive standard grades may form a combination such as Firsts and Seconds Sound Wormy, or No. 1 Common and Better Sound Wormy. However, either in combination or as a standard grade, Sound Wormy must not grade less than No. 1 Common.

The extremes in quality may be illustrated by the difference between Firsts and No. 2 Common. The former must have a minimum width of 6 in., and 91⅔% of the surface measure of a piece must be able to be cut into clear face material. The number of pieces into which a piece can be cut to give the required percentage of clear material depends upon the grade and upon the size of the piece.

For example, a piece with 14 sq. ft. of surface measure can be cut into only 2 pieces, with a minimum size of 4 in. by 5 ft. or 3 in. by 7 ft. if the piece is to be classified as Firsts; but a piece of No. 2 Common of this size can be cut into 7 pieces with a minimum size of 3 in. by 2 ft. There are other requirements which are not mentioned here.

For a summary of hardwood grading, see reference [1].

Hardwood Sizes. Dressed hardwood lumber of less than 8 in. in nominal widths is cut ⅜ in. scant; lumber 8 in. or more is cut ½ in. scant of their nominal width. Hardwoods are cut in lengths from 4 ft. to 16 ft. in one foot increments. These are known as standard lengths. The thickness of hardwood lumber is based upon both a rough lumber and a lumber surfaced on two sides (S2S). Frequently, hardwood lumber is expressed in terms of the number of quarter-inches in the nominal thickness. Thus, 1½ in. thick rough will be expressed as ⁶⁄₄. Rough thicknesses from ⅜ in. to 1½ in. are surfaced down to a thickness ³⁄₁₆ in. less than their rough measurement, and thicknesses of 1¾ in. up to 6 in. are surfaced to a thickness ¼ in. less than the rough thickness.

Lumber Abbreviations. Lumber terms are most frequently abbreviated in commercial usage, grading literature, and in construction specifications and drawings. Table 5–3 lists some of the most frequently abbreviated terms that should be committed to memory [1].

Wood Identification and Use. There are several woods that are known by more than one common or commercial name. This variation may be caused to some extent by the wide geographical distribution of the species. Clarity in construction requires specific terms. Although woods are frequently spoken of by their commercial, or common, name, they should be identified by their botanical name when written in construction specifications.

Table 5-4, which is adapted from ASTM D 1165, lists the more useful domestic woods and some of their construction applications.

28. BUILDER'S HARDWARE

This article considers various materials that fasten the wood parts of a building together. All the devices considered are not only used in framing a structure; many are used on such parts as interior finish. For instance, large nails and spikes are used in framing, whereas finish nails are not. However, it is convenient to consider finish nails in this article.

TABLE 5-3
Lumber Abbreviations

AD—air dried
B1S—beaded one side
B2S—beaded two sides
B&B or B & Btr.—B and better
B&S—beams and stringers
Bd.—board
Bd. ft.—board foot
Bdl.—bundle
Bdl. bk. s.—bundle bark strips
B/L—bill of lading
Bm.—board measure
CB1S—center bead one side
CB2S—center bead two sides
CG2E—center groove two edges
Clr.—clear
CM—center matched
Com.—common
Csg.—casing
CV1S—center V one side
CV2S—center V two sides
DB. Clg.—double-beaded ceiling (E&CB1S)
DB. Part—double-beaded partition (E&CB2S)
DET—double end trimmed
D&CM—dressed (1 or 2 sides) and center matched
D&H—dressed and headed
D&M—dressed and matched
D&SM—dressed (1 or 2 sides) and standard matched
D2S&CM—dressed two sides and center matched
D2S&M—dressed two sides and (center or standard) matched
D2S&SM—dressed two sides and standard matched
Dim.—dimension
Dkg.—decking
D.S., D/S or D Sdg.—drop siding
DT&G—double tongued and grooved
E—edge

EB1S—edge bead one side
EB2S—edge bead two sides
E&CB1S—edge and center bead 1 side; surfaced 1 or 2 sides and with a longitudinal edge and center bead on a surfaced face
E&CB2S—edge and center bead 2 sides; all 4 sides surfaced and with a longitudinal edge and center bead on the 2 faces
ECM—ends center matched
E&CV1S—edge and center V 1 side; surfaced 1 or 2 sides and with a longitudinal edge and center V-shaped groove on a surfaced face
E&CV2S—edge and center V two sides
EG—edge (vertical) grain
EE—eased edges
EM—end matched, either center or standard
ESM—ends standard matched
EV1S—edge V one side
EV2S—edge V two sides
FAS—First and Seconds
FBM—foot or feet board measure
Fcty.—factory (lumber)
FG—flat (slash) grain
Flg.—flooring
FOHC—free of heart center or centers
F.o.k.—free of knots
Frm.—framing
GM—grade marked
G/R or G/Rfg.—grooved roofing
Hdwd.—hardwood
H&M—hit and miss
H or M—hit or miss
Hrt.—heart
Hrtwd.—heartwood
J&P—joists and planks
KD—kiln-dried

TABLE 5-3 (Continued)

K.d.—knocked down

Lbr.—lumber

Lr. MCO—log run, mill culls out

Lth.—lath

M—thousand

MBM—thousand (feet) board measure

MC—moisture content

MCO—mill culls out

Merch.—merchantable

M.G.—mixed grain

Mldg.—molding

MR—mill run

M.s.m.—thousand (feet) surface measure

MSR—machine stress rated

M.w.—mixed widths

N1E—nosed one edge

N2E—nosed two edges

Og.—ogee

PE—plain end

PET—precision and trimmed

Pln.—plain, as plainsawed

P&T—post and timbers

Qtd.—quartered—when referring to hardwoods

R/S, Res.—resawed

Rfg.—roofing

Rfrs.—roofers

Rgh.—rough

R/L—random lengths

Rnd.—round

R. Sdg.—rustic siding

R/W—random widths

R/W&L—random widths and lengths

S&E—surfaced 1 side and 1 edge

S1E—surfaced one edge

S2E—surfaced two edges

S1S—surfaced one side

S2S—surfaced two sides

S1S1E—surfaced 1 side and 1 edge

S2S1E—surfaced 2 sides and 1 edge

S1S2E—surfaced 1 side and 2 edges

S4S—surfaced four sides

S4S&CS—surfaced four sides with a calking seam on each edge

S&M—surfaced and matched; that is, surfaced 1 or 2 sides and tongued and grooved on the edges; the match may be center or standard

S2S&SM—surfaced two sides and standard matched

S2S&CM—surfaced two sides and center matched

S2S&M—surfaced two sides and center or standard matched

S2S&S/L—surfaced two sides and shiplapped

Sap.—sapwood

SB—standard bead

Sd.—seasoned

Sdg.—siding

Sel.—select

SESdg.—square-edge siding

SE&S—square edge and sound

Sftwd.—softwood

SG, S/L, SL—slash (flat) grain

Ship.—shiplap

S.m.—surface measure

SM—standard matched

Snd.—sound

SR—stress rated

Std.—standard

SW—sound wormy

T&G—tongued and grooved

TB&S—top, bottom, and sides

Tbrs.—timbers

V/S—V 1 side; that is, a longitudinal V-shaped groove on 1 face of a piece of lumber

V2S—V on 2 sides; that is, a longitudinal V-shaped groove on 2 faces of a piece of lumber

VG—vertical grain

TABLE 5-4

Standard Nomenclature of Domestic Hardwoods and Softwoods From ASTM D 1165

Commercial Names for Lumber	Official Common Tree Names	Botanical Names	Uses
HARDWOODS			
Red alder	red alder	*Alnus rubra*	interior finishes, sash, doors
Black ash (Brown ash)	black ash	*Fraxinus nigra*	
Oregon ash	Oregon ash	*F. latifolia*	
Pumpkin ash	pumpkin ash	*F. profunda*	plywood,
White ash	blue ash	*F. quadrangulata*	interior finishes
	green ash	*F. pennsylvanica*	
	white ash	*F. americana*	
Aspen (Popple)	bigtooth aspen	*Populus grandidentata*	plywood
	quaking aspen	*P. tremuloides*	
Basswood	American basswood	*Tilia americana*	door frames, finishes, patterns
	white basswood	*T. heterophylla*	core stock for veneer architect models
Beech	American beech	*Fagus grandifolia*	flooring, cabinets
Birch	gray birch	*Betula populifolia*	
designated as	paper birch	*B. papyrifera*	plywood, doors,
Red Birch,	river birch	*B. nigra*	interior finishes
Sap (White)	sweet birch	*B. lenta*	
Birch or Birch	yellow birch	*B. alleghaniensis*	above, and flooring
Butternut	butternut	*Juglans cinerea*	plywood, interior finishes
Cherry	black cherry	*Prunus serotina*	plywood, flooring, interior finishes
Chestnut	American chestnut	*Castanea dentata*	core stock, plywood, interior finishes, poles, posts

TABLE 5-4 (Continued)

Commercial Names for Lumber	Official Common Tree Names	Botanical Names	Uses
HARDWOODS			
Gum (may be designated Red or Sweetgum)	sweetgum	*Liquidambar styraciflua*	plywood, trim, interior finishes
Hickory	mockernut hickory	*Carya tomentosa*	plywood, flooring "The impossibility
	pignut hickory	*C. glabra*	of distinguishing
	shagbark hickory	*C. ovata*	between Hickory lumber and Pecan
	shellbark hickory	*C. laciniosa*	lumber for accurate species
Pecan	bitternut hickory	*C. cordiformis*	identification in
	nutmeg hickory	*C. myristicaeformis*	all cases is
	water hickory	*C. aquatica*	recognized."
	pecan	*C. illinoensis*	[9]
Holly	American holly	*Ilex opaca*	plywood
Hard Maple (If specified as white, requires sapwood.)	black maple	*Acer nigrum*	special floors, i.e. bowling alleys
	sugar maple	*A. saccharum*	and dance halls, interior finishes
Oregon Maple	bigleaf maple	*A. macrophyllum*	flooring, interior finishes
Soft Maple (If specified as white, requires sapwood.)	red maple	*A. rubrum*	flooring, interior finishes
	silver maple	*A. saccharinum*	
Red Oak	black oak	*Quercus velutina*	
	blackjack oak	*Q. marilandica*	
	California black oak	*Q. kelloggii*	
	cherrybark oak	*Q. falcata var. pagodaefolia*	
	laurel oak	*Q. laurifolia*	subflooring,
	northern pin oak	*Q. ellipsoidalis*	flooring,

TABLE 5-4 (Continued)

Commercial Names for Lumber	Official Common Tree Names	Botanical Names	Uses
	HARDWOODS		
	northern red oak	*Q. rubra*	interior finishes
	Nuttall oak	*Q. nuttallii*	
	pin oak	*Q. palustris*	
	scarlet oak	*Q. coccinea*	
	Shumard oak	*Q. shumardii*	
	southern red oak	*Q. falcata*	
	turkey oak	*Q. laevis*	
	willow oak	*Q. phellos*	
White Oak	Arizona white oak	*Q. arizonica*	subflooring, flooring, trestles, paneling, exterior doors, thresholds, frames, interior finishes
	blue oak	*Q. douglasii*	
	bur oak	*Q. macrocarpa*	
	California white oak	*Q. lobata*	
	chestnut oak	*Q. prinus*	
	chinkapin oak	*Q. muehlenbergii*	
	Emory Oak	*Q. emoryi*	
	Gambel oak	*Q. gambelii*	
	Mexican blue oak	*Q. oblongifolia*	
	live oak	*Q. virginiana*	
	Oregon white oak	*Q. garryana*	
	overcup oak	*Q. lyrata*	
	post oak	*Q. stellata*	
	swamp chestnut oak	*Q. michauxii*	
	swamp white oak	*Q. bicolor*	
	white oak	*Q. alba*	
Persimmon	common persimmon	*Diospyros virginiana*	plywood
Poplar	yellow-poplar	*Liriodendron tulipifera*	plywood, siding, weatherboard, interior finishes (painted or stained), frames, sashes

TABLE 5-4 (Continued)

Commercial Names for Lumber	Official Common Tree Names	Botanical Names	Uses
HARDWOODS			
Sycamore	American sycamore	*Platanus occidentalis*	flooring
Walnut	black walnut	*Juglans nigra*	parquet flooring, interior finishes, cabinet work, plywood
SOFTWOODS			
Alaska Cedar	Alaska-cedar	*Chamaecyparis nootkatensis*	interior finishes, cabinets
Incense Cedar	incense-cedar	*Libocedrus decurrens*	split shingles, foundation sills
Eastern Red Cedar	eastern redcedar	*Juniperus virginiana*	
	southern redcedar	*J. silicicola*	plywood, closet linings
Western Red Cedar	western redcedar	*Thuja plicata*	plywood, shingles, interior finishes, frames, siding
Northern White Cedar	northern white-cedar	*T. occidentalis*	construction framing
Southern White Cedar	Atlantic white-cedar	*Chamaecyparis thyoides*	construction
Cypress (Red White, or Yellow)	baldcypress	*Taxodium distichum*	gutters, shingles, siding, millwork, framing, plywood, exterior trim
	pondcypress	*T. distichum var. nutans*	plywood

TABLE 5-4 (Continued)

Commercial Names for Lumber	Official Common Tree Names	Botanical Names	Uses
SOFTWOODS			
Douglas Fir May be specified as Coast Region or Inland Region	Douglas-fir	*Pseudotsuga menziesii*	general construction, plywood, framing, sheathing
White Fir	subalpine fir California red fir grand fir noble fir Pacific silver fir white fir	*Abies lasiocarpa* *A. magnifica* *A. grandis* *A. procera* *A. amabilis* *A. concolor*	plywood, general construction, sash, doors
Eastern Hemlock	Carolina hemlock eastern hemlock	*Tsuga caroliniana* *T. canadensis*	framing, subflooring, sheathing
Mountain Hemlock	mountain hemlock	*T. mertensiana*	framing, subflooring, sheathing
West Coast Hemlock	western hemlock	*T. heterophylla*	framing, subflooring, sheathing, floors, siding, joists, planks
Western Larch	western larch	*Larix occidentalis*	rough dimension, flooring, sash, doors, planks, poles, small timbers
Jack Pine	jack pine	*Pinus banksiana*	posts, poles
Lodgepole Pine	lodgepole pine	*P. contorta*	general construction, poles, siding, flooring

TABLE 5-4 (Continued)

Commercial Names for Lumber	Official Common Tree Names	Botanical Names	Uses
SOFTWOODS			
Norway Pine	red pine	*P. resinosa*	general construction, piles
Ponderosa Pine	ponderosa pine	*P. ponderosa*	piles, posts, doors, panels, trim, general construction
Sugar Pine	sugar pine	*P. lambertiana*	sash, doors, frames, general construction
Idaho White Pine	western white pine	*P. monticola*	general construction, paneling, siding, exterior doors, patterned wood construction, trim, frames
Northern White Pine	eastern white pine	*P. strobus*	exterior doors, trim, frames, panels, patterned siding
Southern Yellow Pine	loblolly pine	*P. taeda*	kraft paper and below listed uses of other Southern Yellow Pines
	longleaf pine	*P. palustris*	general construction, siding, timbers, flooring
	shortleaf pine	*P. echinata*	
	slash pine	*P. eliottii*	
Redwood	redwood	*Sequoia sempervirens*	shingles, plywood, foundations, siding, general construction

TABLE 5-4 (Continued)

Commercial Names for Lumber	Official Common Tree Names	Botanical Names	Uses
SOFTWOODS			
Eastern Spruce	black spruce red spruce white spruce	*Picea mariana* *P. rubens* *P. glauca*	framing, general millwork
Engelmann Spruce	blue spruce Engelmann spruce	*P. pungens* *P. engelmannii*	flooring, sheathing, siding, studs, dimension stock
Sitka Spruce	Sitka spruce	*P. sitchensis*	doors, millwork
Tamarack	tamarack	*Larix laricina*	framing

Connectors are of the utmost importance in building construction. Many failures in building construction can be traced to poor connector use.

Nails. Steel-wire nails are ordinarily used in building construction. They are formed from steel wire of the same diameter as the nails. Nails are also made of copper and aluminum. The usual forms of nails are shown in Figure 5.3.

A nail has a shank, tip, and head. Nail lengths are usually denoted by the shank length. Shank designs may be plain or deformed. Deformed shanks have barbs or annularly threaded, or helically threaded spirals. The threaded shank nails are usually called *drive screws*. Nails that have a grooved shank are called *fluted* nails or *rolled grooved* nails. The nails that have deformed shanks with ridges are called *barbed nails*. These nails have greater resistance to withdrawal than plain shank nails.

Tips are diamond shaped, chiseled, or flat tipped. The process of bending over the tips of nails that have been fully driven is called *clinching*. Some nails have tips specially designed to facilitate clinching. These nails are referred to as *clinched nails*.

Heads of nails are usually flat, although cylindrically shaped heads are used for finish work. See Figure 5.3. *Double headed* nails have two

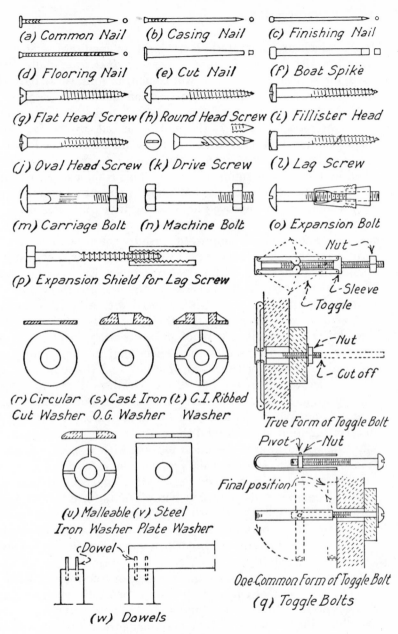

FIGURE 5.3 Nails, screws, bolts, washers, and dowels.

heads which are separated about ⅜ in. to facilitate removal. These nails are also known as *duplex head* or *duel head* nails. They are used for securing temporary work such as formwork, batterboards, and the necessary temporary construction incident to the project.

Nails that have a natural finish or a polished surface are referred to as *bright finished* nails. If the nail is heated, the surface attains a deep prussian blue cast, and the nails are called *blued nails*. Nails that are called *cement coated* nails have a resinous coating that is applied by a dipped or tumbled process. This resinous coating, intended to improve bond between the nail and the surrounding wood, provides only a temporarily improved bond when driven into less dense woods. Cement coated nails driven into dense or hardwoods do not realize lasting or appreciably improved withdrawal resistance. Nails that have a roughened surface, such as sand blasted or chemically etched shank, have increased withdrawal resistance. Zinc coated nails have increased resistance to nail corrosion and stain of the wood; the zinc coating does not contribute to increased withdrawal resistance.

Several factors determine the ability of nails to support their implied loads. The loads may be of short duration, extended, repetitive, or vibratory. Other factors that affect nail holding ability are the physical properties of the wood, the nail, and the manner of driving. The physical properties that affect withdrawal resistance are the configuration of the shank, length, diameter, point design, and surface coating. Wood characteristics that influence withdrawal are the moisture content, density, and the grain configuration. Another factor affecting withdrawal is the grain pattern with respect to the orientation of the nail. Nails driven *perpendicular to grain* have greater holding ability than nails driven *parallel to grain* or into the end grain. Generally, nails that have sharp points have greater resistance to withdrawal than those with blunt points. The sharper point does not destroy the fibers of the nail immediately adjacent to the nail. Likewise, nails driven into predrilled holes, called *pilot holes*, have greater holding ability.

Deterioration of nails and the staining of wood is inhibited by the use of nails made of corrosion resistant materials, such as aluminum alloys, copper alloys, and stainless steel. These materials will inhibit the "bleeding" or staining that may disfigure a finished wall surface. Steel nails that are to be used for exterior applications should be galvanized or zinc coated.

Nail size is designated by *penny* and is symbolized by the lower case d; thus, 2d is called 2-penny, a nail that is 1 in. long. Historically, the penny designation originated with the cost for 100 nails. Nails are sold in smaller quantities by the pound, but in construction they are sold by the keg, which contains 100 lbs. of nails.

Probably the most frequently used nail in general wood construction is the *common nail*. Common nails are used in wood construction framing where there is no objection to the exposed head and where the wide head is desirable. They are used in sheathing, subflooring, and for securing joists, studs, and other wood rough framing members. Common nails may be bright or cement coated. They vary in size from 2d to 60d, the latter being 6 in. long.

Similar to the common nail is the *box nail*, which has a more slender shank for a given length or size. They are available in sizes up to 16d. They are sometimes called *slim nails*. These nails cause less splitting than common nails when driven near the end of a board; however, they have less holding ability. To avoid splitting by nails, predrilled holes somewhat smaller than the nail diameter may be drilled.

Casing nails are used principally with matched flooring, ceiling, and drop siding. *Finish nails* are used with interior and exterior finish applications. Finish nails sizes range from 2d to 20d, which have a length of 4 in. The nail heads, countersunk below the surface with a nail set, form a hole that is filled with putty or plastic wood filler. A barbed flooring nail is illustrated in Figure 5.3 (*d*). *Shingle* and *lath nails* are small nails of the same general shape as common nails. Wire spikes are also of the same general shape, but they may have diamond or chisel points and flat or convex heads. Their size is designated by the length in inches and varies from 6 to 12 in. Wire nails, plain or galvanized, with large heads, are used with prepared roofing; galvanized and zinc-coated nails are used where resistance to corrosion is required. Various other types of wire nails are on the market, and galvanized common nails are available in many sizes.

Cut nails and spikes (Figure 5.3 (*e*)) are stamped from steel plates of the same thickness as the nail. Various sizes and shapes are manufactured to correspond with wire nails and spikes.

The initial holding power of cut nails is greater than that of wire nails, but due to their pyramidal shape, the holding power when they are partly withdrawn is less. Wire nails are more easily driven than cut nails. Cut nails have a longer life when exposed than the more widely used wire nails.

Boat spikes, which are used in heavy timber framing, are made of square bars of steel or wrought iron. They have a wedge-shaped point and a head (*f*). The size of boat spikes varies from ½ in. sq. by 3 in. long to ⅝ in. sq. by 12 in. long.

The various types of nailed joints used in light wood framing and the number and sizes of nails are shown in Figure 5.11 and Table 5-5.

Screws.   Screws are divided into two general classes, finish wood screws and lag or coach screws.

Wood screws are made of steel, brass, or bronze. Steel wood screws may have the natural steel finish called *bright*, or they may be blued, chromium- or nickel-plated, bronzed, lacquered, galvanized, or anodized. Various forms of wood screws are shown in Figure 5.4 (*g*) to (*k*). *Slotted head screws* can be driven with a common screwdriver. The *Phillips head* recess screw has a cross-shaped depression that requires a screwdriver with this shape.

With the exception of the drivescrew, screws have gimlet points. *Drivescrews* have diamond-shaped points and steep-pitched threads so that they may be driven with a hammer. Wood screws are identified by their head shape. A wide variety of head shapes are available, but only a few are used in wood construction.

Flat head screws (*g*) are the most common. *Oval head screws* are used for a better class of finish woodwork. They, like the flathead, are countersunk into the wood surface. *Roundhead wood screws* are most suitable for connecting light gage metals to wood members. They are not countersunk.

The size of wood screws is designated by the gage, length in inches, and head shape, for example: #10 x 1" O.H.W.S. (oval head wood screw). Several gages are available in each length. These lengths vary from ¼ in. to 6 in. Their length is determined by that portion which is embedded in the wood.

Lag screws have a conical point and a square head. *Coach screws* have a gimlet point and a square or hexagonal head, as shown in (*l*), but both forms are usually called lag screws. The size of lag and coach screws is designated by the diameter and length of the shank in inches. The lengths, measuring from the bottom of the head to the tip, vary from 1 in. to 12 in. and the diameters from ¼ in. to 1¼ in. They are used for heavy timber framing.

Screw holes should be prebored to prevent wood from splitting, to make driving easier, and to improve withdrawal resistance. This pilot hole, smaller than the root diameter, is approximately ¾ the root diameter of the thread. However, this hole diameter depends upon the species of wood penetrated.

Bolts.   Bolts used in wood construction may be divided into carriage bolts, machine bolts, and drift bolts.

Carriage bolts have a round head shaped as shown in (*m*) and a square nut. To prevent the bolt from rotating while the nut is turned, the portion of the shank immediately under the head is square, and the remainder of the shank is round. Bolt size is designated by the length of shank and the diameter in inches. Carriage bolts may be obtained in sizes from ¼ in. to 1 in. They are used in bolting steel, cast-iron members, and timber connectors.

A *drift bolt* is a piece of round or square steel rod, with or without head or point, driven as a spike. Drift bolts are used in heavy framing. Before driving, a pilot hole must be bored.

Expansion bolts have many different forms, but in all forms, a special nut or shield is used which is so designed that, after insertion in a hole, the process of turning the bolt will so enlarge or expand the shield that it cannot be withdrawn. One form is illustrated in (*o*). *Expansion shields* shown in (*p*) are often used with lag screws. Lead expansion shields are available for use with ordinary screws. Expansion bolts are used to fasten wood to masonry or concrete which is already in place.

A *toggle bolt* is so arranged that, after the bolt has been inserted head first into a hole on the other side of the piece, it will rotate or open up in such a manner that it cannot be pulled back through the hole. Two forms of toggle bolt are illustrated in (*q*). Toggle bolts are used where bolts cannot be inserted in the usual way because one face is inaccessible.

Washers.   Washers used under the head and below the nut of a bolt provide a larger bearing area and prevent crushing of the wood fibers. Washers are of five types and various sizes to suit the various sizes of bolts. These types are *circular cut washers* (*r*), *cast-iron O.G. washers* (*s*), whose name is derived from the Ogee curve; *cast-iron ribbed washers* (*t*), *malleable iron washers* (*u*), and *steel-plate washers* (*v*), which are specially made.

Dowels.   *Dowels* are steel drift bolts or wooden pins extending into and connecting, but usually not through, two members of a structure. A *tree-nail*, similar to a wood dowel, has one or both ends exposed. Wood dowels are usually glued.

Timber Connectors.   Various types of metal connectors designed to connect individual pieces of wood are called *timber connectors* and consist of metal rings, plates, and disks embedded in the surfaces of two members (Figure 5.4(a) to (e)). The embedment prevents sliding, and thereby stress is transmitted from member to member. Joints with these connectors are much more effective than simple bolted joints. Metal connectors have greatly extended the possibilities of timber construction for trusses and arches.

Many kinds of joints have been devised, but only a few are used to any extent. The more common types are illustrated in Figure 5.5. Many require bolts passing through the centers of the connectors to hold the timbers in contact; with some connectors, the bolts force the

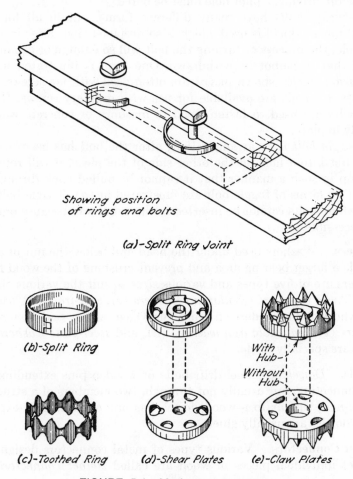

Showing position of rings and bolts

(a)-Split Ring Joint

(b)-Split Ring

(c)-Toothed Ring   *(d)-Shear Plates*

With Hub-

Without Hub-

(e)-Claw Plates

FIGURE 5.4   Modern connectors.

connectors into the timbers. Special tools groove the timbers to receive the connectors when connectors require such preparation.

The *split ring* in (*a*) and (*b*) fits into precut grooves in the timber faces and is used in heavy construction. The *toothed ring* in (*c*) is placed between the surfaces of two timbers without previous preparation and is forced into the timbers by pressure to produce joints in light construction. The *shear plates* in (*d*) are used in pairs, with each unit let into a prepared depression or dap in the timber face until its back is flush with the surface of the timber; or else one unit is inserted into a timber member so that a metal member can be bolted to it (Figure 5.5). In either case, the stress is transmitted between members by shear in the bolt. The *claw plates* shown in Figure 5.4(*e*) are used in the same manner as shear plates, but the dap is not made deep enough to receive the teeth, which are forced into wood below the depth of the dap. These units are used singly or in pairs. One has a flush back, whereas the other has a projection which fits into the hole of the other so that stress can be transferred across the joint by the units themselves without producing shear in the bolt. Either unit can be used to connect a metal member to a timber member. It is desirable to relieve the bolt from the shearing stress; the hole in the metal member is made large enough to receive the projection on the back of the claw.

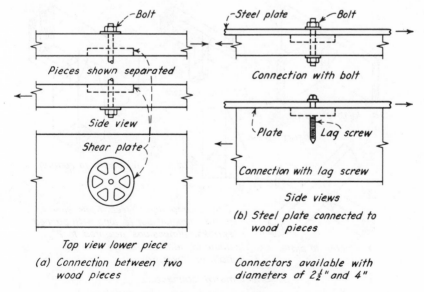

Top view lower piece

(a) Connection between two wood pieces

(b) Steel plate connected to wood pieces

Connectors available with diameters of $2\frac{1}{2}$" and 4"

FIGURE 5.5   Use of shear plate connectors.

Each type of connector is available in several sizes and capacities. The number of connectors required in any joint is determined by the stresses in the members and the properties of the wood. When they first came into use many years ago, timber connectors were called modern connectors, but this designation is rarely used now.

Sheet Metal Connectors. Zinc-coated and galvanized sheet steel connectors of various types are manufactured to replace the usual nailed joint in light wood framing (Figure 5.6). They are attached to the members with special half-length nails designed for driving into prepunched holes provided in the connectors.

Some connectors have several "toothed nails," which are formed when they are stamped out and bent at right angles from the rectangular piece of sheet metal. The connectors are pressed into an assembled wood truss. These lightweight trusses, usually 2 by 4's or 2 by 6's, are used for framing roofs of relatively short span. The connectors usually receive several common nails that are installed for erection and handling stresses.

One of the most widely used sheet metal connectors is the *sheet metal joist hanger* or *beam connector*. These connectors, made of various gages and sizes, accommodate all sizes of wood members and facilitate wood framing. They are usually held in place by half-length

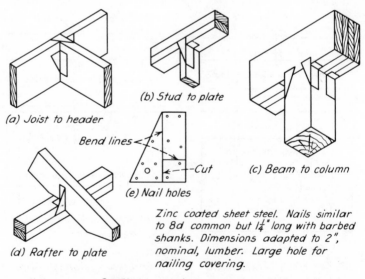

(a) Joist to header

(b) Stud to plate

(c) Beam to column

Bend lines — Cut

(e) Nail holes

(d) Rafter to plate

Zinc coated sheet steel. Nails similar to 8d common but 1¼" long with barbed shanks. Dimensions adapted to 2", nominal, lumber. Large hole for nailing covering.

B - Sheet metal connectors

FIGURE 5.6   Sheet metal connectors. Timber Engineering Company.

nails. There are several designs. Some are simply nailed into the side of the supporting wood members; others hang from the supporting beam. Most have a pocket which receives the member that it is supporting.

Small wood posts may be anchored at top or bottom with galvanized metal framing bases. The bases are anchored into supporting concrete by anchor bolts.

Lightweight gage metal *tie-down* anchors are used to stabilize a structure against lateral and uplift forces. These tie-downs may consist of an assembly of an embedded anchor bolt which is threaded through a steel angle that is nailed to a post or stud. Specially fabricated sheet metal tie-downs are embedded into concrete. They are nailed through prepunched holes to a wood post, column, or stud. These connectors are used at each end of wood stud shear walls that resist seismic and wind loads.

Powder Driven Studs.   Metal studs that are driven through wood, metal, or concrete by controlled explosive shot are called *powder driven studs*. These devices are designed to anchor wood and metals to metals or concrete. For wood construction, powder driven studs are frequently used to anchor wood sill plates to concrete slabs or to a steel anchorage. These studs, which are made of heat treated steel, are from $\frac{1}{4}$ in. to $\frac{1}{2}$ in. in diameter and from $\frac{3}{4}$ in. to 6 in. in length.

The drive studs are loaded into a "gun-like" device that is actuated by a powder charge. The size of the powder charge (or "shell") is determined by the material that is to be penetrated as well as by the material into which the drive stud will be embedded. Excessive loads will tend to split wood plates or penetrate entirely through the member. Charges that are too light will not embed the stud properly into the anchorage material. Heavy charges are capable of penetrating up to $\frac{1}{2}$ in. of steel.

These studs have an assortment of head designs that permit a choice of anchorages. Drive pins have a flat or button head; others that have an "eye" or hole through the head permit metal hangers or wire to be threaded through them. Threaded studs permit nut fasteners on the stud.

These types of fastening devices have the distinct advantage of providing economical and quickly assembled anchorages requiring field assembly. This is especially true where intended embedded concrete anchorages have been omitted, where engineered revisions require field changes, or where embedded anchorages may be awkward or time consuming if set at the time of the concrete placement.

29.   WOOD FRAMING

The structural elements of many buildings are made of wood. According to the classification based on the types of construction given in Article 1, there are three classes of buildings which make extensive use of wood structural members: *Wood Frame Construction, Ordinary Construction,* and *Heavy Timber Construction.* Consideration will be given to each of these types.

Wood Frame Construction

Wood frame construction, frequently called *Frame Construction,* is defined in Article 1 and is used extensively in dwellings. It is that type of construction in which the walls, partitions, floors, and roof are wholly or partly of wood or other combustible material. Our discussion is confined to buildings of this type. The height is usually limited to two habitable floors above the grade line, not including a finished attic or a basement. Wood frame construction is not permitted within certain fire zones.

The structural frame is lumber, most of which has a nominal thickness of 2 in. The center-to-center spacing of the various structural members is a module of 48 in. that is usually 16 in., but occasionally they are spaced at 12 in. and 24 in. Unless this spacing is constant, framing difficulties ensue. The depths of joists and rafters are determined by the loading conditions, span, allowable stresses, and required rigidity. Most studs and plates are 2 x 4's. Types of Wood Frame Construction are illustrated in Figures 5.7 and 5.8.

The exterior wall framing is usually covered with sheathing which is, in turn, covered with finished surfaces of siding or a masonry veneer, as later described. The roof framing is covered with a deck of sheathing or other material to receive roofing, as explained in Article 73. The floor framing is covered with a subfloor on which a finish is laid. The wall and roof sheathing are factors in insulation and, together with the subfloors, provide essential lateral rigidity.

Before the frame is started, termite shields (see Article 27) should be installed on top of all foundation walls. Two by or three by *mud sills* or *plates* are then placed on top of the foundation walls and anchored by bolts previously installed. These sills, which may be four or six inches wide, are the first members of the frame to be placed.

If there are unexcavated spaces below the first floor, there should be a *crawl space* with a clearance of at least 18 in. between the bottoms of the joists and the ground. This space provides for ventilation (to avoid decay) and access for inspection and repairs. If the soil is of

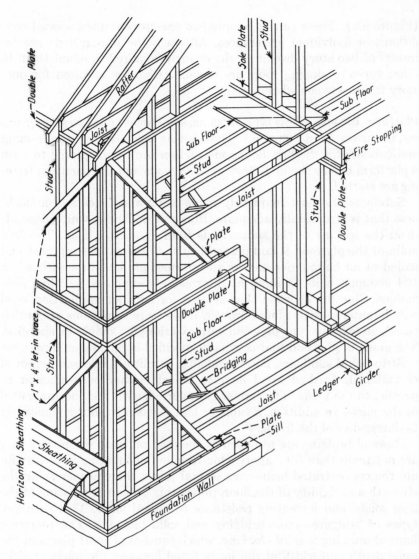

FIGURE 5.7 Platform frame. (Adapted from reference 4).

a type, such as clay, which conducts moisture upward by capillary action, the surface of the unexcavated soil could be covered with a layer of heavy polyethylene film to serve as a vapor barrier and assist in humidity control.

There are two types of wood frames commonly used for dwellings. These are the *platform frame* (Figure 5.7) and the *balloon frame*

(Figure 5.8). These can be adapted to one-story frames, special conditions, or individual preferences. Although the illustrations are for frames of two-story dwellings, the changes required to adapt them to other types of sloping roofs are obvious. Changes required for one-story frames are considered later.

Platform Frame.   This type of frame, sometimes called *western framing*, is illustrated in Figure 5.7. The drawing indicates that the framing and subfloor for the first-floor construction can be completed to form a platform before the first-story exterior walls and the partition framing are started.

Subfloors are most frequently made of plywood sheets of a thickness that is structurally adequate. Plywood thickness, which depends upon the spacing of the supports, is from ½ in. to 1¼ in. The face grain of the plywood is laid perpendicular to the supporting joist and nailed at all boundaries and at all intermediate supports with 8d or 10d common nails. Nailing at intermediate supports is called *field nailing*. Glue for plywood that is used for subfloor application need only be interior type glue; however, if moisture conditions are anticipated, an exterior type plywood with exterior glue should be specified. The grade of plywood that is usually specified is called *standard*.

Strip board subfloors, laid diagonally, permit a tongue-and-grooved or matched, finished wood floor to be laid either perpendicular or parallel to the joists. The former is preferred, with the flooring nailed to the joists. In addition, a subfloor laid diagonally adds significantly to the rigidity of the frame.

Rows of bridging are provided at the joist midspan or on rows that are not more than 8 ft. apart. One function of bridging is to distribute the concentrated loads over several joists, thereby increasing the strength and rigidity of the floor, preventing torsion or twisting of the floor joists, and increasing resistance to lateral loads. There are two types of bridging—cross-bridging and solid blocking. The preferred form of bridging is *solid blocking*, which consists of short pieces of the same depth and width of the joists fitted between the joists at right angles to them and offset to permit end nailing. *Cross-bridging*, also called *X-bracing*, which is usually of one by two's, is nailed in at the top and bottom. Solid blocking may be preferred to X-bracing because it may serve as a solid support and nailer for the side edges of plywood subflooring; it transmits lateral loads and may form a solid support for partitions.

Framing the first-story walls and partitions is started by nailing the sole plates to the subfloor. The first-story studs are then erected, diagonally braced, fireblocked, and capped with doubled top plates.

The frame is now ready to receive the second-floor joists, bridging, and subfloor to complete the second-floor platform.

The framing for the second-story walls and partitions can now be completed, the ceiling joists placed, and the attic subfloor laid if there is to be one.

Double joists must be provided under partitions which are parallel with the joists. If these are bearing partitions unsupported by walls, partitions, or girders, the partition should be trussed by one or more diagonal braces. The 1 × 4 in. or 1 × 6 in. diagonal brace is nailed on temporarily; a power saw is run along the brace, cutting the studs to the depth and width of the brace. The wood is chipped out and the diagonal brace is nailed in flush to the studs, forming a *let-in* brace. Properly installed, let-in braces should extend from the top plate to the sill plate. Bracing arrangements are necessary to avoid partition openings, such as doors and ducts.

Finally, the rafters are erected. They are notched, a *birdsmouth*, to provide bearing on the top plate of the exterior wall studs. The upper ends of full-length rafters are cut to fit against the vertical sides of the ridge board to which they are nailed. To strengthen the roof, horizontal members, called *collar beams*, are often nailed to every third pair of rafters at approximately their upper third points.

The frame is now ready to receive the exterior sheathing. Because wall sheathing, including horizontal wood sheathing, does not provide adequate lateral stability, the frame must be braced laterally by *corner bracing* (Figure 5.7). Such bracing, similar to that described above, consists of 1 × 4 in. boards let-in to notches cut in the outer faces of the studs so that the outer faces of the braces are flush with the outer edges of the studs to which they are nailed.

Often the framing for various portions of the walls and partitions is assembled in a horizontal position on the platform, tilted into the vertical position, and secured, facilitating construction operations.

The frame for a two-story dwelling is identical with the portion of the one-story frame above the foundation sill. This is the usual type of frame for one-story dwellings that do not have masonry veneers or stucco finishes.

Balloon Frame.   The basic difference between the balloon frame and the platform frame is that the balloon frame exterior wall studs extend as one piece from the foundation sill passing through the two floors. The interior partition studs are interrupted only by a single top plate. No header is provided at the exterior ends of the first-floor joists. They are simply nailed to the studs.

The two-story exterior wall studs and their double top plates are

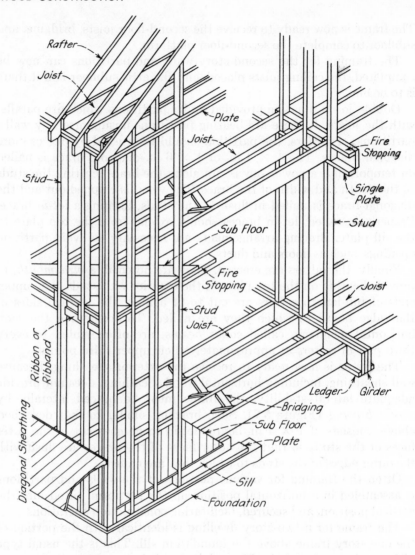

FIGURE 5.8   Balloon frame. (Adapted from reference 4).

erected with the lower ends at the studs bearing on the sill. The ends of the second-floor joists (first-floor ceiling) are supported on a flat 2 × 4 block resting on a 1 × 4 in. *ledger board* let-in to the inside of the notched studs. This board is called a *ribbon* or *ribband*. The joists are placed in contact and nailed to the studs. If there are interior foundation walls under the first-story partitions, the first-story studs are extended to plates on these walls; otherwise, they are supported on girders. The second-story joists are lapped at their interior, which is a single 2 × 4 top plate of the first-story bearing partitions. The relative positions of the first- and second-story bearing partition studs and the second-story joists are shown in Figure 5.8.

The lower portion of the frame for a one-story dwelling is identical to the lower portion of the frame for a two-story dwelling. A double wall plate placed along the tops of the one-story studs supports the ceiling joists and the roof rafters.

The bridging, roof framing, double joists, and corner bracing required for platform framing apply also to the balloon frame. Horizontal sheathing requires that the frame be corner braced, but diagonal sheathing or plywood sheathing does not (see Article 27).

The Merits of Platform and Balloon Frames. The primary advantage of the platform frame over the balloon frame is its relative ease of construction. The primary advantage of the balloon frame is its relatively smaller vertical shrinkage.

The easier erection of the platform frame results primarily from utilizing the platform in each story for wall and partition frame preassembly.

The small vertical shrinkage of the balloon frame results from the continuity of the exterior wall studs extending through the two stories and the approximate continuity of the bearing partition studs. This takes advantage of the negligible longitudinal shrinkage of wood.

The relatively large vertical shrinkage of the platform frame is due to the inclusion of considerable cross-grain wood along the lines of vertical support. The shrinkage of wood in cross-grain is relatively large because the shrinkage of the wood cells in their cross-sectional dimension is large when there is a reduction in moisture content.

The vertical shrinkage produces unequal vertical movements at different points. This causes plaster and glass to crack, joints to open up, openings to be out of plumb and distorted, and doors and windows to bind. Such movements can be minimized in both types of frame by equalizing the total vertical thickness of cross-grain wood along all lines of vertical supports.

Difficulties due to vertical shrinkage cannot be avoided by the platform frame if the building is faced with a masonry veneer. Masonry veneers do not shrink with reduction in moisture content, but the supporting frame does shrink. The relative downward movement of the frame causes stresses which are difficult to avoid even if foreseen, especially at openings. They are, of course, more pronounced in two-floor than in single-floor dwellings. Exterior finishes such as shingles and wood sidings can adjust themselves to vertical shrinkage of the supporting frame, but stucco finishes cannot unless expansion joints are included.

Considering vertical shrinkage in the platform frame, horizontal sheathing is preferable to diagonal sheathing. Diagonal sheathing would restrict the vertical movement of the first-and second-floor construction and preclude the equal vertical settlement of the structure, which is an objective of the platform frame.

If lumber were seasoned to the ultimate in-use moisture content, shrinkage problems would be minimized. This is unfeasible for many reasons, including the seasonal variations in the moisture content of the framing. The most objectionable shrinkage occurs after the finished wall surfaces have been completed.

Framed Openings.   Typical framings around window and door openings are illustrated in Figure 5.9(a) to (c). In general, the jambs at these openings are doubled. The heads consist of two pieces proportioned according to the span and load and supported by the double

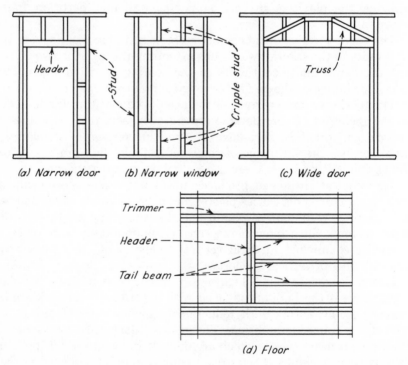

(a) Narrow door   (b) Narrow window   (c) Wide door

(d) Floor

FIGURE 5.9   Framing around openings.

jambs, as shown in (a), (b) and (c). Short studs above and below openings (a) and (b) are called *cripple studs*. A double jamb may be separated (a) if blocks are securely nailed between the two studs. Wide door and window openings may be trussed (c), or deep headers may be used. The framing must support the subjected loads without exceeding permissible stresses and deflections. The framing around an opening in a floor is illustrated in (d), where a header, trimmers, and tail beams are shown.

Fire-stopping. Concealed spaces between studs, joists, and rafters, in the clearance space required between masonry chimneys and wood framing, and in other locations permitting air currents to circulate should be sealed by barriers to impede the spread of fire. Usually 2 in. thick wood members are used for fire-stopping, although noncombustible materials, such as cement asbestos, are permitted. Examples of fire-stopping given in Figure 5.10 are similar to joists supported on steel girders or masonry walls.

Nailed Connections. Nailing is the most common means for joining the 1 in. and 2 in. lumber used in wood frame construction. Nails are driven in various ways with reference to the pieces they connect. In balloon frames, the faces of the joists and studs are placed in contact (Figures 5.7 and 5.8) and nailed together.

The types of nailing (Figure 5.11) are face nailing, end nailing, toe nailing, and blind nailing. In *face nailing*, the nails are driven through the face of one piece into the face or edge of another (a).

In *end nailing*, the nails are driven through the face (perpendicular to grain) of one piece and into the end (parallel to grain) of another piece (b). In *toe nailing*, the nails are driven at about a 30° angle through the face of one end and into the face of another (c). Usually toe nails are driven into both faces of the first members; but sometimes, as in cross-bridging, only one face is accessible. *Blind-nailing* is used for nailing matched flooring to the subfloor. The nails are driven at about a 45° angle at the top of the tongue and into subfloor and joists, as shown in (d). Finishing is not considered a part of the framing. Two-inch matched subflooring is secured by both face and blind nailing to the supporting members.

The type of nailing is determined partially by the accessibility of the nailhead. Nailed joints are strongest when nails driven perpendicular to the grain are subjected to shearing stresses. They are weakest when the withdrawing forces act parallel with the length of the nail. This is especially true for end nailing, in which the nails are driven parallel to the grain in the supporting member.

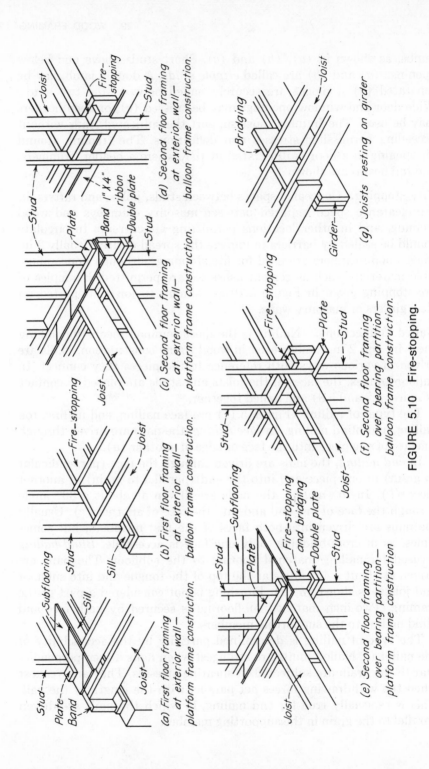

(a) First floor framing at exterior wall—platform frame construction.

(b) First floor framing at exterior wall—balloon frame construction.

(c) Second floor framing at exterior wall—platform frame construction.

(a) Second floor framing at exterior wall—balloon frame construction.

(e) Second floor framing over bearing partition—platform frame construction

(f) Second floor framing over bearing partition—balloon frame construction.

(g) Joists resting on girder

FIGURE 5.10 Fire-stopping.

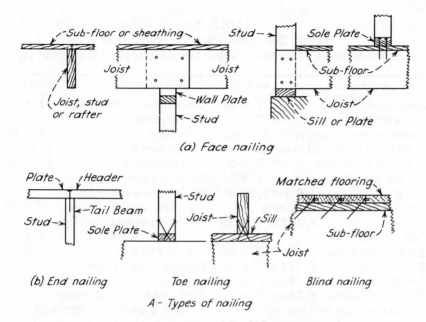

(a) Face nailing

(b) End nailing   Toe nailing   Blind nailing

A - Types of nailing

FIGURE 5.11   Types of nailing. Timber Engineering Company.

The usual minimum recommended nail penetration into the supporting member is one-half the thickness of the attached member.

A *nailing schedule* giving the types of nailing and the sizes of nails suitable for various connections in light wood framing is given in Table 5-5.

Plank and Beam Construction. [6]   A type of framing for dwellings uses 2 in. thick planks for subfloors and roof sheathing, supported by beams and rafters spaced from 6 to 8 ft. apart, which in turn are supported by posts or by trussing the rafters. The planks are perpendicular to the beams. The frames resemble platform frames.

The plank subfloors and the roof sheathing are usually exposed on their lower sides, frequently presenting attractive surfaces. Because of the wide spacing of the beams, tongue-and-grooved planks or planks with grooves on both edges to receive wood strips, called *loose tongues* or *splines*, are used to diminish differential deflection between members and to distribute the floor loads. The finished wood flooring is nailed at right angles to the subfloor planks.

The beams, designed to satisfy structural requirements, may be solid sections, planks placed on edge and spiked together, or planks

TABLE 5-5
Nailing Schedule Using Common Nails
Recommended by National Lumber Manufacturers Association [4]

Location	Nailing	Location	Nailing
Joist to sill or girder	T, 3–8d	Continuous header to stud	T, 4–8d
Cross bridging to joist	T, 2–8d	Ceiling joists	
Girder ledger at each joist	F, 3–16d	To plate	T, 3–8d
Subfloor to each joist or girder		Laps over partitions	F, 3–16d
1 by 6 in. or less	F, 2–8d	To parallel rafters	F, 3–16d
Over 1 by 6 in.	F, 3–8d	Rafter to plate	T, 3–8d
2 in.	B and F, 2–16d	1-in. diagonal brace to	
Sole plate to joist or blocking		each stud and plate	F, 2–8d
	F, 16d at 16 in. c.	Sheathing to each bearing	
Top plate to stud	E, 2–16d	1 by 8 in. or less	F, 2–8d
Stud to sole plate	T, 4–8d	Over 1 by 8 in.	F, 3–8d
Double stud	F, 16d at 24 in. c.	Built-up corner-studs	
Double stud	F, 16d at 24 in. c.		F, 16d at 24 in. c.
Double top plates	F, 16d at 18 in. c.	Built-up girders and beams	
Top plates, laps, and intersections		Along each edge	20d at 32 in. c.
	F, 2–16d		
Continuous header, two pieces			
	F, 16d at 16 in. c.		

Key: F, face nail. E, end nail. T, toe nail. B, blind nail. c. denotes center-to-center spacing.

separated by blocks. Built-up beams can be covered with a single piece of wood to improve appearance. Utility lines can be located in separated beams.

Columns or posts are spaced to correspond to the beam spacing. Columns must be large enough to carry their loads and to provide adequate seating or bearing for the beams. The minimum post size is 4 × 4 in. When the beams abut over a post, the column size should be not less than 4 × 6 in.

Special framing is provided under bearing partitions and heavy concentrated loads. Lateral bracing is provided by solid wall panels with diagonal wood or rigid plywood sheathing or by diagonal corner bracing. Abutting ends of beams should be tied together, preferably with nailed metal straps. The longitudinal joints between the planks of exposed ceilings may be beveled or attractively finished in various other ways.

Ordinary Construction

Ordinary construction is defined in some model codes. Although other codes may not identify and specify a category as ordinary construction, there are types of construction described that meet many of the criteria of ordinary construction.

As defined in the first chapter, ordinary construction requires non-combustible walls, such as masonry or concrete of equivalent structural stability under fire conditions, and a fire resistance rating of not less than 2 hours. The interior assemblies are entirely or partially of wood with sizes less than those required for heavy timber construction. The interior is wood frame as described above in wood frame construction; however, wood trusses and steel beams, girders, and trusses without fireproofing are often used. This class of construction is used extensively for most occupancies when it is not prohibited by fire district regulations.

A cross section of a typical two-story building of ordinary construction with brick and concrete exterior walls is illustrated in Figure 5.12. Some of the features that identify it as representative of ordinary construction are:

(1) The exterior masonry walls conform to the minimum thickness to height ratio specified by the codes.

(2) The parapet wall satisfies requirements for fire protection where required.

(3) The joists bearing on the masonry wall are end cut on an acute angle so that their top edges are free of the masonry, while their bottom edges bear 4 in. into the masonry. This is called a *fire cut*. In the event of collapse caused by fire, it permits the joists to rotate out of the wall without tending to pull over the wall.

(4) Wood joists, beams, and girders supported by exterior masonry walls should have not less than 4 in. of masonry between the ends of the beams and the outside face of the wall.

(5) The exterior ends of the ceiling joists rest on 2 in. thick plates anchored to the top of the masonry wall.

(6) The interior framing is wood frame construction corresponding to the balloon frame illustrated in Figure 5.14. This type of construction, preferable to the platform frame construction, minimizes vertical shrinkage.

(7) Interior joists require a minimum of 3 in. bearing.

(8) The wood columns bear directly on the columns below with the loads transferred through suitable connections.

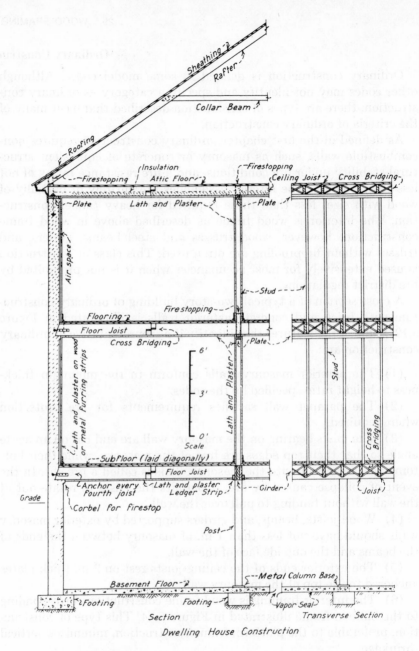

Dwelling House Construction

FIGURE 5.12 Ordinary construction for dwelling.

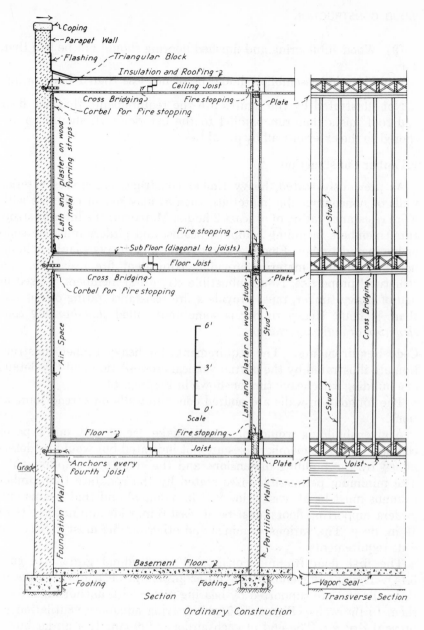

Coping
Parapet Wall
Flashing
Triangular Block
Insulation and Roofing
Ceiling Joist
Cross Bridging
Fire stopping
Plate
Corbel for fire stopping
Lath and plaster on wood or metal furring strips
Fire stopping
Sub floor (diagonal to joists)
Floor Joist
Cross Bridging
Plate
Corbel for fire stopping
Air Space
6'
3'
0'
Scale
Lath and plaster on wood studs
Stud
Floor
Fire stopping
Joist
Plate
Grade
Anchors every fourth joist
Foundation Wall
Partition Wall
Basement Floor
Footing
Footing
Vapor Seal
Section
Transverse Section
Ordinary Construction

Stud
Stud
Stud
Cross Bridging
Joist

FIGURE 5.13 Ordinary construction for light commercial building.

(9) Wood subflooring and finished flooring should not be less than 2 in.

(10) Interior assemblies should be of not less than 1 hour rating.

Not illustrated by the figures are the requirements that the floor and roof joists that run parallel to the exterior wall should be anchored to the exterior wall by metal ties.

Heavy Timber Construction

As previously stated, heavy timber construction requires exterior walls of noncombustible materials, such as masonry or concrete with a fire resistance rating of at least 2 hours. Moreover, the interior structural members including columns, beams, and girders must be solid or laminated timber. Floors and roofs must be heavy plank or laminated wood or a material providing equivalent fire resistance and structural properties. Noncombustible structural members, if used in lieu of heavy timber, must provide a fire resistance rating of not less than 3/4 hour. Heavy timber is sometimes called *slow-burning construction* or *mill construction*.

Code Requirements. The requirements for heavy timber construction are illustrated by the partial vertical cross section and the details of a building with heavy timber shown in Figure 5.14.

The foundation walls are poured concrete walls on spread foundations.

A given building would, of course, make use of only one type of construction. The member sizes are not indicated but would be determined by the building dimensions and the structural requirements. The minimum permissible sizes stated by the code are that timber columns must be at least 8 in. × 8 in. nominal and that beams and girders supporting floors must be at least 6 in. wide and not less than 10 in. deep. The various framings and other details must satisfy the code requirements.

The first floor framing consists of longitudinal girders (Figure 5.14). The exterior ends of the outer girders project into recesses in the walls and are supported by bearing plates with anchor lugs to tie them to the walls (Figure 5.15(*a*)), leaving adequate ventilation to prevent dry rot. The end of each girder is fire-cut. If a girder burns through, the embedded end rotates off the anchor lug and the girder falls free, preventing a collapse of the masonry wall.

The interior ends of girders are supported by wood columns with beveled or *chamfered* corners, which increases their fire resistance. The girder ends are supported on the brackets of steel post caps

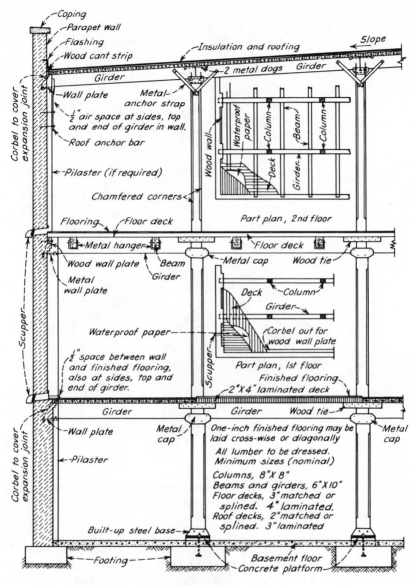

Heavy timber construction

FIGURE 5.14   Building with heavy timber construction.

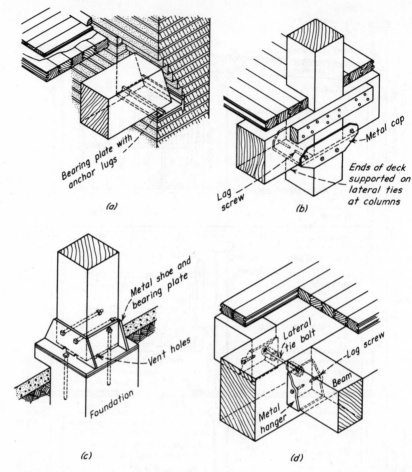

FIGURE 5.15   Details for heavy timber construction. (a) Girder bearing on metal wall plate. (b) Metal column cap with two brackets. (c) Built-up steel column base. (d) Metal beam hanger.

(Figure 5.15(b)), thus permitting second-story columns to bear directly on the post caps. The adjacent ends of the two girders which rest on a single post cap are tied together by lateral ties and lag screws through the cap. The bottom of each column bears on a metal base (Figure 5.15(c)), which is supported by, and anchored to, a concrete footing.

The outer panels of the first floor deck are constructed of splined or tongued-and-grooved planks of a 3 in. minimum thickness between girders. A clearance of ½ in. left between the edge of the deck and the wall surface provides for expansion. This open space is covered by

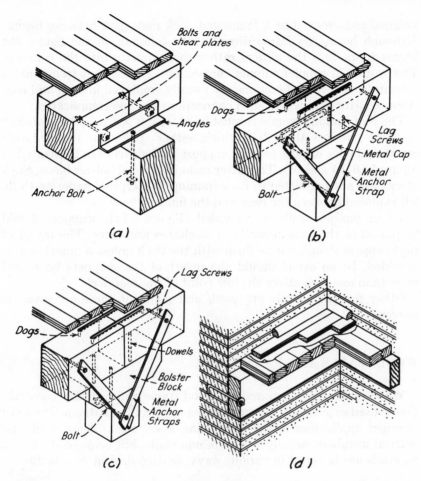

FIGURE 5.16 (a) Beams supported on top of girder. (b) Metal column cap supporting roof girders. (c) Wood bolster block supporting roof girders. (d) Wall beam. National Lumber Manufacturer's Association.

moldings fastened to the wall, or the bottom is covered by a 3 × timber ledger that is bolted to the masonry wall, or the masonry wall is corbeled to support the deck. This ledger or corbel also serves to deter deflection of this one unsplined deck member that may deflect and tear away from the base molding (Figure 5.16(d)).

The second-floor framing of longitudinal girders supported by the wall and the interior columns is similar to the first-floor framing. Transverse beams spanning between girders (Figure 5.14) are supported by metal hangers such as those shown in Figure 5.15(d). A tongued-and-grooved or a laminated deck may span between beams. Although beams do frame directly into columns, in the figure, the beams frame into the girders rather than directly into the columns. This results in simpler column connections. Beams that frame into a column from several directions are usually employed in steel and concrete construction, where such connections are less complicated.

The roof framing members must be at least 6 × 8's, and a roof deck must be not less than 2 in. thick, either splined or tongued-and-grooved. Types of post caps to support roof girders are illustrated in Figure 5.16(b) and (c). The latter includes a wood bolster block. Such blocks are not permitted in floor framing. Waterproof paper or 15 lb. felt is placed between the deck and the finished floor.

Where parapet walls are provided (Figure 5.14), scuppers should be placed in the outside walls to discharge rainwater. The invert of the scuppers should not be flush with the deck unless a gravel stop is provided. In no event should the invert of the scuppers be placed more than one inch above the low point of the roof deck.

Other framing details are used, and many of these are given in reference [7].

30. WOOD WALLS AND PARTITIONS

Wood in the sizes required for wood frame and ordinary construction is used extensively in constructing walls and partitions. The most common application is the supporting wood stud. These studs are vertical members, usually 2 × 4 in. nominally, but occasionally 2 × 6 in. studs are fastened in various ways, as described in Article 29.

Sheathing. To increase the rigidity, lateral stability, and heat-insulating properties of the exterior walls and to form the base for a finish, *sheathing* is nailed to the studs. This sheathing may be plywood ⅜, ½, or ⅝ in. thick, wood boards with a nominal thickness of 1 in., fiberboard ½ and ²⁵⁄₃₂ in. thick, or gypsum board at least ½ in. thick.

One of the most widely used forms of wood is *plywood*. Plywood is an assembly of thin wood sheets, forming a panel that is glued and pressed together. The grains of the plys run alternately at right angles, with an odd number of plys required for a panel. This arrangement equalizes surface tension between plys and deters warping, shrinking, and splitting.

The exposed panel surface is called the *face* veneer; the hidden side is the *back* veneer. The interior plys are referred to as the *core* or *inner* ply. Normally, the face and the back grain run parallel to the long side of the panel; otherwise, the panel is referred to as *crossband* plywood.

The quality, the use, and the structural properties of the plywood are determined by the species of wood, the type of glue, the veneer grades, the use grades, and the thickness of the panels. These we will investigate.

Plywood is broadly classed according to application into *exterior plywood* and *interior plywood*. For exterior plywood, the better grades of veneer are used with a glue that has high resistance to water penetration and delamination. (Table 5-7).

For structural considerations, about 47 softwood species and 8 hardwood species are classified according to their stiffness into the groups listed in Table 5-6. These species are applied as face and back veneers.

The face and back of plywood sheets, which are called the *veneers*, are further classified according to the wood quality and grading characteristics. Splits, knots, and other imperfections according to frequency and severity determine veneer grade. Table 5-7 clarifies.

Interior type grade plywood is usually glued with protein type glue that is moisture resistant. Intermediate glue, used for interior grade, is more resistant to moisture, bacteria, and mold. Where interior plywood is expected to be subjected to excessive moisture, an interior plywood that is assembled with exterior glue may be used. Glues used for exterior grade must be capable of resisting water penetration and repeated wet and dry cycles.

Plywood standard widths are 3, 4, and 5 ft.; and lengths starting at 5 ft. increase by 1 ft. increments up to 12 ft. By far, 4 ft. by 8 ft. is the most common size.

The nominal size of plywood is determined by the number of plys. The face or veneer plys are usually thinner than the interior or core plys. Plywood is manufactured in an odd number of plys: 3, 5, or 7. Plys are usually $\frac{1}{8}$ in. thick, although they may be as thin as $\frac{1}{16}$ in. The thinnest plywood is nominally $\frac{1}{4}$ in. and the thickest is nominally standard $1\frac{1}{4}$ in. Thickness of plywood is usually increased in $\frac{1}{8}$ in. increments.

Veneers to be used on exterior plywood must be classed at least "C" by the criteria listed in Table 5-7.

Plywood Sheathing Grades.   Plywood is classified according to *sheathing grades*. Based upon the species group (Table 5-6) and the veneer classification, there are three sheathing grades. Listed in descending order:

TABLE 5-6
Classification of Species [19]

Group 1	Group 2	Group 3	Group 4	Group 5
Birch	Cedar, Port Orford	Alder, Red	Aspen, Quaking	Fir, Balsam
Yellow	Douglas Fir 2**	Cedar	Birch, Paper	Poplar, Balsam
Sweet	Fir	Alaska	Cedar	
Douglas Fir 1*	California Red	Pine	Incense	
Larch, Western	Grand	Jack	Western Red	
Maple, Sugar	Noble	Lodgepole	Fir, Subalpine	
Pine, Southern	Pacific Silver	Ponderosa	Hemlock, Eastern	
Loblolly	White	Redwood	Pine	
Longleaf	Hemlock, Western	Spruce	Sugar	
Shortleaf	Lauan	Black	Eastern White	
Slash	Red	Red	Poplar, Western***	
Tanoak	Tangile	White	Spruce, Engelmann	
	White			
	Almond			
	Bagtikan			
	Maple, Black		*Douglas Fir 1—Washington, Oregon, California, Idaho, Montana, Wyoming, British Columbia, Alberta.	
	Meranti			
	Mengkulang		**Douglas Fir 2—Nevada, Utah, Colorado, Arizona, New Mexico.	
	Pine			
	Pond		***Also known as Black Cottonwood.	
	Red			
	Western White			
	Spruce, Sitka			
	Sweetgum			
	Tamarack			

Courtesy American Plywood Association, Tacoma, Washington.

TABLE 5-7
Veneer Grades Used in Plywood [19]

N Special order "natural finish" veneer. Select all heartwood or all sapwood. Free of open defects. Allows some repairs.

A Smooth and paintable. Neatly made repairs permissible. Also used for natural finish in less demanding applications.

B Solid surface veneer. Circular repair plugs and tight knots permitted.

C Knotholes to 1". Occasional knotholes ½" larger permitted providing total width of all knots and knotholes within a specified section does not exceed certain limits. Limited splits permitted. Minimum veneer permitted in Exterior-type plywood.

C Improved C veneer with splits limited to ⅛" in width and knotholes and borer holes limited to ¼" by ½".

D Permits knots and knotholes to 2½" in width and ½" larger under certain specified limits. Limited splits permitted.

Structural I is composed of woods only from Species Group 1. *Structural II* is made from the woods that are listed in Groups 1, 2, and 3. These two grades are used for structurally engineered applications, such as sheathing diaphragms, shear walls, gusset plates, box-beams, folded plates, stress skin panels, and other places requiring a quality structural plywood. *Standard (C-D)* sheathing grade is used for general sheathing applications. The C and D indicate the veneer quality. The D veneer grade indicates that such sheathing would be for interior application only. Should standard grade be used for exterior application, both faces must be C veneer grade, and the sheathing grade would be *Standard (C-C)*. These three sheathing grades are stamped with a notation that indicates the recommended spacing of the supports for the sheathing. The plywood is stamped with a number that indicates the recommended span of the plywood with the face grain of the plywood perpendicular to the supports. The indication, for example, may be $^{32}\!/_{24}$; this means that the recommended spacing for supports when the plywood is used for roof sheathing is 32 in. and that supports should not exceed 24 in. when the plywood is used as floor sheathing.

Other Sheathing Materials.   Wood sheathing boards plain, matched, or shiplapped, with a nominal width of at least 6 in. or 8 in., should be nailed to each stud with not less than two 8d common nails. Sheathing may be installed horizontally or diagonally. Horizontal sheathing is more economical than diagonal sheathing but lacks rigidity; therefore, the frame must be diagonally braced at the corners by preferably continuous 1 × 4 in. pieces let-in to the studs in a continuous line. Ducts or window openings may complicate bracing.

Fiberboard is made of cane fiber, straw, or similar fibrous materials pressed into sheets 2 and 4 ft. wide and usually 8 ft. long. The 4 ft. wide sheets, placed with the length parallel or perpendicular to the studs, are nailed with galvanized large-head roofing nails 1½ to 1¾ in. long. Field nailing (which are nails in the central surface) is spaced 6 in., and edge or boundary nailing is spaced 3 in. apart. The 2 ft. wide sheets are edge matched and are placed with the length horizontal. A special board impregnated or coated with bituminous material is available, which should not be vaporproofed.

Exterior Surfaces.   The exterior walls may be covered with horizontal or vertical siding and nailed to the studs with or without an intervening layer of sheathing. Noncorrosive nails, such as zinc-coated steel-wire nails or aluminum nails, are used to avoid rust spots. A heavy, specially treated paper, such as *kraft building paper*, or sheathing paper placed under the siding minimizes air infiltration and weatherproofs the sheathing.

Wood products that are called *patterned, worked*, or *worked to pattern* are machined to a configuration or a designed profile. This includes various shaped sidings and moldings.

There are two general types of wood siding: *bevel siding*, sometimes called *weatherboarding*, which is tapered and laid so that the thinner upper edge laps over the lower edge, and *drop* or *novelty siding*, which has a tongue-and-groove joint or a *rebated* or *shiplap joint* (Figure 5.17(*b*) to (*h*) and is installed flush against the sheathing. Some forms of the latter siding give the same effect as bevel siding. Redwood bevel siding is graded Clear All Heart, VG (Vertical Grain), Clear All Heart, Clear VG, and Clear. Moldings are made from Clear grade, either heart or sapwood.

Siding materials worked to pattern are shiplapped, Centermatched (CM), Dressed and Matched (D & M), and Tongue and Groove (T & G).

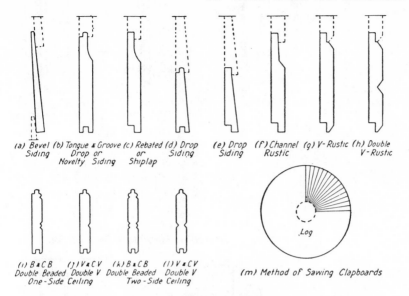

(a) Bevel Siding (b) Tongue & Groove Drop or Novelty Siding (c) Rebated or Shiplap (d) Drop Siding (e) Drop Siding (f) Channel Rustic (g) V-Rustic (h) Double V-Rustic

(i) B & CB Double Beaded One-Side Ceiling (j) V & CV Double V (k) B & CB Double Beaded (l) V & CV Double V Two-Side Ceiling

(m) Method of Sawing Clapboards

.Log

FIGURE 5.17   Types of wood siding and ceiling.

Many other designs, such as *rustic sidings* and *colonial sidings*, are special forms of drop sidings. The common siding widths are 6 in. and 8 in. (nominal), and the thicknesses are $9/16$ and $3/4$ in. (actual); but other widths and thicknesses are manufactured.

Siding is made of Douglas fir, white and yellow pine, spruce, hemlock, redwood, cedar, and cypress.

Figure 5.17 shows representative redwood patterned sidings.

Clapboards, once widely used as siding, are now seldom used.

The use of siding without sheathing, illustrated in Figure 5.18(*a*), is an inferior grade of construction.

Wood Shingles.   *Wood shingles,* applied to solid wood or plywood (*c*), may be placed with more shingle exposure than on roof applications. Shingles with a scored face, called *shakes,* although longer, are laid in the same manner with a considerable length to the weather. Asbestos-cement or asphalt shingles are also used in a similar manner.

Siding.   Square-edge boards attached vertically to horizontal girts, with the joints covered with battens, may be used for siding (Figure 5.18(*j*)). Tongued-and-grooved boards, without battens, may be used in the same manner. In a better type of construction, vertical siding is nailed to 1 in. wood sheathing. Air infiltration is minimized by nailing kraft paper between the siding and the sheathing.

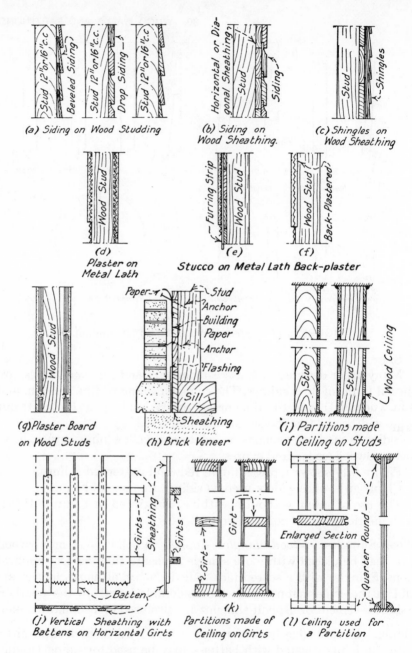

FIGURE 5.18   Types of frame walls.

Veneer. Sheathed exterior stud walls may be veneered with brick (*h*) or with other masonry. The veneer resting directly on the masonry foundation should be tied to the frame at intervals not to exceed 16 in. vertically and horizontally.

The weakest point in veneer construction is the tie between the face brick and the backing. The usual method of tying is illustrated in Figure 5.19(*a*). The masonry is brought up to the elevation at which a corrugated galvanized-steel tie, sharply bent into an L, is placed in the position shown in (1) and attached by a nail driven into a stud. The bend should be placed at the height of the joint, and the nail should be driven into the bend. This practice is not usually followed because it is possible for the veneer to pull away from the frame. This can be avoided by the procedure illustrated in (*b*). The masonry is stopped one course below the elevation of the tie (1); the tie is placed flat along the stud and nailed to the stud (2); another course of brick is laid (3); and the tie is bent sharply over the head of the nail and into the joint (4). If this procedure is followed, the tie cannot yield. Minimum nail size required for anchors is an 8d common galvanized or aluminum nail driven slightly inclined into a stud.

A 1 in. air space between the brick and the sheathing, which is covered by a layer of nonvapor-resistant sheathing paper, increases the insulating properties of the wall. Openings should be carefully flashed to prevent moisture penetration behind the facing. The frame

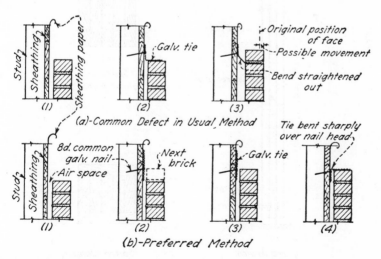

FIGURE 5.19   Tying face brick to wood frame. (*a*) Common defect in usual method. (*b*) Preferred method.

construction should not extend below the first-floor joints. By using the balloon frame (page 264), shrinkage in the timber will be minimized, especially where the window and door frames penetrate the masonry veneer.

Veneered construction resists exposure to exterior fires far better than frame construction. Its resistance to interior fires is about the same as that of frame construction; but if not properly firestopped, it may be difficult to extinguish fires behind the brick facing.

31.   SOLID WOOD AND LAMINATED MEMBERS

Columns and Studs.   Wood columns, as ordinariy used, are square timbers rarely smaller than 4 × 4 in., and usually not larger than 12 × 12 in., although larger timber columns can be obtained. Columns built up of small timbers fastened for unified action of the timbers under load are not as effective as glued or solid columns. One type of built-up compression member is the spaced column (Figure 5.20(a)), which has a load-carrying capacity much greater than that of the individual vertical members when not connected by a spacer block.

Several planks may be spiked together (b) to form a column, but the arrangement shown in (c) results in a more unified action and is

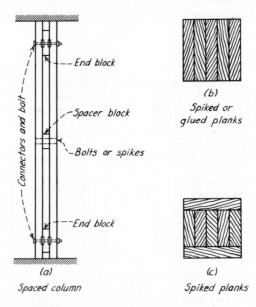

FIGURE 5.20   Built-up wood columns.

more effective. Columns are glued in several laminations as described in the paragraph on glued laminated beams, rather than by spiking (*b*). Glued laminated columns have the same advantages as beams of this type.

Solid Beams and Girders. Several specific types of flexural members are included under the general designation of beams and girders. Among these are floor and ceiling joists, rafters of sloping roofs, roof purlins, and beams and girders in various locations.

In wood frame construction, the closely spaced joists and rafters usually have a thickness of 2 in. and a common depth of from 4 to 8 in. for rafters and 6 to 12 or 14 in. for joists, depending upon the span and loading. Consideration must also be given to the rigidity, as determined by the deflection under load. Often rigidity rather than the allowable stress determines the size. Beams may have a solid section of the required size; but in this type of construction, the required size is often achieved by spiking together 2 in. planks set on edge. Such pieces are readily available and easily assembled, or fabricated, on the job.

In heavy timber construction, the minimum thickness permitted for floor joists, beams, girders, and other members of this general type is 6 in., and the minimum depth is 10 in. This size may not be determined by structural criteria, but it is a fire resistive requirement.

Glued Laminated Beams and Girders. If several pieces of laminations of uniform section laid flat on each other form a beam supported at the ends, their load-carrying capacity and vertical rigidity is relatively low. When loaded, they deflect, slide on each other along the surfaces of contact, and act as individual units. If sliding or relative movement can be prevented, the several pieces will act as a composite or solid unit, and the load-carrying capacity and rigidity will be markedly increased. Various devices have been used satisfactorily, but the most effective procedure is gluing and clamping wood laminations together under pressure. These beams are called *glued laminated beams* and are widely known as *glulams* (Figure 5.21(*a*)). The usual thickness of laminations is 1½ in., or nominal 2 in., although ¾ in. laminations are used for attaining small radius curves. The minimum number of laminations is 2, which produces a 3 in. member, whereas the larger members may have as many as 60 laminations. Glued laminated members are produced in several standardized widths. The usual widths expressed in inches are 2¼, 3⅛, 5⅛, 6¾, 8¾, 10¾, 12¼, and 14¼ [45].

Adhesives, which are known as *water-resistive* glue or *dry-use* adhe-

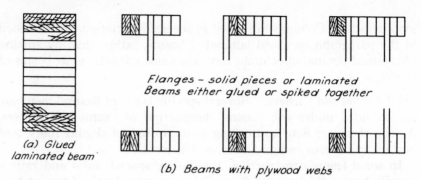

Flanges – solid pieces or laminated
Beams either glued or spiked together

(a) Glued
laminated beam

(b) Beams with plywood webs

FIGURE 5.21   Glued laminated beams and beams with plywood webs.

sives, are used for interior glued laminated members. These glues, which are the casein type, are used on laminated members intended for interior applications where the moisture content of the wood is not expected to exceed 16%. Members which are intended for exterior use, or where interior use will be subjected to extreme moisture conditions (showers, commercial laundries) require a *waterproof glue* known also as *wet-use adhesive*. The waterproof glues are resorcinol, phenol-resorcinol, phenol, and melamine.

Although it is not always practical to apply both types of adhesive on the same member, some structural engineers may require dry-use adhesive and wet-use adhesive on a long beam. Dry-use adhesives are less expensive; hence, the portion of beam that spans a wide interior is specified with dry-use adhesive, and that portion extending out to the exterior may be specified with a wet-use adhesive.

To obtain the desired lengths, laminations are spliced with inclined *scarf joints* made by pregluing overlapping ends with opposite bevels with slopes of about 1 in 10. After a lamination is spliced, it is surfaced to uniform thickness and width throughout its length and is then assembled into the glued beam. Each lamination is usually as wide as the beam, although two laminations may be placed side by side to make up the beam width. The longitudinal vertical joints between adjacent laminations in successive layers must be staggered. Full width top and bottom laminations should be used.

Glued laminated beams and girders are factory produced. They may have a constant cross section, a tapered top, the maximum cross sections at the center or elsewhere, cambering or other profiles. Girders with spans exceeding 100 ft. have been constructed, and girders with 2 in. laminations have been made as deep as 8½ ft.

Glued laminated construction has advantages compared with solid members because small sizes can be utilized, larger members can be constructed, defects can be eliminated; and because the moisture content may be reduced, there is better opportunity for seasoning. It is sometimes called *glulam construction*.

Glued Laminated Appearance Grades.   There are three grades which refer to the appearance of the glued laminated members. They are industrial, architectural, and premium. These appearance grades refer to the lumber characteristics and surfacing and to the structural qualities of the members. *Industrial appearance grade* is surfaced on only two sides and has visible sound knots. This grade is used in applications where appearance is not a prime consideration. *Architectural appearance grade* is used where appearance of the member is a consideration. These members have only slight imperfections. They are surfaced on the exposed sides. *Premium appearance grade* is used where appearance is paramount. This grade consists of lumber made of material that is clear with matching grain characteristics. The finished laminated surface is smoothed on the exposed faces.

Because premium grade is the most expensive, architects most frequently specify that the member will be architectural grade, with the last or bottom exposed member of premium grade.

Plywood Beams.   Beams and girders are sometimes constructed with plywood webs and with built-up flanges (Figures 5.21(*b*)). The possible depth is limited by the width of the plywood because horizontal splicing is not feasible. Because the maximum length of plywood is limited, vertical web splices are required for spans exceeding 12 ft. The flanges may be made up of vertical laminations or solid sections. If shop-fabricated, the member may be pressure-glued, but if field fabricated, the members are nailed or bolted with or without glue between them. This type of construction is rarely used but does have advantages.

Trussed Beams.   Beams, trussed (Figure 5.22) to increase their load-carrying capacity, are used occasionally.

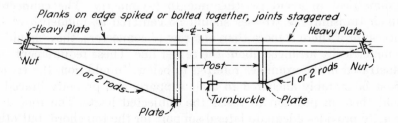

FIGURE 5.22   Half-elevation of trussed beams.

32. WOOD TRUSSES

Types of Trusses.   The various types of trusses are considered earlier and illustrated in Figures 3.4 and 3.5. The most common types of wood trusses for symmetrical roofs are the Howe (Figure 3.4(*i*)), the Pratt (*j*), the Fink (*m*), and the Belgian.

Flat or gently sloping roofs may be supported by the Howe truss, the Pratt truss, or the Warren truss (*p*) to (*s*). These types frequently support column-free floor areas.

Timber Connectors.   Timber connectors have made it possible to use wood members with small cross sectional areas. These connectors are readily available, require little labor to assemble, and result in only small reductions in the effective cross sectional area of the timber. One of the most efficient connectors for wood to wood connections in trusses for buildings is the *split ring connector* (Figure 5.23 and Figure 5.4). With this type of connector, the only function of the bolts is to hold the adjacent wood surfaces in contact with one bolt passing through several connectors.

The stress that a single connector will resist depends upon the size of the connector and the properties of the wood in which it is embedded. The number and size of the connectors required is determined by the total tendency to slide along the contact surfaces.

Another connector is the *shear plate* (Figure 5.4(*d*)). It also makes field connections between two wood members with plates embedded in each member as shown in Figure 5.5(*a*). In such arrangements, the bolts are subjected to the entire shearing stress between the two pieces connected.

If the lumber in a truss is not seasoned to its ultimate expected moisture content, the bolts should be tightened at intervals. However, *lock washers* should be used on all bolts to alleviate this problem, as the spring action of the lock washers will allow expansion and contraction.

General Requirements.   All members, joints, and trusses should be symmetrical to avoid twisting due to asymmetry. The connectors which transmit the stresses should be centered at middepth of the pieces; or if there is more than one row of connectors, they should be symmetrically arranged about the center line. These requirements are illustrated in the examples referred to below. In addition, the trusses must be suitably anchored to their supports and properly braced to hold them in position to resist the subjected loads. The roof deck usually provides adequate lateral support for the top chord, but other bracing is usually required.

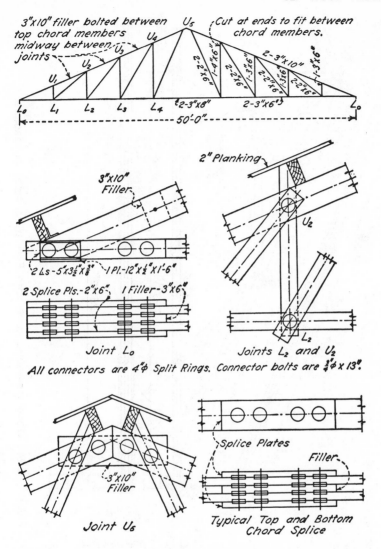

FIGURE 5.23   Joint details of wood Pratt truss using connectors.

A Pratt roof truss and a Fink roof truss with sloping top chords and split ring connector are illustrated in Figure 5.23 and 5.24, respectively. Note that the ends of the Fink truss are attached to wood columns and that *knee braces* are provided by continuing certain truss chords until they connect to the columns. The truss and columns form a *transverse bent* capable of resisting lateral or transverse loads on the roofs and sidewalls. A Fink truss supported by walls is similar to

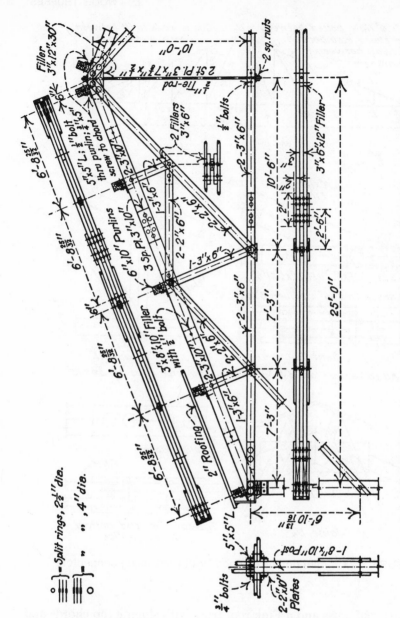

FIGURE 5.24 Wood transverse bent with Fink truss using connectors.

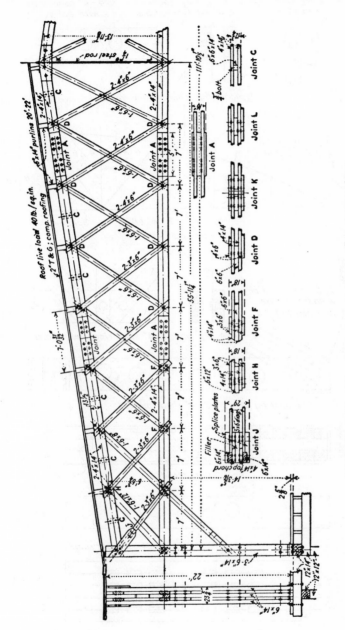

FIGURE 5.25 Wood transverse bent using connectors.

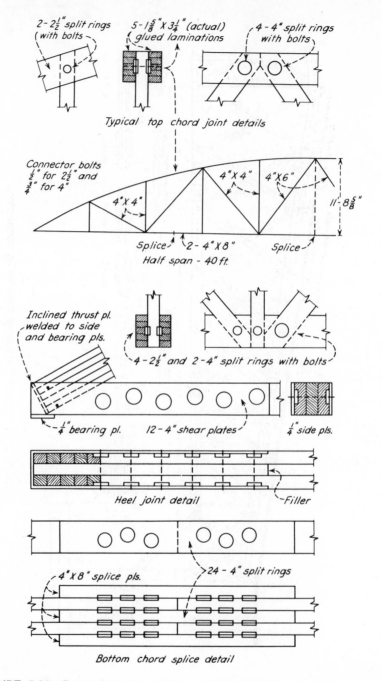

FIGURE 5.26 Bowstring truss with two-piece glued laminated top chord. Adapted from Timber Engineering Co. design.

the one illustrated; but the knee braces are omitted, and the stresses in the truss members are changed.

A relatively long span transverse bent using split ring connectors is shown in Figure 5.25.

A bowstring roof truss with a two-piece glued laminated top chord and split ring connectors at all joints except the heel joints is shown in Figure 5.26. An enlarged cross section of the top chord and an enlarged detail of the heel joint are shown. The web members fit in between the divided chords and are fastened with split ring connectors. Six pairs of shear plates, illustrated in Figure 5.4(d), and 5.5(a), are required at each joint.

A typical bowstring roof truss with single-piece glued laminated chords is illustrated in Figure 5.27. Both chords and all web members have the same width so that the web members can be joined to chords by bolted steel-strap end connections. Shear plates are used with the bolts when the stresses require them. Each chord is spliced at midspan by steel splice plates attached with bolts and passing through shear plates embedded in the members. A detail of the end connection is shown in the figure. This connection is assembled by welding a heavy inclined thrust plate between two side plates at the required angle to receive the end of the top chord, which is cut normal

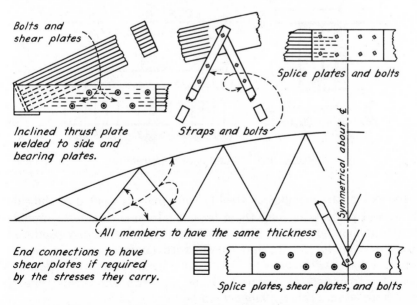

FIGURE 5.27   Glued laminated monochord bowstring truss.

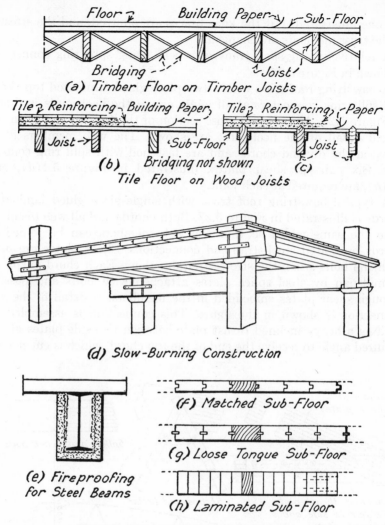

Floor *Building Paper* *Sub-Floor*

Bridging *Joist*

(a) Timber Floor on Timber Joists

Tile *Reinforcing* *Building Paper* *Tile* *Reinforcing* *Paper*

Joist *Sub-Floor* *Joist*

(b) *Bridging not shown* *(c)*

Tile Floor on Wood Joists

(d) Slow-Burning Construction

(e) Fireproofing for Steel Beams

(f) Matched Sub-Floor

(g) Loose Tongue Sub-Floor

(h) Laminated Sub-Floor

FIGURE 5.28   Wood floor construction.

to the axis of the member. A steel bearing plate is welded to the side plates and projects beyond them far enough on each side to provide for the anchor bolt holes. Trusses of this type are factory-produced with spans up to 150 ft., and trusses with longer spans can be obtained. Trusses with spans of 250 ft. have been constructed. Because each chord consists of one unit rather than two units, this type of truss is sometimes called a *monochord bowstring truss*.

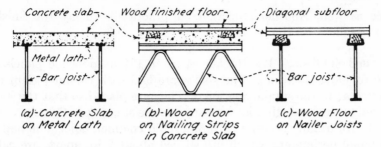

(a)-Concrete Slab on Metal Lath

(b)-Wood Floor on Nailing Strips in Concrete Slab

(c)-Wood Floor on Nailer Joists

FIGURE 5.29 Concrete and wood floors on open-web joists. (a) Concrete slab on metal lath. (b) Wood floor on nailing strips in concrete slab. (c) Wood floor on nailer joists.

33. WOOD FLOOR CONSTRUCTION

Wood Floors on Wood Joists.   Floor construction for wood frame and ordinary construction buildings consists of a matched-wood finished floor that is supported by wood joists. Preferably a layer of building paper of 15 lb. felt is laid between the subfloor and the finished floor as shown in Figure 5.28(a). This insulates and helps to alleviate squeaking. Other finish floor materials may be used in place of matched flooring, with appropriate construction made to receive the various finish materials.

Joists are usually 2 in. wide and from 6 to 14 in. deep. For heavy loads, the joists may be 3 or 4 in. wide. Heavy timber construction, mentioned elsewhere, requires joists not less than 6 in. wide. Usual joist spacings are 12 and 16 in.; 24 in. spacing is too great for lath that is to support plaster. Light metal lath may span up to 12 in., but heavy rib lath permits greater joist spacing. In some cases, 1 × 2 in. furring strips, properly spaced for lath, are nailed at right angles to the underside of the joists. The spacing of the joists is therefore independent of the lath. However, the airspace below the joists may enable rapid spread of fire. Another objection to wood furring strips nailed to the underside of wood joists is that the nails loosen their grip as the joists dry out.

A convenient function of the subfloor is to serve as a platform during the early stages of construction. A subfloor makes the floor system more substantial, more fire-resistant, and decreases sound and heat transmission. When a subfloor is used, there is a tendency to nail the finished floor to the subfloor rather than to the joists. This practice is objectionable because the finished floor does not remain in place as well. Another objectionable practice is to lay the finished floor parallel with the joists.

Plywood Subfloor.   The thicker ⅝ in. to 1¼ in. plywood panels are used for floor sheathing. The sheets are laid with the face grain perpendicular to the supporting joists. They are placed so that the joints formed by the ends that are on supports are staggered. These plywood panels are usually secured with 8d or 10d common nails. The nails at the panel perimeters, which are spaced about 6 in. apart, are called *boundary* nails. The nails which are in the central portion of the panel are called *field* nails. These are usually spaced farther apart but nailed to each intermediate support.

The joint (the long side of the abutting plywood panels) that spans between joists should have solid blocking beneath. This will prohibit differential deflection between adjoining plywood panels. This problem is sometimes alleviated by the use of a metal, H-shaped clip placed between adjacent panels.

The sheathing grades are frequently used for this application. They are stamped with a number, in inches, that is the recommended spacing of the supporting members.

Underlayment.   Frequently, in order to attain a smooth finish floor surface, a second layer of thinner plywood is placed over the thicker supporting plywood sheathing. This thinner plywood is called an *underlayment*. This underlayment ply is nailed to the supporting panel. All joints of the underlayment are staggered so as not to align with the joints of the supporting panel. The purpose of this underlayment is to provide a smooth, deflection-free surface that will in turn support some finished floor covering. This type of construction is used for underlaying carpets and various resilient floor tile installations.

Strip Board Subfloors.   A strip board subfloor should be laid diagonally; or if strips are placed on the subfloor and over the joists, the finished floor and subfloor may both be laid at right angles to the joists, the strips overcoming the uneven places in the subfloor. The subfloor may be of ordinary sheathing, matched boards, shiplap, or plywood sheathing. Sometimes the subfloor is omitted entirely to decrease the cost, and if so, the finished floor would probably have to be laid before plastering, during which operation it must be protected from dirt and water. It will absorb moisture given off by the plaster. This will cause swelling, followed by shrinking and opening of the joints.

Some form of layer between the subfloor and the finished floor should always be used. Asphaltic felt, asbestos paper, or gypsum board increase fire-resistance and sound-deadening and warrant the additional expenditure, although kraft building paper is more economical and is insulating.

The joists are bridged as shown in Figure 5.28(a).

If a wood subfloor is laid before the building is enclosed from the elements, provisions should be made for the swelling of the subfloor, or the exterior walls may be cracked and pushed out. To prevent this, spaces $\frac{1}{4} \times \frac{1}{2}$ in. may be left between the boards, or every tenth or twelfth board may be omitted at first and placed after the building is under cover.

If a tile floor is supported by wood joist construction, the joists may be designed to carry a reinforced-concrete slab (Figure 5.28(b)), which is a base for the tile. The construction shown in (c) is also used but is more likely to produce cracks than the construction in (b). A thin-setting bed for ceramic floor tile which does not require special construction is described on page 520.

This type of floor construction is used in residences and other buildings of wood frame and ordinary construction and may be used in buildings with steel frames. Inexpensive and light, it is sufficiently strong for heavy loads but very combustible unless protected on the underside by plaster or metal lath.

Heavy Wood Subfloor on Wood or Steel Beams.   Wood subfloors vary in thickness from 3 to 10 in., depending upon the loaded span. They are supported directly by girders running between columns, or by beams which are supported by the girders, as shown in Figure 5.15 and described in Article 29. In the first arrangement, the lateral spacing of columns commonly is not over 10 or 12 ft. because of the heavy subfloors required for longer spans. In the second arrangement, the beams are spaced 4 ft. or more apart, and the column spacing is not restricted by the length of the subfloor.

Heavy wood subfloors are either *matched*, (Figure 5.28(f)), *loose-tongue* (g), or *laminated* (h). The matched floor is used for thicknesses of 3 or 4 in. However, greater thicknesses produce excessive waste in matching; therefore, the hardwood loose-tongue may be more economical. This loose-tongue is also called a *slip tongue* or *spline*.

Laminated floors 4 in. or more in thickness are constructed by laying 2 in. lumber on edge and nailing the adjacent pieces together with spikes spaced about 18 in., 2 × 4's are used for a 4 in. floor, 2 × 6's for a 6 in. floor, and so on. A laminated floor is easier to lay than a heavy loose-tongue floor. The pieces are smaller, more easily handled and drawn together; however, more feet, board measure, are required because 2 in. material is only 1½ in. thick. The cost of the loose-tongue is saved in the laminated floor.

34. WOOD RIGID FRAMES, ARCHES, AND DOMES

As explained previously, there is no clear distinction between rigid frames and arches. The evaluation made in that article (page 167) will be followed in this discussion, although many authorities would not make such a distinction. The classification based on the number of hinges, as has been explained, is always followed.

Glued laminated construction, discussed in Article 31, is used extensively in constructing wood rigid frames and arches. The frames are factory-produced and transported to the site. Because it is not feasible to transport completed frames of the usual sizes, they are assembled and connected by field joints or splices.

For rigid frames, which include portions with relatively small radii of curvature, it is necessary to use laminations which are thinner than the 1½ in. thickness ordinarily used for beams. For example, the minimum permissible radius for 1½ in. laminations varies from 30 to 40 ft.; and for ¾ in. laminations, from 7 to 12 ft., depending upon the species of wood.

Rigid Frames.   Typical forms of glued three-hinged laminated rigid frames are illustrated in Figure 5.30(a), for which the slopes of the top members are relatively small, in (b), where the slopes are steep, and in (c), in which the top member is curved. A more detailed illustration of the frame in (a) is shown in (d). In all of these, outer surfaces of the legs are vertical, but the inner surfaces may be vertical instead. Frames may be curved at the eaves, as shown by the dashed line on the left side of (d), rather than angular. Various types of connections are made at the bottom of each leg and at the crown, some of which are shown in Figure 5.30(e) to (h). Even though the joints at the footings and the crown are considered to be hinged, the usual types of connection develop some resistance to rotation.

The two units of the frame are usually raised to position separately, and the leg and crown joints are then secured.

In the crown connections shown in Figure 5.30(f) and (g), a steel dowel passing through shear plates embedded in adjacent ends of the two pieces are provided to resist the shear caused by unbalanced loading. To hold the ends together horizontally, splice plates are bolted to the sides of the member and across the joint in (f) and on the top and bottom surfaces in (g). These plates are not provided to develop resistance to rotation. Shear plates are not required around these bolts. In the crown connection shown in (h), a bolt with countersunk ends may be used instead of a dowel because of the acuteness of the angle. Therefore, splice plates are not required to hold the ends together horizontally.

A simple connection of a leg to a footing is illustrated in Figure 5.30(e). If the soil has inadequate resistance to carry the horizontal components of the end thrust of the frames, a steel rod may be provided to tie appropriately designed bases of the two legs together and thus to resist the equal and opposite horizontal components. This tie is placed below the floor level and is embedded in concrete to protect the rod from corrosion. This member is called a *tie beam*.

A two-hinged glued laminated rigid frame is illustrated in Figure 5.31. It is similar in construction to the three-hinged frame in Figure 5.30(a) and (d), but it is constructed in three sections, as shown in the figure. These are spliced together on the job. The splices are located at points where bending moments have minimum values. However, since various combinations of dead load, unbalanced snow load, and wind loads from any direction must be provided for, a splice

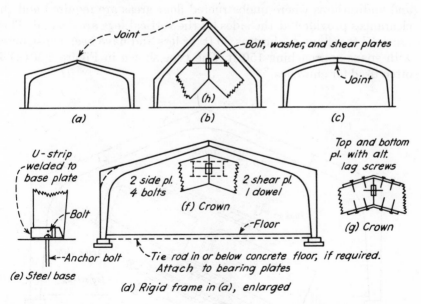

FIGURE 5.30 Three-hinged glued laminated rigid frames. Partially from Unit Structures, Inc.

in any location must always be designed to resist bending moments of considerable magnitude.

One effective form of splice is shown in Figure 5.31. The bent plate is nailed to the leg at the factory. Top and bottom splice plates receive the lag screws which attach these plates. Shear plates are embedded in the members, and holes are bored to receive the screws. These operations are accomplished at the factory. The shearing resistance of the splice may be increased by means of dowels and shear plates (Figure 5.30(g)). The three sections may be assembled flat on the floor, or they may be placed in position separately. If the latter procedure is used, the legs are erected first. After they are in position, the center beam is hoisted and lowered into place on the projecting legs of the bent plates. It is called a *drop-in beam.* It is supported while the lag screws in the splice plates are driven and the splice is secured.

The roof decks are usually constructed of heavy matched planks spanning the distance between frames or between purlins supported by the frames. The underside of the deck is usually exposed and is appropriately finished.

Glued laminated rigid frames are attractive in appearance. They are used for many kinds of buildings such as gymnasiums, churches, and auditoriums where unobstructed floor areas are required and the clearances provided at the sides by the vertical legs are desired. They are used for spans as short as 30 ft. or less and have been constructed with spans approaching 150 ft. The type shown in Figure 5.30(b) is often used for churches.

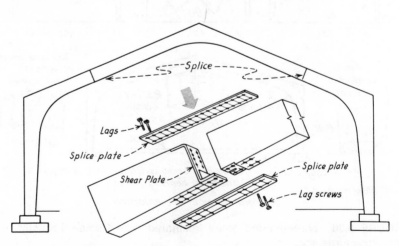

FIGURE 5.31   Detail of splice. Unit Structures, Inc.

Arches. In contrast to rigid frames, arches are so proportioned that the stresses produced by the loads are primarily compressive and the shears and flexural stresses are relatively small. Flexural stresses must be considered, however, in the design of the arch section, splices, and supports. The arch ring commonly has a constant radius and constant cross section, although rings with parabolic profiles and variable cross sections are sometimes used. Glued laminations are usually ¾ or 1½ in., the actual thickness depending upon the species of lumber used and the curvature. The arches may be fixed, two-hinged, or occasionally three-hinged for the longer spans.

A glued laminated arch supported on piers is diagrammed in Figure 5.32(a). Vertical sidewalls may be provided as shown by the inner dashed lines. If such sidewalls are constructed, the projecting ends of the arches may be exposed and treated to resist the weather, or they may be incorporated in building space. An arch may be elevated (b) and supported on rigid reinforced concrete buttresses or on walls or columns with the horizontal thrusts at the ends carried by steel tie

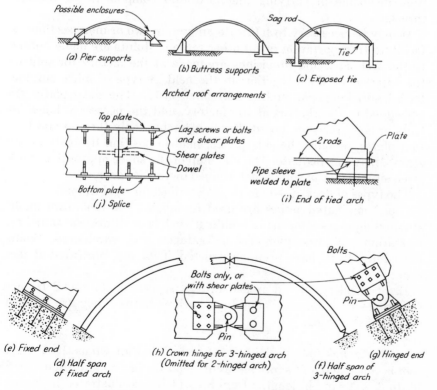

FIGURE 5.32   Glued laminated timber arches.

rods (c). Sagging of the rods is prevented by vertical *sag rods*, or straps. The arches shown in (a) and (b) may have fixed or hinged ends, but the ends of the tied arch in (c) are always considered to be hinged.

An enlarged drawing of half the arch shown in Figure 5.32(a) or (b), with fixed end supports, is illustrated in (d), and a detail of the support is shown in (e). A hinged end support for these arches is illustrated in the half-arch in (f) and the detail in (g). They are usually built up of steel plates welded together. If these arches are three-hinged, there is a hinge at the crown as shown in (f) and the detail in (h). An end support for the tied arch in (c) is illustrated in (i). This would be attached to a column or bearing wall with appropriate provision made for stability. The horizontal components of the end thrusts are carried by the *tie rods*. To prevent them from sagging under their own weight, *sag rods* may be used as shown in (c). Tied arches may also be used if the end supports are very near the ground level. Such an arrangement would only be used if the foundation soil were incapable of carrying the horizontal components of the end thrusts.

It may be necessary to fabricate an arch in two or more sections to facilitate transportation or for other reasons. Joints are made normal to the arch axis, and provisions are made at the factory for splicing the adjacent sections together in the field. A type of splice suitable for a heavy long-span arch is illustrated in (j). The shear plates are embedded in the timbers at the factory, and the necessary holes for bolts or lag screws are bored. For light arches, only the top and bottom plates, or only the side plates, held in position by bolts or lag screws, may be adequate. The make-up of the splice appropriate for a given situation is, of course, determined by the designer.

The types of decks used are those described for rigid frames.

Glued laminated arches are used to support the roofs over unobstructed floor areas for many buildings such as auditoriums, churches, recreation buildings, gymnasiums, garages, and warehouses. Spans exceeding 240 ft. have been constructed. They are fabricated at factories.

Light glued laminated arched rafters, spaced as close as 2 ft., are sometimes used for small buildings such as barns. They are covered with 1 in. sheathing.

Lamella Arches.   A special form of arched roof known as the *lamella roof* is constructed of short pieces of wood varying in size from 2 × 8 in. to 3 × 16 in., in lengths from 8 to 14 ft., as illustrated in Figure

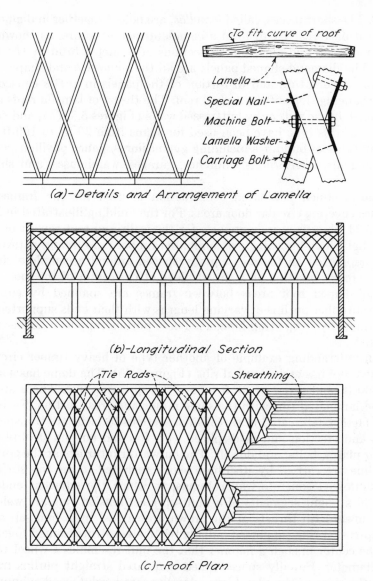

To fit curve of roof

Lamella
Special Nail
Machine Bolt
Lamella Washer
Carriage Bolt

(a)—Details and Arrangement of Lamella

(b)—Longitudinal Section

Tie Rods   Sheathing

(c)—Roof Plan

FIGURE 5.33   Wood lamella arch roof. (*a*) Details and arrangement of lamella. (*b*) Longitudinal section. (*c*) Roof plan.

5.33. The short pieces, called *lamellas*, are bolted together in diamond-shaped patterns (*a*) to form a complete roof structure, as shown by the roof plan and longitudinal sections of a simple form in (*b*) and (*c*). The diamond-shaped panels are all the same size and shape. The apparent curvature and distortion of the panels in (*c*) is caused by the changing slope of the arched roof. The thrust of lamella roofs may be taken by tie rods or by buttressed walls (Figures 5.32(*b*), and (*c*)). Roofs of this type have been used for spans from 25 ft. to 150 ft. on many types of buildings including gymnasiums, dance pavilions, exhibition halls, auditoriums, churches, garages, warehouses, and sheds.

Domes.   Glued laminated construction is used for the frames of domes covering circular floor areas. For the building illustrated in Figure 5.34, the structural supports for the walls and roof consist of sixteen half three-hinged, glued laminated timber rigid frames with their legs supported on footings equally spaced around a complete circle, their crowns meeting on a thrust block at the center of the roof. The wedge-shaped roof areas between frames are spanned by equally spaced timber purlins of varying lengths with their ends supported on the frame by steep strap hangers. These purlins support a timber roof deck.

An outstanding example of another type of heavy timber circular domed roof has arched radial ribs (Figure 5.35). The dome has a span of 300 ft. and a rise of 51 ft. above the springing line. There are 36 glued laminated ribs with the same radius and cross section. Each rib was factory-fabricated in three equal sections to facilitate transportation and erection and field-spliced with top and bottom steel plates, shear plates, bolts, and dowels. The outward rib thrusts at the springing line are carried by steel base shoes anchored to the top surfaces of reinforced concrete wall columns, to which the adjacent ends of the 36 straight segments of a built-up steel tension ring are welded. The inward rib thrusts at the crown are carried by a built-up steel compression ring supported radially by 18 steel struts framed together at the center in such a manner that the unit resembles a wheel 18 ft. in diameter. Equally spaced glued laminated straight purlins frame into the sides of the ribs. Light, closely spaced, solid wood subpurlins are attached to the tops of the purlins and are normal to them. These support 3 in. wood-fiber concrete roof deck panels. Each of the 36 sectors between ribs has sets of diagonal steel-strap cross bracing, with one system located immediately above and one immediately below the subpurlins. The ends of the sets of cross bracing in adjacent segments and in each layer are welded together to form steel nets

FIGURE 5.34 Domed roof with glued laminated rigid frame supports. Span 96 ft. Cocoa High School, Cocoa, Florida. Unit Structures, Inc., Designers and Manufacturers.

enveloping the entire dome. The 36 columns are connected by horizontal reinforced concrete girts, and brick panel walls are constructed between columns to provide lateral support to the columns and the structure as a whole. For a more detailed description and the erection procedure, see reference 3. This type of construction is suitable for smaller structures.

A dome with wood lamella construction is illustrated in Figure 5.36. The outside diameter is 142 ft., and the rise is 18½ ft. A reinforced concrete tension ring is located around the perimeter of the dome and

FIGURE 5.35   A 300-ft diameter dome with 36 glued laminated wood 7 by 16¼ in. ribs. Montana State University Field House. Architects: Wilson and Berg, Jr. Structural Engineer: Ben F. Hurlbutt. Fabricator and Erector: Timber Structures, Inc.

on top of reinforced columns spaced about 12 ft. apart. Alternate panels between columns are filled with masonry walls. A crown block is located at the center of the dome to carry the thrusts of the lamella which frame into it. The lamella vary in cross section from 4 by 16 in. for those framing into the tension ring to 2 by 12 in. at the crown. The lamella are connected in the same manner shown for the arch roof in Figure 5.37. The roof-deck is matched sheathing varying in thickness from 2 in. for the lower portion of the dome to 1 in. for the portion surrounding the crown.

The pattern of the lamella in Figure 5.36 is shown in Figure 5.37(b). In this pattern, the number of lamellas for each concentric passing through the lamella intersections remains constant. The lamellas terminate into collective radial ribs at the crown.

Another pattern is shown in (a). For this pattern, the 360° angle at the crown is divided into from 4 to 12 units, permitting all the lamellas and ring purlins to have the same radius and cross section.

FIGURE 5.36 A 140 ft. wood lamella dome roof. Ham, Hanover, and Williams, Architects. Roof Structures, Inc., Webster Groves, Missouri, Manufacturers and Erectors.

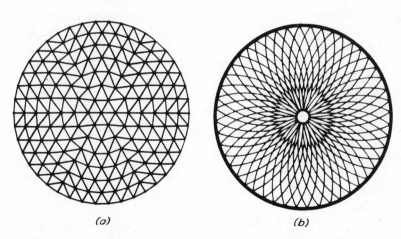

(a)   *(b)*

FIGURE 5.37 Lamella dome patterns. Lamella Roof Associates.

REFERENCES AND RECOMMENDED READING

1. *Wood Handbook,* Agriculture Handbook 72, Forest Products Laboratory, U. S. Department of Agriculture, Washington, D. C., 1955.
2. *Typical Designs of Timber Structures,* Timber Engineering Co., Washington, D. C.
3. "Biggest Dome Spans 300 Feet," *Engineering News-Record,* January 10, 1957, p. 32.
4. *Manual for House Framing,* National Lumber Manufacturers Association, Washington, D. C.
5. *Plank-and-Beam Framing for Residential Buildings,* National Lumber Manufacturers Association, Washington, D. C.
6. *Plank-and-Beam System for Residential Construction,* Construction Aid 4, Housing and Home Finance Agency, Division of Housing Research, Washington, D. C., 1953.
7. *Heavy Timber Construction Details,* National Lumber Manufacturers Association, Washington, D. C.
8. *National Building Code,* National Board of Fire Underwriters, 1955.
9. *ASTM 1971 Annual Book of Standards, Part 16,* American Society for Testing and Materials, Philadelphia, Pennsylvania.
10. *Sweet's Catalog,* 1971 Edition, F. W. Dodge Corp., New York, New York.
11. *Standard Specifications for Grades of California Redwood Lumber,* Redwood Inspection Service, San Francisco, California, 1970.
12. *Grading Rules for West Coast Lumber #16,* West Coast Lumber Inspection Bureau, Portland, Oregon, 1970.
13. *Timber Construction Manual,* First Edition, American Institute of Timber Construction, John Wiley and Sons, 1966.
14. *American Softwood Lumber Standard, NBS Voluntary Product Standard, PS 20–70,* U. S. Department of Commerce, National Bureau of Standards, U. S. Government Printing Office, Washington, D. C., 1970.
15. *Mosaic-Parquet Slat Flooring, NBS Voluntary Product Standard PS 27–70,* U. S. Department of Commerce, National Bureau of Standards, U. S. Government Printing Office, Washington, D. C., 1970.
16. *Use and Abuse of Wood in House Construction,* Miscellaneous Publication #358, U. S. Department of Agriculture, Washington, D. C., 1939.

17. *Wood Decay in Houses*, Home and Garden Bulletin No. 73, U. S. Department of Agriculture, Washington, D. C., 1967.
18. *Floors*, U. S. Post Office Department, 1962.
19. *Guide to Plywood Sheathing for Floors, Walls and Roofs*, American Plywood Association, Tacoma, Washington, 1966.
20. *Cellon*, A New Pressure Process to Preserve Lumber and Plywood, Koppers Company, Inc., Pittsburgh, Pennsylvania, 1961.
21. *Characteristics of Hardwood*, Appalachian Hardwood Manufacturers, Inc., Cincinnati, Ohio.
22. Richard J. Preston, Jr., *North American Trees*, Second Edition, The Iowa State University Press, Ames, Iowa, 1961.
23. Charles G. Ramsey, and Harold R. Sleeper, *Architectural Graphic Standards*, Sixth Edition, John Wiley and Sons, 1970.
24. *Rules for Measurement and Inspection of Hardwood Dimension Parts, Hardwood Interior Trim and Moldings, Hardwood Stair Treads and Risers*, Fifth Edition, Hardwood Dimension Manufacturers Association, Nashville, Tennessee, 1961.
25. *The Fundamentals of Good Lumber Manufacturing and Grading Practice*, West Coast Lumber Inspection Bureau, Portland, Oregon, 1971.
26. *Rules for the Measurement and Inspection of Hard and Cypress Lumber*, National Hardwood Lumber Association, Chicago, Illinois, 1971.
27. *Standard Grading Rules for Southern Pine Lumber*, Southern Pine Inspection Bureau, Pensacola, Florida, 1970.
28. *Grading Rules for Northeastern Lumber*, Northeastern Lumber Manufacturers Association, Inc., Glens Falls, New York, 1970.
29. Catalog: *Products for Building and Construction*, Koppers Company, Inc., Pittsburgh, Pennsylvania.
30. *Timber Design and Construction Handbook*, Timber Engineering Company, National Lumber Manufacturers' Association, McGraw-Hill, 1956.
31. Virginia Polytechnic Wood Research Laboratory, *Better Utilization of Wood through Assembly with Improved Fasteners*, Blacksburg, Virginia, 1959. Sponsored by Independent Nail and Pack Company.
32. German Gurfinkel, *Wood Engineering*, Southern Forest Products Association, New Orleans, Louisiana, 1973.
33. *Uniform Building Code*, 1970 Edition, Vol I, International Conference of Building Officials, Whittier, California.
34. *The National Building Code*, 1967 Edition, Recommended by the

American Insurance Association, New York, New York.

35. *U. S. Product Standard PS 1–66 for Softwood Plywood* together with DFPA grade-trademarks, American Plywood Association, Tacoma, Washington, 1970.

36. *Plywood Construction Guide for Residential Building*, American Plywood Association, Tacoma, Washington, 1967.

37. *Plywood Construction Systems*, American Plywood Association, 1970.

38. *Plywood Design Specification*, American Plywood Association, 1969.

39. *Plywood Specification Guide*, American Plywood Association, 1970.

40. *Guide to Plywood Underlayment*, American Plywood Association, Tacoma, Washington, 1969.

41. *Plywood Excerpts from Minimum Property Standards of the Federal Housing Administration*, American Plywood Association, Tacoma, Washington, 1967.

42. *Guide to Plywood Grades under U. S. Product Standard PS 1 for Softwood/Plywood Construction and Industrial*, American Plywood Association, Tacoma, Washington, 1970.

43. *Guide to Plywood for Siding*, American Plywood Association, Tacoma, Washington, 1970.

44. *Plywood Components*, American Plywood Association, Tacoma, Washington, 1967.

45. *Inspection Manual AITC 200–63* and "Standard Specifications for Structural Glued Laminated Timber of Douglas Fir, Western Larch, Southern Pine and California Redwood AITC 203–70," American Institute of Timber Construction, Englewood, Colorado, 1970.

46. *Douglas Fir Use Book*, West Coast Lumbermen's Association, Portland, Oregon, 1958.

47. *Plywood Concrete Forms*, American Plywood Association, Tacoma, Washington.

6

STEEL CONSTRUCTION

35. STEEL MATERIALS

Compared with Other Ferrous Products.   The primary object of the various steel manufacturing processes is to reduce the amount of carbon present in the pig iron and to control the amounts of phosphorus, sulfur, manganese, silicon, and other ingredients of steel manufacture.

Steel differs in physical properties from pig iron and from cast iron by possessing more ductility and malleability and less brittleness. It contrasts with malleable cast iron by having malleability without treatment after casting. Steel differs in chemical composition from pig iron, cast iron, and malleable iron principally by having a much lower percentage of carbon but also by the smaller amounts of manganese, silicon, and phosphorus present. Low carbon steel and wrought iron have similar chemical composition; but in wrought iron, the iron itself contains a smaller percentage of impurities because a part of the impurities are in the slag, which is mechanically mixed with iron.

Low-carbon steel differs from wrought iron chiefly by the process of manufacture and structure, rather than any great difference in other physical properties. Wrought iron is no longer produced.

Effect of Composition on Properties.   Carbon has the most pronounced effect on the physical properties of steel. The amount of carbon present in steel may vary from almost 0 to about $1\frac{1}{2}\%$. Increasing the amount of carbon increases the strength, hardness, and brittleness of steel but decreases its ductility. Steel may be classified according to carbon content approximately as follows.

	Carbon Content,%
Low-carbon steel	0.06–0.30
Medium-carbon steel	0.30–0.50
High-carbon steel	0.50–0.80

No clear demarcation exists between the various grades so these limits are subject to considerable variation.

Silicon, in the amounts usually found in steel, has little effect on its properties; but when present in amounts as high as 0.3 or 0.4%, it increases the strength without a sacrifice in ductility.

Sulfur has little effect on the strength or ductility of steel, but it makes steel brittle and likely to crack when worked in red heat. This property is called *red shortness*. The maximum amount of sulfur permitted by specifications for structural steel is about 0.05%.

Phosphorus is very objectionable because it causes brittleness at ordinary temperatures, or *cold short*. The maximum amount of phosphorus permitted by specifications for structural steel is about 0.06%.

Manganese in amounts ordinarily present is beneficial to steel, but its action is too complex to consider here.

Effect of Mechanical Working.   Hammering, pressing, and rolling steel while it is hot tends to eliminate flaws. If the working is continued while the metal is cooled past a certain critical temperature, the steel will be *fine grained*. This results from the breaking up of the crystals and deterring their reformation.

The cold-working of steel increases the strength and elastic limit but decreases the ductility.

Effect of Heat Treatment.   Heating, annealing, and sudden cooling have very marked effects on the strength, ductility, and grain size of steel, but this subject is too complex for consideration here.

Processes of Manufacture.   The principal processes used in manufacturing steel are the *basic oxygen* process, the *electric furnace* process, the *vacuum melting* process, and the *open-hearth* process, which produce steel by purifying pig iron. The open-hearth process has replaced the Bessemer process.

Open-Hearth Process

The open-hearth process for manufacturing steel is carried out in an open-hearth furnace (Figure 6.1.). There are two open-hearth processes, the *acid* and the *basic*. The basic process will be discussed first.

The Basic Process.   The furnace consists essentially of a hearth in the central portion and two openings, called *ports*, at each end. The hearth is lined with calcined magnesite, with anhydrous tar as a binder. The tar burns to coke and becomes hard and firm. The capacity of the furnaces in use varies from 100 to 200 tons in 12 hours.

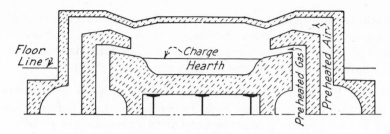

FIGURE 6.1 Open-hearth furnace.

Operation. In charging the furnace, pig iron, steel scrap, iron ore, and limestone are placed in the hearth through doors on one side of the furnace. The furnace is heated by introducing preheated gas through one port at one end of the furnace and preheated air through the other port at the same end. The gas and air ignite at the ports. The exhaust gases are drawn through both ports at the other end by natural draft. These gases pass through regenerators filled with checker brick and heat the brick to a high temperature. In about 20 minutes, valves are turned so that the direction of flow of the gases is reversed. The gas and air are preheated by passing through the regenerators that were heated by the exhaust gases. Regenerators are provided at each end of the furnace so that one set is always being heated by the exhaust gases while the other is heating the incoming gas and air. The direction of flow is changed at intervals of 15 or 20 minutes. The regenerators make it possible to heat the furnace to a temperature high enough to melt the charge in 4 or 5 hours. The process requires 6 to 12 hours to remove the impurities. Then the furnace is tapped through a tap hole, and the molten steel is run into ladles.

Removal of Impurities. The oxygen in the iron ore of the charge is the oxidizing agent that combines with the silicon, manganese, and phosphorus present in the charge, forming corresponding oxides which, with the limes of the charge, form the *slag*. The carbon present in the materials charged into the furnace is oxidized by the oxygen of the iron ore, forming carbon dioxide gas, which leaves the furnace with the other exhaust gases. The elimination of sulfur is uncertain.

Sources of Heat. Some heat is supplied by the oxidation of the impurities; but most of the heat is obtained from the combustion of the fuel, which may be gas, fuel oil, or powdered coal.

Recarburizing.   The process is always continued until the amount of carbon present is less than the amount desired in the finished steel. The additional carbon is supplied by adding ferromanganese with coal, charcoal, and coke to the metal in the ladle.

The Acid Process.   The process just described is the basic process because the slag formed is basic in character and requires a basic lining for the hearth in order to prevent any chemical action between the slag and the lining. Phosphorus can be removed by the basic process but not by the acid process; because all the ores in the U.S. contain phosphorus in objectionable amounts, the acid process is of very little use.

Electric Furnace.   There are two types of electric furnace, the *electric arc furnace* and the *induction furnace*. The former produces an electric arc between two electrodes which melt the charge. The charge is usually scrap iron. This method of steel manufacture, which can be highly controlled, is used to produce steel that requires special alloys, such as stainless steel, surgical steel, spring steel, and other quality or specialty steels.

The induction furnace has an electrically charged copper coil which surrounds a round converter. The charge is melted and purified by the heat produced by a high frequency current. For a descriptive illustration of this process, see Figure 6.2.

Vacuum Melting and Degassing Process.   Steels which are exceptionally pure and of extremely high grade are produced by placing an electric furnace (either arc or induction) in a vacuum chamber. The gaseous impurities are removed by the vacuum, leaving a steel of exceptional physical qualities. This steel is used for the aerospace industry. See Figure 6.3 for an illustrative flow diagram of this process.

Basic Oxygen Process.   This process, which originated in Austria, employs oxygen of high purity. Oxygen is shot under extremely high pressure through a water-cooled, retractable lance that is lowered into the converter to a position just above the molten steel. The impurities of the iron are burned off. This process greatly accelerates the steel-making process. As much as 300 tons of steel may be processed in less than an hour, as compared to the open hearth process, which may require up to 8 hours to produce a comparable quantity and quality of steel. This process is shown in Figure 6.4.

Properties and Uses.   As stated, the strength, ductility, and other properties of steel are affected by its composition, the mechanical treatment it receives when it is hot and cold during processing, and many other factors. In general, it may be classed as a relatively high-strength and low-cost metal. Its most significant defect is its lack of resistance to corrosion.

Steel is the most used structural metal in building construction. Its uses are so numerous and so well known that it seems unnecessary to mention them here, but many of them are considered in detail in subsequent articles.

Alloy Steels

Various alloying elements are introduced into steel to change the properties, such as increasing the strength and resistance to rust. Included in these elements are silicon, manganese, copper, nickel, chromium, tungsten, molybdenum, and vanadium. They may be introduced singly, or two or more may be used. Such steels are called alloy steels.

A common steel used for curtain walls, decorative purposes, sinks and counter tops, and other building purposes is stainless steel, which resists corrosion, takes a high polish, and is attractive. Some types are made by adding from 10 to 30% of chromium to steel. In others, from 5 to 15% of nickel also is included.

The melting points of various steels vary from 2600° to 2800°F. Their weight is 490 lb. per cu. ft. Small amounts of copper, ranging from 0.15 to 0.30%, introduced in steel increases its resistance to atmospheric corrosion without having significant effect on its other properties. Copper-bearing steel is extensively used for sheet-metal products.

Structural Steels. Structural steels are identified by the ASTM designation number that specifies the steel. The most commonly used structural steel for buildings and general construction is designated *A 36* steel. It is referred to as a *carbon steel.*

There are several steels which have greater structural abilities that are known as *high strength steels.* A structural steel which is known as *A 242* is known as *High Strength Low Alloy* structural steel. This steel is used where smaller members are required, or where a savings in the lesser weight would realize a more economical structure. Additionally, this steel has increased resistance to corrosion. It may be riveted, bolted, or welded. Its use is confined to members that have a thickness 2 in. or less.

Another exceptionally high strength steel, identified as *A 440,* has

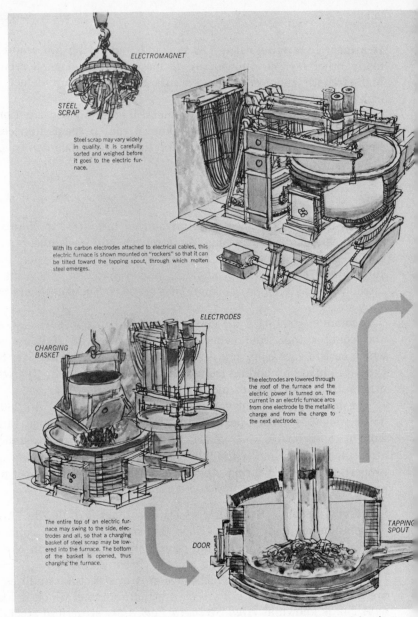

ELECTROMAGNET

STEEL
SCRAP

Steel scrap may vary widely
in quality. It is carefully
sorted and weighed before
it goes to the electric furnace.

With its carbon electrodes attached to electrical cables, this
electric furnace is shown mounted on "rockers" so that it can
be tilted toward the tapping spout, through which molten
steel emerges.

ELECTRODES

CHARGING
BASKET

The electrodes are lowered through
the roof of the furnace and the
electric power is turned on. The
current in an electric furnace arcs
from one electrode to the metallic
charge and from the charge to
the next electrode.

The entire top of an electric furnace may swing to the side, electrodes and all, so that a charging
basket of steel scrap may be lowered into the furnace. The bottom
of the basket is opened, thus
charging the furnace.

DOOR

TAPPING
SPOUT

FIGURE 6.2 Electric furnace process. American Iron and Steel Institute.

ELECTRIC FURNACE STEELMAKING

...ng-deserved reputation for producing alloy, stainless, tool, and ...r specialty steels belongs to America's electric furnaces. Opera- ...have also learned to make larger heats of carbon steels in these ...aces; this development helps account for the record tonnage ...uts of recent years.

...heat within the electric furnace is intense and rigidly controlled. ...ern electric furnaces have top sections that can be moved away ...hat special containers can charge scrap into them from above. ...etimes pig iron is also charged and prereduced iron ore, in vari- ...forms, is rich enough in iron to be used as an electric furnace ...making charge.

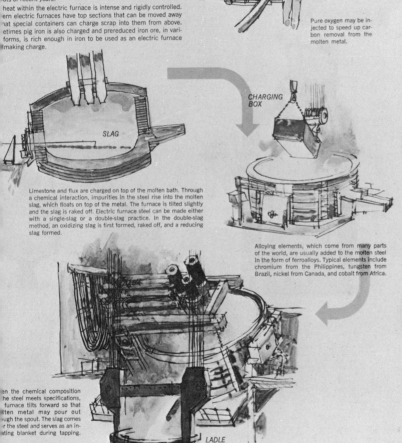

Pure oxygen may be in-jected to speed up car-bon removal from the molten metal.

SLAG

CHARGING BOX

Limestone and flux are charged on top of the molten bath. Through a chemical interaction, impurities in the steel rise into the molten slag, which floats on top of the metal. The furnace is tilted slightly and the slag is raked off. Electric furnace steel can be made either with a single-slag or a double-slag practice. In the double-slag method, an oxidizing slag is first formed, raked off, and a reducing slag formed.

Alloying elements, which come from many parts of the world, are usually added to the molten steel in the form of ferroalloys. Typical elements include chromium from the Philippines, tungsten from Brazil, nickel from Canada, and cobalt from Africa.

...en the chemical composition ...he steel meets specifications, ...furnace tilts forward so that ...lten metal may pour out ...ugh the spout. The slag comes ...r the steel and serves as an in-...ating blanket during tapping.

LADLE

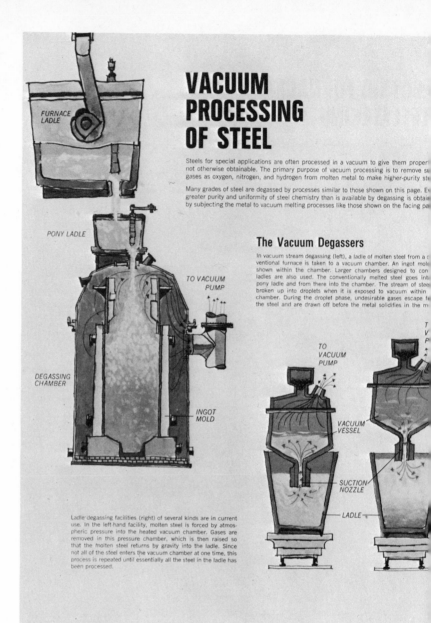

VACUUM PROCESSING OF STEEL

Steels for special applications are often processed in a vacuum to give them proper[?] not otherwise obtainable. The primary purpose of vacuum processing is to remove su[?] gases as oxygen, nitrogen, and hydrogen from molten metal to make higher-purity ste[?]

Many grades of steel are degassed by processes similar to those shown on this page. Ev[?] greater purity and uniformity of steel chemistry than is available by degassing is obtai[?] by subjecting the metal to vacuum melting processes like those shown on the facing pa[?]

The Vacuum Degassers

In vacuum stream degassing (left), a ladle of molten steel from a c[?] ventional furnace is taken to a vacuum chamber. An ingot mol[?] shown within the chamber. Larger chambers designed to con[?] ladles are also used. The conventionally melted steel goes int[?] pony ladle and from there into the chamber. The stream of stee[?] broken up into droplets when it is exposed to vacuum within [?] chamber. During the droplet phase, undesirable gases escape f[?] the steel and are drawn off before the metal solidifies in the m[?]

FURNACE LADLE

PONY LADLE

TO VACUUM PUMP

DEGASSING CHAMBER

INGOT MOLD

TO VACUUM PUMP

VACUUM VESSEL

SUCTION NOZZLE

LADLE

Ladle degassing facilities (right) of several kinds are in current use. In the left-hand facility, molten steel is forced by atmospheric pressure into the heated vacuum chamber. Gases are removed in this pressure chamber, which is then raised so that the molten steel returns by gravity into the ladle. Since not all of the steel enters the vacuum chamber at one time, this process is repeated until essentially all the steel in the ladle has been processed.

FIGURE 6.3 Vacuum melting and degassing process. American Iron and Steel Institute.

320

The Vacuum Melters

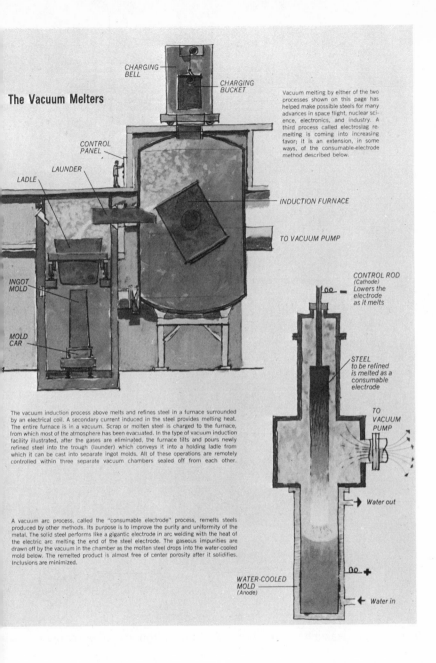

CHARGING BELL

CHARGING BUCKET

CONTROL PANEL

LAUNDER

LADLE

INGOT MOLD

MOLD CAR

INDUCTION FURNACE

TO VACUUM PUMP

CONTROL ROD
(Cathode)
Lowers the electrode as it melts

STEEL
to be refined is melted as a consumable electrode

TO VACUUM PUMP

Water out

WATER-COOLED MOLD
(Anode)

Water in

Vacuum melting by either of the two processes shown on this page has helped make possible steels for many advances in space flight, nuclear science, electronics, and industry. A third process called electroslag remelting is coming into increasing favor; it is an extension, in some ways, of the consumable-electrode method described below.

The vacuum induction process above melts and refines steel in a furnace surrounded by an electrical coil. A secondary current induced in the steel provides melting heat. The entire furnace is in a vacuum. Scrap or molten steel is charged to the furnace, from which most of the atmosphere has been evacuated. In the type of vacuum induction facility illustrated, after the gases are eliminated, the furnace tilts and pours newly refined steel into the trough (launder) which conveys it into a holding ladle from which it can be cast into separate ingot molds. All of these operations are remotely controlled within three separate vacuum chambers sealed off from each other.

A vacuum arc process, called the "consumable electrode" process, remelts steels produced by other methods. Its purpose is to improve the purity and uniformity of the metal. The solid steel performs like a gigantic electrode in arc welding with the heat of the electric arc melting the end of the steel electrode. The gaseous impurities are drawn off by the vacuum in the chamber as the molten steel drops into the water-cooled mold below. The remelted product is almost free of center porosity after it solidifies. Inclusions are minimized.

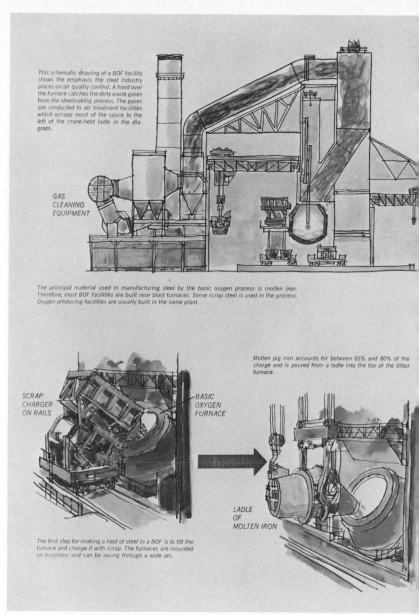

This schematic drawing of a BOF facility shows the emphasis the steel industry places on air quality control. A hood over the furnace catches the dirty waste gases from the steelmaking process. The gases are conducted to air treatment facilities which occupy most of the space to the left of the crane-held ladle in the diagram.

GAS
CLEANING
EQUIPMENT

The principal material used in manufacturing steel by the basic oxygen process is molten iron. Therefore, most BOF facilities are built near blast furnaces. Some scrap steel is used in the process. Oxygen producing facilities are usually built in the same plant.

Molten pig iron accounts for between 65% and 80% of the charge and is poured from a ladle into the top of the tilted furnace.

SCRAP
CHARGER
ON RAILS

BASIC
OXYGEN
FURNACE

LADLE
OF
MOLTEN IRON

The first step for making a heat of steel in a BOF is to tilt the furnace and charge it with scrap. The furnaces are mounted on trunnions and can be swung through a wide arc.

FIGURE 6.4   Basic oxygen process. American Iron and Steel Institute.

SIC OXYGEN STEELMAKING

...ca's capability to produce steel by the basic oxygen process has grown ...ously from small beginnings during the middle 1950's. The high ton-... f steel now made in basic oxygen furnaces—commonly called BOF's— ...es the consumption of large amounts of oxygen to provide operational ...d to promote the necessary chemical changes. No other gases or fuels ...ed.

...sic oxygen process produces steel very quickly compared with the other ...methods now in use. For example, a BOF may produce up to 300-ton ...s in 45 minutes as against 5 to 8 hours for the older open hearth process. ...grades of steel can be produced in the refractory-lined, pear-shaped ...es.

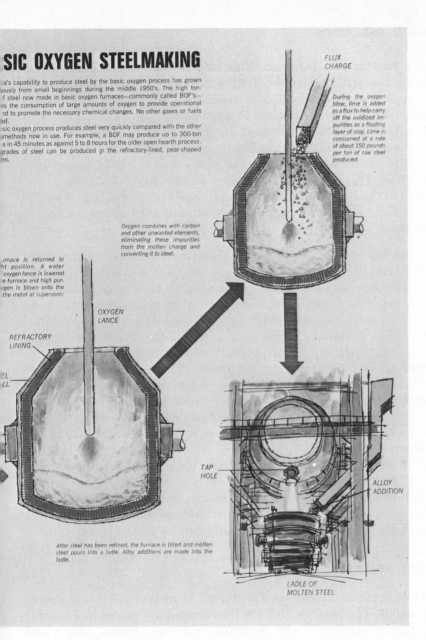

FLUX
CHARGE

During the oxygen blow, lime is added as a flux to help carry off the oxidized impurities as a floating layer of slag. Lime is consumed at a rate of about 150 pounds per ton of raw steel produced.

Oxygen combines with carbon and other unwanted elements, eliminating these impurities from the molten charge and converting it to steel.

...urnace is returned to ...ht position. A water ... oxygen lance is lowered ... furnace and high pur-...ygen is blown onto the ...the metal at supersonic

OXYGEN
LANCE

REFRACTORY
LINING

...EL
...LL

TAP
HOLE

ALLOY
ADDITION

After steel has been refined, the furnace is tilted and molten steel pours into a ladle. Alloy additions are made into the ladle.

LADLE OF
MOLTEN STEEL

a high resistance to corrosion. This steel is used where weight reduction in the structure is required while maintaining high stresses. Although defects may be welded (welding is not recommended), this steel should be riveted or bolted at connections.

Similar to A 440 in physical qualities is *A 441, High Strength Manganese Vanadium Steel*. This material, which has comparable strength qualities to A 440, is used where weldability is a chief consideration. This steel also has high resistance to deterioration.

Two other high strength structural steels are identified as A 572 and A 588. The latter steel is primarily intended where welding is required.

Weathering Steel. For years the corrosion deterrence of structural steel, especially at inaccessible places, has been a maintenance problem. This has been alleviated by the use of a bare, unprotected steel which acquires a protective oxide coat that deters rust and corrosion. This steel, called *weathering steel*, soon acquires a sepia-toned patina that eventually ripens into a deep brownish-mauve. During this aging, there is a small amount of the oxide residue which is carried off by rainwater and wind erosion. This residue deposits upon surfaces directly below the steel, leaving a stained coat upon those materials that are stainable, such as stucco and concrete. Weathering steel should not be used above permeable surfaces where appearances are important unless provision is made for protection and drainage of the run-off.

This steel is ideal for use in bridges, towers, and other structures where inaccessibility for maintenance is a factor. For the effect of its rich color, its maintenance-free quality, and its forthright honesty of structure, entire buildings of architectural character have been built using this steel exposed on both interior and exterior.

36. STEEL SHAPES

After the iron ore has been converted to molten steel, it is poured into molds. The ingots which are thus formed are placed in a *soaking pit*, then processed in the *blooming mill* into blooms, billets, and slabs. *Blooms* are square-like members having an area exceeding 36 sq. in. A *billet* is somewhat flatter, with a cross-sectional area between 16 and 36 sq. in. The flat-like *slab* has a thickness not less than 1½ in. For a description of the process, see Figure 6.5.

After leaving the blooming mill, the steel is sent to the rolling mill, where the blooms are shaped into structural shapes; the billets, into rods, bars, and tubes; and the slabs are rolled into plates, sheets, and strips.

In the rolling mill, the blooms are run through a series of rollers which gradually shape the hot steel into its finished form. Figure 6.5 indicates the steps that would be required for shaping a wide-flange member.

Various rolled shapes or sections form columns, beams, girders, and other structural units shown in Figure 6.6. The parts of these shapes are designated as flanges, webs, and stems.

The *wide-flange*, called *W-shapes* or *sections*, vary in depth from a maximum of 36 in. to a minimum of 4 in. and weigh from 7½ lbs. to 730 lbs. per linear foot. The inner faces of these flanges are parallel to the outer face, or the inner face may have a slope of 1 to 20, depending upon the manufacturer. There are two general forms of wide-flange shapes; those relatively deep and narrow (*a*) are more suitable for beams and girders, and those more nearly square (*b*) are desirable for column sections.

The *American standard beams*, called *I-beams* and identified as *S-shapes* (*c*), vary in depth and in weight per foot from a maximum of 24 in. and 120 lb. to a minimum of 3 in. and 5.7 lb. The inner face of their flanges has a slope of 1 in 6.

The *American standard channels* and miscellaneous channels, called *channels* (*d*), vary in depth and in weight per foot from a maximum of 18 in. and 58 lb. to a minimum of 3 in. and 4.1 lb. The inner face of the flanges of these sections has a slope of 1 in 6.

In addition to the wide-flange and standard shapes mentioned, light beams, joists, columns, and standard mill beams, called *miscellaneous M-shapes*, are used.

Angles are divided into those with equal legs and those with unequal legs (*e*). Equal-leg angles vary in size, thickness, and weight per foot from a maximum 8 × 8 × 1⅛ in. and 56.9 lb. to 1 × 1 × ⅛ in. and 0.80 lb. Unequal-leg angles vary in size from 8 × 6 × 1 in. and 44.2 lb. to 1¾ × 1¼ × ⅛ in. and 1.23 lb.

Tees (*f*) are commonly made by splitting the webs of wide-flange or American standard beam shapes, and thus are available in sizes which correspond to the sections from which they are cut. These are called *structural tees*, indicated as *WT*, and are the ones ordinarily used in structural framing.

Square and rectangular tubular shapes are rolled in several sizes and wall thickness. Square tubes are available with 2 × 2 in. through 10 × 10 in. outside dimensions. The wall thicknesses vary from about ³⁄₁₆ in. for the lightest 2 in. tube to ⅝ in. for the heaviest 10 in. tube.

The raw materials of steelmaking mus
brought together, often from hundr
of miles away, and smelted in a b
furnace to produce most of the iron
goes into steelmaking furnaces. Air
oxygen are among the most impor
raw materials in iron and steelmak

PELLETS

IRON ORE

SINTER

BLAST
FURNACE

LIMESTONE

CRUSHING

SCRAP OR
PREREDUCED ORE

BASIC
OXYGEN
FURNACE

MOLTEN IRON
TRANSFER CAR

COAL

COKE OVENS

SLAG

OPEN
HEARTH
FURNACE

COKE OVEN
BY-PRODUCTS

CASTING
PIG IRON

MIXER

ELEC
FURN

A FLOWLINE ON STEELMAKING

This is a simplified road map through the complex world of steelmaking. Each stop along
the routes from raw materials to mill products contained in this chart can itself be charted.
From this overall view, one major point emerges: Many operations—involving much equipment and large numbers of men—are required to produce civilization's principal and least
expensive metal.

FIGURE 6.5   The steel-making process. American Iron and Steel Institute.

Molten steel must solidify before it can be made into finished products by the industry's rolling mills and forging presses. The metal is usually formed first at high temperature, after which it may be cold-formed into additional products.

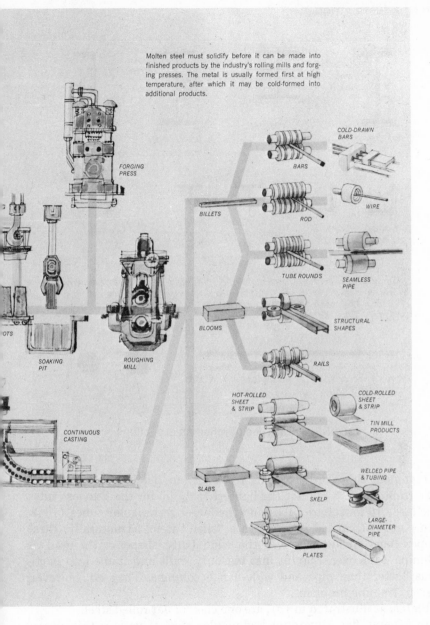

FORGING PRESS

COLD-DRAWN BARS

BARS

BILLETS

WIRE

ROD

TUBE ROUNDS

SEAMLESS PIPE

BLOOMS

STRUCTURAL SHAPES

SOAKING PIT

ROUGHING MILL

RAILS

HOT-ROLLED SHEET & STRIP

COLD-ROLLED SHEET & STRIP

TIN MILL PRODUCTS

CONTINUOUS CASTING

SLABS

WELDED PIPE & TUBING

SKELP

LARGE-DIAMETER PIPE

PLATES

INGOTS

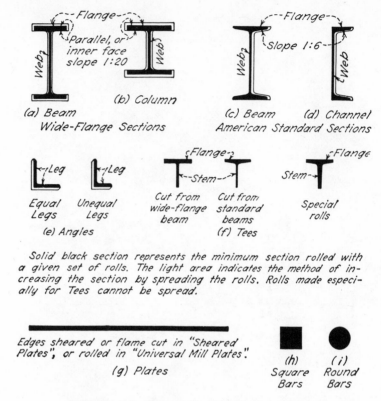

(a) Beam　　(b) Column
Wide-Flange Sections

(c) Beam　　(d) Channel
American Standard Sections

Equal　Unequal
Legs　Legs
(e) Angles

Cut from　Cut from
wide-flange　standard
beam　beams
(f) Tees

Special
rolls

Solid black section represents the minimum section rolled with a given set of rolls. The light area indicates the method of increasing the section by spreading the rolls. Rolls made especially for Tees cannot be spread.

Edges sheared or flame cut in "Sheared Plates", or rolled in "Universal Mill Plates".
(g) Plates

(h)　　(i)
Square　Round
Bars　Bars

FIGURE 6.6　Types of rolled-steel sections.

Rectangular tubes are 3 × 2 in. to 12 × 8 in. Their wall thicknesses vary from about $\frac{3}{16}$ in. for the lightest to $\frac{1}{2}$ in. for the heaviest tube. The large 10 in. square tubes and the larger rectangular tubes (12 × 8 in.) are not carried in stock as a "shelf" item. Structurally, these shapes have advantages over the wide-flange shapes when used as columns. Moreover, they fit into masonry walls and frame into partitions better than pipe and wide-flange columns. They do, however, present framing problems.

Plates, as illustrated in (g), are one class of flat rolled steel.

In general, flat, square, or rectangular steel of various thicknesses and widths are referred to as *plate* if thick and wide, as *bar* if thick and narrow, as *sheet* if thin and wide, and as *strip* if thin and narrow. There are, however, definite dimensions of these members which classify them. For structural steel, the classifications made by the AISC usually apply. See Table 6-1.

Plate is indicated by PL, and when billed or referred to on draw-

TABLE 6-1

Thickness (Inches)	Width (Inches)					
	To 3½ Inclusive	Over 3½ to 6	Over 6 to 8	Over 8 to 12	Over 12 to 48	Over 48
0.2300 and thicker	Bar	Bar	Bar	Plate	Plate	Plate
0.2299 to 0.2031	Bar	Bar	Strip	Strip	Sheet	Plate
0.2030 to 0.1800	Strip	Strip	Strip	Strip	Sheet	Plate
0.1799 to 0.0449	Strip	Strip	Strip	Strip	Sheet	Sheet
0.0448 to .00344	Strip	Strip				
0.0343 to 0.0255	Strip	Hot rolled sheet and strip not generally produced in these widths and thicknesses				
0.0254 and thinner						

*Courtesy of American Institute of Steel Construction, Inc.

ings, it is properly indicated by thickness × width × length, with all dimensions in inches—using fractions or decimals. If the width is 8 in. or less, they are usually classed as *bars*; if more than 8 in. they are *plates*. Plates which are more than 2 in. thick are called *bearing plates* because they are used to provide bearing areas for the ends of various members. Plates are available in thicknesses from ½ in. to 2 in., and widths from 8 in. to lengths of 15¼ ft. Bearing plates are available in sizes up to 56 in. and in thicknesses up to 10¼ in. For thicknesses up to 2 in., plates are sufficiently smooth and flat to use as bearing plates to receive accurately finished milled ends of columns, and plates up to 4 in. can be straightened in presses to the necessary flatness. The top surfaces of plates thicker than 4 in. should be planed over a sufficient area to receive milled-column sections if the plates are grouted on concrete foundations, but both faces should be planed if they rest on steel. Plates on which the edges are rolled when the plates are being rolled are called *universal mill plates*. Plates with sheared edges are called *sheared plates*.

Square and *round bars* (*h*) and (*i*) are used for ties, lateral bracing, sag rods, and hangers in structural-steel framing and as reinforcing in reinforced concrete structures. Many sizes are available.

The various weights available for the shapes (Figure 6.5) are obtained by spreading the rolls that produce the shapes. This process adds to the area of the shapes. The thickness of the flanges, as well as the depth of the wide-flange shapes, increases as the rolls are spread, but all the shapes in a given series are designated by a single nominal depth regardless of the actual depth. An increase in the flange thickness is desirable because a given amount of material added to the flanges increases the strength and rigidity of a member much more than the same amount of material added to the web. American standard beams and channel sections of a given depth are increased in weight and cross-sectional area by spreading the rolls in such a manner that all the additional material goes into the web, where it is not very effective in increasing the strength and rigidity of the section. The actual depth of all the shapes in a given depth series is the same and is equal to the depth by which the series of sections is designated. Spreading the rolls to increase the thickness of an angle also increases the lengths of the legs, but these increases are not taken into account in designating the size of the angle.

Structural-steel shapes are manufactured at steel mills. The operation of cutting the various shapes to the required size and fastening them together to form columns, girders, trusses, and other structural members is called *fabrication*. This work is done by fabricating plants. The individual units are shipped to the building site in sizes that can be transported and handled on the job, and are placed in position, or erected.

The individual sections are fastened together, in both fabrication and erection, by riveting, welding, or bolting, as described in Article 37.

Designation of Shapes.   The standard designations for structural steel shapes as adopted by the American Institute of Steel Construction are shown in Table 6-2.

The number of pieces of a given type, size, and length is always shown by a numeral in front of these designations; and the length, by a dimension in feet and inches following the designation. For example, $2 - L\,3 \times 3 \times \frac{1}{4} \times 12' - 6''$ would indicate 2 angles with 3 in. equal legs that are $\frac{1}{4}$ in. thick and a length of $12' - 6''$. The symbols for inch and pound are not used in these standard designations except in lengths, but they are ordinarily used elsewhere.

Reference.   Reference [18] gives complete information on all significant properties of all structural-steel shapes available and the companies which roll each shape; estimating and detailing; allowable loads on beams, columns, and their connections; standard specifications and codes; and many other pertinent subjects.

37.   STEEL FABRICATION

The various structural steel shapes and plates of which built-up members are composed are fastened together or fabricated in the shop

TABLE 6-2
Hot Rolled Structural Steel Shape Designations*

New Designation	Type of Shape	Old Designation
W24 × 76	W shape	24 WF 76
W 14 × 26		14 B 26
S 24 × 100	S shape	24 I 100
M 8 × 18.5	M shape	8 M 18.5
M 10 × 9		10 JR 9.0
M 8 × 34.3		8 × 8 M 34.3
C 12 × 20.7	American standard channel	12 20.7
MC 12 × 45	Miscellaneous channel	12 × 4 45.0
MC 12 × 10.6		12 JR 10.6
HP 14 × 73	HP shape	14 BP 73
L 6 × 6 × ¾	Equal-leg angle	∠ 6 × 6 × ¾
L 6 × 4 × ⅝	Unequal-leg angle	∠ 6 × 4 × ⅝
WT 12 × 38	Structural tee cut from W shape	ST 12 WF 38
WT 7 × 13		ST 7 B 13
ST 12 × 50	Structural tee cut from S shape	ST 12 I 50
MT 4 × 9.25	Structural tee cut from M shape	ST 4 M 9.25
MT 5 × 4.5		ST 5 JR 4.5
MT 4 × 17.15		ST 4 M 17.15
PL ½ × 18	Plate	PL 18 × ½
Bar 1 □	Square bar	Bar 1 ⌑
Bar 1¼ o	Round bar	Bar 1¼ φ
Bar 2½ × ½	Flat bar	Bar 2½ × ½
Pipe 4 Std.	Pipe	Pipe 4 Std.
Pipe 4X—Strong		Pipe 4X—Strong
Pipe 4XX—Strong		Pipe 4XX—Strong
TS 4 × 4 × 0.375	Structural tubing: square	Tube 4 × 4 × 0.375
TS 5 × 3 × 0.375	Structural tubing: rectangular	Tube 5 × 3 × 0.375
TS 3 OD × 0.250	Structural tubing: circular	Tube 3 OD × 0.250

*Courtesy of American Institute of Steel Construction, Inc.

by riveting and welding. The pieces are cut to the required shapes and sizes and fastened together to form a completed member.

The members which are placed in position or erected in the field to form the structural frame are connected with field rivets, unfinished bolts, or high-strength bolts and by welding, as described in subsequent paragraphs. Field riveting is done with the riveting gun, and field welding is done with manually operated equipment.

Types of Rivets and Bolts. The various types of fasteners used in shop or field connections are rivets, unfinished bolts, high-strength bolts, and welds. Turned bolts, which were formerly used, have largely been replaced by high-strength bolts.

Drift Pins. Before rivets or bolts are inserted, the holes in a connection are brought into matching positions by one or more *drift pins*. A drift pin is a tapered steel rod which is driven into a hole. Specifications prohibit enlarging mismatched holes by driving the drift pins. Such holes must be reamed to make them match, but inspectors reject poor connections with poorly matching holes.

Field Bolts. After the holes of a connection have been matched, an adequate number of temporary unfinished bolts hold the connected parts in position until the final fasteners are installed. The temporary bolts are removed when a sufficient number of the final fasteners are in place.

Rivets. Rivets that conform to ASTM 502 are made in two grades. *Grade 1*, which is made of carbon steel, is used for general construction. *Grade 2* rivets are made of carbon-manganese steel. They are used for connections in high-strength carbon steels and the high-strength, low-alloy steels.

Rivet sizes are identified by their shank diameter. Diameters range from $\frac{1}{2}$ to $1\frac{1}{4}$ in. in $\frac{1}{8}$ in. increments. Lengths of rivets are from $1\frac{1}{4}$ in. to $8\frac{7}{8}$ in.

Rivets are driven in holes $\frac{1}{16}$ in. larger than the shank diameter. They usually consist of a *buttonhead* and a cylindrical shank long enough to provide material to form the other head (Figure 6.7(a)).

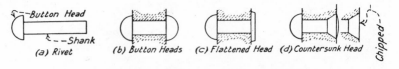

FIGURE 6.7   Rivets and rivet heads.

A rivet with two buttonheads is shown in (*b*), with a *flattened* head in (*c*), and with a *countersunk* head and with a *countersunk and chipped* head in (*d*).

The buttonhead is the usual type. Countersunk and flattened heads may be required for clearance, and chipped countersunk heads are required on bearing surfaces.

The methods and equipment used in driving rivets are considered later. Rivets are heated to a cherry red before driving. As they cool, they contract and exert a clamping force on the included material. This action results in considerable friction between surfaces of the material connected, which augments the shearing and bearing resistance of the rivets. This clamping force is not considered in determining the number of rivets required.

The conventions used on structural drawings to represent rivets of the various types are shown in Figure 6.8. The heads of shop rivets are shown in plan by circles the size of the head, but field rivets are indicated in plan by blackened circles the size of the hole. The longitudinal sections of *shop* rivets, on the extreme left of Figure 6.8, are not shown in the drawings, but the blackened sections of field rivet holes, on the extreme right, are shown even though the view of the member is not a section through the rivets.

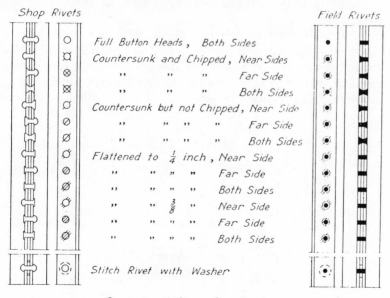

Shop Rivets   Field Rivets

Full Button Heads, Both Sides
Countersunk and Chipped, Near Sides
     ,,   ,,   ,,   Far Side
     ,,   ,,   ,,   Both Sides
Countersunk but not Chipped, Near Side
     ,,   ,,   ,,   ,,   Far Side
     ,,   ,,   ,,   ,,   Both Sides
Flattened to $\frac{1}{4}$ inch, Near Side
     ,,   ,,   ,,   ,,   Far Side
     ,,   ,,   ,,   ,,   Both Sides
     ,,   ,, $\frac{3}{8}$,,   Near Side
     ,,   ,,   ,,   ,,   Far Side
     ,,   ,,   ,,   ,,   Both Sides

Stitch Rivet with Washer

Conventional Signs for Rivets

FIGURE 6.8

The line passing through the centers of a row of rivets, in plan view, is called a *gage line*. The center-to-center spacing of rivets is called the *pitch*. The distance from the center of a rivet to the nearest edge is called the *edge distance*. The length of a rivet, determined by the total thickness of the material it penetrates, is called the *grip*.

Unfinished Bolts.   *Unfinished bolts*, or *A307 bolts*, are made of low-carbon steel and have rough unfinished shanks and square, unmarked heads and nuts (Figure 6.9(a)). They are used in holes $\frac{1}{16}$ in. larger in diameter than the shank. The permissible carrying capacity is lower than that of rivets of the same diameter. Their maximum diameter is 4 in. They are relatively inexpensive. They may be tightened with hand wrenches; but when bolts are used, power wrenches may prove more economical and probably give more uniform results.

According to the American Institute of Steel Construction (AISC) Specifications, unfinished bolts may be used in tier building field connections providing they are not used for column splices on buildings more than 200 ft. in height, nor for column splices on buildings with certain height to width ratios.

They are not permitted in roof truss splices and connections of trusses to columns, column splices, column bracing, knee braces, and crane supports in structures carrying cranes of more than 5-ton capacity.

They are permitted in column splices, with the limitations just given, because the abutting ends of column sections are milled and the splices so arranged that the compressive stresses are transmitted directly across the splice. Within the specified limits, unfinished bolts are considered satisfactory to resist the flexural stresses in the col-

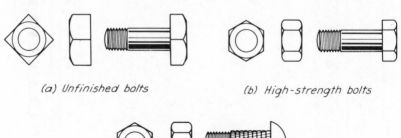

(a) Unfinished bolts   (b) High-strength bolts

(c) High-strength body bolts

FIGURE 6.9   Types of bolts.

umns caused by lateral loads. Splice plates are required, regardless of flexural stresses, to hold the column section in alignment. Unfinished bolts are not permitted for connections of the members of structures that carry large cranes because the repeated action of the cranes tends to work the joints loose and permit the structure to rock back and forth with the movement of the crane.

A sufficient number of unfinished bolts are used temporarily in erecting members with riveted field connections to hold the members in true position while the field rivets are being driven.

High-Strength Bolts. Some bolts which are used for connecting structural steel, especially high-strength structural steel, are identified as *high-strength structural bolts*. These bolts, which are identified by the ASTM number stamped on their heads, are designated as A 325, A 449, and A 490. Under Specification A 325, there are three types of bolts. *Type 1*, which is of medium carbon steel, is made in sizes from ½ in. to 1½ in. The heads of the bolts, in addition to the ASTM mark, are marked with three short radial lines 120° apart. *Type 2* bolts are of low-carbon steel with diameters from ½ in. to 1 in. The heads of the bolts are marked with short radial lines 60° apart. ASTM A 325 *Type 3* bolts are made of weathering steel and are used to connect members that are of weathering steel. They may be identified by the *underlined* ASTM A 325 on the head.

High-strength bolts A 490 are made of alloy steel. A 449 bolts are made of medium-carbon steel; two A 449 bolts are illustrated in Figure 6.9(*b*) and (*c*). The type shown in (*b*) is designated as the *high-strength bolt*, A 325 or A 490, and is inserted in a hole with a diameter ⅟₁₆ in. larger than that of the shank of the bolt. The type shown in (*c*) is designated as the *high-strength bearing bolt* or *high-strength interference-body bolt*. On the shank, it has a raised pattern consisting of straight or spiraled ribs or knurls. Since the outside diameter of the shank is greater than the diameter of the hole, the bolt must be driven into the hole with a maul or pulled in with the nut. This process causes the shank to bind in the hole, and the hole is more completely filled than it is when the plain high-strength bolt is used. This is referred to as an *interference fit*, which gives the bolt one of its names. It also gives better bearing between the bolt and the sides of the hole, which is recognized in the name bearing bolt.

According to the assembly specifications, the type of bolt, and the method of tightening, high-strength bolts may or may not require *hardened washers* under either the head or the nut or both. The head is kept from turning initially by gripping with a hand wrench. The high-strength bearing bolt binds in the hole and does not tend to turn when being tightened. The head is button-shaped, as shown in the figure. A hardened washer is generally required under the nut.

The nuts on both types of bolts are tightened with pneumatic impact wrenches. Because of the high strength of the bolts, they are generally tightened so that the clamping forces are sufficient to insure, under most conditions, that the friction between the surfaces of the various parts connected is adequate to prevent any sliding movement due to the applied loads and to make sure that the nuts will not be loosened. In addition, the high tensile strength of the bolts and the initial tension to which they are subjected make them effective in resisting forces that subject them to tension. Under conditions where the tendency to slip is unusually high and where there are abnormal vibrations, bearing bolts are superior to bolts with plain shanks.

The allowable carrying capacities of high-strength bolts are equal to and, under ordinary conditions, greater than those of rivets of the same nominal diameter.

Both types of bolts are used for field connections under the same conditions as rivets. They have the advantage of being cheaper and quieter to install than rivets.

Welding Processes.   Many processes are used to weld pieces of metal together. Those used in structural welding are divided into two main groups: *pressure processes*, in which the weld is completed by applying pressure after the pieces to be welded have been placed in contact where the weld is to be formed and have been heated to the required temperature; and the *nonpressure* or *fusion process*, which requires no pressure to complete the weld.

Nonpressure or fusion processes used in structural welding are divided into arc welding and gas welding according to the source of heat.

In *arc welding*, the heat is provided by an electric arc formed between the work to be welded and an electrode held in the operator's hand with a suitable holder or in an automatic machine. The electrode may be a metallic rod, as in *metal-arc welding*, or a carbon rod, as in *carbon-arc welding*.

In *gas welding*, the heat is provided by a gas flame produced by burning a mixture of oxygen and a suitable combustible gas. The flame is formed at the tip of a *blowpipe* or *torch*, which is held in the operator's hand or in an automatic machine. The gas used in structural welding is acetylene, which gives this process the name of the *oxyacetylene process*.

In both the arc and gas processes, the pieces to be welded are placed in contact; the edges are melted so that metal from the two pieces flows together; and, when cooled, the pieces are joined by the weld. In order to make a satisfactory joint, additional metal must be supplied.

It is provided by the metallic rod used as the electrode in the metal arc process, but in the carbon-arc and gas processes a metal rod called a *filler* or *welding rod* is used. The end of this rod is melted off into the joint as the joint is being formed.

The only pressure-welding process used in structural work is the *spot-welding process*. In this process, a small area or spot on the surfaces to be joined is heated by placing electrodes against the outer surfaces of the pieces and passing an electric current through the pieces and across the contact surface. The heat required to raise the temperature of the pieces to a welding temperature at the spot where the weld is to be made is generated by the resistance offered by the metal between the electrodes to the flow of electric current. This is, therefore, a *resistance process*. Since the weld is consummated by exerting pressure across the spot, this is a *pressure process*. If spot welds are formed progressively in a continuous overlapping row, the process is known as *seam welding*. Disk electrodes are used to apply the pressure.

Types of Welded Joints.   Various types of welded joints used in structural work, with the names applied to the parts of welds, are illustrated in Figure 6.10. The types of welds illustrated are the *single-* and *double-fillet lap weld* in (*a*), the *single-* and *double-vee groove weld* in (*b*), the *groove weld* with double-fillet welded backing strip in (*c*), the *double fillet T weld* in (*d*), and the *plug* or *rivet weld* in (*e*)—all of which are fusion welds—and the *spot weld* in (*f*), which is a resistance-pressure weld. The electrodes are shown in this figure.

Fabricating Procedures.   The following paragraphs on cutting, punching, drilling and reaming, riveting, milling and straightening, bending, and rolling are quoted, with permission, from *Structural Shop Drafting* published by the American Institute of Steel Construction [3].

These procedures apply specifically to riveted members, but the comments on cutting, milling and straightening, bending, and rolling apply also to welded members.

"*Cutting.* There are several types of cutting operations, each of which is particularly adapted to the class of work involved. Plates are cut in a guillotine type machine, called a *shear*. Angles are cut on a

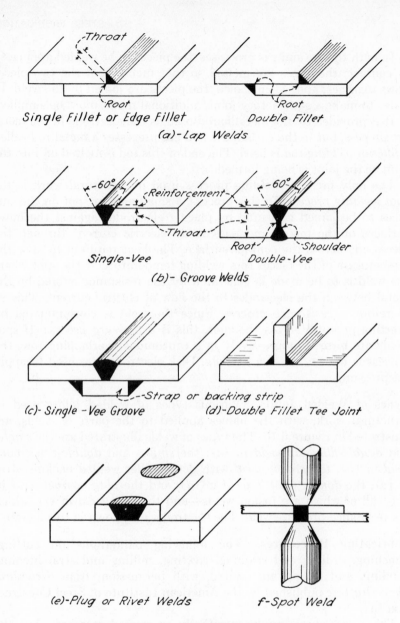

FIGURE 6.10 Types of welded joints.

machine with the shear blade shaped and set at an angle to cut both legs with one stroke. Beams, channels, and light columns are usually cut on high-speed saws. There are several types of saws for this purpose, the most common one being the *friction saw*, which cuts through steel as fast as a power crosscut saw passes through timber. Bars are cut by a shear of the guillotine type, the dies and shear blades of which are made in the shape of the material to prevent its distortion. Large bars are cut on a *power hack saw*.

Cutting of material which is over the capacity of the various machines and most cutting of curved and complex forms is done with the acetylene torch and is known as *flame-cutting* or *gas-cutting*. The cutting torch is the most useful and flexible means of cutting steel ever devised. [See *Explanatory Comments.*]

Punching, Drilling and Reaming. Punching is the most economical common method of making rivet and bolt holes in steel. Generally the material is punched rather than drilled, where the size of the hole and the thickness of the metal comes within the range of the machines and when punching is permitted by the specifications. Normally, mild carbon steel up to a thickness one-eighth of an inch greater than the diameter of the rivet can be punched without undue stress on the machine or excessive breakage of the punches.

Except in special cases, all holes are punched with a diameter $\frac{1}{16}$ of an inch larger than that of the rivet or bolt to be used. Some tolerance is required to provide for less than perfect matching of holes in adjacent parts when they are lined up for rivet driving.

Drilling of structural steel is largely confined to making holes in material thicker than the capacity of the punch machines and dies, as in slabs and the heavier wide flange beams and columns, or to meet specification requirements.

Specifications for bridge work and certain other specialized work require all holes to be punched at least one-eighth of an inch smaller than the final size. Then, after the piece has been assembled, it is placed under a battery of reamers operating like jibs or gantries, and the holes are reamed to the desired size. The procedure is usually referred to as *sub-punch and ream*. Reaming produces a clean hole with all burrs and distorted metal removed. Reamed holes are also used for connections which will be field-bolted with close-fitting *turned* or *finished bolts*.

Riveting. Riveting machines of both the pressure and impact type are operated by compressed air at 80 to 100 lb. per sq. in. pressure. The *pressure-type riveter*, frequently known as a *bull riveter*, has a

C-shaped frame with a throat depth sufficient to permit a machine to reach halfway across an average girder. Small-size pressure riveters are hung from jib cranes while the larger machines are generally supported from a power-operated gantry or even from an overhead traveling crane. Pressure riveters complete the driving operation with one stroke, the larger machines exerting a pressure of 80 tons.

Impact riveters, generally referred to as *riveting guns,* are portable hand tools which can be connected to the end of a compressed air hose. A rapid succession of blows is delivered to the rivet by means of a plunger moving up and down in the gun barrel much as the piston in a reciprocating steam engine moves back and forth. These tools are also used by the fitter for driving tapered *drift pins* through holes to align the several parts of a member. The gun drives just as satisfactory a rivet as the pressure riveter but, being slower, its shop use is limited to driving rivets inaccessible to the larger machines and scattered rivets where the amount of driving does not warrant taking the work to the machines.

Standard-size rivets used in a structural shop are almost always heated before driving. The heating in most shops is done in either oil or gas-fired furnaces and then raised to a minimum temperature color of light cherry red. In some shops, electric rivet heaters are used. [See *Explanatory Comments.*]

Milling. The ends of columns, struts, and members in compression chords of trusses which are required to bear evenly against another or against supporting base plates are *milled* to a smooth even surface in a *milling machine.* The machine consists of one or more rotating cuting heads fitted with teeth or blades, and a bed on which to securely hold the work in proper alignment during the milling operation.

Column base plates over 4 in. in thickness must be milled over the area in contact with the column shaft. This milling is usually done on a *planing machine.*

Straightening, Bending, and Rolling. All material which is bent or distorted during shipping, handling, or in the punching operation must be straightened before further fabrication is attempted. The machine generally used for straightening beams, channels, angles, and heavy bars is commonly known as a *bulldozer.* This machine has a heavy cast steel frame with a horizontal plunger or ram centered between two lugs about 2 ft. apart set on an anvil or head. With the piece to be straightened placed against the two lugs, pressure is applied by the ram at points along the length of the bent member until it is in alignment once more.

The bulldozer also is used to make long-radius curved beams, channels and angles and to make minor angle bends in heavy pieces and to make other bends within its capacity. Short-radius curves, where material must be heated, are handled by the blacksmith. [See *Explanatory Comments*.]

Long plates which are slightly curved or cambered out of alignment in the direction of their longitudinal center line are usually straightened by cold rolling. By passing these plates between a pair of rolls bearing against their top and bottom faces, with more pressure exerted along the shorter or concave side than along the convex side, the thickness of the plates on the concave side is reduced ever so slightly. The metal thus displaced by the rolls increases the length of this side enough to bring the plate back into true longitudinal alignment."

Explanatory Comments.   Certain explanations of the procedures in the foregoing paragraphs seem desirable.

In driving buttonheaded rivets with the pressure riveter, the heated rivet shown in Figure 6.7(a) is placed in the hole provided for it. The member is then moved into position between the jaws of the C-frame. A depression is provided to receive the formed head. The end of the plunger is provided, with a depression in which the other buttonhead (b) is formed as the pressure is applied to the projecting end of the rivet. If a rivet is driven with a riveting gun, the formed head is supported in a heavy *dolly*. The other head is formed by several blows from the plunger of the gun, the impact of the plunger being resisted by the dolly, which acts as an anvil. Rivets with one head flattened or countersunk ((c) and (d)) are formed in a similar manner.

Flame cutting or gas cutting is also known as *oxygen cutting*. It includes a group of processes in which the cutting is accomplished by the chemical reaction of oxygen with the base metal at elevated temperatures. These temperatures are attained by means of gas flames resulting from the combustion of oxygen and one of several kinds of gas, the most commonly used gas being acetylene, as in *oxyacetylene cutting*. Steel is also cut by *arc cutting*, a group of processes in which metals are cut by melting them with the heat of an electric arc between an electrode and the base metal [4]. Accurate cuts can be made by these procedures regardless of the thickness of the metal.

At least one plant, the Commercial Shearing and Stamping Company, Youngstown, Ohio, is equipped to cold-bend S, W, or wide-flange shapes, with depths as great as 30 in., in the plane of the web. Bends in the plane perpendicular to the plane of web are made more easily. The minimum radius of curvature for bending in the plane of the depth is about fourteen times the depth. A single piece can be bent to various radii along its length. The maximum length of piece which can be handled is 30 ft.

Welding.   The following comments on welding procedures are quoted, with permission, from *Structural Shop Drafting*, published by the American Institute of Steel Construction [3]:

"The process most commonly used in the welding of structural steel is known as *metal-arc welding*. In this process energy is supplied by an electric arc established between the *base* or *parent* metal (the parts being joined) and a metal *electrode*. The instant the arc is formed tremendous heat is concentrated at the point of welding, which is located at the junction formed by the two parts to be welded together. The parent metal melts in a small pool; additional metal supplied by the electrode is transferred through the arc in the form of tiny globules and deposited in the pool. As the electrode (and hence the arc) is moved along the joint, either manually or automatically, the molten metal left behind solidifies as a uniform deposit or *weld*, which joins the parts solidly, thus forming the desired connection.

Electrodes may be either *bare* or *coated*, although most welding is done today with coated electrodes. Bare manually operated electrodes produce welds which are not as reliable as those made with electrodes that are heavily coated. In the case of the coated electrode, the heat from the arc produces from the coating a gaseous and slag flux shield, as shown in Figure 6.11, which completely envelops the arc and the molten metal, thus preventing contact with the oxygen and nitrogen of the air, contamination from which tends to produce brittle welds. The slag coating which is formed floats on the molten pool thus protecting it from the atmospheric elements. This slag is easily removed after the weld has cooled.

In *manual welding*, the welding operator manipulates the electrode by hand. An insulated gripping device connected by a flexible cable to the source of current is used to hold the electrode. The operator wears proper covering and a helmet which affords protection against *weld spatter* (globules of metal thrown off during welding), heat, the blinding light so characteristic of the process and harmful rays emanating from the arc."

Note: Welds in any accessible position can be made by manual welding because the globules of metal from the electrode are forced across the arc almost independently of the action of gravity.

"When a weld is terminated abruptly, a *crater* is formed in the base

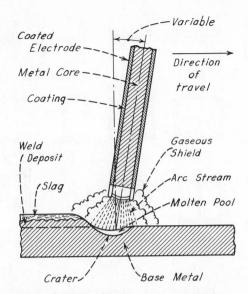

FIGURE 6.11 Shielded metal-arc welding process.

metal by the blast of the arc. It is required that this crater be filled with weld metal so that the finished appearance and the cross section of the weld are uniform.

In *submerged-arc automatic welding*, the electrode is fed by a machine which has a reel for carrying a coiled supply of uncoated wire. With this method, the arc is shielded by a granular flux automatically deposited on the joint ahead of the arc. The welding wire pushes through the flux; the arc is established; part of the flux melts to form a slag flux shield; and the customary slag floats on the molten pool. Unfused flux is recovered for re-use."

38. STEEL COLUMNS

Types of Column Sections. Structural steel columns may consist of a single piece, such as the wide-flange shape (Figure 6.12(a)), or they may be built up of various shapes (b) to (n). The individual shapes are riveted or welded together or welded to form the built-up sections.

Wide-Flange Columns. The most commonly used column section is the wide-flange column shown in Figure 6.12(a), with nominal depths of 8 to 14 in. Various weights are available for each nominal depth, but it should be noted again that the spreading of the rolls to pro-

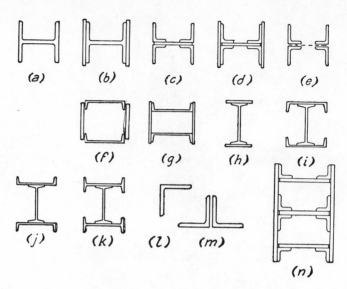

FIGURE 6.12   Steel columns.

duce sections with several areas of the same nominal depth increases the actual depths of the sections. For example, the wide-flange shape which has a nominal depth of 14 in. and weighs 87 lb. per ft. has an actual depth of 14 in. and a flange thickness of 0.688 in.; whereas the maximum 14 in. section, which weighs 730 lb. per ft., has an actual depth of 22.44 in. and a flange thickness of 4.910 in.

Cover-Plated Wide-Flange Columns.   Column sections heavier than a maximum wide-flange shape can be obtained by riveting or welding cover or angle plates to such a shape (Figure 6.12(b)). There is a special 14 in. core shape weighing 320 lb. per ft. for this purpose. It has a relatively thick web, $1\frac{7}{8}$ in., to support the thick flanges formed by cover plates. Cover plates, each with thicknesses of 1 in. or more, might be riveted or welded to each flange to give a total thickness 3 in. or more for each flange.

Plate and Angle Columns.   Columns are built up of four angles and a web plate (c), with cover plates also (d). Columns of almost any size desired can be built up in this way.

Other Types.   The types of sections shown in Figure 6.12(e) to (k) are not used to any extent. The angle sections (l) and (m) may be used in very light corner columns or wherever their form is advantageous. These angle sections are extensively used, however, as members of trusses.

39. STEEL BEAMS AND GIRDERS

The general term *beam* may include various flexural members such as beams, girders, joists, lintels, purlins, and rafters, as explained in Article 19, which are given specific names because of the manner in which they are used.

Types of Beams. The cross sections of various types of beams are shown in Figure 6.13. The wide-flange shape shown in (*a*) and the American standard S-beams shown in (*b*) are the most commonly used sections. They are available in a great variety of sizes. The channel shown in (*c*) is extensively used for roof purlins. The wide-flange shape or the S-beam may be provided with *cover* or *flange plates* (*d*). Beams may be built up of plates and angles (*e*) and (*f*) to obtain almost any size desired. The built-up sections shown in (*g*) and (*h*) are used occasionally.

Separators. In (*i*) and (*j*), two beams have been fastened together with *separators* to hold them in position and to make them act somewhat as a unit. The *plate-and-angle separator* in (*i*) is more effective in this respect than the *pipe separator* in (*j*). If the beams are not too far apart, two overlapping angles, with legs riveted together, can be used as an *angle separator*. Separators spaced 5 or 6 ft. apart are usually required when beams are used in pairs.

Lintels. The sections shown in (*k*) to (*n*) are lintels arranged to carry brick or stone masonry over openings and can be largely concealed. The single angle shown in (*n*) carries the outer layer of masonry; the remainder of the wall may be carried by a *relieving arch*, or another angle may be used.

Plate Girders. A built-up plate girder is illustrated in (*o*). The section consists of four flange angles, a web plate, and cover plates, as shown in (*f*). The web plate is usually made quite thin, a ratio of unsupported depth to thickness of 170 being permitted if *web stiffeners* are provided, to keep the web plate from buckling. Since these stiffeners must project over the flanges, it is necessary to bend or *crimp* them around the vertical legs of the flange angles in order to place them against the web (*p*); or *fill plates*, equal in thickness to the flange angles, may be placed between the stiffeners and the web

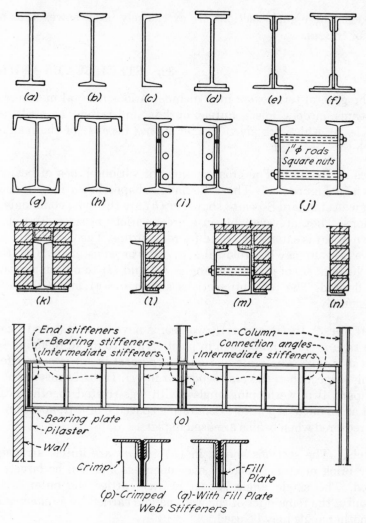

FIGURE 6.13   Steel beams.

plate (q). If the unsupported depth of the web does not exceed 70 times its thickness, *intermediate stiffeners* are not required; but *end stiffeners* are always required. Stiffeners are also required at all points where concentrated loads are supported. The spacing of stiffeners is determined by the shearing stresses. It is often required that the thickness of web plates be at least ¼ in.

A girder may be built up of two flange plates and a web plate welded

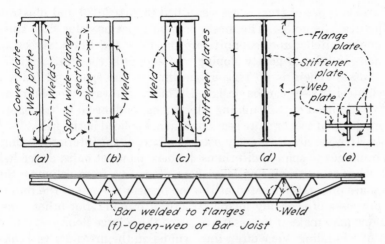

(f)-Open-web or Bar Joist

FIGURE 6.14 Welded plate girders.

together (Figure 6.14(*a*)) or of a wide-flange section split along the center of the web and separated to receive a narrow plate which increases the depth (*b*), the three parts being welded together. Stiffeners are provided by welding plates between the flanges and normal to the web (*c*) to (*e*). Welded girders are lighter than riveted girders with the same carrying capacity.

Joists. When the loads are rather heavy and joists are widely spaced, S-beams may prove economical, but special lightweight sections are available for light loads and closely spaced joists. One type of lightweight joist is the *M-shape beam*, a rolled section of the same form as the standard S-beam, but much lighter because the metal is rolled thinner. The sizes vary from a 6 in. depth weighing 4.4 lb. per ft. to the 14 in. depth weighing 17.2 lb. per ft. The web thickness of the smaller section is less than $\frac{1}{8}$ in., and of the larger section about $\frac{3}{16}$ in. Other sections lighter than the American standard sections are available.

Another form of joist for light loads is the *open-web* or *trussed joist*, one type of which is illustrated diagrammatically in Figure 6.14 (*f*). The flanges of joists of this type are made up either of two light angles, two bars, or a tee, with web members consisting of flat bars welded to the flanges, or of a continuous round bar bent back and forth to form the diagonals and welded to the flanges. Some types are called *bar joists* because they are made up largely of bars.

40. STEEL TRUSSES

Various types of trusses are described in Article 20 and illustrated in Figures 3.3 and 3.4. The most common types used for steel trusses for double-pitched roof supports are the Fink and Pratt trusses and, for floor and flat or gently sloping roof supports, the Pratt and Warren trusses. There is no reasonable limit to the span and loads for which steel trusses can be designed. They are always shop-fabricated and transported to the building site. Spans too long to be transported are made up of two or more sections and are spliced in the field.

Steel trusses are frequently used as a part of the interior framing of tall buildings to span auditoriums, lobbies, banquet halls, dance halls, and other rooms requiring large floor areas free from columns. Such floors are preferably located at the top or near the top of the building for purposes of economy in the structural framing, but other considerations may make it necessary to place these large floor areas in any part of a building. Very often they will be on the first floor with many stories above (Figure 6.15(a)). The depth of the truss may occupy one or more stories. It may not be possible to use one of the common types of truss because of doors or hallways, for the floor or floors immediately above may have to pass through the truss and will determine its form. Such trusses have to be of very heavy construction. Some typical cross sections of truss members are given in Figure 6.15(b).

Another common use for steel trusses is the support of the balconies

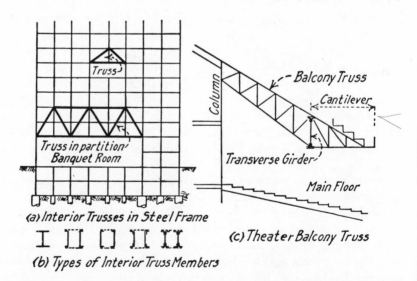

(a) Interior Trusses in Steel Frame

(b) Types of Interior Truss Members

(c) Theater Balcony Truss

FIGURE 6.15   Trusses located in partitions and balcony truss.

of theaters. Columns are, of course, objectionable and they are eliminated whenever possible. A common form of construction consists of
cantilever trusses overhanging a transverse supporting truss or girder
(Figure 6.15(*a*)).

The various members of which a truss is composed are connected
at the *joints*, or points where they meet, to form a truss by riveting
or welding, as described in Article 37.

Riveted Trusses.   The members meeting at a *joint* or *panel point*
may be connected by attaching them to *connection* or *gusset plates*
¼ in. or more thick and large enough to provide for the number of
rivets required by the stresses in the members, and, occasionally, by
the stresses in the plates themselves. The rivets may be as small as
⅝ in. in diameter for light trusses but usually are ¾ in. and may be
larger for long spans or heavy loads.

For trusses of ordinary spans, there is usually one gusset plate at
each joint (Figures 6.16(*b*) and 6.17(*d*)). To avoid twisting action,
the pieces of each principal member connected to a gusset plate
should be symmetrically located with reference to the plate. To
accomplish this objective, the principal members, such as the top
and bottom chords, are made up of two units as illustrated by the
two angles, placed back to back, of the top chord in Figure 6.16(*a*)
and the two channels, placed back to back, for the bottom chord,
shown in (*b*) and in Figure 6.17(*c*). These are spaced apart far
enough to receive the connection or gusset plate between them (Figure 6.16(*b*)). Members carrying minor stresses may be composed
of a single small angle which does not cause significant twisting
action because of the low magnitude of the stresses.

If the top chord of a truss is subjected to bending stresses by members such as roof purlins or floor joists not located at the joints, it
may be composed of two angles and a plate (Figure 6.17 (*a*)), or two
channels (*c*). Similarly, a bottom chord subjected to such stresses
may be composed of two angles and a plate (*b*) or two channels (*c*).
If the member includes a plate, that plate may also serve as a connection plate at the joints.

Truss members consisting of two angles or channels, without a
plate, should be provided with *stitch rivets* at intermediate points
between the connection plates (Figure 6.16(*b*)). The distance between the backs of the angles or channels is maintained by washers
through which the rivets pass. Stitch rivets are provided to make the
individual parts act together more nearly as a unit.

Wood or steel purlins may be fastened to the top chords of steel

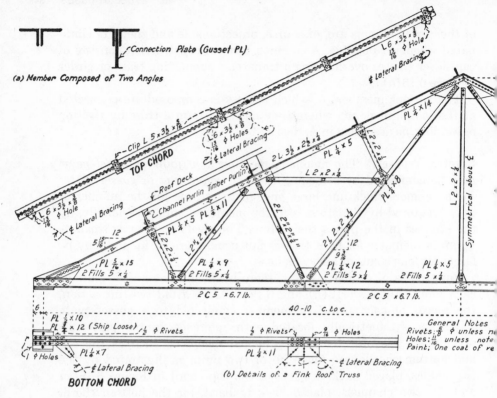

(a) Member Composed of Two Angles

(b) Details of a Fink Roof Truss

BOTTOM CHORD

FIGURE 16.16 Fink steel roof truss.

roof trusses by short lengths of steel angles, called *clip angles* (Figures 6.16(*b*) and 6.17 (*d*)).

Heavy riveted trusses are made up of large angles and many other types of members such as those shown in Figures 6.12(*a*) to (*d*).

Welded Trusses. The connection or gusset plates in riveted trusses serve as indirect paths through which stresses are transferred from one member to another member connected to the same plate. A significant advantage of most welded trusses over riveted trusses is the direct transmission of stress from one member to another, which eliminates gusset plates. To accomplish this objective, appropriate types of members must be used.

The ends of members to be connected at a joint are so cut and so shaped, by slotting or otherwise, that they will provide suitable opportunities for welding. Most structural welding is done by the electric-arc welding process described earlier. The basic types of welds used

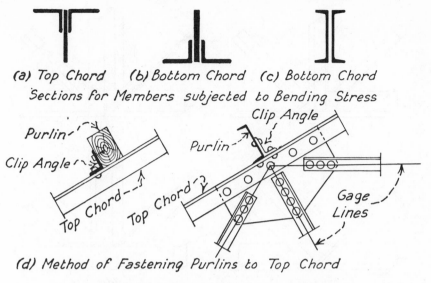

(a) Top Chord *(b) Bottom Chord* *(c) Bottom Chord*
Sections for Members subjected to Bending Stress

(d) Method of Fastening Purlins to Top Chord

FIGURE 6.17   Simple steel roof truss details.

are the *groove weld* by means of which abutting edges or ends of plates and shapes are joined (Figure 6.10(*b*)), and the *fillet weld*, formed along the junction of two surfaces which meet at an angle (Figure 6.10(*a*) and (*d*)).

The following comments concerning the structural shapes and arrangements of members at the joints are quoted, with permission and minor changes, from the eleventh edition of the *Procedure Handbook of Arc Welding Design and Practice,* published by the Lincoln Electric Company [4].

"Arc-welded trusses may be designed in various ways using T-shapes, H-shapes, or U-shapes for chords. The web members are generally angles or channels. (Various types of arc-welded truss connections are illustrated in Figure 6.18.)

(1) Perhaps the simplest type of truss connection is made of angle sections for web members (Figure 6.18(*a*)). This is easy to fabricate and weld because the sections lap each other and fillet welds are used.

(2) For a heavier truss, the vertical member can be an I-beam or H-beam. The web of this member is slotted to fit over the stem of the T-section. The T-section is used for both the top and bottom chord members. The diagonal members are made of a double set of angles. (Fillet welds are used.)

(3) Some trusses make use of T-sections for their diagonal mem-

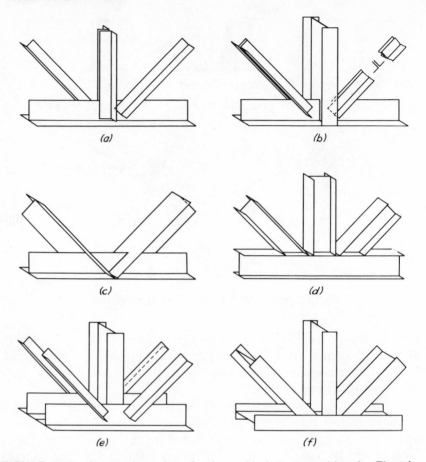

FIGURE 6.18   Connections for simple welded trusses. Lincoln Electric Company.

bers (*c*). The flanges of the diagonal members must be slotted to fit over the stem of the T-section used for the top chords as well as the bottom chords. The stem of the diagonals is also cut back and is butt-welded to the stem of the top and bottom chords. (Fillet welds are used elsewhere.)

(4) Quite a few trusses are made completely of H-sections (*d*). Both top and bottom chords are made of H-sections, as are the web members. This allows loads to be placed anywhere along the top and bottom chords because of the high bending strength of the H-section. With the conventional truss design, loads must be placed only at

points where diagonal or vertical members connect to the chord members. Almost all the welds are on flanges of the top and bottom chords and, since they are flat surfaces, there is no unnecessary filling of the members to make these connections. (The welds are all fillet welds.)

(5) In the connection shown in (*e*), two T-sections are intermittently welded together at their flanges to form top and bottom chord members. The vertical member is an H-section set in between the stems of the T-sections. Each diagonal member is made of a set of angles. All the welding consists of fillet welds.

(6) Sometimes the flanges are made of wide-flange beams and H-sections with the webs of the top and bottom chords placed horizontally (*f*). The welding of these members consists mainly of butt-welding the flanges together. Under severe loading, gusset plates are added in between the flange connections in order to strengthen the joint and reduce the possibility of concentrated stresses."

The types of connections shown in Figure 6.18 can be adapted to many forms of truss, including the bowstring roof truss, for which the top chord can be cold-formed with the desired curvature.

The end panels of a heavy riveted Warren roof truss with 12 panels and a span of 150 ft., connected to a column, are illustrated in Figure 6.19. Because of the large stresses involved, the types of connections in Figure 6.18 are not suitable and gusset plates are required.

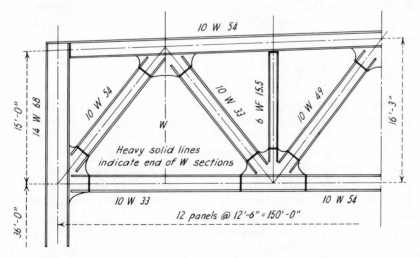

FIGURE 6.19   Welded truss with gusset plates. Lincoln Electric Company.

The edges of the gusset plates are curved between members, to improve the stress distribution within the welds. This also improves the appearance. They are also made straight, however.

Except at the top of the column, the structural section, of which each truss member is composed, ends at a gusset plate. At each end, its web is butt welded to the gusset plate. The flanges are butt welded to ends of the stiffener plates, which are welded normal to each side of the gusset plate.

Instead of ending the structural section of the gusset plate, as just described, the web of the section may be cut off at this point and the projecting flanges slotted to form a forked end which fits closely over the gusset plate. The web is butt welded to the edge of the gusset plate, and the projecting flanges are fillet welded to the sides of this plate. No stiffening plates are required. The latter arrangement has the advantage of not breaking the continuity of the flange material, but it requires a longer main section and does not utilize the web material which is cut out in forming the slots.

The end top chord member which connects to the top of the column carries only the relatively small stresses due to lateral loads, such as the wind load. It is butt welded to the column. The gusset plate at the end of the bottom chord is fillet welded to the column. A suitable end connection can be devised for a truss, simply supported by a column, ending at the outer end of the lower chord; the end member of the top chord is omitted. The sizes of the gusset plates are determined by the stresses they must carry and by the lengths of the welds required to transmit the stresses from the members to the plates.

41. STEEL FRAMING

General Comments.   The preceding articles in this chapter have been concerned with the structural steel shapes available and their use in constructing the various types of members used in the *frames* or *framing* of buildings. The means employed for fastening pieces together have been considered. Examples of the use of gusset plates in forming the points of riveted and welded trusses and of welded joints without gusset plates have been given. This article is concerned primarily with column, beam, and girder splices and connections and with lateral bracing for the frames of tier buildings and mill or industrial buildings.

A tall building with steel frame is often referred to as a *skeleton construction* building because its frame resembles a skeleton; it is called a *tier building* because its stories constitute tiers and is known as a *skyscraper* because of its height. Buildings of any height which enclose undivided floor areas that can be subdivided to suit the tenants are often called *loft buildings*.

Riveted Construction. Various details used in connections with rivets as fasteners are explained in the following paragraphs.

Lap and butt joints. Joints provided by riveting plates or parts of members which lap over each other (Figure 6.20(*a*)) are called *lap joints*. Those formed by butting the ends of two parts together and fastening them by rivets passing through a splice plate or connection plate (*b*) are called *butt joints*.

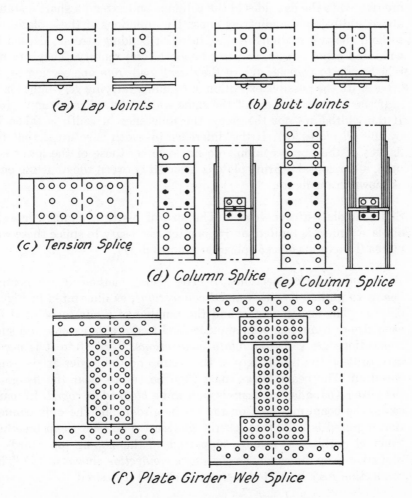

(a) Lap Joints   (b) Butt Joints

(c) Tension Splice

(d) Column Splice   (e) Column Splice

(f) Plate Girder Web Splice

FIGURE 6.20   Riveted splices in steel members.

Column Splices.   Steel column sections are usually made constant for two-story heights. Because of the change in the load on a column at each floor, it would be possible to save column material by reducing the column section in each story. This practice would be undesirable because the cost of the splices and the increase in erection costs would probably offset any saving in column section. In order that they will not interfere with the beam and girder connections, the splices are usually made about 2 ft. above the floor line.

The abutting ends of the columns at the splice are accurately *milled* so that the compressive stresses can be transferred directly from the upper column to the lower column by bearing. Splice plates are riveted to the flat sides of the columns and extend a short distance above and below the abutting ends. The functions of these plates are to hold the two sections in line, to resist bending stresses caused by wind and other causes, and to provide lateral rigidity. They are not relied upon to transfer any of the direct load from one section to the other, with the possible exception of columns carrying very light loads.

If the two sections are of the same width, the splice is simple (*d*). If the widths are not the same, the difference in width is taken up by the *fill plates* (*e*). If the difference in width is so great that the flanges of the upper columns do not bear on those of the lower column, a horizontal bearing plate is inserted between the abutting ends as shown in the figure.

Splices in plate-girder webs.   The size of plates available for webs of plate girders is limited, and it is often necessary to splice these web plates. Two forms of *web* splices are shown in (*f*).

Beam and girder connections.   The usual method of connecting beams to girders is with *framed connections*, as illustrated in Figure 6.21(*a*). The size of angles and the number of rivets to be used for each size of beam has been standardized to quite an extent, and such connections are called *standard connections*. Very often it is necessary to keep the top flanges of the beam and the girder at the same elevation. The beam flange must then be cut to clear the flange of the girder (*b*), and the beam is said to be *blocked* or *coped*. In many cases, the beam may rest on top of the girder, and the only connection required is bolts through the flanges (*c*) to hold them together. Channel purlins are usually supported on the sloping top chords of roof trusses by means of the *clip-angle connection* shown in (*d*). This connection may be bolted instead of riveted if desired.

Connection of beams and girders to columns.   There are two types

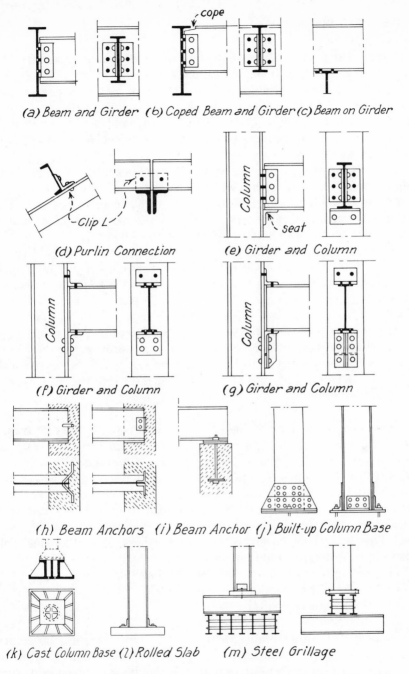

(a) Beam and Girder (b) Coped Beam and Girder (c) Beam on Girder

(d) Purlin Connection   (e) Girder and Column

(f) Girder and Column   (g) Girder and Column

(h) Beam Anchors (i) Beam Anchor (j) Built-up Column Base

(k) Cast Column Base (l) Rolled Slab   (m) Steel Grillage

FIGURE 6.21   Riveted beam end connections and column bases.

of connections used between beams or girders and columns: the *framed-connection type* and the *seated type*.

The framed-connection type, illustrated in (*e*), consists of two angles riveted to the beam or girder in the shop, the rivets between the angles and the column being driven in the field. A *shelf angle* may be provided to support the beam or girder during erection. This is called an *erection seat*.

The seated type of connection shown in (*f*) consists of an angle at the bottom side of the beam which is shop-riveted to the column and an angle at the top which is field-riveted to the beam and column. For large reactions, it is necessary to provide one or two *stiffener angles* to support the outstanding leg of the bottom angle (*g*).

The shop work is simpler on the seated type, and fewer field rivets are required; but this type may project through the fireproofing if used in connecting to column flanges, and then the framed connection is used.

Wall supports for beams and girders.   Steel beams and girders may be built into masonry walls. *Bearing plates* are provided to distribute the reaction over a larger area, and anchors (*h*) tie the beam and wall together. Where beams and girders are not built into the wall, *anchor bolts* (*i*) are provided.

Column bases. The load at the lower end of a column must be transferred to a concrete footing or pier which in turn transfers the load to the ground. If the end of the steel column were permitted to rest directly on the concrete, the concrete would be crushed where the two came in contact because the working stress in the steel is much greater than the strength of the concrete in bearing. It is therefore necessary to distribute the column load over a large area of the footing. This is done by means of the *column base*. Bases for steel columns may be divided into four classes.

(1) Built-up bases made from structural sections (*j*).

(2) Cast bases of steel or cast iron (*k*).

(3) Rolled-steel slabs (*l*).

(4) Steel grillages (*m*).

Built-up bases are suitable for light loads, but where it is necessary to distribute the column load over a considerable area, cast-iron or cast-steel bases may be used, cast-steel being much stronger and more reliable than cast-iron. These are rarely, if ever, used. Rolled-steel slabs have come into general use. They are more economical and reliable than the cast-iron bases and more economical than the cast-steel bases. The end of the column is milled and bears directly on the steel slab. A simple connection is made between the column end and the slab by means of two angles. Slabs are available up to 12 in. thick, but slabs over 6 in. thick are not usually economical.

Welded Construction.   Typical details for welded beam, girder, and column connections are illustrated in Figure 6.22, together with the pertinent standard welding symbols.

Connections for Rods or Bars.   The ends of round or square bars may be threaded to receive nuts to form an end connection. Several connections making use of nuts on *threaded ends* are shown in Figure 6.23(*a*). When long bars are used, it is usually economical to enlarge the end of the bar so that the area of the section at the root of the threads is somewhat greater than the area in the body of the bar. A smaller bar can then be used, for the threads do not reduce the section area. These enlarged ends are called *upset ends* (*b*).

The *clevis* shown in (*d*) is a convenient form of end connection for round and square bars. Upset ends are usually provided. The pin in the clevis is a *clevis pin*, as shown, or an ordinary bolt and nut. The small split pin inserted in the larger pin is called a *cotter pin* or *cotter key*.

It is often desirable to make tension rods and bars adjustable so that they may be tightened. Two devices are used for this purpose: the *turnbuckle* (*b*) and the *sleeve nut* (*c*). In both devices, a right-hand thread is used at one end and a left-hand thread at the other so that the abutting ends of the bar may be drawn together or pushed apart by turning the turnbuckle or sleeve nut in the proper direction.

Tier Building Frame.   The riveted frame of a tier building formed by assembling the various beams, girders, and columns is illustrated in Figure 6.24. In this frame, no special provision was required for wind bracing. The following points should be noted:

(1) The outside wall, called a *panel* or *enclosure wall*, is supported at each story by the spandrel beams which transmit their load to the outside columns.

(2) The panel walls shown are veneered with stone ashlar 3¾ in. thick, which is bonded to the backing by a bond course 7½ in. thick every third course and by galvanized anchors at the intermediate joints. The backing may be of brick or of structural clay tile. One bond course rests on the spandrel beam.

For office buildings and some other occupancies, panel walls with large glass areas, as described in Article 84, are used extensively. Such walls are usually called *curtain walls* or *window walls*.

(3) The floor beams are placed at the third points of the girders

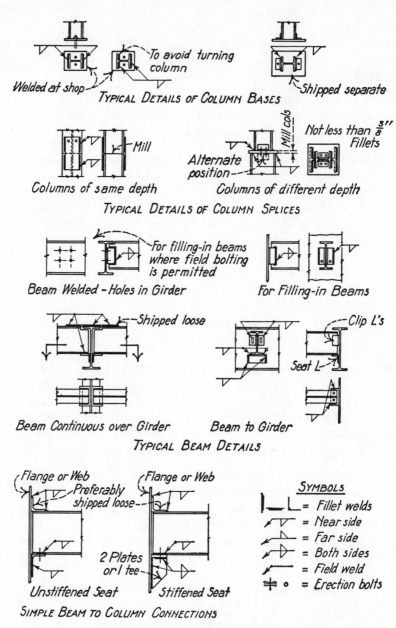

FIGURE 6.22   Typical welded connections.

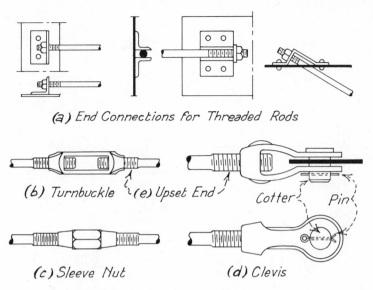

(a) End Connections for Threaded Rods

(b) Turnbuckle   (e) Upset End   Cotter   Pin

(c) Sleeve Nut   (d) Clevis

FIGURE 6.23   Connections for rods or bars.

where they cause the least moment. They are connected to the girders by framed connections.

(4) The floor beams are attached to the columns by seated connections. Framed connections can be used if there is room for them.

(5) Framed connections are used between the girders and columns. Seated connections can be used if the stiffener angles can be small enough so that they will not project through the fireproofing.

(6) Wide-flange section columns are used. The columns are continuous through two stories, or sometimes continuous for three stories.

(7) The column splices are located about 2 ft. above the floor.

(8) Reinforced concrete floor slabs supported by steel beams are shown. Many other types of floor, as described in Article 59, might have been used.

(9) The supports for the columns are not shown. The columns would rest on slabs or grillages which would spread the column load over concrete footing or piers.

(10) All steel members are fireproofed.

Wind Bracing for Industrial Buildings and Arches.   Buildings must be designed to resist the horizontal forces caused by wind as well as the vertical forces from the weight of the buildings and their contents. Provision must also be made for the lateral thrust of cranes and

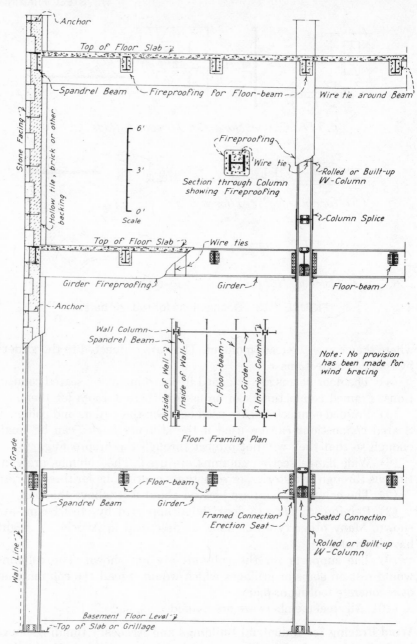

Anchor

Top of Floor Slab

Spandrel Beam Fireproofing for Floor-beam Wire tie around Beam

Stone Facing

Hollow tile, brick or other backing

6'

3'

Fireproofing

Wire tie

0'

Scale

Section through Column showing Fireproofing

Rolled or Built-up W-Column

Column Splice

Top of Floor Slab Wire ties

Girder Fireproofing Girder Floor-beam

Anchor

Wall Column

Spandrel Beam

Outside of Wall

Inside of Wall

Floor-beam

Girder

Interior Column

Note: No provision has been made for wind bracing

Floor Framing Plan

Grade

Floor-beam

Spandrel Beam Girder

Framed Connection

Erection Seat

Seated Connection

Rolled or Built-up W-Column

Wall Line

Basement Floor Level

Top of Slab or Grillage

FIGURE 6.24 Tier building construction without special wind bracing.

other equipment. Bracing provided to resist all lateral loads is usually called *wind bracing*.

A simple industrial or steel mill building frame with lateral bracing omitted is shown diagrammatically in Figure 6.25 (*a*). The steel frame for a building with three-hinged arches is shown in (*b*). The arches are braced in pairs, as shown by broken lines. Several types of steel industrial building frames are shown in (*c*). These are provided with lateral bracing as described in the following paragraphs. Three types of wind bracing are used in steel mill buildings. The simplest type makes use of *knee braces* to brace the columns rigidly to the trusses (Figure 6.26(*a*)), or trusses with considerable depth at the ends are rigidly fastened to the columns (*b*). This type of construction provides bracing to resist wind forces on the sides of the buildings, but wind forces against the ends of the buildings are resisted by *cross bracing* in the plane of the sides or by a rigid lattice girder just below the eaves.

Another method of bracing steel industrial buildings to resist wind forces consists of providing cross bracing in the plane of the bottom chord of the truss to make the building rigid from end to end in this plane. In addition to such bracing, the sides and ends are made rigid by *diagonal bracing* so that the whole structure acts like a rigid box (see Figure 6.28). It is not necessary to provide diagonal bracing in all the *bays* (spaces between columns) on the sides. Bracing is also provided in the plane of the top chord to hold the tops of the trusses in position, but this bracing is not essential to the wind bracing system, and therefore it is not shown in the figure. It is evident that this system might interfere seriously with the windows and doors.

A third system provides lattice girders in the sides and ends to make them rigid. The bracing in the plane of the bottom chord is the same as in the previous method (*c*). Various other arrangements are used.

Wind Bracing for Tier Buildings.   In tall buildings of skeleton construction such as office buildings, providing adequate bracing to resist wind is an important feature of the design. If a frame as illustrated in Figure 6.27(*a*) is subjected to lateral forces such as wind forces, it would collapse by distorting as in (*b*). In low buildings of considerable width, the stiffening effect of the panel walls and partitions and the rigidity of the joints between the girders and columns may be sufficient to provide satisfactory resistance to wind forces and earthquake shocks, but this is not true of tall buildings.

Code requirements for wind loads and earthquake shocks were con-

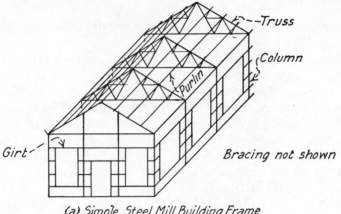

(a) Simple Steel Mill Building Frame

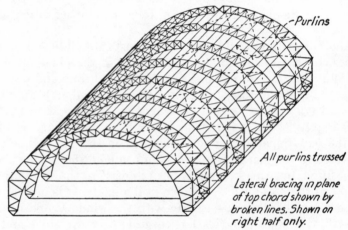

All purlins trussed

Lateral bracing in plane
of top chord shown by
broken lines. Shown on
right half only.

(b) Steel Frame with Three-Hinged Arches

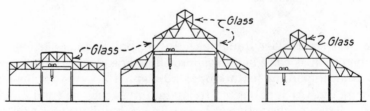

(c) Types of Steel Mill Building Frames

FIGURE 6.25 Framing for steel industrial buildings.

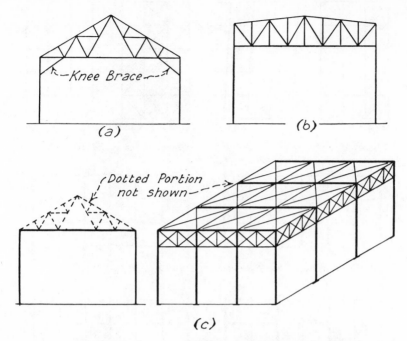

FIGURE 6.26   Wind bracing for steel industrial buildings.

sidered in Article 2. No further mention will be made of bracing to resist earthquake shock except to note that the methods employed are similar but more extensive than those employed for wind bracing.

In designing the wind bracing for a building, the stresses in the structural frame must not exceed allowable values, and the deflections and vibrations of the building must be controlled to prevent damage to the structure. The maximum permissible lateral deflection at the top of a building is sometimes limited to one one-thousandth of the height.

Several types of wind bracings have been devised. The most direct type consists of diagonals crossing the vertical panels between the columns and the floor girders (Figure 6.28). It is evident, however, that this method is limited in its application because of its interference with the use of the building and with the locating of windows and doors, even though it is necessary to brace only a relatively small number of panels, as will be explained later.

If sufficiently rigid connections are provided between the girders and columns (Figure 6.27(d)), a skeleton frame will be able to resist the lateral forces. Obviously, there is a tendency to bend the columns

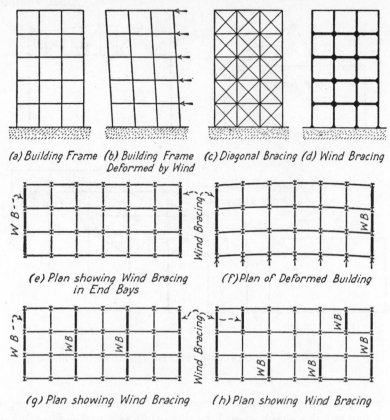

(a) Building Frame   *(b) Building Frame*   *(c) Diagonal Bracing*   *(d) Wind Bracing*
Deformed by Wind

(e) Plan showing Wind Bracing
in End Bays

(f) Plan of Deformed Building

(g) Plan showing Wind Bracing

(h) Plan showing Wind Bracing

FIGURE 6.27   Wind bracing for steel office or tier buildings.

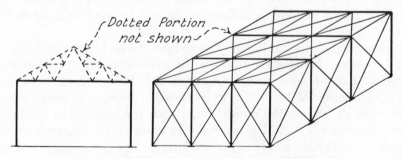

FIGURE 6.28   Bracing for steel mill buildings.

and girders when this type of bracing is used, and therefore it is necessary to consider the bending stresses in the design of these members. This type of wind bracing can be arranged to interfere very little with the design and the use of a building.

The group of braced vertical panels in a single vertical plane designed to resist wind stresses is called a *wind bent*. It is not necessary to make all panels of a building rigid, although this may be desirable as it reduces the size of the bracing. Sometimes wind bents can be placed in the outside walls (*e*). With this arrangement, the floors of a building tend to deform (*f*) under the action of wind forces. The floors used in modern building construction are usually rigid enough to carry the wind load to the wall bents; but if they are not, special bracing can be provided in the plane of the floors. It may be undesirable or impossible to place all the required wind bracing in the outside bents of a building; if so, some of the interior bents must be utilized. It is desirable but not necessary for these to be continuous across the building (*g*), but wind bents may be distributed throughout the building (*h*), each designed to carry its part of the wind load. They should be so placed that there is no twisting effect in the frame caused by greater rigidity on one side than on the other, and due consideration must be given to relative lateral deflections.

In buildings that diminish in size in the upper stories, wind bents located in the exterior walls may be continued down through the interior of the lower part of the building, or sometimes it may be desirable to transfer the wind loads to wall bents in the lower stories (Figure 6.29(*a*)). The horizontal effect of the wind on the upper section may be transferred to the wall bents or other bents in the lower stories by means of a heavy concrete floor slab where the building changes section, or by special bracing in the floor. The vertical reactions of the columns of the wind bents are transferred directly down through the corresponding columns in the lower section.

In buildings with towers projecting above a relatively low and broad main building, the towers are usually provided with wind bracing which is independent of the main structural frame (*b*). The tower bracing may be in the exterior bents only or in the interior bents also.

Buildings of irregular shape and buildings which change in section require special study. The system of bracing shown in (*c*) has been used on such buildings. No wind bracing is required for the upper floors. From sections *M–N*, *P–Q*, and *R–S*, it is seen that the wind bracing is placed entirely in the outside walls, but all outside walls do not contain wind bracing. The horizontal thrust on the portion of

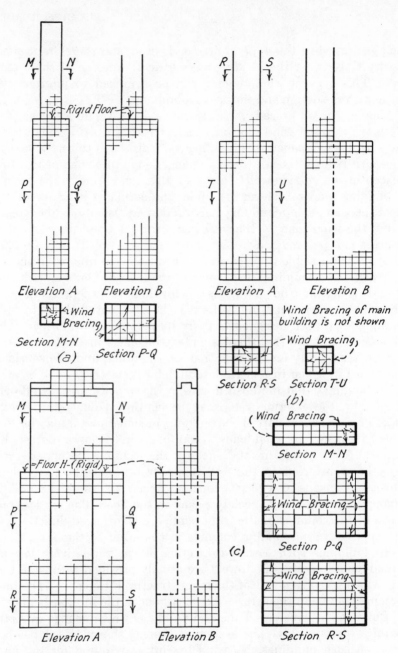

FIGURE 6.29 Wind bracing for steel office or tier buildings.

the building above floor H is transmitted to the wind bents in the outside walls below floor H. In order to transmit this horizontal thrust, floor H must be specially designed. The vertical wind loads in the columns above floor H are transmitted directly down the same columns below floor H.

Walls enclosing elevator shafts and stairways or other walls without openings sometimes can include diagonal wind bracing, and sometimes vertical bents have been converted into vertical cantilever wind trusses. In addition, instead of using diagonal bracing or vertical trusses, reinforced concrete walls may be installed in such locations. These are called *shear walls*. They are relatively rigid.

For further comments, see paragraph entitled lateral bracing in Article 48.

Wind-Connection Details. Various types of connections between wind girders and columns, using shop rivets and field rivets or high-strength bolts, are illustrated in Figure 6.30 (a) to (f). The *simple connection* in (a) may be used when the moment is small. It can be used on interior connections as well as on exterior connections, since it does not occupy any space. The *knuckle connection* (b) will develop considerable resisting moment and occupies very little space. The connecting members are structural tees cut from wide-flanged shapes. A *framed connection* can be used with this detail in place of the *seated connection* shown. The connection shown in (c) is sometimes used. The brackets are wide flange or S-beams which have been cut diagonally. A triangular plate and a pair of angles can be substituted for each wide flange or S-beam bracket, and a plate girder can be substituted for the wide flange or S-beam. The connection shown in (d) can be used to develop large resisting moments. A single plate is used to form both brackets and to replace a part of the web of the girder. This requires the use of splice plates as shown. The connection shown in (e) is similar to that in (d), except that stiffener angles are provided along the edges of the bracket if necessary to prevent buckling. The connections shown in (d) and (e) can be altered to have only a bottom bracket, and all three types may be changed so as to have top brackets instead of bottom brackets. The connections in (d) and (e) are called *gusset-plate* connections.

In tall slender buildings, the wind moment may place tension in some of the columns. These stresses must be provided for in the design of the column splices, in anchoring the columns to the foundations, and in the design of the foundations.

Many other types of riveted connections have been used, and con-

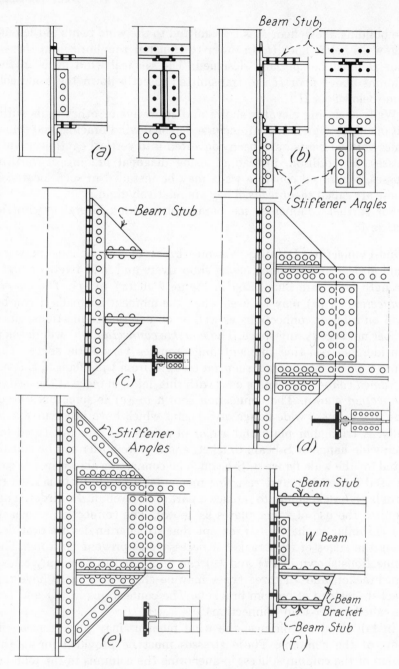

FIGURE 6.30 Types of wind connections.

nections which are suitable for welded construction have been developed.

Before about 1950, all the taller skeleton construction buildings were field-riveted. Since that time, high-strength bolts have come into extensive use to replace rivets with no restriction on the building height. Shop and field welding is also being used for buildings of considerable height and its use will doubtless continue to expand.

Lateral Bracing of Rigid Frames, Arches, and Domes. The lateral bracing for structures of these types is considered in Article 42.

Fire Protection of Structural Steel. The classification of buildings according to construction was considered in Article 1. The required fire ratings of the various parts of buildings in each class is given in Table 1-2, these ratings being based upon the behavior of materials in a *standard fire test*, which is explained. The ratings must be sufficient to withstand the hazard involved.

The capacity of stuctural steel to carry stresses is considerably reduced at the high temperatures that may be expected to prevail during a severe fire. For that reason, it is necessary to protect structural steel members with fire-resistant materials. According to the 1955 Building Code of New York City, the thicknesses of fire-resistive materials, exclusive of air spaces, required to give various fire-resistive ratings are given in Table 6-3.

Methods used in protecting structural steel against the action of fire are shown in Figure 6.31.

42.   STEEL RIGID FRAMES, ARCHES, DOMES AND CABLE-SUPPORTED ROOFS

Rigid Frames

General Comments. Rigid frames, as defined in Article 21, constructed of steel, are used extensively for auditoriums, gymnasiums, armories, field houses, industrial plants, and buildings with many other classes of occupancy. Steel rigid frames normally are not advantageous for spans of 40 ft. or less. They are widely used for spans up to 100 ft. and have been constructed for spans exceeding 200 ft. For the longer spans, however, steel arches are usually used.

Rigid frames are constructed by connecting structural steel shapes and plates either by riveting or welding them together. Welding is particularly well suited to this type of structure, however, and is usually employed.

TABLE 6-3
Thickness of Fire-Resistive Materials
For Protection of Structural-Steel Members for Various Fire Ratings, According to the 1955 Building Code of New York City.

Fire-Resistive Materials	Inches Required for Rating			
	4 hr.	3 hr.	2 hr.	1 hr.
Brick, burned clay or shale	3¾	3¾	2¼	2¼
Brick, sand lime	3¾	3¾	2¼	2¼
Concrete brick, block, or tile, except cinder-concrete units	3¾	3¾	2¼	2¼
Hollow or solid cinder-concrete block and tile having a compressive strength of at least 700 lb. per sq. in. of gross area	2½	2	2	1½
Solid gypsum block (to obtain 4 hr. rating must be plastered with ½ in. of gypsum plaster)	2	2	1½	1
Gypsum poured in place and reinforced	2	1½	1½	1
Hollow or solid burned clay tile or combinations of tile and concrete	2½	2	2	1½
Metal lath and gypsum plaster	2½	2	1½	⅞
Cement concrete, Grade I*	2	2	1½	1
Cement concrete, Grade II†	4	3	2	1½
Cement concrete, Grade II with wire mesh	3	2	2	1½
Hollow gypsum block (to obtain 4 hr. rating must be plastered with ½ in. of gypsum plaster)	3	3	3	3
Metal lath and vermiculite-gypsum plaster provided that, to obtain a 4 hour rating for columns, a backfill of loose vermiculite shall be employed. For the 3 and 2 hour ratings for floors, the thickness may be ¾ in. Thickness shown includes finish coat of plaster.	1	1	⅞	¾

*Grade I concrete has aggregate consisting of limestone, traprock, blast-furnace slag, cinders, calcareous gravel.

† Grade II concrete has aggregate consisting of granite or siliceous gravel.

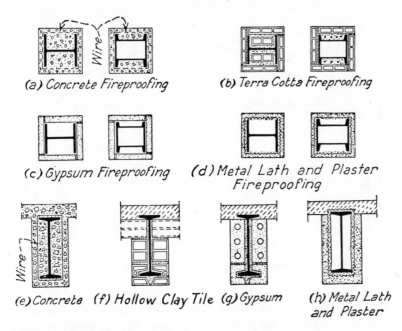

(a) Concrete Fireproofing

(b) Terra Cotta Fireproofing

(c) Gypsum Fireproofing

(d) Metal Lath and Plaster Fireproofing

(e) Concrete   (f) Hollow Clay Tile   (g) Gypsum   (h) Metal Lath and Plaster

FIGURE 6.31   Fire protection for steel members.

The Common Type. The most common type of welded steel rigid frame is illustrated in Figure 6.32. The vertical legs or columns are provided with flat bases which bear on the tops of the foundations. Actual hinges are never provided at these points, and thus there is always some resistance to rotation. If the column bases are so anchored to the foundation that this resistance is nominal and is not considered in the foundation design, the connections are considered to be hinged in preparing the frame designs. If the bases are securely anchored to the foundations so as to resist the moment fully, and the foundations are designed accordingly, the connections are considered to be fixed. Between these two extremes, the condition of partial fixity is often considered.

Fixing the column bases increases the rigidity of the frame but may increase the cost of the foundations significantly. For that reason, rigid frames are usually designed on the assumption that the joints at these points are hinged. In most, if not all, frames, full fixity is provided at the crown. Therefore, the common type of single-span rigid frame is usually considered to be two-hinged for design purposes.

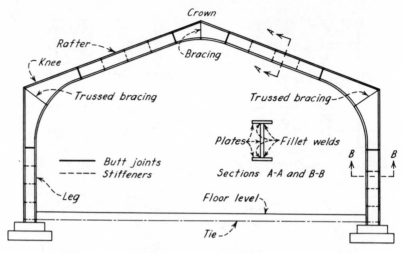

FIGURE 6.32   Steel rigid frame.

Advantages of Fixing to Foundations.   The lateral wind loads on a high, narrow, rigid frame may cause it to deflect laterally enough to damage rigid enclosing walls. In addition, a rigid frame whose columns support a traveling crane may have frequent and excessive lateral deflections caused by the operation of the crane. These might damage rigid enclosing walls and the frame itself and interfere with the operation of the crane. Since fixing the column bases reduces lateral deflections, fixed bases may be essential and fully justify the greater cost of the foundations [4].

Tie between Foundations.   The horizontal component of the thrust of each leg at the top of its foundation must be considered in the design of the foundation. The soil may be adequate to resist this component; but if not, the column bases at the two sides may be tied together by steel rods, embedded in concrete to resist corrosion, to form a tie beam located below the floor, as shown in the figure; or the tie rods may be embedded in a concrete floor. Turnbuckles (Figure 6.23(*b*) are provided on the tie rods to adjust them accurately to the required length.

Basic Components of a Rigid Frame.   The long-span frame (Figure 6.32) includes two *knees* and a *crown* with H-shaped cross sections, each made up of flange plates fillet-welded to a web plate. The portions of the frame included between the ends of the crown and knees, often called *rafters*, may be wide-flange shapes or, to obtain greater depths, they may be built up of a web plate to which flange plates are fillet welded, as for the knees and crown. The various sections are groove welded together, as indicated by the heavy lines in the figure. The bottom of each leg is fillet welded to a heavy baseplate with holes to receive bolts that anchor the frame to the foundations.

Stiffeners. Stiffeners consisting of plates fitted between the flanges, normal to the webs, and fillet welded to the flanges and the web are used in various locations. They must be provided at all points where either of the flanges changes direction to prevent the flanges and the web from buckling. Such conditions exist at the groove welded joints between the various sections indicated by the heavy lines on the figure. They also exist at the center of each knee and the crown, as shown by the light solid lines in the figure. Stiffeners may be required at intermediate points along the length of each rafter and leg, as shown by the dashed lines. This condition may prevail if the web is relatively thin or if there are heavy roof purlin loads along the rafter.

Lateral Bracing. Effective lateral bracing is required between adjacent frames, at the ridge, and in the diagonal plane through the knee where the stiffeners are located. The latter is called an *eaves strut*. Such bracing often consists of light trussed members.

Each frame and all its parts must be supported laterally to remain in its true position. Roof purlins, combined with diagonal cross bracing, consisting of angles in the plane of the top flange and located in alternate panels between frames, usually constitute the principal bracing for this part of the structure. In addition, lateral support must be provided for the legs.

Other Types. Many other types of knees and other parts may be used on rigid frames of the general type illustrated in Figure 6.33. The top member may be horizontal, steeply pitched, segmental, or curved, as shown in Figure 3.10.

Two-story buildings have been constructed, with a rigid frame for each story, and multiple-span frames (Figure 3.10) are rather common. For such frames, the knees of the interior columns are replaced by Y-shaped units. Rigid frames may be used to support sawtooth roofs (Figures 3.10 and 6.34).

Rigid frames are often used for craneways by supporting the crane girders on brakets welded to the legs. In addition, balconies may be cantilevered out from the legs of rigid frames (Figure 3.10).

FIGURE 6.33 Rigid frame in Field House at University of Colorado.

FIGURE 6.34 Sawtooth roof with welded steel rigid frames.

Arches

Use.   Steel arches are used extensively to support roofs covering large unobstructed floor areas in structures such as hangars, field houses, and exhibition halls with spans that may exceed 300 ft. Often, rigid frames would be preferred for intermediate spans.

Classification according to Hinges.   As explained in Article 21 and illustrated in Figure 3.7, arches may be three-hinged, two-hinged, one-hinged, and fixed or hingeless. The most common type is three-hinged. Hingeless arches are often used, two-hinged arches are occasionally used, and one-hinged arches are never used.

Comparisons of Classes.   The stress computations for three-hinged arches are relatively simple, whereas those for two-hinged and fixed arches are more complex. The stresses in three-hinged arches are not affected significantly by minor movements of the abutments. Horizontal movements of the abutments can be prevented by tying the two abutments together with tie rods (Figures 6.35 and 6.36) if the soil supporting the foundations cannot provide adequate lateral support. The stresses in two-hinged arches are unaffected by minor vertical movements of the abutments if the span remains constant. The stresses in fixed arches are affected by vertical or horizontal movements or rotation of the abutments. In general, the three-hinged arch is usually preferred, except when significant savings in material can be achieved by the other types, especially the fixed arch. The latter requires favorable foundation conditions.

Types of Ribs.   The arched members are ordinarily called ribs. Arches may be *girder arches*, whose ribs have solid H-shaped cross sections such as those of plate girders or wide-flange shapes, or they may be framed in a manner similar to trusses, in which case they are said to be *trussed arches*, as explained in Article 21 and illustrated in Figure 3.7.

Lateral Bracing.   Since both chords or flanges of arch ribs are in compression, they must be braced laterally to prevent buckling in that direction. This is usually accomplished by using trussed purlins whose depth commonly is equal to that of the arch rib. If the spacing of the arches is relatively small, rolled sections shallower than the ribs are sometimes used for purlins, in which case the bottom chord or flange may be supported by knee braces located near the ends of the purlins. Other lateral bracing is required to give stability to the structure as a whole. This consists of diagonal bracing between the

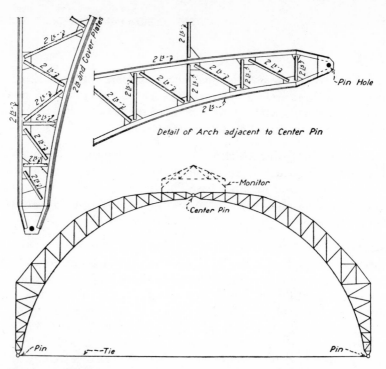

FIGURE 6.35 Riveted three-hinged trussed or framed steel arch with 2 L chords and web members.

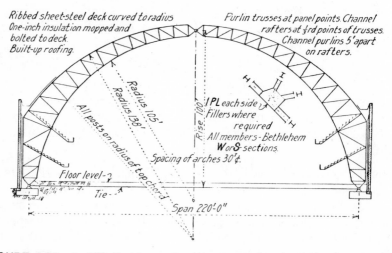

FIGURE 6.36 A 220 ft. span, riveted three-hinged trussed steel arch, with W-section chords and W-section and S-beam members. University of Minnesota Field House.

arches in the curved surface tangent to the top chord and sometimes the bottom chord also, a simple example of which is illustrated in Figure 6.25(*b*).

Methods of connecting.   The various parts of an arch may be either riveted or welded together. The method used for this purpose determines, to a large extent, the structural shapes of which the members are composed.

Girder Arch Ribs.   A riveted plate-girder arch rib is built up in a manner similar to a riveted plate girder (Figure 6.13(*f*)) with a web plate and two flanges, each consisting of two angles placed back to back and cover plates as required. A welded plate girder arch rib consists of a web plate to which top and bottom flange plates are welded, as for the welded plate girder section shown in Figure 6.14(*a*).

Riveted or welded plate girder ribs may be made up of straight segments spliced together or they may fabricated to conform to the curves of the ribs. The flange angles and flange plates for ribs of the latter type are cold-bent to the curve of the arch, and the webs are cut with this curvature. If wide-flange shapes are used for arch ribs, they may be made up of straight or cold-bent segments groove-welded together.

Trussed Arch Ribs.   A determining factor in selecting the flange or chord sections of a trussed arch rib is the fact that provisions must be made for connecting the ends of the web members to the chords. In riveted trussed arches, these connections are usually made by gusset plates. Two angles placed back to back with gusset plates between them (Figure 6.35) is a common form. The cross section of the flange can be varied, without changing the angles, by the use of appropriate cover plates. The web members used with chords of this type are usually composed of two angles placed back to back with the gusset plates between them.

Another common arrangement for riveted trussed arches utilizes W-sections for both chord and web members (Figure 6.36). The webs of all members are placed perpendicular to the plane of the truss. Sections of the same depth, or of such a size that the differences in depth can be compensated for by the use of filler plates, are used. By this arrangement, the members meeting at a joint can be connected by two gusset plates each riveted to a flange of the W-sections, as shown in the figure. The chord members may be straight segments or they may be cold-bent to the required curvature. S-beams are also used.

The ends of members meeting at a joint of a trussed rib may be welded together in the same manner as truss members are welded, described in Article 40 and illustrated in Figure 6.18(*f*).

Riveted Three-Hinged Constant-Depth Trussed Arch.   A long-span, riveted, three-hinged trussed arch with a constant depth, except at the hinges, is illustrated in Figure 6.37. The spacing of the arches is 40 ft. [5]. The chords are composed of 14 in. wide-flange shapes with the webs normal to the plane of the arch. They are cold-bent to the desired curvature. The same wide-flange section is used throughout, and the section area is increased, where required, by riveting plates to the webs rather than the flanges, as is the common practice. Each web member consists of two angles, one riveted to the flange on each side of the rib. Only a few gusset plates are required. Web members are not provided with lacing or battens except near the center of the arch.

The chords are braced laterally by trussed purlins with depths equal to that of the rib. Other bracing is provided in the surfaces tangent to the chords. The roof deck is composed of the V-beam type of asbestos-protected sheet metal, with glass-fiber insulation applied to the underside of the deck.

Welded Constant-Depth Hingeless Trussed Arch [6].   A long-span welded constant-depth trussed arch without hinges is illustrated in Figure 6.38. The fixed ends are supported by reinforced concrete abutments. Each flange consists of a structural T-section, and each web member consists of two angles placed back to back and spaced far enough apart to fit over the stems of the flanges to which their ends

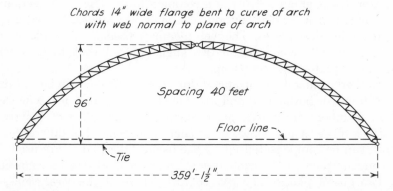

FIGURE 6.37   A 360 ft. span, riveted constant-depth, three-hinged steel trussed arch with wide-flange chords. Hangar at Edwards Air Force Base. Van Dyke and Barnes, Architects and Engineers.

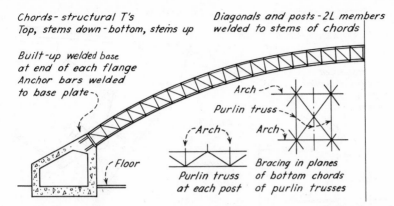

Chords- structural T's   Diagonals and posts - 2L members
Top, stems down - bottom, stems up   welded to stems of chords

Built-up welded base
at end of each flange
Anchor bars welded
to base plate

Arch

Purlin truss

Arch   Arch

Floor   Bracing in planes
Purlin truss   of bottom chords
at each post   of purlin trusses

FIGURE 6.38   A 300 ft. span, fixed arch with constant-depth rib-welded trussed and structural T chords. U.S. Navy hangar, Patuxent River, Md. Arranged from *Engineering News Record.*

are welded. The chord members are made up of segments straight for two panel lengths. The ends of adjacent segments are groove-welded together.

Since both chords are compression members and tend to buckle laterally under stress, they are braced effectively in this direction. The arches are spaced 25 ft. apart. Purlin trusses located between the arches are framed into each radial arch truss member or post to provide lateral support. They are in the same planes as these members, are as deep as the arch ribs, and are welded trusses with T-section chords and angles for web members. Diagonal bracing, consisting of T-sections, is located in surfaces tangent to the top and bottom chords and between the chords of the trussed purlins.

For ease in erection, the chords of each arch are divided into five segments of about equal length and are spliced by field welding. The roof is wood sheathing spanning the distances between the trussed purlins.

Riveted Three-Hinged Girder Arch. A long-span, riveted three-hinged girder arch is shown in Figure 6.39. The ribs have I-shaped cross sections built up of four flange angles, a web plate, and cover plates riveted together as shown in the figure. Trussed purlins provide lateral support for the arch ribs as well as for the light-gage-steel roof deck. Other bracing is provided to increase the lateral support of the arch ribs and the structure as a whole. Each arch is erected

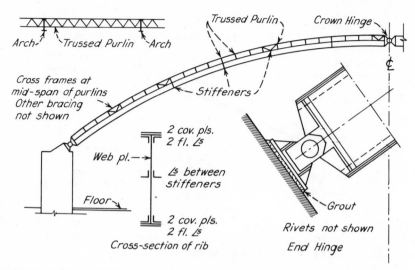

FIGURE 6.39   A 330 ft. span, three-hinged plate girder rib, riveted arch. Hangar at Kennedy International Airport, Port of New York Authority, Arranged from *Engineering News-Record*.

in four sections, locating the field splices near the quarter points of the arch [7].

Welded Three-Hinged Girder Arch.   The framing for the Field House at the University of Vermont, shown in Figure 6.40, consists of 28 welded three-hinged arches with a span of 153 ft. covering a floor 486 ft. long. The arch ribs consist of 26 × ⅜ in. web plates and 15⅝ × ⅝ in. flange plates shaped to conform to the curvature of the arch and welded together.

The roof deck consists of tongue-and-groove planks 4 in. thick, spanning between arches. They are anchored to the arches by 4 × 10 in. wood nailing strips attached to the top flanges of the arches at the fabricating shop. The roof deck is covered with a built-up roofing.

Each arch was shipped to the site in four sections, assembled on the ground in two halves, and pin-connected to the foundations. The two halves were then raised into position by truck-mounted cranes and pinned at the crown. The two sections of each half arch rib were field-connected with high-strength bolts.

Domes

General Comments.   Coliseums, field houses, exhibition halls, auditoriums, sports arenas, and other large floor areas are often roofed

FIGURE 6.40   A 153 ft. span, welded three-hinged girder arch. University of Vermont Field House. Architect: Freeman-French-Freeman. Structural Engineers: Severud, Elstad and Krueger Associates. Steel fabricator and erector: Vermont Structural Steel Corporation. Structural steel furnished by the Bethlehem Steel Corporation.

over with domes that have supporting members of steel. Such domes have been constructed with diameters of more than 400 ft.

A dome is an integral or self-contained unit. Its perimeter may be supported directly by the foundation, which carries the vertical and horizontal thrusts of the ribs. It is usually provided, however, with a steel tension ring which carries the outward thrusts of the ribs and is supported on steel or reinforced concrete columns located at the bottom ends of the ribs. They may be vertical or their tops may be tilted outward. They are braced together laterally to form a stable structure.

Patterns of Framing. Many arrangements or patterns of steel frames have been used for the support of domes. Most of them provide for a *tension ring* around the perimeter of the dome. Usually there are radial *ribs*, with bottom ends framing into the tension ring and top ends framing into a *compression ring*, with its center at the center of the dome.

Three types of patterns are illustrated in Figure 6.41. The tension and compression rings and the radial ribs are indicated by heavy

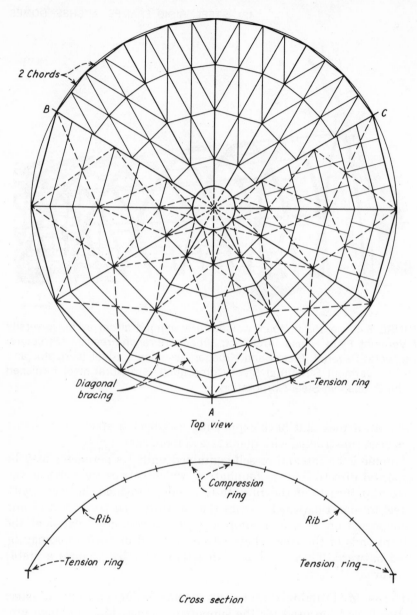

2 Chords

B

C

Diagonal bracing

Tension ring

A
Top view

Compression ring

Rib

Rib

Tension ring

Tension ring

Cross section

FIGURE 6.41   Patterns of dome framing.

lines. The rings are usually made up of straight segments. The number of radial ribs varies from 12 to 48, depending upon the diameter of the dome, the framing patterns, and other factors.

The tension ring is made up of straight segments between ribs. The rim of the compression ring is supported by radial members (shown by the dashed lines), which may be located at every second or third rib, causing the entire assembly to resemble a wheel.

For the pattern shown in the sector *AB* of the figure, the roof deck is supported directly by equally spaced straight purlins spanning from rib to rib. For the pattern in the sector *AC* of the figure, the roof deck is supported directly by rafters normal to straight purlins, which are in turn supported by the ribs. An interior view of framing with this pattern would resemble Figure 6.42. The pattern in the sector *BC* includes secondary framing members parallel to the main ribs, extending from the tension ring until they meet the ribs to form diamond-

FIGURE 6.42 Steel framed dome with wide-flange radial ribs. Field House, Syracuse University. Span 300 ft.; 36 precast reinforced concrete columns. Seating capacity 2,000. Architects: King and King. Structural Engineers: Eckerlin and Kleeper. General Contractor: R. A. Culotti Construction Co.

shaped patterns, as in the *lamella* pattern. The Houston, Texas, stadium has a 642 ft. clear-span dome of this pattern with 12 trussed main ribs and secondary ribs dividing the tension ring into 72 segments all supported on 72 steel columns [17].

Types of Ribs. The ribs may be fabricated by welding or riveting. They may be of the girder or the trussed type, as described under arches. The ends of the ribs may be fixed or hinged at their junctions with the tension and compression rings.

Lateral Bracing. For all the patterns shown in Figure 6.41, the ribs receive lateral support from the purlins which frame into them. In the patterns in sectors *AB* and *AC*, additional lateral support is provided by diagonal cross bracing, as shown by the dashed lines. The pattern in sector *BC* is made up of diamond-shaped panels which are subdivided into triangles by the circumferential members or purlins. Since triangles are stable units, each sector is stable against lateral forces.

As has been stated, a dome is usually supported on steel or reinforced concrete columns located at the bottom ends of the ribs. The dome itself, framed as described, is stable against lateral forces, but the lateral forces against the dome are transmitted to the tops of the columns. In addition, provisions must be made to resist the lateral forces on the portion of the structure below the perimeter of the dome. These may consist of rigid framing or of diagonal bracing in the panels between columns.

Dome with Wide-Flange Ribs. The roof framing for the Field House at Syracuse University (Figure 6.42) consists of a dome with thirty-six 18 in. wide-flange ribs spanning between a steel tension ring 300 ft. in diameter and a steel compression ring 18 ft. in diameter. It is supported on 36 tapered precast reinforced concrete columns 30 ft. high. The domed roof is surrounded with an overhang which projects 18 ft. beyond the exterior walls of the building.

The ribs consist of segmental sections groove-welded together. There are five intermediate circumferential rings of steel purlins, spanning between the ribs and bulb-tee rafters or subpurlins with a constant spacing of 2 ft. 9 in. spanning between the purlins, as shown in the figure. These are shimmed to conform to the spherical roof surface. There are three sets of cross-rod bracing in each of the 36 segmental panels of the dome.

Between the tension and compression rings, the roof deck of the dome is made up of 3 in. wood-fiber-cement planks spanning between rafters. These are lightweight and noncombustible and contribute significantly to heat insulation and sound absorption. The planks are covered with a 1 in. rigid insulation, over which is a conventional 5-ply built-up roofing with a marble surface. A 7 in. slab of lightweight concrete 40 ft. in diameter covers the opening within the compression ring and is elevated to provide ventilation. The 18 ft. overhang is constructed of vermiculite concrete on a sheet-metal roof deck with an asbestos-cement board underside. The ventilator slab and the overhang are covered with a conventional 5-ply built-up roofing [15].

Cable-Supported Roofs

General Comments. As stated in Article 35, cold-drawing of steel bars through dies to form wire markedly increases the strength of the steel. For this reason, cables made up of steel wires are efficient members for carrying loads that subject them to tension. Such action occurs in various forms of roofs which are supported by wire cables and are called *cable-supported* or *cable-suspended roofs*, the former designation being less common but probably preferable because it is more general. Because of the flexibility of cables, special consideration must be given to the effects of unbalanced loads on deflections and the tendency of gusts of wind to cause vibration and flutter which, if excessive, may cause failure.

According to Roebling [16], a group of wires twisted together (Figure 6.43(a)) is called a *strand* and a group of strands twisted together, as shown in (b), is called a *wire rope*. Resistance to corrosion is markedly increased by using galvanized wire.

Bridge strand and rope are *prestretched* by the manufacturer to

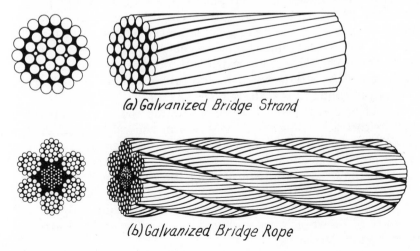

(a) Galvanized Bridge Strand

(b) Galvanized Bridge Rope

FIGURE 6.43   Wire strand and rope.

remove constructional looseness which is present when they come from the stranding or closing machines. The process is often called *prestressing*, but this term must not be confused with other meanings explained in Article 52.

Normally, the cables for cable-supported roofs consist of *galvanized bridge strand*. It is considered superior to *galvanized bridge rope*, which is also available, because it has higher strength for a given diameter and because it elongates less when subjected to a given unit tensile stress.

Galvanized bridge strand is available, from one manufacturer, in diameters varying from $\frac{1}{2}$ in., made up of 7 or 14 wires, to $3\frac{5}{8}$ in., made up of 223 wires. Galvanized bridge rope is available in diameters varying from $\frac{5}{8}$ in., with 6 strands and 7 wires per strand, designated as 6 x 7, to 4 in., with 6 strands and 43 wires per strand, designated as 6 x 43. Other manufacturers have corresponding strands and wire ropes differing somewhat in available diameters and numbers of wires. Adjustable connections are available for anchoring the ends of cables.

Single-Layer, Cable-Supported Roofs. The single-layer, cable-supported roof in Figure 6.44(a) and (b) shows the basic structural features of a covered stadium, 310 ft. in diameter, with cylindrical exterior walls 85 ft. high, located in Montevideo, Uruguay. The rainfall in this area is very low, there is no snow, and earthquakes do not occur. Such favorable conditions rarely exist. The roof support consists of a reinforced concrete *compression ring* located on top of the 4 in. reinforced concrete wall surrounding the stadium, a steel *tension ring* at the center, and 256 galvanized wire cables radiating between the two rings. The cables are made of strands 0.6 in. in diameter. The roof deck consists of thin reinforced concrete slabs precast to the dimensions of the trapezoidal space each occupies in the completed roof. These slabs are anchored to the cables and to each other by means of projecting ends of the reinforcing rods.

After all the roof slabs were in place, the roof was temporarily loaded with brick to increase the stress in the cables. Loading increased their length and widened the joints between the slabs. The joints were then filled with cement mortar. After the mortar had hardened, the brick were removed and the cables tended to return to their previous length. Their action was resisted by the slabs and joints, which were compressed, and the joints remained tight. This procedure is known as prestressing. The prestressing also increased the stiffness of the roof.

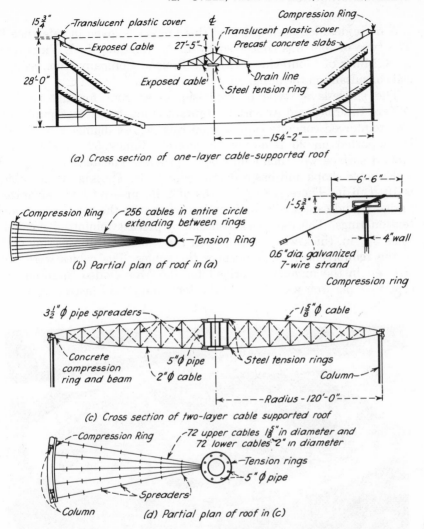

(a) Cross section of one-layer cable-supported roof

(b) Partial plan of roof in (a)

(c) Cross section of two-layer cable supported roof

(d) Partial plan of roof in (c)

FIGURE 6.44 Single-layer and double-layer cable-supported roofs.

Slabs were not placed in a circular area 65 ft. in diameter at the center of the roof. This was covered with a ventilated skylight (Figure 6.44(a)). Because of the absence of snowfall, it was not necessary to provide for an unbalanced snow load.

The tops of the slabs were waterproofed. Rainwater that falls on the roof is collected in a trough around the central opening and conducted through four pipes, hung from the ceiling, to downspouts on the outside wall and into a storm sewer.

A more detailed description of this project is given in reference 8. The type of roof was conceived by L. A. Mondino, L. I. Viera, and A. S. Miller of Montevideo with the Preload Company, Inc., as special consultant on the design of the roof system.

The single-layer type of cable-supported roof for the Chicago O'Hare International Airport Restaurant is illustrated in Figure 6.45. The reinforced concrete compression ring, whose diameter is 190 ft., is supported on 26 reinforced concrete columns 58 ft. above the ground surface with diamond-shaped profiles. The tension ring is a wide-flange shape and has a diameter of 13 ft. The sag of the cables is 10 ft. 6 in. There are 52 lengths of 2 in. prestretched galvanized bridge strand cables with their ends anchored to the compression and tension rings. The roof deck consists of precast reinforced concrete slabs 3½ in. thick which fit between the cables with projecting ends of the reinforcement hooking over the cables. The cables were prestressed in the manner described for the Montevideo Stadium, except that pig iron was used for the temporary load instead of brick.

FIGURE 6.45   Single-layer, cable-supported roof. Chicago O'Hare International Airport Restaurant. Architects and Engineers: C. F. Murphy Associates. Roof Contractor: E. H. Marhoefer, Jr. Cables furnished by John A. Roebling's Sons Division, Colorado Fuel and Iron Corp.

A cable-supported roof similar to that of the Montevideo Stadium was constructed for the Assembly Building at San Antonio, Texas. The span is much smaller, only 132 ft. There are 200 cables attached to a welded built-up steel ring. This ring is designed to carry the compressive and flexural stresses to which it is subjected because it is supported by 20 vertical steel columns spaced about 20 ft. apart around the perimeter. The inner ends of the cables are attached to a welded-steel tension ring 40 ft. in diameter. Steel trusses radiating out from a hub at the center of the roof to the tension ring support the roof over the area surrounded by the tension ring.

The remainder of the roof consists of small trapezoidal shaped precast concrete slabs arranged in concentric rings, supported on the cables, with open joints which were filled with mortar after the cables had been prestressed by loading the roof with brick. All operations follow procedures similar to those used in building the Montevideo Stadium. This comment also applies to the drainage system [13].

Double-Layer, Cable-Supported Roof.   A roof supported by a double layer of cables is illustrated in Figure 6.44(c) and (d). This roof covers the Memorial Auditorium at Utica, New York, a circular building 240 ft. in diameter. The supporting system consists of two 72-cable layers. The outer ends of the cables in both layers are anchored to a reinforced concrete compression ring, with a section area of about 20 sq. ft., supported on twenty-four 2-ft.-square reinforced concrete columns equally spaced around the perimeter of the building and supported laterally. The two welded-steel tension rings are 18 ft. apart vertically and are held apart by equally spaced steel pipes 5 in. in diameter. Vertical pipe struts are inserted at intervals of about 14 ft. between each pair of cables in the same vertical plane and located on concentric circles.

The two tension rings were supported in contact on a temporary scaffold at the center of the building while the cables were being placed and anchored to these rings and to the compression ring. At this stage, the tension rings were at a predetermined elevation. Hydraulic jacks were inserted between the rings and gradually forced them apart. Downward movement of the lower ring was made possible by removing timber blocking which was provided on top of the scaffold. Jacking was continued until the two layers were 18 ft. apart, vertically, at the center. Since the ends of the cables were anchored to the tension and compression rings, the vertical movement produced by the jacks subjected the cables to tension. The operations were so controlled that the stresses in the cables reached predetermined values. The jacks were replaced by permanent pipe separators. The tensions in the cables were then increased to desired values by inserting the vertical struts between the upper and lower cables, as shown in the figure. Diagonal cross bracing, consisting of rods, was placed in each panel between struts. This bracing distributed the effects of concentrated loads due to mechanical equipment and stiffened the roof. Final adjustments were made to insure that the tension in the cables was in close agreement with the desired values.

Inducing tensile stresses in the cables before the loads to be carried are applied is called *pretensioning* or *prestressing*. Pretensioning controls the deflection of the roof and prevents objectionable vibrations which may develop in cable supported structures. Both cables remain in tension under all conditions of loading.

The roof deck is constructed of light-gage steel supported directly on the upper cables except near the perimeter where intermediate channel purlins are required.

A more detailed description of this project is given in reference 9. The structural system was conceived and designed by Lev Zetlin, consulting engineer.

A cable-supported roof, which resembles the roof just described, was provided for the United States Pavilion at the Brussels Worlds Fair. The span is 302 ft. There are two layers of cables, but they are not prestressed [10].

Cable-Supported, Cantilever Roofs.   Cable-supported cantilevers are used to support the roofs of hangars. They may be of the single- or double-cantilever types. Portions of the structures other than the cables may be constructed of structural steel, reinforced concrete, or various combinations of these materials.

One example of a cable-supported, single-cantilever hangar is illustrated in Figure 6.46(a). The principal roof support consists of a series of heavy wide-flange steel girders, each hinged at one end to a rigid side section of reinforced concrete and supported near the other end by a cable passing over a steel mast erected above the innerside of the side section and anchored to the outer side of that section. The roof deck consists of precast reinforced concrete channel slabs, as described in Article 53, supported on rolled-steel purlins spanning between the cantilever girders. The relative weights and dimensions of the cantilever and side sections and the anchorage provided must insure stability [11].

A cable-supported, double-cantilever hangar is illustrated in Figure 6.46(b). The steel cantilevers project outward from both sides of a center section. The inner ends of the cantilevers are hinged to the

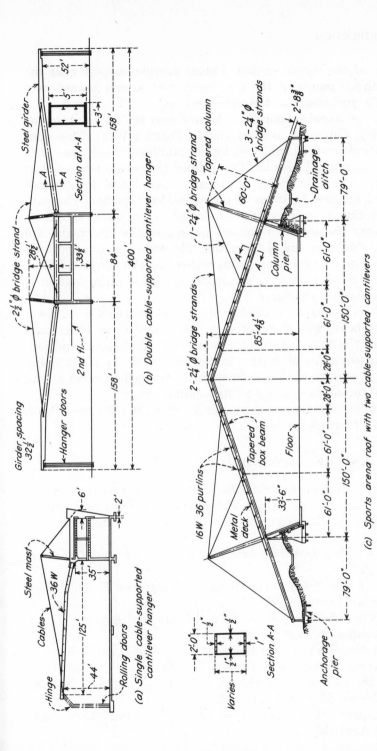

(a) Single cable-supported cantilever hanger

(b) Double cable-supported cantilever hanger

(c) Sports arena roof with two cable-supported cantilevers

FIGURE 6.46 Cable-supported cantilever roofs.

393

steel frame of the center section. Cables provide support near the outer two-thirds points of the cantilevers and extend inward over steel masts to anchorages on the center section.

Principal structural supports for the roof over the Sports Arena at Squaw Valley, California, is illustrated in Figure 6.46(c). They consist of two independent cable-supported cantilevers with free ends meeting at the ridge of the roof. The inclined girders are tapered-steel box sections. The columns or masts are tapered-steel four-flanged cross-shaped sections built up of plates welded together. The cantilever projection of each girder is supported at two points by cables that pass over the top of a column normal to the girder and are anchored to the lower end of the girder. The end is in turn anchored to a pier, the uplift at this point being resisted by the pile foundation. Each column is supported by a column pier. The entire system is arranged as shown in the figure.

The roof deck consists of cellular sheet-steel panels, 4½ in. deep, supported on heavy wide-flange steel purlins spaced about 12 ft. apart and spanning the 32 ft. between the cantilevers, all designed and connected so as to require no further lateral bracing in the plane of the roof. The top surface of the roof panels is arranged to serve as the finished roof surface to receive a plastic coating.

Further information on this project is given in reference 12, on which these comments are based.

REFERENCES AND RECOMMENDED READING

1. *Steel Construction,* American Institute of Steel Construction.
2. *Specifications for the Design, Fabrication and Erection of Structural Steel for Buildings,* American Institute of Steel Construction.
3. *Structural Shop Drafting,* American Institute of Steel Construction, 1950.
4. *Procedure Handbook of Arc Welding Design and Practice,* Lincoln Electric Company, 1957.
5. "Steel Trussed Arches Span 360 ft. in Hangar," *Engineer News-Record,* August 9, 1956, p. 36.
6. Arsham Amirikian, "Navy Builds 300-Ft. Welded Arch Hangar," *Engineering News-Record,* January 26, 1950, p. 24.
7. "Long Steel Arches for Hangars Erected Fast," *Engineering News-Record,* May 25, 1950, p. 40.
8. M. Schupack, "Cable-Supported Roof Cuts Cost," *Civil Engineering,* April 1958, p. 52.

9. "Prestressing Stabilizes Unusual Cable Roof," *Engineering News-Record*, April 28, 1960, p. 36.
10. "U.S. Pavilion at Brussels Features Cable Supported Roof," *Civil Engineering*, July 1955, p. 95.
11. Boyd Anderson, "Planning Hangars for Jet Aircraft," *Civil Engineering*, May 1959, p. 57.
12. "Cable Supported Roof for Olympic Arena," *Civil Engineering*, September 1959, p. 47.
13. W. E. Simpson, "Cable-Suspended Roof for San Antonio Assembly Building," *Civil Engineering*, November 1960, p. 63.
14. Howard Seymour, "Suspension Structures," *Architectural Record*, September 1960.
15. Edward C. Kurt, "Steel Frame Dome Spans Arena," *Engineering News-Record*, May 10, 1962, p. 39.
16. *Suspension Bridge Technological Data*, John A. Roebling's Sons Division, Colorado Fuel and Iron Corp., Trenton, New Jersey, p. 1.
17. "Record-Span Dome Roofs Air-Conditioned Stadium," *Engineering News-Record*, February 27, 1964.
18. *Manual of Steel Construction*, Seventh Edition, American Institute of Steel Construction, Inc., New York, New York, 1970.
19. Brochure: *Steel-Making Flow Charts*, American Iron and Steel Institute, Washington, D. C., 1970.
20. *USS Stainless Steel, Architectural Design Details*, United States Steel Corp., San Francisco, California.
21. *The Picture Story of Steel*, American Iron and Steel Institute, Washington, D. C.
22. *Exposed Steel, Architectural Design Details*, United States Steel Corporation, San Francisco, California, 1966.
23. *ASTM 1971 Annual Book of Standards, Part 14*, American Society for Testing and Materials, Philadelphia, Pennsylvania.
24. *Specification for the Design, Fabrication and Erection of Structural Steel for Buildings*, American Institute of Steel Construction, New York, New York, 1969.

7

REINFORCED CONCRETE CONSTRUCTION

43. CONCRETE MATERIALS

Concrete is a mixture of portland cement and water combined with inert materials called aggregates—sand, gravel, and crushed rock. The cement and water combine chemically, forming a paste that sets and hardens, binding the aggregates into a rocklike mass.

Cement Manufacture. The materials necessary to the manufacture of portland cement are lime, silica, and alumina. These ingredients are obtained by processing a mixture of limestone with impure limestone that contains considerable clay, or with clay or shale, or with blast furnace slag. They must be mixed in the proper portions as determined by analysis.

At the risk of oversimplification, the process is as follows. Quarried materials, such as limestone, are introduced to the primary crusher, where they are reduced to about 6 in. material. Further crushing and grinding in the secondary crusher brings the material down to about ¾ in. Measured portions of raw materials, mentioned above, are then blended. This blend is subjected to fine grinding and further mixing in *ball mills* and *tube mills,* a step accomplished by a *dry process* or a *wet process* in which water is added to the raw materials, forming a slurry.

The raw materials are then submitted to a rotary kiln for about four hours, where they are subjected to temperatures of about 3000°F. The resultant *clinker* is cooled. Gypsum, to control the time of set, is then added, and the mixture is pulverized by a series of grindings to an extremely fine powder—cement.

The cement is then cured and prepared for shipment in bulk or in the familiar 1 cu. ft. paper bags weighing 94 lbs.

Types of Cements. Portland cement should conform to ASTM C 150. All cement types are not alike in behavior and properties. Special portland cements satisfy specific requirements.

There are five types of portland cement that are generally used for construction. *Type I* cement is called *standard* or *normal*, and is used for constructing columns, beams, floor slabs, and other general construction purposes. *Type II* cement is *modified*. Similar to Type I, it has a lower content of tricalcium of aluminate and is intended for the same usage as Type I but in localities where a light sulfate resistance is required. This cement, which has a lower heat of hydration (see below), is used in structures of larger dimension and massive pours, such as dams. *High Early Strength* cement is *Type III* and attains an accelerated strength, as is desired for precast concrete. The increased tricalcium silicate content accelerates the heat of hydration. This makes Type III cement most suitable for cold weather concreting. This type of cement should not be used in pouring members of large dimension because the high heat of hydration tends to cause expansion and subsequent cracking. *Type IV* cement, termed *low heat*, is used in massive pours to diminish cracking. *Type V* cement is *sulfate resistant*. It is used in areas where excessive sulfate activity is expected. Type V cement should be used for sewage treatment plants and for concrete drainage structures that are expected to carry waters with a high sulfate content.

Air entrained cements produce concrete that has innumerable microscopic air bubbles. This permits the concrete effectively to resist *scaling* and *spalling*, which are deterioration caused by the contraction and expansions of freezing and thawing cycles. Air entrained concrete, identified as *Type IA, IIA,* and *IIIA,* is also characterized by improved workability, reduced segregation, and negligible bleeding. In addition to their air entraining properties, Type IA possesses the usual properties associated with standard; Type IIA, those associated with modified; and Type IIIA, those associated with high early strength cements [20].

A special type of cement is white portland cement, which is produced in Types I, III, and IA. This nonstaining cement is used effectively with coloring agents or as plain white cement in terrazzo, precast concrete, masonry mortar, and stucco or in any other building component suitable for gray concrete.

A warm buff-tone, Type I portland cement is now manufactured. The adobe-colored concrete produced offers a relief to the monotony of the usual dead gray concrete.

Mixing Water.   Water should be free from excessive amounts of oils, acid, sugar, alkali, organic matter, or other deleterious substances, but *potable* water is usually satisfactory for making concrete. Seawater can be used for concrete, but requires a richer mixture. *Surface* water, which is nearly always present on aggregates even though they appear dry, must be considered in determining the amount of mixing water. Also, if dry aggregate is used, volume corrections must account for the water that the aggregate will adsorb. In any event, the quantity of water required is the net amount that enters the cement paste.

Setting.   When water is added to portland cement, a chemical reaction, *hydration*, transforms the materials into a product that possesses cementatious properties. As this mortar sets, it releases heat, known as the *heat of hydration*. The process of setting is an *initial set*, followed by the *final set*. The rate of the setting is measured by the penetration of weighted Gillmore needles, or by the Vicat device. Initial set should not occur before 45 minutes, nor should final set require more than 10 hours.

Infrequently, a premature setting, known as *false set*, occurs in a few minutes. Its cause is thought to be due to improper manufacturing and curing of the cement, with gypsum as the culprit.

Aggregates.   The inert material, called *aggregate*, is usually divided into two classes according to size. Usually, all aggregate passing a sieve with $\frac{1}{4}$ in. openings is *fine aggregate* and that retained is *coarse aggregate*, but other grading criteria are in use. Aggregates graded so that the finer particles progressively fill the voids surrounding coarser particles have a smaller percentage of voids than aggregates consisting of uniform particles. Therefore, the graded aggregates are more economical because a smaller amount of cement paste is required to fill the voids. Specifications ordinarily state the size grading of each aggregate.

Fine aggregate is usually sand, and coarse aggregate is gravel or crushed stone; but when sand is unavailable, fine aggregate can be made by crushing rock. Crushed, air-cooled, *blast-furnace slag* is sometimes used.

Some aggregates produce lightweight concrete. *Cinders* and *coke breeze*, because of their sulfur content, tend to corrode steel reinforcing and embedded steel pipes. *Burned clays* and *shales*, which expand, become vitrified during the burning process. *Vermiculite* ore is micaceous aluminum. *Magnesium silicate* greatly expands by sudden heating, causing the water between its thin layers to vaporize, forming greatly expanded lightweight cellular granules. *Perlite* is a volcanic siliceous rock crushed and heated to a high temperature, causing combined water to vaporize and form innumerable tiny bubbles inside the lightweight glassy particles. Others are natural lightweight stones such as lava, pumice, and tuft. *Nailable concrete* is produced by using asbestos fiber, sawdust, and cinder aggregates.

Sometimes aggregates are natural mixtures of sand and gravel just as they come from a gravel pit, without grading or washing. This is called *pit-run gravel, bank gravel,* or *bank run.* It is not a satisfactory aggregate because of the uncertain quality of the resultant concrete. Better and more economical concrete can normally be produced by carefully graded aggregate because a saving in cement will be realized.

Aggregates should be rounded, insoluble, clean, strong, hard, and durable particles, which are not coated with dust, clay, silt, or other substances. They should not expand or contract when wet with water or react chemically with the cement. Shale and some cherts are particularly objectionable because of their volume change and unsoundness. Flat, elongated, and angular particles, if present in large quantities, reduce the workability of concrete. Aggregates are *screened* to secure the desired gradings and are washed to remove impurities.

The maximum usable size of aggregate depends upon the character of the work. The size should not exceed one-fifth the width or thickness of the form in which it is being placed or three-fourths the clear space between reinforcing bars. A maximum size of 3 in. is often specified. Questionable aggregates should be tested because there is the possibility of producing inferior concrete, regardless of the proportions used. The tests may include a sieve analysis (ASTM C 136), analyses for organic impurities (ASTM C 40) and soundness (ASTM C 88), and the Los Angeles Rattler test (ASTM C 131) to determine aggregate hardness. The compressive strength of concrete is a good index of the quality of the aggregates used and of the durability, water tightness, and other desirable qualities of the concrete itself.

Heavyweight aggregates are used in concrete for nuclear structures requiring radiation shielding and other purposes requiring dense concrete. Such natural occurring heavyweight aggregates are *barite, limonite,* and *magnitite.* Steel products such as steel chips, nuts, bolts, shot, filings, and punchings are also a type of heavyweight aggregate. This type produces the densest concrete, weighing nearly 300 lbs. per cu. ft.

Admixtures.   *Admixtures* are materials added into the standard concrete mixture for the purpose of controlling, modifying, or imparting some particular property to the concrete mix.

Some additives have more than one effect on the concrete. These effects may be desirable or detrimental, depending upon conditions and requirements. For example, an admixture may retard the set, permitting improved finishing while reducing the water content, or an additive that accelerates the time of set may have adverse effects by producing excessive cracking.

Functionally, admixtures are implemented for a number of reasons, which are clearly summarized by ACI committee 212 [10], as follows:

(1) To increase workability without increasing water content or to decrease the water content at the same workability.

(2) To accelerate the rate of strength development at early ages.

(3) To increase the strength.

(4) To either retard or accelerate initial setting.

(5) To retard or reduce heat evolution.

(6) To modify rate of or capacity for bleeding, or both.

(7) To increase durability or resistance to severe conditions of exposure including application of ice-removal salts.

(8) To control expansion caused by the reaction of alkalies with certain aggregate constituents.

(9) To decrease capillary flow of water.

(10) To decrease the permeability to liquids.

(11) To produce cellular concrete.

(12) To improve penetration and pumpability.

(13) To reduce segregation particularly in grout mixtures.

(14) To reduce or prevent settlement, or to create slight expansion, in concrete or mortar used for filling block-outs or other openings in concrete structures, and in grout for seating machinery, columns or girders, or for filling post-tensioning cable ducts or voids in preplaced aggregate.

(15) To increase bond of concrete to steel.

(16) To increase bond between old and new concrete.

(17) To produce colored concrete or mortar.

(18) To produce fungicidal, germicidal, and insecticidal properties in concrete or mortars.

(19) To inhibit corrosion of embedded corrodible metal.

(20) To decrease the unit cost of concrete.

An *accelerating admixture* increases the rate of set and the rate of strength gain. Accelerated strength development, a decided advantage in precast concrete work, is desirable where early form removal and erection are necessary for economy. A further advantage of accelerators is a reduction of the curing time. Although there are several other accelerating admixtures, calcium chloride is the most widely used.

Air-entraining admixtures are certain detergents, as in the salts of wood resins, petroleum acids, sulfonated lignum, proteinaceous materials, and fatty and resinous acids. These admixtures cause numerous air bubbles about 1 mm in diameter to form in the concrete. This produces a concrete with an improved workability that permits a reduction in the quantity of water required for placement. With the improved water-cement ratio that results, some increase in concrete strength is realized. The microscopic bubbles result in improved resistance to freezing and thawing cycles and decreased spalling and deterioration caused by the weathering cycles. Air-entrainment is particularly desirable in concrete used for highway work in colder regions.

Admixtures are further classified as *water-reducing, retarding, water-reducing and retarding,* and as *water-reducing and accelerating.*

These water-reducing admixtures reduce the water requirement while producing a concrete that retains its workability and consistency. Retarders and accelerators modify the setting times of concrete. This permits increased working time for finishing operations, especially exposed aggregate finishes. Retarders, such as ligno sulfonate, are also useful in obtaining crack-free surfaces, particularly in pours of large volumes. Sugar and citric acids have a similar effect on concrete. They should not be used as admixtures, but precautionary measures should avoid the use of mixing waters or form boards that have any contact with these materials.

Finely divided mineral admixtures are known as relatively chemically inert materials, cementitious materials, and pozzolans. The *inert materials* include ground quartz, ground limestone, bentonite, hydrated lime, and talc. Some of the *cementitious materials* are natural cements, hydraulic limes, slag cements, and granulated iron blast furnace slag. ASTM C 219 defines *pozzolan* as "a siliceous or siliceous and aluminous material, which in itself possesses little or no cementitious value but will, in finely divided form and in the presence of moisture, chemically react with calcium hydroxide at ordinary temperatures to form compounds possessing cementitious properties." Fly ash, a diatomatious earth, volcanic glass, and some shales or clays are pozzolanic materials.

These admixtures produce a concrete that has excellent workability. If properly proportioned, they are beneficial in concrete that is pumped through pipe lines. Mineral admixtures vary in their effect on concrete. Generally, they may improve sulfate resistance, effect a lower temperature rise in the mix, prevent excessive expansion, and reduce permeability. However, these properties are more or less marked in their effect depending upon the particular mineral admixture, the proportions used, and the cementitious materials used. These admixtures supplement or improve the proportions of finer aggregates required for a mix. They are also used for special concrete mixtures, such as preplaced aggregate construction.

There are several other admixtures that are occasionally used to produce special effects. *Gas-forming admixtures*, such as hydrogen peroxide and powdered aluminum, release gas bubbles in fresh concrete. Aggregate settlement and bleeding are thus minimized, and increased bond to reinforcing may be realized.

Expansion-producing admixtures, such as granulated iron, produce a grout that has a nonshrink property.

Bonding admixtures, which are frequently two-component epoxy compounds, are mostly used for patching and bonding for anchorages.

Various *color admixtures* provide black, red, brown, buff, green, and white concretes. The earthen colors are reliable and stable and are preferred choices among architects. Blues and greens are elusive.

Certain copper and chemical compounds have been used, although with mixed results, as an admixture for *fungicidal, germicidal,* and *insecticidal* deterrents. More research is definitely needed in this area.

Soaps, stearate, mineral oils, asphalt emulsions, and certain cutback asphalts are used as *dampproofing admixtures*. They do have side effects detrimental to concrete strengths. The effectiveness of dampproofing admixtures may be gravely questioned, as results do not substantiate expectations. Architects frequently prohibit any dampproofing admixtures.

Admixtures are made in liquid, paste, or powder. Accuracy in batching concrete proportions is essential, especially the additions of admixtures. Admixtures that are mixed in as liquids are air-entraining agents, water reducing agents, retarders, and those that have a two-fold function, such as water reducing retarders and water reducing accelerators. Pozzolans, gas forming agents, coloring materials, grouting, and some retarding admixtures are produced in powdered form. These types should be mixed in a dry state with the cement.

Admixtures should be of only the finest quality. When using additives, caution should be taken by using the product of only one manufacturer and by following their printed instructions without deviation.

Proportioning. The proportioning of concrete ingredients has been the subject of much research, and several methods for proportioning have been devised. A method in vogue for years called for arbitrary proportions that neglected the characteristics of the materials or the amount of water used. For example, a 1:2:4 concrete consisted of 1 volume of cement, 2 volumes of fine aggregate, and 4 volumes of coarse aggregate. This is not an entirely satisfactory method and is not recommended for engineered construction.

Emphasis is now placed on the *water-cement ratio method* of proportioning. The strength of the cement paste that binds the aggregate particles together to make concrete is determined by the proportion of water to cement used in making the paste. In order to obtain a workable plastic paste, more water is used than is actually necessary to satisfy the required chemical changes while the paste sets and hardens. However, by increasing the proportion of water (beyond that required to produce the plastic mixture), the paste is diluted and its strength and durability reduced. The result is a strength reduction in the concrete produced from dilute paste because the strength of the aggregates is usually greater than the paste and is therefore not the determining strength factor. Restated, and to paraphrase the maxim, the concrete is only as strong as its weakest ingredient. The weakest link in the "concrete chain" is the paste, and the strength of the paste is determined by the proportion of water used.

The ratio of the quantity of water to the quantity of cement is called the *water-cement ratio*—a ratio expressed in terms of U.S. gallons of water to a sack of cement. Increasing the proportion of cement paste beyond that required to produce a workable mix does not increase the strength or durability of concrete. The proportion of cement paste required to produce a workable mix depends upon the aggregate gradation. For a given volume of aggregate, coarse particles present less total surface to be covered by the cement paste than finer particles do. It is therefore desirable to use the lowest proportion of fine aggregate that will fill the voids in the coarse aggregate. However, an excess of coarse aggregate causes the mixture to be harsh and difficult to place. To correct this situation, there is a tendency to increase the amount of water, thus reducing the strength and durability. The proper procedure is to increase the workability by better grading of the aggregates, but if this is not possible, the proportion of cement paste must be increased. For a given water-cement ratio, rounded aggregates produce a more workable concrete. Angular shaped aggregates produce a concrete difficult to place. In other words, for the same workability, concrete composed of rounded aggregates requires a paste with a lower water-cement ratio and therefore produces a higher strength than concrete made from angular aggregates.

As stated, the strength and durability of concrete are dictated primarily by the water-cement ratio. The proportioning of concrete requires a water-cement ratio that will produce a strong and durable concrete meeting the required conditions of exposure. The next step is to determine the amount of cement paste to produce a workable mix. The consistency, or state of fluidity, that gives concrete the required workability depends upon the placement conditions. Concrete required for a large mass without reinforcing bars can be stiffer than concrete placed in confined, reinforced members. Also, vibrating the concrete in the forms, by special equipment, permits stiffer mixtures.

The water-cement ratio that produces concrete of the required strength can be determined by tests made with the materials used.

Permissible Water-Cement Ratios

Minimum compressive strength, lb. per sq. in. at age of 28 days	2000	2500	3000	3500	4000
Gallons of water per sack of cement	8	7¼	6½	5¾	5

The consistency required is indicated by the *slump test*, which is a convenient measure of plasticity or workability. In this test, a representative sample of concrete from the mixer is placed in an open-ended sheet-metal mold shaped like the frustum of a cone. It is 12 in. high with a top diameter of 4 in. and a bottom diameter of 8 in. and rests on a flat, stable surface. The concrete is placed through the open top in three layers, each layer is rodded 25 times, and the top surface struck off with the top of the mold. The mold is immediately lifted. The measurement which the top surface of the concrete drops is called the *slump*. A stiff mixture obviously will have a small slump and a fluid mixture a large slump. A supplementary test for consistency is the *Ball Penetration Test*, ASTM C 360. A 30 lb., 6 in. diameter hemispherical weight, sometimes called a *Kelly Ball*, is placed on a flat, freshly placed concrete surface. The depth that the ball sinks into the fresh concrete surface is recorded. This penetration, which indicates the consistency of the concrete, corresponds to one-half of an equivalent slump recorded by the slump test.

Bulking. The volume of a given quantity of damp sand is much greater than the same volume of dry sand. This volume increase is due to an increase in the moisture content and is called *bulking*. When the weight of the moisture reaches about 6% of the weight of the sand, the bulking may be as high as 20 to 30%. Further moisture additions tend to decrease the amount of bulking until the sand is completely inundated when there is practically no bulking. The bulking of fine material is more for a given percentage of moisture than that of coarse material. Coarse aggregate bulks very little. Methods of measuring sand should be such that bulking can be avoided.

Mixing. Concrete is mixed in batches by power-driven mixers with capacities from 2 cu. ft. to 8 cu. yd. or more. Although the strength and uniformity of concrete increases with the mixing time, the improvement after the first minute or two after all the materials are in the mixer is not considered to justify the longer mixing times. The speed of rotation of the mixing drum, within reasonable limits, has little effect on the quality of the concrete. The minimum mixing time permitted is $1\frac{1}{2}$ minutes.

Concrete is usually purchased from central mixing plants which deliver mixed concrete to the job in dump trucks. Where considerable delay in placement is likely, the concrete may be placed in an agitator where it can be kept workable for an hour or more. No water should be added in the agitator. The dry ingredients for concrete may be weighed at a central batching plant and dumped into transit mixer trucks with water added; the mixing will then be completed enroute to the job.

Transporting. Concrete is transported from the mixer to the forms by wheelbarrows, two-wheeled buggies, power buggies, bottom-dump buckets, dump cars, dump trucks, and chutes; or it may be pumped through hoses and steel pipes. Concrete placed by pumping is called "pumpcrete." Before selecting the mode of transport, factors that must be considered are the quantity, the job layout, and the equipment available. The method employed must prevent the separation of the materials, called *segregation*, and insure quality concrete. The use of long chutes has diminished during recent years because segregation occurs, but short chutes do not have this failing. Concrete is transported vertically in wheelbarrows and buggies by elevators; however, bottom-dump buckets lifted by cranes are used to place concrete in inaccessible places. The use of aluminum chutes and contact between concrete and aluminum should be avoided.

Placing. Before concrete is placed, the forms should be carefully cleaned; and except in freezing weather, the forms should be thoroughly sprinkled unless oiled forms are used. The concrete should be placed to avoid segregation. Dropping concrete above a height of five feet is objectionable because it causes segregation and entrapped air. The concrete next to the forms should be spaded and tamped to prevent honeycombing and to improve the appearance of the exposed surface, but excessive spading or tamping will result in segregation.

Concrete is *honeycombed* when it contains pockets where the coarse aggregate is not surrounded with fine aggregate and mortar.

The quality of concrete can be improved by consolidating it by vibration. Some types of vibrators are attached to forms, others are placed on the surface of the freshly deposited concrete, such as slabs and pavements, and still others are inserted into the concrete. They may be driven by electricity or compressed air. If vibrators are used, a stiff, low slump, low water-cement ratio concrete can be placed satisfactorily. Thus, stronger and more durable concrete can be formed from the same materials. Vibration also facilitates the flow of concrete around closely spaced reinforcement and into places that would be difficult to reach by ordinary methods, although, as a practice, concrete should not be moved laterally by vibrators.

Partially hardened concrete or concrete that has attained initial set should never be used even though its plasticity is restored by *retempering*, which means adding water and remixing.

Underwater concrete placement should be avoided. However, if concrete must be placed under water, it must not be allowed to fall through the water because the mortar would wash out. Two devices for depositing concrete under water are in use: the *bottom-dump* or *drop-bottom* and the *tremie*. The drop-bottom bucket is constructed with a bottom that opens when it touches the surface on which the concrete is to be deposited. The tremie is a steel pipe long enough to pass through the water to the required depth of placement. In starting operations, the pipe bottom is plugged to exclude water and to retain the concrete with which it is filled. It is then positioned at the point where the first concrete is to be deposited. The plug is then forced out, and concrete flows out of the pipe to its place in the forms without passing through the water. Concrete is supplied at the top of the pipe at a rate sufficient to keep the pipe continually filled. The flow of concrete in the pipe is controlled by adjusting the depth of embedment of the lower end of the pipe in the deposited concrete. The upper end of the pipe may have a funnel shape or a hopper which facilitates feeding concrete to the tremie. Care must be exercised to keep the tremie from losing its charge, otherwise it will fill with water, necessitating a recharge with the plugged tremie. If a richer concrete is used, however, some engineers permit the operation to continue while the water is forced from the tremie by the concrete; the additional cement takes the place of that washed out of the concrete by the water in the tremie. After the concrete is placed, it is not spaded or puddled.

Curing. The chemical reaction that transforms cement into hardened concrete continues for a long period if conditions remain favorable. This results in a gradual increase in the strength of the concrete and an improvement in other properties. For these reactions to continue, moisture must be present and the temperature favorable. An excess of water is always present when the concrete is placed. If the evaporation of this water is prevented or reduced, the water necessary to continue the chemical operations will be present. Evaporation is reduced by covering the surface with wet burlap or plastic film, by sprinkling, by coating the surface with a waterproof paint, spraying with various liquids, or by covering slabs with sand which is kept moist. This protection against evaporation should be continued for 7 days or longer.

Cold Weather Concreting. Concrete must be protected from freezing at least during the first 7 days after placement. The heat generated by the chemical reactions is an important factor in maintaining desirable temperatures. The mixing water may be heated and, if more heat is required, the aggregates may also be heated, especially to remove the frost and ice. Type III cement may be used. Specifications often require that concrete be maintained above 50°F for the first 7 days after placing, but the materials should not be heated enough to raise the temperature of the concrete above 70°F. After concrete has been placed, it can be protected by enclosing it in a plastic film enclosure and maintaining heat in the enclosed space. Covering a temporary framework with canvas tarpaulins is a common method. The heat may be supplied by steam which is allowed to escape into the enclosed space to provide moisture for curing. Salamanders, which are commonly oil-burning stoves, are also used to provide heat. The freezing point of concrete can be lowered by the use of salt or other chemicals; however, this practice is objectionable.

Concrete should never be placed on frozen ground because settlement may occur when the ground thaws. All ice and frost should be removed from forms before concrete is placed.

Hot Weather Concreting. Precautionary measures must be taken during hot weather. Concrete that generates excessive heat may lose strength and exhibit excessive plastic shrinkage cracking. These detrimental effects may be alleviated by lowering the temperature of the concrete. Aggregates are cooled by sprinkling them with cold water, and chilled water or ice is used in the mixing water. Retarding admixtures, judiciously used, are beneficial in lowering the heat of hydration while permitting a longer curing time. Proper concreting practice will assure a measure of success in hot weather, especially proper and prolonged curing, protection from direct sunlight and warm breezes, and scheduling that will assure afternoon placements.

Pneumatically Applied Concrete. Concrete or cement mortar can often be advantageously placed by a pneumatic procedure to coat concrete or masonry surfaces, to fill small spaces without using forms, or to patch or repair defects in concrete surfaces. In this procedure, a dry mixture of cement and sand is forced through a hose by compressed air. At a hand-held nozzle, water is supplied to the mixture in a mixing chamber. Directing the nozzle deposits the mortar in the desired place. Because the apparatus resembles a gun, the product is often called *gunite* and, because it is "shot" into place, it is also called *shotcrete.*

If properly proportioned mixtures are skillfully applied, the finished product will be strong, dense, and highly impermeable to water penetration.

Dry Packed Mortar. It is often necessary to place cement mortar beds which will be strong and have a minimum shrinkage in small, relatively inaccessible spaces. Examples of such spaces are: the construction clearances required between the bottoms of column bases and the tops of footings; open spaces provided at construction joints so that overlapping bars which project from the ends of adjacent members can be welded to establish continuity; and other small spaces provided for any purpose. Bolt holes in concrete and surface defects caused by improper placing must also be filled or patched.

The mortar used to fill such spaces usually consists of a relatively rich mixture of sand and cement with a minimum moisture content of putty-like consistency which will not develop surface moisture when pressed nor shrink when setting. It is commonly placed by hand tamping and protected against premature drying. It is often referred to as *dry pack*, although some moisture is present.

Cast-in-Place Reinforced Concrete. Cast-in-place concrete, as its name implies, is cast in its permanent position and location in contrast to precast concrete, which is cast at some other location. Prestressed concrete is reinforced concrete in which the reinforcement has been stressed, before the concrete is loaded, to counteract partially the expectant stresses caused by the anticipated loads. Most prestressed concrete is precast; precast and prestressed are considered together in Chapter 8.

Concrete with a minimum of reinforcement, that is, with only enough reinforcement to resist shrinkage and temperature changes, is

classed as *plain concrete*. Reinforced concrete contains enough reinforcement to act effectively with the concrete in resisting forces.

Preplaced Aggregates. Heavyweight concretes are placed in the
conventional manner, or by the *preplaced aggregate method*, which is
also known as the *prepacked method, grout intrusion*, or *grouted concrete method*. In this method, the coarse aggregate is placed in watertight forms, and grout is then pumped in to permeate through the
aggregate. Proper vibration of the formwork assures a smooth surface
free from honeycomb.

Well suited for the heavyweight aggregates, this system has been
employed for surface and underwater heavy foundations, and for wall
panels, especially those that require exposed rock or aggregate surfaces. Rock aggregates that possess the qualities mandatory for
exposed surfaces (page 196) may be set as form liners that are held
in place with wire mesh, or by wire mesh assemblies.

Heavy, well constructed watertight formwork is required to contain the grout mixture and to resist the imposed hydrostatic pressures.
Reinforcement is conventional. Grouting pipes, spaced at about 5 ft.
intervals, are then placed in the formwork. Wet, clean, coarse aggregate is then consolidated into the formwork by tamping.

Grout, specially prepared, is usually mixed with a pozzolan and a
fluidity agent that contributes to a fluid grout, while acting as a water
reducing and set retarding agent. The grout, mixed to a creamy consistency by high-speed mixers, is then pumped through the piping
into the aggregate mass. The level of the grout, which must be moni-
tored, should be maintained above the pipe outlet. Grout level is
ascertained by several methods, of which a *sounding well* is one simple
method. This contains a float that continually indicates the grout
level.

This method of construction is one of the newer evolutionary developments in concrete technology. Its proper implementation requires
expertise. The system is well outlined in a report by the American
Concrete Institute (ACI), "Preplaced Aggregate Concrete for Structural and Mass Concrete,' from which this description is drawn.

Laitance. When excess water is used, fine inert particles from the
cement aggregate, called *laitance*, accumulate in the water on the top
surface of concrete. As the water evaporates, this fine inert material
forms a residue and, if unremoved, forms a plane of weakness. Because
of the excess water or *water gain*, the concrete below the thin layer
of laitance is usually porous and lacks durability. The formation of
laitance should be prevented by avoiding excess water rather than

trying to remove the laitance after it has formed. Laitance may be removed by wire brushing or sandblasting.

Reinforcement.   Concrete has a low tensile strength and therefore is an ineffective material for members subjected to loads that tend to cause bending. Steel reinforcement, which has high tensile strength, is embedded in the concrete to resist these tensile stresses. In addition, vertical rods, commonly called *longitudinal rods* in columns, reduce the column size, or cross-sectional area, required to carry a given load because an area of steel is many times more effective in resisting compressive stresses than an identical area of concrete. The usual type of reinforcement is steel rods or bars.

Round reinforcement is classed as plain and deformed. *Plain bars* are relatively smooth and are made in $\frac{1}{4}$ in. diameters only. *Deformed bars* have raised surfaces that facilitate an effective bond with concrete. These deformations are spaced uniformly along the bars, following patterns that bond tests have proven most effective. Each steel mill rolls its own unique deformation pattern into the bars along with the mill identification mark. A numerical grade mark on each standard length identifies the tensile quality of the steel [15]. The deformations are controlled by ASTM specifications. Most reinforcing bars are of this type, but deformed bars are not available with diameters less than $\frac{3}{8}$ in.

Bars are classed as axle, rail, and billet. They are further graded as structural, intermediate, hard, regular, special, and high strength. Intermediate billet steel is the most commonly specified.

The available sizes of deformed bars have nominal diameters varying by $\frac{1}{8}$ in. increments from $\frac{3}{8}$ in. to about $1\frac{3}{8}$ in. Each bar diameter is designated a number equal to the number of $\frac{1}{8}$ in. increments included in the diameter. For example, a No. 7 bar is $\frac{7}{8}$ in. in diameter. The standard mill lengths of bars are 40 and 60 ft. Two special size bars S14 and S18 are made in billet steel. They have 1.69 and 2.25 in. diameters, respectively.

Reinforcing is also available in the form of electric welded wire fabric with a rectangular or square mesh. The wire is of cold-drawn longitudinal or carrying wires and transverse wire with the two sets of wires welded at the intersections. The longitudinal wires are spaced from 2 in. to 6 in., and the transverse wires are spaced from 8 in. to 16 in. The wires, sized by gages, range from 10 gage to 7/0 gage, which is 0.49 in. They are also made in #4 bar grids. Various combinations of wire sizes and spacings are available. They are purchased in rolls 5 ft. wide by 100 ft. long.

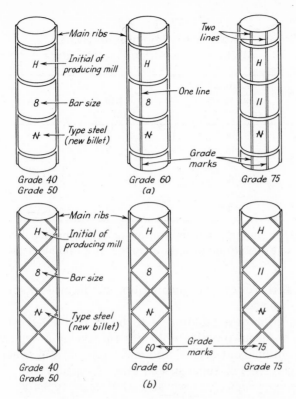

FIGURE 7.1 U.S. standard bars—identification marks. *Manual of Standard Practice*, 1970 edition, Published by the Concrete Reinforcing Steel Institute, Chicago, Illinois, [15].

Code Requirements. *The Building Code Requirements for Reinforced Concrete*, prepared and adopted by the American Concrete Institute [1], have been included in many building codes and represent the best concrete practice. Specific comments about design in this treatise are based chiefly on these requirements.

Plain and Reinforced Concrete Walls

Uses. Concrete has largely replaced stone and brick for foundation walls, for it is generally more economical, more substantial, and more watertight than brick or stone masonry. However, brick masonry will withstand uneven settlement without serious cracking better than plain concrete. Bearing walls above grade may be constructed of concrete. Attractive buildings with concrete exterior walls are being constructed. In some, the exposed surfaces are left as they come from

TABLE 7-1
Stock Size of Reinforcing Bars
All Bars are deformed except No. 2

Bar Number	Nominal Diameter (in inches)	Bar Number	Nominal Diameter (in inches)*
2	¼	8	1
3	⅜	9	1.128 (1⅛)
4	½	10	1.270 (1¼)
5	⅝	11	1.410 (1⅜)
6	¾	14	1.69　(1⅝)
7	⅞	18	2.25　(2¼)

*Diameters in parentheses are approximate. Bar number equals number of whole one-eighth increments included in its nominal diameter.

the forms, with very little touching up; but in others, special surface treatments are used.

If the interior of a building is to be of reinforced concrete construction, the usual practice is to use the skeleton type of building with wall columns and beams, and enclosure walls of concrete, brick, structural clay tile, or some special type of curtain wall. This article will deal only with concrete walls; the other types are considered elsewhere.

Concrete is used to an extent for bearing partitions and fire walls, but it is unsuited for nonbearing partitions because of its weight, the cost of forms, and the difficulty of installation after the floors are built. Concrete partitions cannot be poured much thinner than 4 in.; therefore, other forms of construction, some hollow and some thinner, have distinct advantages in weight and cost and are at least as satisfactory in other respects.

Concrete Foundations or Basement Walls. Foundation or exterior basement walls are of several types. The simplest is that shown in Figure 7.2, which supports the vertical load and withstands the lateral pressure of the earth. Ordinarily, the earth pressure is not considered in this type of construction because its effect is small compared to the vertical loads; but for low buildings with deep basements, the earth pressure is a factor in design. Walls of this type are frequently unreinforced except in the footings, but longitudinal reinforcement is desirable to reduce objectionable cracking produced by temperature changes, shrinkage, and uneven settlement. This reinforcement is

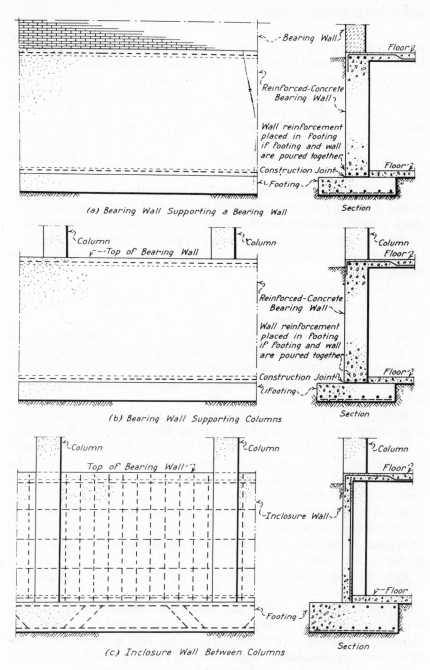

FIGURE 7.2   Reinforced concrete foundation walls.

placed near the top of the wall and near the bottom just above the footings. Longitudinal reinforcement placed in the footings is not fully effective in the beam action of the wall because of the construction joint between the footing and the wall. Special provision made for earth pressure may require a thicker wall than would otherwise be necessary. Vertical steel reinforcement placed near the inner surface of the wall will cause the wall to act as a vertical slab supported at the bottom by the basement floor and at the top by the first floor constructions.

The type shown in Figure 7.2(b) is designed to carry the concentrated loads of columns instead of the uniform load of a wall. Since there is now a definite beam action, the wall must be designed as a continuous reinforced concrete beam. The effect of earth pressure and the provisions for such pressure are the same as in the walls of the first type.

Instead of resting the columns on a bearing wall, they may be carried to a continuous footing (Figure 7.2(c)). The walls are required to carry only the lateral earth pressure and possibly a load contributed by the first floor. The lateral earth pressure may be provided for by reinforcing the wall as a vertical slab supported by the basement and first floors, as shown in the figure, or the main reinforcement may be placed horizontally; the necessary support is provided by the columns.

In Figure 7.3 (a), the columns are supported on independent footings, and the wall carries only the lateral earth pressure and possibly a load contributed by the first floor. The wall is designed as a vertical slab supported at the top by the first floor and at the bottom by the basement floor; the main reinforcement is placed vertically near the inner face. In (b), the wall is supported by the columns, and the main reinforcement is placed horizontally near the inner face. Windows or other openings in the wall will determine the most desirable method of support, or it may be economical to design the wall as a slab supported on four sides.

Deep basement walls of larger buildings are described in Article 14.

Concrete Walls.   All walls constructed of concrete should contain some reinforcement to provide against cracks caused by temperature changes, shrinkage, or unequal settlement. This is particularly true at corners and around openings. Building codes divide concrete walls into two classes, plain and reinforced.

According to the Building Code of the ACI 318-71 [11], *reinforced concrete walls* are required to have a horizontal reinforced area equal

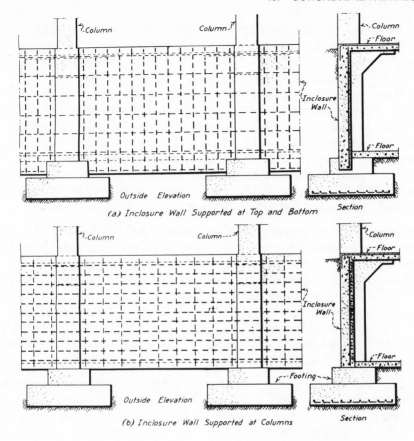

Outside Elevation   Section

(a) Inclosure Wall Supported at Top and Bottom

Outside Elevation

(b) Inclosure Wall Supported at Columns   Section

FIGURE 7.3   Reinforced concrete foundation walls.

to at least 0.0020 of the vertical cross-sectional area of a wall and a vertical reinforced area equal to at least 0.0012 of the horizontal cross-sectional area. Walls with less reinforcement, or none, are classed as *plain concrete*.

Plain Concrete Walls above Grade.   In exterior walls classed as plain concrete, horizontal reinforcement is much more essential than vertical reinforcement. It is frequently desirable to place a band consisting of one or two bars all around the building just above the window openings and another band just below these openings. It is also desirable to place a similar band of steel near the bottom of the wall.

Reinforced Concrete Walls above Grade.   Reinforced concrete is extensively used for panel walls in skeleton construction and to a lesser degree for bearing walls. In skeleton construction, the exterior walls rest on wall beams or grade beams and are carried by the structural frame. They are called *panel walls* and are usually installed after the structural frame is completed.

Panel walls are required to carry only the wind loads, and thus their thickness is usually determined by the minimum thickness which can be poured and which will be watertight and still satisfy requirements. The thickness should not be less than 4 in. and will rarely be more than 12 in. Precast *enclosure walls* are reinforced with wire fabric or small rods. A weathertight joint between the column and the wall is secured by a metal diaphragm or by a polysulphide sealant. A turned-up spandrel beam may serve as a panel wall.

Structural Requirements for Reinforced Concrete Walls. The design requirements for reinforced concrete bearing walls may be satisfied by certain empirical requirements [11] These requirements apply when the wall dimensions are determined by other than structural analysis. Moreover, these empirical requirements apply only when the loads are reasonably concentric with the walls. They are adapted with permission from Section 14.2, "Building Code Requirements for Reinforced Concrete (ACI 318-71)," a copyrighted publication of the American Concrete Institute.

(1) The length of the wall to be considered as effective for each concentrated load shall not exceed the center-to-center distance between loads, nor shall it exceed the width of the bearing plus four times the wall thickness.

(2) Reinforced concrete bearing walls shall have a thickness of at least $\frac{1}{25}$ of the unsupported height or width, whichever is the shorter.

(3) Reinforced concrete bearing walls of buildings shall be not less than 6 in. thick for the uppermost 15 ft. of their height; and for each successive 25 ft. downward, or fraction thereof, the minimum thickness shall be increased 1 in. Reinforced concrete bearing walls of two-story dwellings may be 6 in. thick throughout their height, provided that the permissible load is not exceeded.

(4) The area of the horizontal reinforcement of reinforced concrete walls shall be not less than 0.0025 and that of the vertical reinforcement not less than 0.0015 times the area of the wall. These values may be reduced to 0.0020 and 0.0012, respectively if the reinforcement is not larger than $\frac{5}{8}$ in. in diameter and consists of either welded wire fabric or deformed bars with a specified yield strength of 60,000 lbs. per sq. in. or greater.

(5) Walls more than 10 in. thick, except for basement walls, shall have the reinforcement for each direction placed in two layers parallel with the faces of the wall. One layer consisting of not less than one-half and not more than two-thirds the total required shall be placed not less than 2 in. nor more than one-third the thickness of the wall from the exterior surface. The other layer, comprising the balance of the required reinforcement, shall be placed not less than ¾ in. and not more than one-third the thickness of the wall from the interior surface. Bars, if used, shall not be less than #3 bars, nor shall they be spaced more than 18 in. on centers. Welded wire reinforcement for walls shall be in flat sheet form.

(6) In addition to the minimum as prescribed in (4), there shall be not less than two #5 bars around all window or door openings. Such bars shall extend at least 24 in. beyond the corner of the openings.

(7) Reinforced concrete walls shall be anchored to the floors, or to the columns, pilasters, buttresses, and intersecting walls.

(8) Panel and enclosure walls of reinforced concrete shall have a thickness of not less than 4 in. and not less than $\frac{1}{30}$ the distance between the supporting or enclosing members.

(9) Exterior basement walls, foundation walls, fire walls, and party walls shall not be less than 8 in. thick.

(10) Where reinforced concrete bearing walls consist of studs or ribs tied together by reinforced concrete members at each floor level, the studs may be considered as columns.

As an example of the reinforcing that would be required, the number and spacing of rods given in Table 7.2 satisfy the requirements (4) of this code. They should be placed as specified.

TABLE 7-2
Minimum Wall Reinforcement
#3 Bars Assumed

Thickness (in inches)	Spacing of Horizontal Bars	Spacing of Vertical Bars
6	7¼	12
8	5½	9
10	4½	7¼

44.   CONCRETE FORMWORK

Concrete, a plastic material, must be formed. Formwork means the containment or support of plastic concrete that allows the concrete to shape and attain its final form.

Formwork must have several qualities beyond the required or intended shape. It must be economical and strong enough to support the imposed dead, live, and erection loads. The forms should be reusable and easily erected and removed. A versatile formwork will meet any jobsite conditions or special design requirements. The essential problem in formwork design is to assemble a temporary structure that can be disassembled in a relatively short period of time, that is economical, and which will support the excessive dead and live loads. It may be uneconomical to assemble the formwork structure by means which ignore the fact that the structure may be disassembled to permit reuse of most of the materials.

Class I formwork is generally regarded as temporary, or for the support of light loads requiring a plywood of an inferior quality compared to the plywood required for Class II formwork. Class II formwork is for the type of structure that will be supporting the heavier loads, and formwork of concrete is for the larger, more permanent structures.

Economy in formwork is achieved by implementing a number of factors. Foremost is overall jobsite planning and control. This should include the designing of forms and a planned sequential coordinated movement of large sectional components or assemblies. Preconstruction planning should include a careful selection of forming materials. This may include panelized materials or sheathing boards, plastic-faced plywoods, hardboards, steel panels, or other materials that will assure a high reuse factor.

The proper supervision and the effective employment of experienced workmen while erecting efficiently constructed forms will permit fast erections and form removal, which is called *stripping*. The vertical members that support horizontal formwork are called *shores*; however, the vertical bracing members are generally referred to collectively as the *falsework* or *centering*. After forms are stripped from floor slabs, the slabs are *reshored* by replacing the shore braces against them, thus preventing deflection and cracking.

There are several systems available commercially that assure economical erection and stripping. Specially designed metal clamps and metal wedges and assortments of these may be used for assembly.

Formwork, to be safe and economical while molding the concrete to the desired shape, should be structurally adequate. Formwork failures by collapse are frequent enough to warrant the attention of engineered design. Usually collapse failures are caused by one or a combination of the following: localized overloading of placement equipment, lack of lateral support, defective form materials, poorly assembled and misaligned materials, and undersized and overspaced supporting members.

Collapse of poorly planned but adequately constructed formwork may be initiated by bumping a vertical supporting member with a loading buggy, crane booms, or buckets. A chain reaction on poorly planned formwork layouts may result in eventual complete collapse.

One of the more troublesome aspects of concrete form construction is locating and installing embedded items. *Embedded items* are usually materials that penetrate into the concrete, although this includes materials that pass through the concrete, such as pipes. Requirements frequently include blockouts for various equipment, pipe sleeves, passage openings, inserts and anchorages, waterstops, electrical boxes and conduits, and other mechanical equipment. The proper location and alignment, termed *interface*, is essential with mechanical equipment. If provision is not made for these items, or if they are mislocated in the formwork, embedment by drilling and cutouts into the hardened concrete will be required—often at great expense. Coordination between the formwork designer-constructor with the various crafts is mandatory. A thorough examination of all drawings is essential.

Materials. Form materials that are in contact with the concrete must be supported by frames, shores, braces, studs, and walers, which are usually made of wood or steel. The usual forming materials are strip boards (sheathing), plywood, hardboard, steel, and reinforced plastic. In some cases, compacted earth, formed or sculptured sand, and even other concrete are used as forms for reinforced concrete.

Most forming systems employ wood in some degree. Generally, the most economically available wood is used. Frequently, Douglas fir or Southern yellow pine serve this purpose, although Western hemlock, ponderosa, Northern white sugar, and Idaho pine are used. Wood that contacts the concrete should be nonstaining and free from organic substances that may impart detrimental effects to the concrete.

Lumber. Sheathing boards are either centermatched Tongue and Groove (T&G), shiplapped, or square edged and surfaced on all sides

(S4S). Square edged sheathing leaves a rough surface that effects a texture.

Early efforts at formwork used mostly solid wood, which usually consisted of 1x, S4S, or centermatched T&G sheathing in contact with the concrete. In wall forms, these are nailed to vertical members called *studs*, which in turn are supported by a double row of horizontal *walers* that are spaced about 3 ft. on center. This formwork usually impresses the grain of the form boards on the surface of the finished concrete. If S4S material is used, a rougher texture that is characterized by bulges, depressions, deflections, knot holes, grain, and imprints of the form imperfections is impressed in the finished concrete surface. Formerly, these surfaces were frequently ground smooth and given a grout coat of cement that covers all traces of the formwork. Today, this type of rough textured surface is desired for special effects, and some architects make a special effort to gain this effect.

Plywood.   Plywood is the most common forming material. Plywood for formwork is classed according to usage as Class I and Class II. The latter class includes that which is intended for heavy construction, that which entails great expense, and that which is planned for high reuse. These classes of plywoods are usually D.F. (Douglas fir), and Southern yellow pine. They are made with faces of grade B materials in $5/8$ in. and $3/4$ in. thicknesses in the standard plywood dimensions.

Plywood that is cohered by a highly water-resistant glue may be used for concrete forms reusable 10 to 15 times.

Plastic coated plywood is called *High Density Overlaid (HDO)*. This hard, impervious surface increases the usability of these plywood forms up to about 200 times. The smooth, laminated surface may be thermosetting resin of phenol or melamine.

Form plywoods may be specified as *mill oiled*, that is, the surfaces are impregnated with a coat of oil at the mill.

Generally, plywood is oriented with the face of the grain of the wood spanning across its principal supports.

Hardboard.   Hardboard, if tempered, is a suitable forming surface. Plastic coatings have been applied to tempered hardboard that improve its ability to resist water penetration, and to prolong the forming life of the hardboard surface. Usually hardboard is about $1/4$ in. thick in 4 ft. × 8 ft. sheets. Hardboard is frequently used as a form liner when supported by solid wood forms. Various surface effects are obtained by patterned hardboards. The screen effect that is usually on the backside of the smooth face is frequently used for special architectural effects. Hardboard forms, when properly backed by framework, lend themselves to forming curved surfaces.

Metals.   Steel has been used in various parts of forms for some time. Steel members, both cold rolled and hot rolled, have been used as framing members of the form panel, for support of the formwork frame, and for the form liner itself. Other metals such as magnesium and aluminum have been used for formwork. Because aluminum has a detrimental effect on the strength of the finished product, it should not be allowed direct contact with concrete. These lightweight metals do have the advantages of lighter shipping and handling and a rust-free, prolonged life.

Plastics.   Glass-fiber reinforced plastic forms are used for the forming surface. This material is braced and framed with steel or wood framing materals. The plastic forming materials are glass-fiber reinforcement bound by a plastic polyester resin. The principal advantage of this forming material is its ability to take any designed shape and finish. Forms for architectural concrete in any finish or shape are obtainable by applying the plastic resin to a designed shape of plaster, wood, or other suitable forming. Plastic forms are seldom jobsite manufactured items; they are usually manufactured under controlled shop conditions.

Form Liners.   Should a special finish be desired, it may be obtained by implementation of a designed surface called a *form liner*, which is attached to the forming surface that is to form the concrete surface. Plastic form liners made of various plastics, such as polyvinyl chloride and polystyrene, impart a design in the concrete surface. These forms are usually reusable for numerous pours. These patterned forms, more properly called moulds, give a variety of designs in high relief or low relief, which consist of depressed and raised surfaces. Designed surfaces may present problems of entrapped air on the face of the form liner. Vent holes and proper vibration can alleviate this problem.

Other special effects have been achieved by ingenious designers with the use of patterned neoprene form liners, sand castings (or more properly, sculptured sand surfaces), designed forms of foamed plastic, various insulation boards, and, of course, wood blocks.

If left in place, foamed polystyrene has the advantage of serving as an insulation material as well as a patterned form liner.

Formwork Hardware. Connecting devices for formwork consist mainly of nails, bolts, or lag bolts of various sorts. Usually double-headed common nails provide adequate holding ability. Withdrawal is not too difficult because of the double-headed feature. Box nails are often used for attaching the materials that contact the finished concrete because the smaller head leaves a less objectionable impression in the finished concrete surface.

The ability of nails to hold wood members together depends upon several factors, most of them variable—the diameter, length, nail and wood surface finish, wood quality, density and species, wood moisture content, edge or end distance, and the orientation of the wood grain to the axis of the load. Cement coated nails offer more resistance to withdrawal, at least initially. Other fastening devices are wood screws, lag bolts, and through bolts alone or in combinations with various metal connecting devices, such as split-ring and toothed ring connectors.

Concrete in the plastic state exerts lateral pressure in wall forms. To counteract this pressure, forms on each side of the wall are held in place by various form hardware devices called *ties*. There are generally two types with respect to designs: *single member* and *internal disconnecting*. Generally, the single member ties are equipped with form spreader devices that keep the forms from collapsing inward. These types usually feature a crimped or weakened breaking point that permits their separation and removal. The internal disconnecting types are characterized by threaded male and female portions that can be separated by unscrewing. Some kinds, such as she-bolts, are similar to sex bolts and are reusable. Their application is usually for Class II forms.

If special architectural finishes are required, the specific form ties and formwork liners, if used, should be specified. The problem of patching or deliberately exposing tie holes depends upon the function of the structure or the desired aesthetic effect. Frequently architects have required the bolt holes made by the form ties, especially she-bolt holes, to remain as a decorative feature. In such cases, these bolt holes should be spaced on a designed pattern.

Square or rectangular column forms are held together by steel column clamps or steel strapping, or wood two-by members that are nailed together. Some column clamps have rigid angles at each corner; others have hinged or threaded devices to hold them together. Some are held together with wedge shaped pins or threaded devices.

Prefabricated Forms. Prefabricated forms of various types are available for rental or purchase. They usually consist of welded light gage metal frames that feature a plywood panel. These forms, which are designed for a high reuse factor, come in various sizes and are economical if reused to the optimum. Prefabricated forms are identified by two general classes according to usage. *Ready-made* forms are for general construction where forming consists of ordinary beams, columns, and walls. These forms are made in a modular dimension permitting some flexibility. *Custom-made* forms, as the name implies, are made for special designs and may be for a single job and for the more unusually formed concrete such as would be found in dams and tunnels. These forms may be used in forming the precast wall panels or for special curtain wall designs. Their pattern design is thus used only for the project at hand. Some of these forms may be made of heavy cardboard or treated paper.

Prefabricated forms may have a provision for an adequate number of form ties to eliminate the need for wales or walers.

Generally, prefabricated forms consist of a combination of light gage metal frames, wood frames, and plywood panels. All-metal panels with metal frames are also available.

Form Surface Preparation. The interior surfaces of forms usually receive an application of a coating by brushing, spraying, rubbing, or dipping. These substances, which are frequently some type of oil, are intended to deter adhesion between the fresh concrete and the form surface, leaving an intended finish.

In any event, form liners should be of firm, rigid material that will not fold, crumble, or sag under the influence of the concrete. Otherwise, a misshapen surface finish is apt to result.

Well oiled impervious form surfaces, even those of plywood, frequently produce a concrete surface that is riddled with minute holes, sometimes called *bug holes*, on the surface of the concrete. These slight imperfections, which can be eliminated by a coat of rich grout, are the result of air pockets that have adhered to the well oiled surface and have not risen to the surface in spite of persistent vibration.

Concrete Placement. Various procedures produce suitable finishes on exposed surfaces of concrete walls. If a smooth finish with a minimum of form marks is desired, steel or plywood forms or forms lined with hardboard are used. The forms are oiled to protect them for reuse and to avoid adherence of mortar to the form surface; however, excessive oil is likely to stain the concrete. Tight forms avoid mortar leakage and the consequent formation of surface defects. The same brand of cement and the same aggregates and proportions should reduce color variations. The consistency of the mix will insure complete filling of the forms, including corners and irregularities. Drier mixtures can be placed if concrete is vibrated properly. The concrete should be placed in one continuous operation between determined expansion joints, construction joints, or between other appropriate locations. The coarse aggregate should be worked well back from the forms with spades or other suitable tools, permitting the mortar to contact the forms and eliminating the formation of voids and aggregate pockets known as *honeycomb.*

With curing by water continuing, the forms must be removed as soon as feasible to permit any necessary surface repairs. Undesirable projections should be removed; and after wetting the surface, voids should be filled with grout matching the color of the concrete.

After defects have been repaired, the surface should be saturated and coated with cement grout and immediately scrubbed with a wood or cork float. Finally, after the grout has hardened sufficiently so that it will remain in the small air holes, the excess grout should be removed with a sponge-rubber float.

Various other methods are used for finishing concrete wall surfaces, including a rubbed finish which is formed by rubbing with carborundum stones. The concrete must have hardened sufficiently so that the aggregate will not be disturbed with or without the application of mortar while rubbing.

Slip Forms.   Moving forms which shape the concrete as they slide over the fresh concrete are called *slip forms.* Horizontal slip forms shape the concrete for canals, tunnels, and highways. Vertical slip forms are used for forming concrete grain storage bins or silos, chimney stacks, and the central service core for buildings. These service cores usually house elevator shafts, ducts, pipe and conduit spaces, and stair wells. Some service cores of buildings which have been constructed by slip form methods act as the principal structural element —the floor is suspended from the central service core or the solid concrete walls serve as shear walls.

The principal components of a vertical slip form are a yoke, sheathing, wales, and a working deck. The *yoke,* usually made of steel, supports the formwork sheathing and a jacking assembly which raises the entire form and working deck assembly.

The jacking system, either manually or hydraulically controlled, lifts the entire formwork assembly by jacking vertically upon steel rods that are embedded in the fresh concrete. In some systems, these rods are removable. Form lifting operations proceed at increments of about 1 in. every 5 or 10 minutes, or about 6 to 12 inches an hour.

Concrete of a stiff consistency is required for this type of construction. Careful preconstruction planning is essential before concreting operations are undertaken. Usually slip form concreting procedures are a continuous, round-the-clock operation.

In the event that slip form construction operations are curtailed for any length of time, a *cold joint* will result. Such a concrete joint must be treated as a preplanned construction joint. Construction preplanning should anticipate such occurrences and provide for the resultant joint.

45. CONCRETE COLUMNS

Unreinforced concrete columns will crack from bending stresses induced by differential settlement, temperature changes, or unbalanced loads on the surrounding floors. Consequently, unreinforced columns are not used except where the least width is large compared to the length; then they would be classed as *piers* rather than as columns. Vertical bars placed near the surface of concrete columns provide resistance to bending; however, the principal load on columns causes compressive stresses that will cause the vertical bars, if used alone, to tend to kick out or buckle and spall off the surrounding concrete. To avoid this, circumferential or lateral reinforcements are used as *ties* spaced 8 to 12 in. apart (Figure 7.4) or in the form of closely spaced *hoops*.

Because it is difficult to make individual hoops, a single rod is bent into a helix (*b*) that acts in the same manner as individual hoops. It is called a *spiral hoop* or simply a *spiral*, although it is really a helix. Columns with ties are called *tied columns*, and those with spiral hoops, usually called *spirally reinforced columns*, are sometimes termed *hooped columns*. The spiral reinforcement serves another purpose. By confining the concrete *core* within the spiral, it prevents sudden and complete column failure, therefore making it more dependable. Ties are not as effective as spirals; the allowable load on a tied column is taken as 85% of the allowable load on a spirally reinforced column with the same area and vertical reinforcement.

Other types of columns composed largely of concrete are the composite column and the combination column. A *composite column* is a steel or cast iron structural member completely encased in concrete with spiral and vertical reinforcement (Figure 7.4(*l*) and (*m*)). A *combination column* is a structural steel member which supports the principal load and which is wrapped with welded wire mesh and encased in concrete that supports some additional load. This type of column resembles the composite column shown in Figure 7.4(*l*).

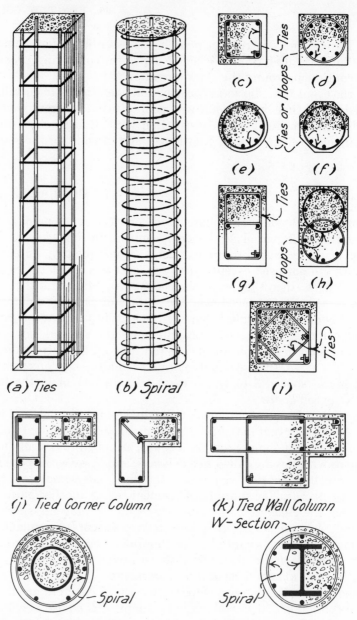

(a) Ties   (b) Spiral

(c)

(d)

(e)

(f)

Ties or Hoops — Ties

(g)

(h)

Ties

Hoops

(i)

Ties

(j) Tied Corner Column

(k) Tied Wall Column

W-Section

(l) Composite Column-Cast-Iron Core   (m) Composite Column-Steel Core

Spiral

Spiral

FIGURE 7.4   Reinforced concrete columns.

Another type of combination column is the concrete filled steel *pipe column.*

The load on a column is considered to be distributed between the concrete and the vertical reinforcement—structural steel, steel pipe, or cast iron included in the column in accordance with specified relationships. Some of the specific requirements for the various types of columns are considered in the following paragraphs.

Spirally Reinforced Columns. This type of column is illustrated in Figure 7.4(*b*), (*d*), (*e*), and (*f*). They are formed with wood or metal and filled from the top; or in the case of long columns, they are filled at pocket openings in the form. The vertical reinforcement, with an area between 1% and 8% of the overall or gross cross-sectional area of the column, must be at least four No. 5 bars. The vertical bar spacing requirements are the same as for spirally reinforced columns. Lateral ties must be at least #2 bars spaced not more than 16 bar diameters, 48 tie diameters, or the least dimension of the column. Columns that require more than four vertical bars require additional ties so that each bar is held firmly in position. The minimum cross sectional dimension of a rectangular column is 8 in., with a gross area of not less than 96 sq. in.

Composite Columns. This type of column, illustrated in Figure 7.4(*a*), (*c*), (*d*), (*e*), (*f*), and (*i*), must not have a metal core cross-sectional area exceeding 20% of the gross area of the column. Hollow metal cores must be filled with concrete. All reinforcement values including spacing, splices, and protective cover must conform to the limitations required of spirally reinforced columns. A minimum 3 in. clearance between the spiral and the metal core is required, except 2″ clearance is permitted when a structural steel wide-flange (W) column is used.

Metal cores must have milled splices that are in alignment. Other requirements include the unified action between the concrete and its reinforcement with the metal core. The metal core must support erection loads or any other imposed loads before it is encased in concrete.

Combination Columns. A combination column requires a minimum of 2½ in. of concrete cover over all metal parts except rivet or bolt heads. The concrete must have reinforcement equal to 4 × 8 / #10 × #10, with the 8 in. wires parallel to the column axis. The mesh, which surrounds the column, must be lapped at splices and covered with a minimum of 1 in. of concrete. The steel column, like the composite column, is designed to carry any erection loads before it is encased in concrete. This type of column, similar to Figure 7.4(m), is used in lower floors of reinforced concrete buildings, thereby reducing the required column size.

Another form of combination column is a steel pipe filled with concrete. Concrete filled steel pipes that are unprotected do not have a fire rating. However, they may have a concrete protective covering that will permit a fire rating. The concrete is frequently covered with a finish for aesthetic expression.

Irregularly Shaped Section. Column cross sections are usually square, circular, or octagonal. However, conditions may require columns with other cross sections, such as the section in Figure 7.4(h) with interlocking spirals and the sections in (g), (j), and (k). The ties are arranged to support the longitudinal rods, including the vertical rod in the middle of each side of the square section.

46.   CONCRETE BEAMS AND GIRDERS

Concrete is strong in compression but weak in tension. This deficiency is overcome by placing steel bars in the concrete on the tensile side of a beam (Figure 7.5(a)) so that they will resist the tensile stresses; a very efficient form of beam results. A sufficient amount of steel reinforcement usually makes the beam as strong in tension as is required.

The simplest form of beam is rectangular in section (b), but because the concrete on the tensile side of the beam is not considered to resist any tensile stress, some of this concrete may be omitted in wide beams, leaving only enough to carry the steel rods and to provide for the shearing stresses. This forms the T-beam shown in (c). The corresponding rectangular beam is shown by the dashed lines. Usually some of the concrete on the compressive side near the neutral axis is also omitted (d). By referring to Figure 3.2, it is evident that the stress resisted by material near the neutral axis is small so that its omission only slightly reduces the strength of the beam.

Frequently, it is necessary to design a beam of given strength in a limited depth which is smaller than would be required for an ordinary rectangular or T-beam. The tensile strength is provided by the required amount of tensile reinforcement, but the amount of concrete available for compression is limited by the space to be occupied. The additional compressive strength necessary is secured by placing steel bars on the compression side of the beam, forming a double-reinforced rectangular beam (Figure 7.5(e)) of balanced design. Steel is about

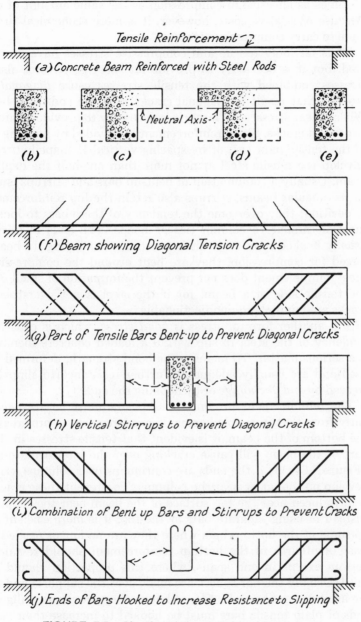

Tensile Reinforcement

(a) Concrete Beam Reinforced with Steel Rods

Neutral Axis

(b)   (c)   (d)   (e)

(f) Beam showing Diagonal Tension Cracks

(g) Part of Tensile Bars Bent-up to Prevent Diagonal Cracks

(h) Vertical Stirrups to Prevent Diagonal Cracks

(i) Combination of Bent up Bars and Stirrups to Prevent Cracks

(j) Ends of Bars Hooked to Increase Resistance to Slipping

FIGURE 7.5 Simple reinforced concrete beams.

15 times as effective in compression as the same amount of concrete. Because of relative costs, however, it is more economical to use concrete to carry compressive stresses.

Thus far, only tensile and compressive stresses have been considered, but it is necessary to provide for shearing stresses also. These stresses combined with the tensile stresses cause *diagonal tension stresses* that result in diagonal cracks near the ends of a beam (*f*). Where shear stresses are large, it is necessary to provide reinforcement to prevent cracks. Shear reinforcement is provided by bending up some of the tensile bars (*g*), or by spacing vertical U-shaped *stirrups* that envelop the tensile steel at not more than one-half the depth of the beam. Usually a combination of bent-up bars and stirrups is used (*i*). In rectangular beams, stirrups also retain the top reinforcement steel in position; they overcome the tendency of these bars to kick out—a function similar to column ties or hoops. The ends of stirrups are usually hooked to increase their resistance to slipping. In beams reinforced for compression, they are bent around the compressive steel. Steel reinforcement does not prevent the formation of small cracks on the tensile side of a beam, for if the steel is not overstressed, the cracks are minute and unobjectionable.

Still another form of stress is *bond stress*, which is caused by the tendency of the steel to slip within a loaded concrete member. This is a serious matter. To increase the bond strength, deformed bars are used and are usually adequate. Further resistance to slippage is increased by *end anchorage* or by hooking the ends (*j*).

The reinforcement in continuous beams must be arranged differently from that in simple beams. If reinforcement is provided only at the bottom of the beam, it is evident that tensile stresses in the upper part of the beam will cause cracking over the intermediate supports (Figure 7.6(*a*)); if the ends are continuous over columns, cracks will develop on top, at or near the columns. To prevent these cracks, steel is introduced near the top surface of the beam near the supports (*b*). Instead of using separate bars at the top, it is more efficient to bend up some of the bottom bars where they are no longer needed (*c*). At least one-fourth of the bottom reinforcement should remain in the bottom except in end spans, where this proportion should be one-third. Usually the bar splices are made over the supports and the bars are arranged as shown in (*d*), although they are usually aligned. The ends of plain tensile bars must be hooked to increase their resistance to slipping; however, hooks are usually not required on deformed bars. The inclined portions of the bent-up bars effectively resist diagonal tension stresses, although it is usually necessary to use stirrups where

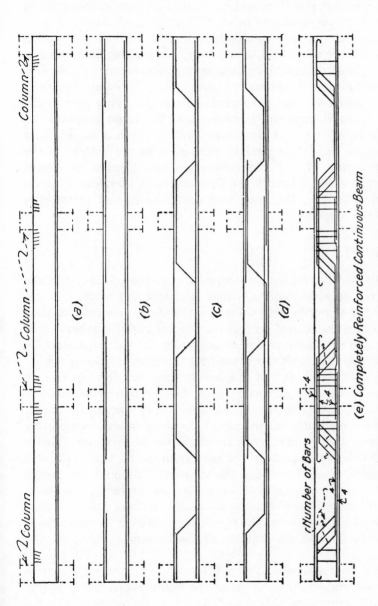

(a)

(b)

(c)

(d)

(e) Completely Reinforced Continuous Beam

FIGURE 7.6　Continuous reinforced concrete beams.

shear reinforcement is not provided by bent-up bars. It may be desirable to have several points at which the bottom bars individually or in pairs are bent up as shown in (e). This figure illustrates a completely reinforced concrete beam built monolithic with the supporting columns, which are reinforced to resist any induced bending stresses.

In building construction, floor slabs are usually cast monolithic with the beams and girders, as shown in Figure 7.7(c). This slab effectively resists the compressive and flexural stresses in the upper part of the supporting beam at midspan; however, near the beam supports the compressive stresses are in the lower part of the beams. At these points, the bars which run straight through near the bottom of the beam carry a part of the compressive stresses. Continuous beams which are cast monolithic with the floor slabs are therefore T-beams in the central part of the span and double-reinforced rectangular beams over the supports.

47. CONCRETE SLABS

Ordinary Slabs.   For design considerations, reinforced concrete slabs are considered as wide shallow beams. As floors and roofs, they are supported by masonry bearing walls or by reinforced concrete or steel frames. Building codes as a general rule do not permit timber beams to support concrete slabs except as nonstructural slabs; however, in composite construction, timber is used for structural slab support.

The simplest example is that of separate slabs simply supported on beams (Figure 7.7 (a)), although such construction is uncommon. Usually slabs are continuous over the beams (b). It is necessary to provide steel reinforcement near the top of the slabs over the supporting beams because the tensile stresses are at the top at this point. This is provided by bending up some of the steel from the bottom of the slab or by installing short bars over the supports. Although no tensile stresses exist at the bottom of the slab over the supports, it is desirable to continue at least one-fourth of the bottom steel straight though. The steel required at the top of the slab over the supports is about equal to the amount required at the bottom at midspan. See Figure 7.8.

Reinforced concrete slabs, monolithic with reinforced concrete beams and girders (c), span between the beams while acting as the flange of the supporting T-beams. The slabs, continuous over the beams, require tensile steel near the top over the beams. The code permits the portion of the slab extending a distance eight times its thickness each side of the beam to be considered as the flange of the

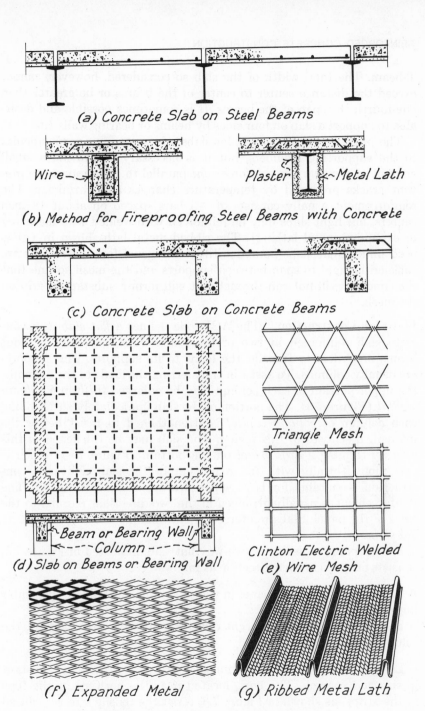

(a) Concrete Slab on Steel Beams

Wire—   Plaster   Metal Lath

(b) Method for Fireproofing Steel Beams with Concrete

(c) Concrete Slab on Concrete Beams

Triangle Mesh

Clinton Electric Welded

Beam or Bearing Wall
Column

(d) Slab on Beams or Bearing Wall   (e) Wire Mesh

(f) Expanded Metal   (g) Ribbed Metal Lath

FIGURE 7.7   Reinforced concrete slabs.

T-beam. The total width of the slab so considered, however, cannot exceed the distance center to center of the beams or be greater than one-fourth the span of the beams. It is sometimes possible and desirable to support a slab on four sides by beams or bearing walls (d).

The principal reinforcement for slabs, of course, runs perpendicular to the supporting members, but it is necessary to provide a small amount of *temperature reinforcement* parallel to the supports to prevent cracks produced by temperature changes and shrinkage. The reinforcement usually consists of #3 bars spaced on about 18 inch centers, but light slabs may be reinforced with welded wire mesh (e) or expanded metal lath (f). The ribbed metal lath shown in (g) is used under light slabs poured without the usual forms. The ribs are sufficiently rigid to span between supports and the mesh so fine that the concrete will not run through but will form a substantial grip on the mesh.

Flat-Slab Construction.   The ACI code defines a *flat slab* as a concrete slab reinforced in two or more directions, generally without beams or girders to transfer the loads to its supports. The supports are usually columns. To assist in transferring the loads to the columns, the upper portion of each column may be enlarged to form a *column capital* (Figure 7.8). The portion of the slab immediately surrounding each column or *column capital* is strengthened to resist better the intensified stresses in this region. The slab may be thickened in this area by forming a *drop panel* on the bottom (Figure 7.8), which is cast monolithically with the remainder of the slab. Flat-slab construction is suitable for bays which are approximately square. The reinforcement is usually arranged in the two directions parallel to the sides of the panel. Slabs reinforced in this manner are called *two-way flat slabs*.

For purposes of design, a two-way flat slab, indicated in Figure 7.8, consists of strips in each direction as follows:

A *middle strip*, one-half panel in width, symmetrical about the center line of the panel,
A *column strip* of two adjacent quarter panels on either side of the column center line.

Longitudinal reinforcing within each strip is usually made constant in size and spacing. The bars, located near the bottoms or the tops of the strips, as shown in Figure 7.8, resist the tensile stresses caused by the loads.

The minimum permissible thickness of slabs with drop panels is

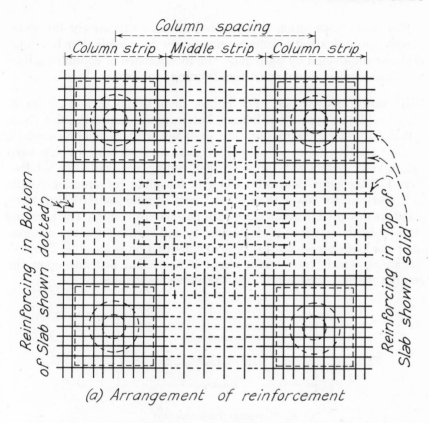

Column spacing

Column strip | *Middle strip* | *Column strip*

Reinforcing in Bottom of Slab shown dotted

Reinforcing in Top of Slab shown solid

(a) *Arrangement of reinforcement*

Drop Panel | *Reinforcing in* | *Column*
Column Capital | *Bottom of Slab dotted* | *Column*

(b) *Section through middle strip*

FIGURE 7.8   Two-way flat-slab construction.

one-fortieth of the longest span but in no case is it to be less than 4 in. The side of a drop panel must be at least one-third the parallel span. The maximum effective central angle of a column capital is 90°. The bottom portion of flat slabs may include recesses or pockets formed by permanent or removable fillers between reinforcing bars, as described later and illustrated in Figure 7.11(f). Many other factors, such as openings and length of reinforcement, are considered in the regulations.

Flat-slab construction as described is used extensively for ware-houses and industrial buildings with heavy floor loads. For buildings that are subdivided by partitions, the interference of column capitals and drop panels presents partition framing problems.

Flat-Plate Construction.   For lighter floor loads, such as those in office buildings, apartment houses, hotels, and dormitories, flat slabs, called *flat-plate construction*, are designed without column capitals or drop panels (Figure 7.9(a)). Flat-plate construction, which is used extensively, gives a flat, uninterrupted ceiling, well suited to subdivision. For the same clear ceiling height, flat-plate construction results in a story height lower than other forms of floor construction. Such slabs are designed in accordance with flat-slab principles. The shearing stresses in the portions of slabs close to and surrounding the columns limit the loads for which this type of construction is feasible. The slab thickness, which varies from about 6 to 10 in., depends upon the column spacing, the loading, and the allowable stresses.

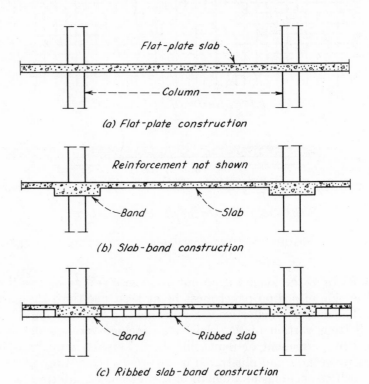

(a) Flat-plate construction

(b) Slab-band construction

(c) Ribbed slab-band construction

FIGURE 7.9   Flat-plate and slab-band floor construction.

Slab-Band Construction. *Slab-band construction* consists of wide shallow beams, called *bands,* running continuously along each longitudinal row of columns and supporting one-way slabs spanning between bands and cast monolithically with them (Figure 7.9(*b*)). The space under the bands may be devoted to closets, washrooms, etc., locating the principal rooms under the slabs to achieve flat ceilings. To accomplish this, the band is sometimes located off center with the columns. A column may also be set somewhat off center from the other columns in a row. Rectangular columns located in partitions may have a width equal to the width of the band and a thickness sometimes as small as 8 in. The slab thickness, which varies from about 5 to 8 in., depends upon the column spacing, the band width, the loading, and the allowable stresses.

A combination of ribbed slabs, described in the following paragraph, and slab-band construction (Figure 7.9(*c*)) may be used to form a flat ceiling. This type of construction is adaptable to longer spans and heavier loads than would be possible with flat-plate construction.

Ribbed Slabs.   In reinforced concrete beams and slabs, the concrete between the neutral axis and the tension face contributes nothing to the flexural strength, but it is effective in resisting a part of the shearing stresses, as explained earlier. The shearing stresses in slabs are usually low; therefore, all this concrete is not necessary. To save concrete and to reduce the weight of the slab, a large part of the concrete on the lower side of the slab is eliminated, leaving only the ribs or joists (Figure 7.10(*a*)). The bottom of the corresponding solid slab is at the bottom of the ribs. These ribs are made wide enough to resist the shearing stresses and to carry the necessary tensile steel. Practically the same amount of steel is required for a solid slab, except for the saving in steel resulting from a reduction of the dead load. The remaining flange may extend down to the neutral axis, but usually it does not. This results in a reduction in the compressive resistance, but this is small because the concrete near the neutral axis carries very little stress. The solid slab would be reinforced with relatively small bars closely spaced, but in the ribbed slab, two larger bars are ordinarily used in each joist. This results in increased bond stresses, but such stresses are not often a controlling factor. Some conditions require solid bridging between joists, spaced not more than 15 ft. apart, for some types of joist floors. This bridging is commonly 4 in. wide and the full depth of the joists. It is reinforced with one rod near the top and one near the bottom. The primary function of bridging is to distribute heavy loads into several joists. The minimum permitted width

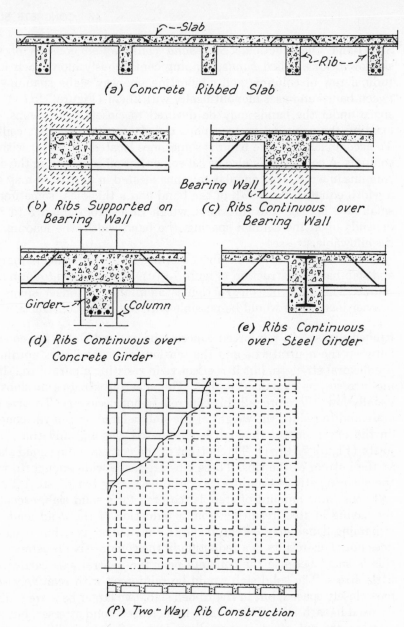

(a) Concrete Ribbed Slab

(b) Ribs Supported on
Bearing Wall

(c) Ribs Continuous over
Bearing Wall

(d) Ribs Continuous over
Concrete Girder

(e) Ribs Continuous
over Steel Girder

(f) Two-Way Rib Construction

FIGURE 7.10   Ribbed slabs of reinforced concrete.

of joists is 4 in., and the depth below the bottom of the slab is some-times limited to three times the width. The clear spacing between the joists is limited to 30 in. The thickness of the top slab varies from $1\frac{1}{2}$ to $3\frac{1}{2}$ in., depending upon the span and loading.

The detail of a ribbed slab supported at a bearing wall is shown in Figure 7.10(b). The details used for slabs continuous over bearing walls, concrete girders, and steel girders are shown in (c) to (e).

Occasionally, it may be advantageous to support a slab on four sides and provide ribs in two directions, as shown in (f).

Ordinary wood forms would be so expensive for ribbed slabs that their cost would be prohibitive. For this reason, various formwork types have been devised. The sides of the joists and the bottom of the slab are formed by structural clay tile, hollow gypsum tile, or sheet-steel cores (Figure 7.11(a) to (c)), or special forms of concrete block called *tile fillers* or simply *fillers*. Wood forms are constructed for the bottoms of the joists. These are usually made of 2 in. material and are sufficiently wider than the joists to support the edges of the clay or gypsum tiles or the steel pans. The formwork is therefore very simple.

The fillers provide a plaster base for the ceiling formed by the under-side of the slab. Special forms of tile or block cover the bottoms of the joists, providing a uniform plaster base over the ceiling. The ends of the end tile may be closed by a thin slab made for that purpose by using pieces of sheet-metal or wire screen.

The sheet-steel cores may be made of heavy material, which can be removed after the concrete has set and reused several times. They may also be thinner material to be left in place. If removable forms are used and the ceiling is plastered, metal fastened to the underside of the joists serves as a plaster base for the ceiling below. Various types of anchors can be cast in the underside of the joists to receive the metal lath. If the sheet-steel cores are to remain in place, metal lath is laid over the formwork before the cores are set. The metal lath is wired to the reinforcing bars in the joists. In one type of steel core, the lath is fastened to the core as shown in Figure 7.11(d) before the core is placed. Sheet-metal closers or end caps are made for the end cores. Because of excessive shearing stresses, it is sometimes necessary to widen the joists at the ends by using tapered cores. The most com-mon widths of metal core are 20 and 30 in., used with joist widths of 4, 5, and 6 in. Core depths vary by 2 in. increments from 6 to 14 in. They are usually corrugated to increase their stiffness and are lapped one or one-half corrugations or more if necessary to provide the re-quired lengths. The cores are supported on forms that form the bottom

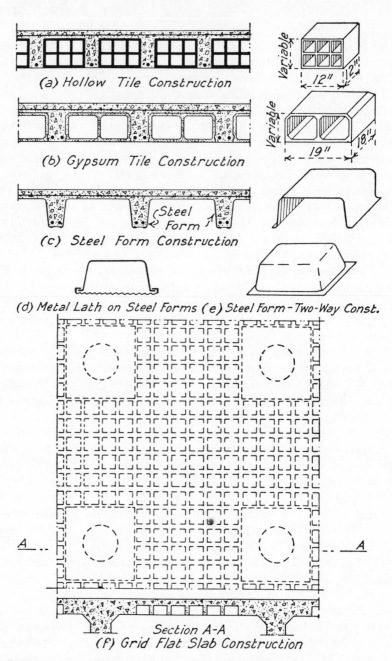

(a) Hollow Tile Construction

(b) Gypsum Tile Construction

(c) Steel Form Construction

(d) Metal Lath on Steel Forms (e) Steel Form–Two-Way Const.

Section A-A
(f) Grid Flat Slab Construction

FIGURE 7.11   Cores for ribbed slabs and grid flat-slab construction.

of the joists. No forms are necessary between the joists. The cost of removing cores must be considered, as must the greater cost of placing the lath. The type most suitable for a given case can be determined only by studying all of the factors involved.

For all types of core, the reinforcement in the top slab must be adequate to resist the loads on the slab and to provide for temperature changes and shrinkage of the concrete. Temperature reinforcement usually consists of wire mesh or $\frac{1}{4}$ or $\frac{3}{8}$ in. rods normal to the joists and spaced from 6 to 12 in. No temperature reinforcement is required, however, parallel with the joists because the reinforcement in the joists is adequate.

Ribbed slabs are suitable for spans varying from 10 to 35 ft. In constructing long-span slabs, camber should be provided to offset the deflections caused by elastic deformation and time yield or *plastic flow,* called *creep.*

Special types of tile and sheet-steel cores are available for two-way construction. A steel core for two-way construction is shown in Figure 7.11(*e*).

A form of flat slab construction has been devised to make use of the ribbed slab in place of the flat slab. This is known as the *grid flat slab* and is illustrated in Figure 7.11(*f*). The reinforcing is not shown in this figure, but would be similar to that used in the two-way system shown in Figure 7.8 with the rods placed in spaces between the cores.

48.   CONCRETE FRAMING

In framing a reinforced concrete building, the forms are first constructed, the reinforcing steel is placed, and finally the concrete is poured, as described in Article 43. After the concrete sets, the forms are removed, and curing procedures continue. It is obviously impossible to pour an entire building of any magnitude in one operation; therefore, construction joints cannot be avoided. These should be located and constructed so as not to impair the strength or appearance of the building. The joints in slabs, beams, and girders should be, preferably, vertical and at the center of the span where the shearing stresses are small. The columns should be poured to the underside of the floor girders for beam and girder construction so that the shrinkage of the concrete columns may precede the placement of the floor above.

Forms are usually constructed of wood, but steel forms are extensively used, especially for buildings of flat-slab construction. The reinforcing steel is held in position by wiring the bars together at their intersections. The bars for each beam, girder, and column are usually wired together forming a frame which is set in position as a unit. Various devices such as *chairs*, *high chairs*, and *spacers* have been designed to hold reinforcing steel in position.

Reinforced concrete buildings may be of bearing-wall construction or of skeleton construction. The bearing walls may be constructed of brick, stone, structural clay tile, concrete block, plain concrete, or reinforced concrete, described earlier.

The cross section of a reinforced concrete building with flat-slab construction is shown in Figure 7.12 and one with beam and girder construction in (*b*).

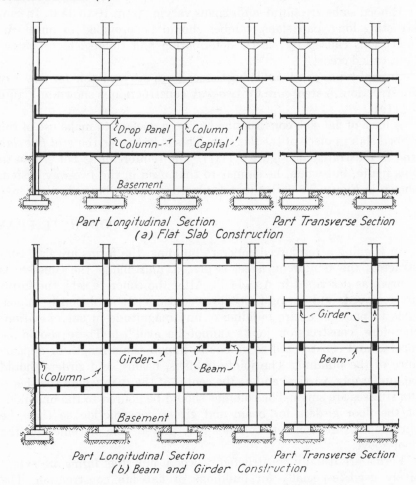

Part Longitudinal Section   Part Transverse Section
(*a*) Flat Slab Construction

Part Longitudinal Section   Part Transverse Section
(*b*) Beam and Girder Construction

FIGURE 7.12   Types of reinforced concrete framing.

Typical details for a building of the beam and girder type are shown in Figure 7.13. All reinforcing is deformed bars. The features which should be noted in this figure are as follows:

(1) The reinforcing in the slab is at the bottom near the center of the span and at the top where the slab crosses the beams. Some of the bottom steel continues through the bottom.

(2) The reinforcing in the beams is similar to that in the girders. The horizontal bars in the bottom of the girder are bent up from the bottom to the top in the region near the supporting columns.

(3) Shear reinforcing is provided by bent-up bars and vertical stirrups.

(4) Dowels extend from the spread footings into the columns. These dowels, equal in size and in number to the longitudinal bars in the column, transfer the stress in the column bars to the footing. They extend into the column and into the footing a sufficient distance so that their bond stress will not be excessive.

(5) The longitudinal bars in the interior columns are arranged around the edge of the columns and are surrounded with closely spaced spiral reinforcing. To avoid confusion, all the longitudinal bars are not shown in the section, but they are shown in the plan view.

(6) The longitudinal bars of the wall columns are held in position by lateral reinforcement ties which are not as closely spaced as the spirals of the interior columns.

(7) The columns are spliced by extending the longitudinal bars from one column upward into the column above. To accommodate the smaller size of the upper columns, the bars are bent inward in the part of the column occupied by the floor construction. They are again made vertical at the floor level.

(8) The slope of the roof is obtained by a lightweight concrete fill that receives the built-up bituminous roofing. The lightweight concrete fill also serves as heat insulation. An expansion joint provides space between the filling and the parapet wall so that the fill can expand as it becomes heated. If this provision were not made, the topping would either buckle or push the parapet wall out. Similar expansion joints should be provided at other points. It is usually desirable to slope the roof for drainage, although good results have been secured with decks which do not slope to drains. Such decks require an impervious surface. Dead level roof decks should be avoided because shallow pools of standing rainwater which may accumulate will bring eventual damage to a built-up roof.

(9) The panel walls consist of aluminum sash which occupy the

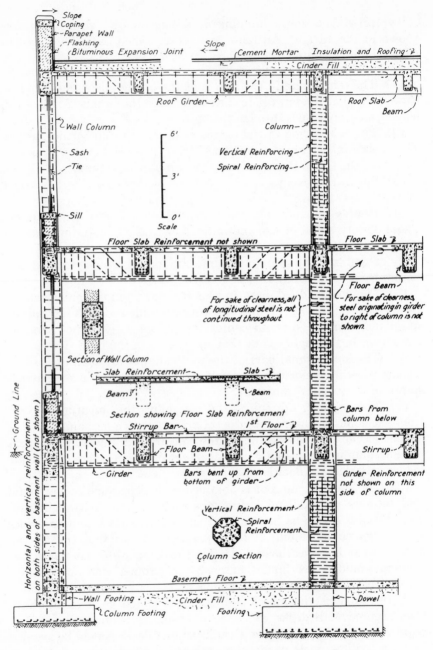

FIGURE 7.13 Beam and girder construction.

entire width between columns and below which a spandrel masonry wall is placed.

(10) The parapet wall is capped with a coping, the top of which slopes and drains onto the roof and not over the face of the building.

Typical details of a building of the flat-slab type are shown in Figure 7.14. The features which should be noted in this figure are as follows:

(1) The substitution of flat-slab construction for the beam and girder construction shown in the previous figure.

(2) The drop panels and the column capitals of the interior columns.

(3) Brackets and drop panels of the exterior columns.

(4) The spiraled interior columns. For clarity, only a part of the longitudinal bars are shown in the section but they are all shown in the plan view.

(5) The tied wall columns. Since these columns are rectangular, they are provided with intermediate ties.

Other features are the same as the corresponding features of the beam and girder type which have just been explained.

Lateral Resistance. The problems involved in designing tall buildings with steel frames to resist the lateral forces caused by wind and earthquake shock were considered in Article 41. As with steel frame buildings, nonbearing masonry walls and partitions contribute to the lateral rigidity of a reinforced concrete building, but the magnitudes of such contributions are uncertain and they are usually neglected in design. The only resisting elements that are considered are the integral structural parts. Members subject to stresses produced by wind and earthquake forces combined with other loads may be proportioned for unit stresses one-third higher than the allowable stresses for dead and live loads only, but the size of a member must not be less than is required for these loads.

The horizontal deflections from lateral loads tend to distort the rectangular partitions constructed between the columns and floor systems and cause them to crack. A similar phenomenon occurs in door and window openings with objectionable results. Therefore, even though a frame may be structurally adequate as far as stresses are concerned, it may not be satisfactory. For that reason, horizontal deflections of a frame as well as stresses must be considered.

A significant factor in determining the special structural requirements for resisting lateral loads of given intensities is the ratio of the

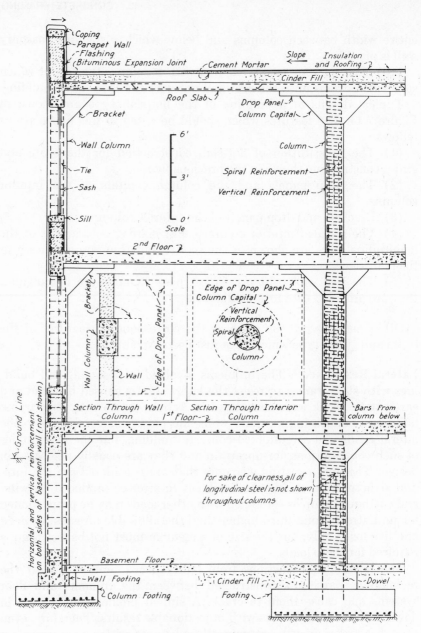

FIGURE 7.14 Flat-slab construction.

height, or the portion of a building above the elevation being considered, to the width, or smallest lateral dimension. Another significant factor is the type of floor construction. Construction with girders framing into the columns and running the width of a building offers considerable resistance to lateral forces without exceeding allowable stresses in the columns or girders, but construction with flat-plate floors are less effective in this respect.

Usually, tier buildings only a few stories high or the upper few stories of tall buildings do not require special provisions for wind unless they are very narrow in at least one dimension. It will be obvious to experienced designers that under some conditions no special provisions are required and that under other conditions they are certainly necessary. Between these two extremes, each case must be given specific consideration.

Tall buildings with long narrow plans require special provisions to resist wind loads acting against the broad side of the building but do not normally require such provision to resist the wind load against the narrow side. This condition results from the smaller wind load and the greater lateral strength of the building in the direction parallel to the longer dimension. When the usual types of construction prove inadequate to resist the lateral forces (after permissible increases have been made in the allowable stresses), a common solution includes reinforced concrete *shear walls* in some of the vertical bents bounded by continuous columns and the floor systems. Shear walls are made integral with the other structural members by tying them together with reinforcing bars. They are designed to resist the lateral forces on the portions of the building that contribute to the lateral loads.

A simple arrangement of shear walls to provide transverse rigidity is illustrated in Figure 7.16. Pairs of columns in every third row in each story are connected by shear walls forming *wind bents*. To provide for a longitudinal corridor, shear walls are not placed between the two interior columns in each transverse row.

The lateral loads are transmitted horizontally to the wind bents by the floor systems as explained for steel tier buildings in Article 41. The usual types of reinforced concrete floor systems transmit these loads, but sometimes special provisions require additional reinforcement or increased sections.

Preferably, shear walls are continuous from the foundation to the maximum height at which they are required. When this is not possible, the lateral loads are transferred horizontally between discontinuous

FIGURE 7.15 Reinforced concrete framed office building. Bank of Georgia Building, Atlanta, Georgia. A reinforced concrete building with conventional framing—30 stories and 390 ft. 10 in. high. Architect-Engineer: Wyatt C. Hedrick. General Contractor: Henry C. Beck Co.

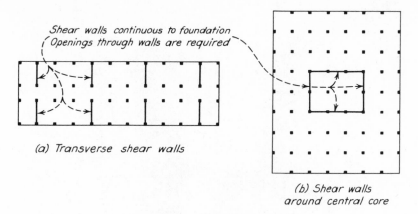

Shear walls continuous to foundation
Openings through walls are required

(a) Transverse shear walls

(b) Shear walls
around central core

FIGURE 7.16   Shear walls.

shear walls by the floor systems where the discontinuities occur. The vertical loads are transmitted directly downward.

The simple patterns of shear walls illustrated in Figure 7.16 may not be possible in all buildings. Shear walls are frequently located in walls that enclose elevator shafts and stairwells, or they are located in other positions which do not interfere with usable floor space.

Shear walls vary from 6 to 12 in. thick depending upon structural conditions. As has been stated, their function is to resist horizontal shearing forces produced by lateral loads, which in many regions may include seismic shocks as well as wind. Generally, shear walls are not considered to support dead and live loads. Reinforced concrete bearing walls do resist lateral forces in a manner similar to shear walls. Such walls carry the dead and live loads in addition to the lateral loads and, for that reason, must be thicker than shear walls. Usually, such construction is uneconomical.

Another means of increasing lateral rigidity and strength is by orienting the longer dimension of rectangular columns parallel with the short dimension of the building plan and framing deep floor girders into them.

Under any conditions, the provisions to resist lateral forces must be such that the resultant resistance at any elevation is colinear with the resultant lateral force at that elevation. Otherwise, the building tends to twist.

Restated, lateral resistance is often provided by a rectangular core bounded by shear walls. In each building of the twin Marina City Towers in Chicago (Figures 7.17, 7.18, and 7.19), a cylindrical rein-

FIGURE 7.17 Model of Marina Towers in Chicago. Architects and engineers: Bertrand Goldberg and Associates. Consulting Engineers: Severud-Elstad-Krueger Associates. Foundation Consultants: Moran, Mueser and Rutledge and R. B. Peck. Sponsors: Building Service Employees International Union.

FIGURE 7.18   Marina Towers in Chicago during construction.

forced concrete core provides resistance to lateral forces. Each core, which has an internal 32 ft. diameter to its full height of 64 stories, rises 588 ft. above the circular reinforced concrete pad or mat on which it bears. The thickness of the core walls varies from 30 in. at the bottom to 15 in. at the top.

The diameter of each foundation mat is 58 ft. Originally, the design thickness of each mat was 3 ft.; but because of construction difficul-

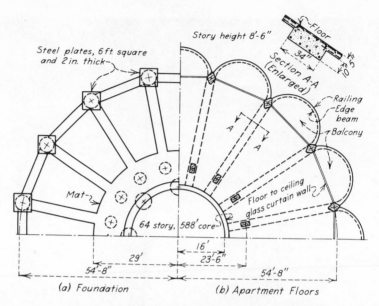

FIGURE 7.19   Partial plans of Marina Towers.

ties, one mat was made 6 ft. thick. Each mat is supported on two concentric rings of piers passing through soil and hardpan to rock about 115 ft. below. The inner ring includes 8 piers located directly beneath the core wall, and the outer ring includes 16 piers, each directly beneath a column located as will be described and illustrated in Figure 7.19.

The structure surrounding the core has a height of 60 stories. Supported primarily on two rings of 16 columns each, column centers are located on circles concentric with the core. One ring has a radius of 23 ft. 6 in.; the other, 54 ft. 8 in. The inner ring of columns bears on the mat, as has been stated, and each column in the outer ring is supported on a foundation pier extending to rock. Each pier in the outer ring is anchored to the mat by a radial tie beam, and the piers are tied together by circumferential beams.

The upper portion of each pier is reinforced, some throughout their entire depth, with vertical bars. A 6-ft. square steel bearing plate 2 in. thick on top of each pier in the outer ring facilitates the transfer of the column load to the caisson. The top ends of the reinforcing bars in the column are welded to the top of the plate. Full bearing is assured by placing the concrete in the top of the pier through a hopper extending through a hole in the center of the plate providing a hydrostatic head.

The lower 4 stories of each building are occupied by shops and services. The next 16 stories provide parking for 448 cars with access to the building. The next 40 stories provide 448 apartments of varying sizes, and the upper 4 of the 64 stories in the core of the building house the elevator machinery, air conditioning equipment, and other mechanical equipment. The elevators, stairways, and various other services are located in the core, which is surrounded at each floor level by a corridor.

The core walls, columns, and foundations are of ordinary stone concrete. The ramps and the floors of the apartments are lightweight concrete slabs integral with wide shallow beams radiating from the core and over the columns in the two rings (Figure 7.19). The parking ramps occupy the space between the core and the outer row of columns and cantilever out 9 ft. 2 in. beyond their centers. On the apartment floors, a semicircular balcony cantilevers out in each bay beyond the columns in the outer row. The apartments are enclosed between exterior columns with floor-to-ceiling glass curtain walls which include sliding glass doors providing access to the balconies. The balconies are open except for railings.

The maximum height of the parapet wall above street level is 581 ft. When completed in 1963, these buildings were the highest reinforced concrete buildings in the U. S. [10] and [11].

Expansion Joints. The folowing material on expansion joints is quoted from the Joint Committee Report [6].

"*a*. Expansion joints are expensive and in some cases difficult to maintain. They are, therefore, to be avoided if possible. In relatively short buildings, expansion and contraction can be provided for by additional reinforcement. No arbitrary spacing for joints in long buildings can be generally applicable. In heated buildings joints can be spaced farther apart than in unheated buildings. Also, where the outside walls are of brick or of stone ashlar backed with brick, or where otherwise insulated, the joints can be farther apart than with exterior walls of lower insulating value.

b. In localities with large temperature ranges, the spacing of joints for the most severe conditions of exposure (uninsulated walls and unheated buildings) should not exceed 200 ft. Under favorable conditions buildings 400 to 500 ft. long have been built without joints even in localities with large temperature ranges.

c. In localities with small temperature ranges, the spacing of joints for unheated buildings or with uninsulated walls should not exceed 300 ft. In such localities buildings up to 700 ft. long have been successfully built without joints where other conditions were favorable.

d. In roof construction, provision for expansion is an important factor. The joints in the roof may be required at more frequent intervals than in the other portions of the building because of more severe exposure. In some cases expansion joints spaced 100 ft. apart have been provided in roofs and not in walls or floors.

e. Joints should be located at junctions in L-, T-, or U-shaped buildings and at points where the building is weakened by large openings in the floor construction, such as at light wells, stairs, or elevators. Joints should provide for a complete separation from the top of the footings to the roof, preferably by separate columns and girders."

Maximum Heights.   The tallest tier buildings in the world have structural steel frames.

Since World War II, the height of buildings with reinforced concrete framing has steadily increased. The tallest buildings of that type in the U. S. in 1962 were the twin Marina City apartment buildings in Chicago, described above, with a height of 60 stories totaling 581 ft. [2]). In 1971, the tallest completed reinforced concrete building in the U.S. was the Shell Building in Houston, Texas. However, construction is proceeding on a taller reinforced concrete building in Chicago.

49.   CONCRETE RIGID FRAMES, ARCHES, AND DOMES

Rigid Frames.   A single-span two-hinged rigid frame is shown in Figure 7.20. The structure, supported on spread footings, has a tie rod under the floor that carries the outward horizontal components of the thrusts on the footings. The roof deck is wood plank sheathing supported by reinforced concrete purlins. Haunched ends improve the lateral support they provide for the frame. Exterior walls, constructed of concrete reinforced for temperature changes and shrinkage, are supported by the frame. Concrete slabs might have been used instead of the wood roof sheathing shown.

Usually, rigid frames can be constructed more economically by precasting rather than casting in place. Rigid frames of that type are considered in Article 56.

Arches.   Long-span reinforced concrete arches support the roofs of hangars, auditoriums, field houses, and other structures requiring unobstructed floor areas.

The cross section of a reinforced concrete thin-shell arched roof with stiffening and carrying ribs supported on cantilever abutments for the War Memorial of Onondaga County in Syracuse, New York,

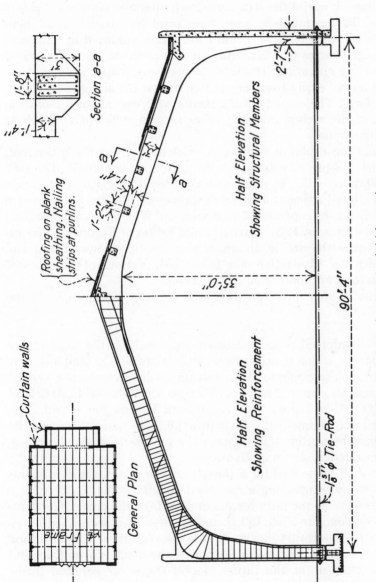

Section a-a

Roofing on plank
sheathing. Nailing
strips at purlins.

1'-2"

1'-4"

Half Elevation
Showing Structural Members

2'-7"

35'-0"

90'-4"

Half Elevation
Showing Reinforcement

1⅝" φ Tie-Rod

Curtain walls

General Plan

¼ Frame

FIGURE 7.20   Reinforced concrete rigid frame in an armory.

455

is shown in Figure 7.21. The arch rib spans 160 ft., and the cantilever projects 27 ft. on the abutments. Twelve identical bays spaced 19½ ft. center to center and arranged in pairs with the dimensions as shown in the figure comprise the structure. Each alternate rib is divided into two half-ribs separated by cane fiber insulation board. This joint, which provides for expansion and contraction produced by temperature changes, serves as a construction joint, enabling two bays to be poured in one operation. The same forms were so arranged that they could be moved easily from bay to bay to cast the six pairs of identical arch bays. The cantilever abutments were cast about a month in advance of the arches, allowing them to gain sufficient strength to support the arches.

A wood fiber insulation board 1 in. thick deck covers the entire roof, over which a 4-ply asbestos-felt built-up roofing is applied. The ceiling was sprayed with 1 in. of acoustical material. Edgarton and Edgarton, Architect-Engineer Associates, prepared the design and were responsible for the supervision. Ammann and Whitney were consulting structural engineers. F. S. Merritt, Senior Editor of *Engineering News-Record*, wrote the article on which this description was based and from which the illustration was taken [8]. Various other types of shell arch roofs are considered in Article 50.

An exterior view of a thin-shell arch coliseum is shown in Figure 7.22.

Domes.   Reinforced concrete is used extensively in the construction of roof domes, which usually cover circular areas. The simplest form of dome would be generated by revolving a solid arch about a vertical axis through its center. Arches of this type are described in Article 21 and illustrated in Figure 7.27. A common form is the ribbed arch, illustrated diagrammatically in Figure 3.6. The spaces between ribs are spanned by purlins that support the roof deck. A simple folded-plate domed roof is shown in Figure 7.30.

A thin-shell dome roof for a theater in San Diego, California, was constructed by compacting a man-made mound of earth 36 ft. high shaped to serve as the form for the underside of the dome. The dome is a pierced concrete shell 190 ft. in diameter. It is supported on its periphery by five thrust blocks. Arches between these supports form openings 9 ft. high. The major portion of the dome is 4 in. thick but increases to 16 in. at the thrust blocks; the edges between thrust blocks are stiffened with edge beams.

The compacted fill was covered with a skim coat of concrete that

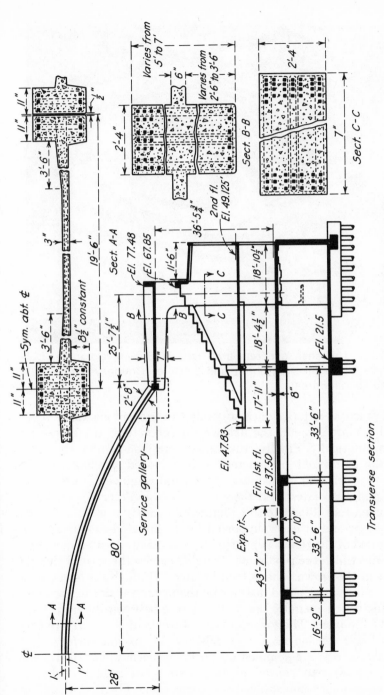

Transverse section

FIGURE 7.21 Thin-shell concrete arched roof supported on cantilever piers. *Engineering News-Record.*

457

FIGURE 7.22 Thin-shell concrete arch roof of Denver, Colorado, Coliseum. Structural Engineers: Roberts and Schaefer. Courtesy Denver Convention and Tourist Bureau.

formed a smooth surface which was oiled to prevent bonding. After placing the reinforcement, the lightweight concrete of the dome was placed pneumatically. Finally, after the concrete had cured, the earth fill and the skim coat of concrete were removed. Architects were Richard Wheeler and Associates; structural engineers were S. J. Blaylock and Associates; and the contractor was Peter Kiewit Sons.

The domed roof illustrated in Figure 7.23 consists of 32 arch ribs with the outer end of each supported by an inverted L-shaped pier. The inner end of the ribs is supported by a centrally located compression ring into which the ribs frame. The ring carries the inward thrusts of the ribs at the crown. The outward thrusts, through the lower ends of the ribs which are carried directly by the piers, transfer to an upper tension ring which encircles the roof structure between the horizontal portions of the piers. The lower outward thrusts of the piers are carried by a lower tension ring located below the ground surface. The spaces between ribs are spanned by purlins spaced about 10 ft. concentrically. Bulb tees spaced about 33 in. span radially between purlins and support noncombustible compressed fiberboard roofing plank 3 in. thick. The roof plank serves as an excellent heat-insulating and acoustic material when used as decking.

The dome, with a span of 250 ft. and horizontal legs of each pier 35 ft. long, has an overall diameter of 320 ft. The ribs and purlins of the dome are constructed of lightweight concrete, and the remainder of the structure is constructed of regular concrete. The piers and the columns supporting the seating area have pile foundations. The facade is a solar screen of perforated brick, and the interior walls are glass and concrete block. The building encloses basketball, ice hockey, trade shows, and other events. [9].

0. CONCRETE SHELL STRUCTURES

y Milo S. Ketchum, Ketchum and Konkel, Consulting Engineers, Denver, 'olorado.

Folded Plates.   The folded plate, shown in Figure 7.24, is the simplest of the shell structures. Its principal advantages are the ease of forming and its simple lines. The structure acts as a concrete slab across the short dimension. At the supports of the plates, it is necessary to provide a stiffener to pick up the reactions of the plates and

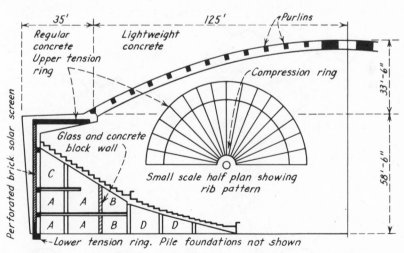

A-Concourse. B-Toilets and concessions. C-Machinery. D-Storage

FIGURE 7.23 Concrete ribbed dome, Jacksonville, Florida, Coliseum. Architects: George R. Fisher and A. Eugene Cellar. Structural Engineer: Grumar E. Kraus.

deliver them to the columns. A horizontal tie is also necessary to hold the horizontal forces. At an edge, the arrangement of the plates must start with a short plate. A two-element plate with an edge member turned up at an angle is shown in (a). A three-element folded plate with a turned-down edge member is shown in (b). The advantage of this shape is that the width of individual plates may be less for the same column spacing. A three-element "Z" shell suitable for buildings having clerestory, north-light windows between the edges of adjacent units is shown in (c). A five-element folded plate which resembles a barrel shell is shown in (d).

The thickness of a folded plate is primarily dependent on the thickness necessary to carry the slab (short) span. This may range from 3 in. for a 5-ft. width to 4½ in. for a 12-ft. width. For structures greater than 12 ft., the slab should be haunched. The slope of the plates should not be greater than 45°, and the ratio of depth to span should be about 1 to 12 for simple spans and 1 to 15 for continuous spans.

Barrel Shells.   Sketches of long-barrel shells are shown in Figure 7.25. In this type of cylindrical structure, the approximate ratio of span to

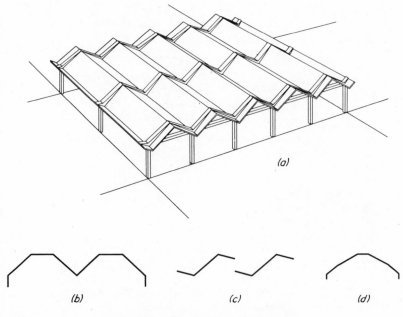

FIGURE 7.24   Folded plate roofs.

width is assumed to be greater than 2 to 1. Short-barrel shells are described in the next paragraph.

The barrel shell acts as a beam in the long direction and as an arch in the short direction. The arch, however, is not supported at its ends, as it is in the usual highway bridge arch, but is supported by the internal shears.

Several cross sections are shown. Figure 7.25(a) is a circular cylindrical barrel with a turned-up edge. The usual section is a circle, but other shapes are satisfactory. A continuous series of curves, as in a corrugated section, is shown in (b). Discontinuous units suitable for a north-light structure are shown in (c). Edge members at the junction of each barrel are shown in (d). The depth of the structure may be increased by this method and longer spans may be used.

The thickness of the barrel is dependent in most cases on the amount of cover over the reinforcing bars and varies from three to four inches. It is customary to thicken the shells slightly at the valley. The ratios of span to depth are about the same as for folded plates. However, the widths may be considerably greater for the same thickness.

Short Shells.   The width of a short shell is large in comparison to the span as sketched in Figure 7.28(a) and (b). The barrels are picked up by arches or frames which may be either above or below the shell. The arches may have many different forms, two of which are shown in (a)

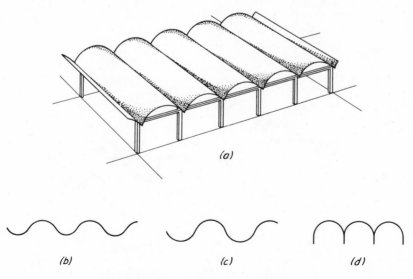

(a)

(b)   (c)   (d)

FIGURE 7.25   Long-barrel shell roof.

and (b). The short shell acts as an arch in the upper part of the curve and delivers its thrusts to a beam element at the lower end of the shell. This area must be braced by an upturned or downturned edge member. In (a), the shell is turned down. The thickness of short shells is usually a minimum, about 2.5 to 4 in., except at an edge which is not supported by an edge plate.

Short shells are often used for quite wide openings. A typical width is 250 ft. with the arch elements spaced 25 to 35 ft. The arch is the predominant structural element.

Domes of Revolution. A typical dome of revolution is shown in Figure 7.27(a). This is a segment of a sphere. The shell acts as a membrane, and all stresses are direct compression or tension. There is

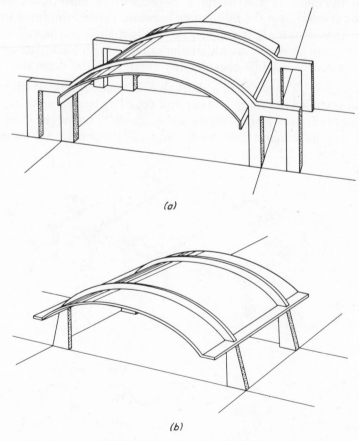

(a)

(b)

FIGURE 7.26 Short-shell roofs.

very little bending in the shell except as a secondary stress near the tension ring at the lower edge. The ring holds the shell mainly in compression and may be prestressed by high-strength steel wires wrapped around the ring.

Many cross sections may be used for a dome of revolution, as shown in the figure. In (d), a central column is used.

The thickness of domes is usually a minimum, and tanks with a diameter of 100 ft. have been built with a roof thickness of 2 in.

The acoustical problems associated with domes may be very difficult to handle.

The Hyperbolic Paraboloid.   The hyperbolic paraboloid is the name of a mathematical surface created by twisting a plane surface, and for that reason these structures are often called *twisted surfaces*. The

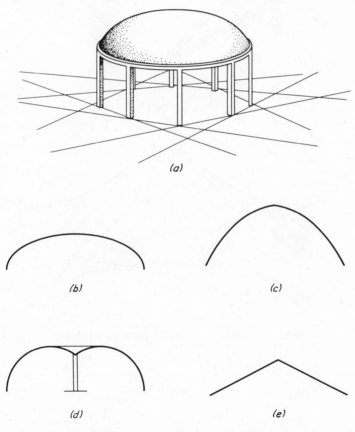

FIGURE 7.27   Domes of revolution.

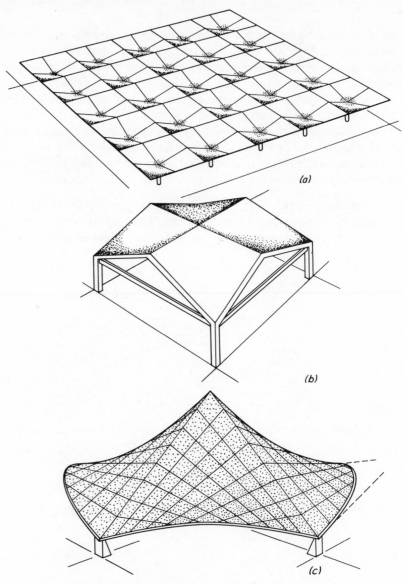

FIGURE 7.28 Hyperbolic paraboloid roofs.

shell has a double curvature, convex in one direction and concave in the other.

If properly arranged, the stresses in a twisted surface are mostly membrane, that is, there is tension in one direction and compression in the other. Ribs are usually necessary at the exterior edges and at the junctions of the surfaces.

Several structures are shown in Figure 7.28(a) to (c). A series of umbrella shells is shown in (a). The support is a column at the center of each square unit. The shell in (b) is a dome arranged with four square or rectangular surfaces supported by triangular tied arches. A circular dome with the corners clipped off is shown in (c). The thrusts from the shell are taken by curved arches. There are an infinite variety of shapes possible using the basic twisted surface.

The thickness of the shell may be a minimum. If two layers of ⅜-in. bars are used and ¾ in. of cover is required, the thickness should be 2¼ in.

Intersection Shell. The dome shown in Figure 7.29(a) is made by the intersection of cylindrical units. This structure acts essentially as a short shell, and the intersections form a rib to stiffen the dome. In (b), the axis of the barrels is reversed, and the shell must be supported by arch ribs at the edges.

Folded-Plate Dome. A dome made with triangular folded plates is shown in Figure 7.30. The plates exert a thrust in the horizontal ring.

Shell Arch. An arch with the cross section of a folded plate is shown in Figure 7.31.

Other Types. The structures shown in this article are only a few of the many possible structures available for shell structures; this field requires the creative genius of the architect and the engineer.

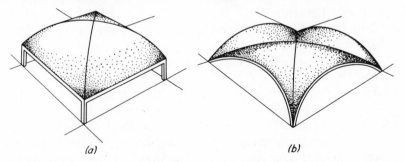

(a)   (b)

FIGURE 7.29   Roofs formed by the intersection of cylindrical units.

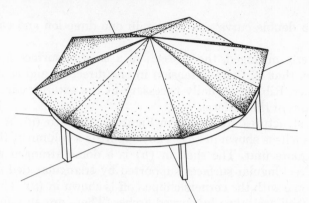

FIGURE 7.30   Round folded-plate domed roof.

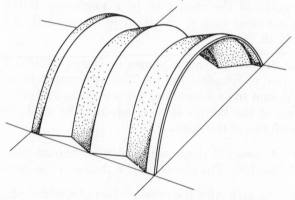

FIGURE 7.31   Arch with folded-plate cross section.

REFERENCES AND RECOMMENDED READING

1. *Building Code Requirements for Reinforced Concrete*, American Concrete Institute.
2. "Tall Concrete Towers for Chicago," *Engineering News-Record*, March 10, 1960.
3. *Reinforced Concrete Floor Systems*, Portland Cement Association, 1956.
4. William Cohen, "How to Stop Partitions from Cracking," *Engineering News-Record*, September 9, 1954, p. 35.
5. J. DiStasio and M. P. Van Buren, and Fred N. Severud, "Flat Plate Floors Designed as Continuous Frames," "Slab Band Floors Are Economical for Apartment Buildings," *Modern Developments in Reinforced Concrete*, No. 20, Portland Cement Association, 1947.
6. *Recommended Practice and Standard Specifications for Concrete and Reinforced Concrete.* Published by American Concrete Insti-

tute, 1940, seventh printing, September, 1950.
7. Phil M. Ferguson, *Reinforced Concrete Fundamentals*, John Wiley and Sons, 1958.
8. F. S. Merritt, "War Memorial Serves Many Purposes," *Engineering News-Record*, November 16, 1950, p. 39.
9. "Domed Roof for Coliseum," *Engineering News-Record*, December 8, 1960, p. 26.
10. *ACI Manual of Concrete Practice, Part 1, 1970*, American Concrete Institute, Detroit, Michigan.
11. *Building Code Requirements for Reinforced Concrete* (ACI 318-71), American Concrete Institute, Detroit, Michigan, 1971.
12. *The Making of Portland Cement*, Brochure HE 108-2, Portland Cement Association, Skokie, Illinois, 1964.
13. *ASTM 1971 Annual Book of Standards, Part 10*, American Society for Testing and Materials, Philadelphia, Pennsylvania.
14. *ACI Manual of Concrete Practice, Part 3, 1968*, American Concrete Institute, Detroit, Michigan, 1968.
15. *Manual of Standard Practice*, 20th Edition, Concrete Reinforcing Steel Institute, Chicago, Illinois, 1970.
16. Catalogs: *Concrete Topics and Trends*, Kaiser Cement, Oakland, California.
17. *Design and Control of Concrete Mixtures*, Eleventh Edition, Portland Cement Association, Skokie, Illinois, 1968.
18. *Plywood Concrete Forms*, American Plywood Association, Tacoma, Washington.
19. *Sweet's Catalog*, F. W. Dodge Corp., New York, New York.
20. *ACI Manual of Concrete Inspection*, Publication SP-2, American Concrete Institute, Detroit, Michigan, 1967.
21. *Concrete Construction, Compilation No. 2*, American Concrete Institute, Detroit, Michigan, 1968.
22. *Cement and Concrete Terminology*, Publication SP-19, American Concrete Institute, Detroit, Michigan, 1967.
23. *Formwork for Concrete*, Publication SP-4, Second Edition, American Concrete Institute, Detroit, Michigan, 1969.
24. *Handbook for Concrete and Cement*, Corps of Engineers, U. S. Army Waterways Experiment Station, Vicksburg, Mississippi.
25. *Cement and Concrete Reference Book*, Portland Cement Association, Skokie, Illinois, 1964.
26. *Concrete Manual*, Seventh Edition, U. S. Department of the Interior, Bureau of Reclamation, Denver, Colorado, 1966.
27. Journal of the American Concrete Institute, Sept. 1971, Guide for Use of Admixtures in Concrete Report by ACI Committee 212.

8

PRECAST AND PRESTRESSED CONCRETE CONSTRUCTION

51. PRECAST CONCRETE

The term *precast concrete* applies to individual concrete members of various types cast in separate forms before placement. In contrast, the term *cast-in-place* or *site-cast* concrete refers to concrete members formed and cast in the positions they are to occupy in the finished structure.

Precast members cast at the building site or at a casting yard remote from the structure are transported by truck or by other means to the site of the structure and positioned by cranes. Concrete block, as described in Article 26, is the simplest form of precast unit, although it is not commonly given that designation.

To improve quality by using a low water-cement ratio and insuring that the forms are completely filled, the concrete is vibrated during placement, as described in Article 42. Lightweight concrete is often employed. In addition to the advantage of light weight, such concrete has a relatively high heat insulation value.

Precast and Tilt-Up Construction.   Concrete walls and partitions are often made of precast units, rather than of cast-in-place concrete, and then positioned as will be described in Article 53. Concrete wall slabs of considerable size are often cast at some distance from the building site and then placed into position. In tilt-up construction, they are rotated about their bottom edges from a horizontal position into their final positions, as described in Article 53.

Types of Members.   Although it may not always be advantageous, nearly every concrete member which can be cast in place can be precast. Included in these types are floor and roof slabs, wall panels, bearing walls and partitions, joists, beams, girders, columns, rigid frames, arches, domes, and piles. Trusses have been precast but not often advantageously constructed of concrete, either precast or cast-in-place. Precast piles are considered in Article 11.

Building Heights.   Except for lift-slab construction in which the floor and roof slabs are precast as described in Article 54, most buildings whose main structural members are precast are one-, two-, or three-story buildings; however, buildings have been constructed of precast concrete to 16 stories. Precast beams, purlins, floor and roof panels, and curtain walls are used, however, irrespective of height.

Prestressing.   Precast concrete members may or may not be pre-stressed, as described in Article 52. In general, prestressing may be advantageous for members subjected to high flexural stresses, such as long-span or heavily loaded slabs, beams, and girders. A few examples of precast construction are illustrated in the following articles, but this type of construction is advantageous for a great variety of members and structures.

Framing.   The superstructure of a building may be assembled in the following ways:

(1) All the members precast.

(2) Precast floor and roof decks supported by cast-in-place concrete of steel girders, columns, or rigid frames.

(3) Cast-in-place floor and roof slabs supported directly by precast joists or purlins.

(4) Precast wall panels supported by cast-in-place concrete or steel columns or rigid frames.

(5) Heavy timber roof sheathing supported directly by precast purlins.

(6) Precast arches supporting precast concrete purlins and cast-in-place concrete slabs or heavy timber sheathing.

(7) Masonry exterior walls and various types of partitions with precast floor and roof decks.

Connections.   A common connection between precast members consists of anchoring steel plates in the precast members so that they will contact when the members are in position (Figure 8.1(a)). These are called *matching plates*. After the members have been positioned, the matching plates are welded together. Occasionally they are bolted, and sometimes angles are used instead of plates.

At the ends or edges of adjacent members, reinforcing bars may be extended so that they overlap (b). They may be joined by welding. The intervening space is then filled with concrete. Instead of welding,

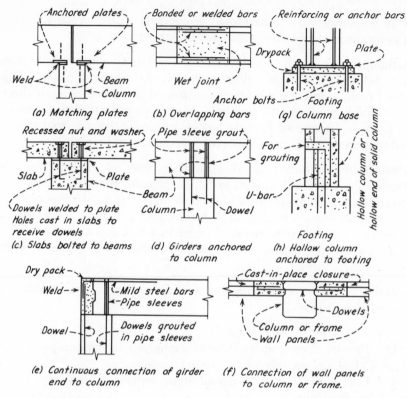

FIGURE 8.1   Typical connections for precast members.

the connection may be established by the bond of the overlapping bars. This is called a *wet joint*.

Pipe sleeves or holes may be cast in members and so located that dowels or bolts cast in the supporting members (*c*), (*d*), and (*e*) can be inserted or threaded into the holes. Some form of anchorage is provided on the ends of the dowels which project beyond the end of the sleeve or hole, or the dowels are grouted into the sleeves.

Various methods are employed to connect beams to columns, one of which is shown in (*d*). A plate welded to the ends of embedded column reinforcing anchors a column base plate which in turn is bolted to its footing (*g*). In (*h*), a hollow column, positioned on its footing, is threaded over a looped anchor bar embedded in the top of the footing. An opening in the side of the column provides a temporary port through which concrete is poured, filling the hollow portion and surrounding the anchor bar protruding from the footing. A hollow end may be cast in a solid column for anchoring in this same manner. Wall panels are connected to columns as shown in (f).

Numerous other types of joints are used; some are explained in subsequent articles.

Advantages.   Precast concrete is more economical when many identical members are to be cast because the same forms can be used many times. In addition to economical formwork, precast concrete has other advantages over cast-in-place concrete:

(1) Concrete quality can be controlled more carefully.

(2) Exposed surface control can be achieved, and plastering is not required to produce a finished surface.

(3) Less jobsite storage space is required.

(4) Casting can proceed under all weather conditions as long as suitable protection is provided.

(5) Curing can be done more effectively.

(6) Erection is less restricted by weather conditions.

(7) Greater erection speed is possible.

(8) Lower cost is possible under conditions favorable to precasting.

52.   PRESTRESSED CONCRETE

Prestressed concrete is concrete subjected to compressive stresses by inducing tensile stresses in the reinforcement. The purpose of prestressing is to counteract tensile stresses in the concrete caused by external loads before the external loads are applied. The term *conventional reinforced concrete* applies to members that are not prestressed. To make prestressing feasible, both the concrete and the reinforcement must have much higher strengths than are required for conventional reinforced concrete members. As will be explained later, the reinforcement consists of high-strength steel wires or cables of high-strength alloy steel bars, all of which are called *tendons.*

The design procedures for prestressed concrete members are the same as those for conventional reinforced concrete members, except that they are extended to provide for the effects of prestressing.

Prestressed concrete is advantageous only under special conditions because of cost. Its use is confined largely but not exclusively to precast members, although only that use is considered in this chapter.

Basic Principle.   To understand the structural action of prestressed concrete members, it is helpful to review this action in homogeneous beams, explained in Article 19 and illustrated in Figure 3.2(e), and in conventional reinforced concrete beams, considered in Article 46. As explained in these articles, the lower portion of a horizontal simple beam carrying downward vertical loads is subjected to tensile stresses. Now the tensile strength of concrete is low, and a homogeneous concrete beam has very little flexural strength. To offset this deficiency, steel reinforcement provided near the bottom of simple beams carries the tensile stresses. The tensile strength of the concrete is neglected in the computations for the flexural strength of reinforced concrete beams.

By subjecting the tensile reinforcement of a beam to tensile stresses before the external loads are applied (described later), compressive stresses are induced in the concrete of the beam. Usually the tensile stresses in the concrete caused by the external load are nearly offset or absorbed by the compressive stresses in the concrete resulting from prestressing the reinforcement which, in turn, prestresses the concrete. The concrete, therefore, is effectively resisting tensile stresses produced by external loads rather than being neglected as it is in conventional design.

Flexural Stresses in Prestressed Concrete. The computed fiber stresses at the section A-A of a homogeneous simple beam with a symmetrical cross section for various designated conditions are shown in Figure 8.2(a) to (e), and those in a prestressed beam in (f).

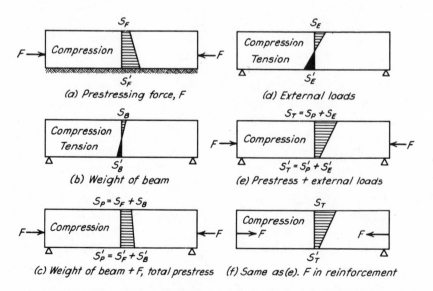

FIGURE 8.2   Fiber stresses caused by various loading conditions.

Before the external forces F are applied, as shown in (a), the beam would normally be supported by forms throughout its entire length and there would be no flexural stresses. As soon as the forces F are applied, the beam deflects upward because of the uneven distribution of the compressive fiber stresses in the concrete (a), and it is supported only at the ends (b). In (b) and the remaining figures, the member acts as a simple beam.

It is not feasible to prestress the beam by applying equal and opposite external compressive forces F at the ends. Substantially the same effect is obtained simply by inserting tendons and anchoring them at the ends of the beam so that a force F is applied to these ends (a) and (e). This procedure is called *posttensioning* because the prestressing force is applied to the tendons after the concrete hardens.

A similar objective can be accomplished by prestressing the tendons in the forms before the concrete is poured. After the concrete sets, it bonds to the tendons, which therefore retain most of the prestress after the prestressing force ceases to act. The member is then removed from the forms. This procedure is called *pretensioning*.

A significant portion of the initial prestress in the tendons is lost because various factors cause the concrete to shorten. Concrete shrinks as it dries. Concrete also shortens elastically as compressive forces are applied and shortens still further over a long period of time because of a phenomenon called *creep, plastic flow,* or *time yield.* The length of each tendon increases elastically while the prestress is applied. This does not reduce the prestress. Steel increases gradually in length, or *creeps,* at a decreasing rate when subjected to a constant unit stress. However, if the length remains constant, as in the tendons of prestressed concrete, the unit stress gradually decreases at a decreasing rate. This is called *relaxation,* and it is the significant factor in prestressed concrete design. Finally, there is a loss of prestress in posttensioning because of the friction between the tendons and the surrounding material. This is especially true if the tendons are curved or *draped,* as explained later. Appropriate allowances are made for these factors which decrease the initial prestress. This is also true for different external loads if the stresses do not exceed those caused by working loads.

The discussions in this article refer to the flexural stresses in a prestressed simple beam. Similar procedures apply to prestressed continuous beams and to columns and piles if they are subjected to loads that would produce significant flexural stresses if they were not prestressed.

Types of Tendons.   As stated, effective prestressing requires high-strength steel for tensile reinforcement. This is because the loss of prestress from shrinkage, elastic deformation, and creep in the concrete and the relaxation and frictional losses in the steel during pre-stressing offset most of the allowable stress in ordinary steel bars.

The individual units, or tendons, which comprise the total tensile reinforcement (corresponding to the bars used in conventional rein-forced concrete) are usually two or more parallel wires about $\frac{1}{4}$ in. in diameter, or they are *strands* of such wires "spiraled" around a straight center wire. Seven-wire strands with diameters of $\frac{1}{4}$ to $\frac{1}{2}$ in. are ordinarily used in pretensioned members. Large-diameter strands with 7, 19, 37, or more wires are used extensively in post-tensioned members.

The wires of the desired diameter are formed by drawing steel rods, while cold, through dies of smaller diameter than the rods. This process of *cold-drawing* markedly increases the tensile strength of the steel. After drawing, the wires may be galvanized to resist corrosion.

High-strength alloy steel bars are also used for tendons. Their strength is further increased by heat-treating or cold-stretching. Diameters vary from $\frac{1}{2}$ to $1\frac{1}{8}$ in. They are used principally in post-tensioned members.

Casting and Prestressing.   Prestressed concrete members, usually precast, are divided into two classes, pretensioned or posttensioned, according to the sequence of casting and prestressing.

For pretensioned members, the tendons are prestressed in the forms before the concrete is poured. The *long-line process* is ordinarily used if several members of the same cross section and identical ten-dons are cast. The casting bed on which the forms are placed may be several hundred feet long and arranged as illustrated diagrammati-cally in Figure 8.3 (a).

The bed includes two end abutments. In the long-line process, tendons long enough to provide the tendons for several members to be cast in line are placed on the casting bed. They are anchored at one end to the abutment; at the other end, to hydraulic jacks operat-ing against the other abutment. Various jacking arrangements have been devised. The tendons pass through a template provided at each end. Holes in each template hold the tendons in the desired posi-tions. Bulkheads are positioned along the long-line form, subdividing it into forms for several individual members of the desired lengths. These bulkheads are notched so that they can straddle the tendons, or the tendons may be threaded through them. The long tendons are

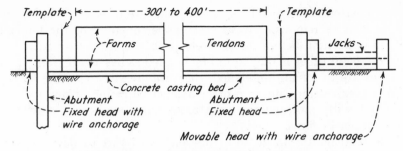

(a) Pretensioning bed for long-line process
Arranged from Portland Cement Assh publication

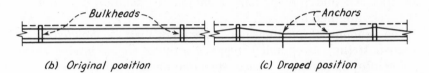

(b) Original position   (c) Draped position

FIGURE 8.3   Pretensioning beds and draping of tendons.

prestressed the desired amount by operating the jacks. The ends of the tendons are anchored and the jacks removed.

The concrete is then placed in the forms. When the concrete has gained its initial set, the forms are covered to retain moisture, or steam curing proceeds. After the concrete hardens sufficiently to carry the prestress, the bulkheads between the ends of the members are removed—a process requiring about 24 hours. The members are separated by cutting the exposed portions of the tendons with a welding torch or by some other method. The beams are removed from the forms by cranes and stored for shipment.

Stirrups or other steel embedded items to be cast in the members are, of course, placed in the forms before the concrete is poured.

Frequently, it is desirable to drape some of the tendons of each member so that they will anchor near the top at the supports and near the bottom in the central portion of a span. The long tendons to be draped are placed near the top of the members and prestressed. They are then pulled downward at the desired points and anchored as shown in Figure 8.3(b) and (c). Care must be taken to avoid overstressing the tendons by this operation. Draping improves the distribution of the prestress in the concrete along cross sections near the support, and the inclined portions of the tendons help carry the shearing stresses.

Individual members can be pretensioned without using the long-line process.

For posttensioned members, the tendons are inserted in holes that are formed by removable cores cast in the concrete; or the tendons, placed in position before the concrete is poured, are surrounded with flexible metal tubes if they are to be bonded as described later. If the tendons are to remain unbonded, they are greased to reduce friction during prestressing and to protect against corrosion. They are then wrapped with mortar-tight sheathing, such as heavy waterproof paper or plastic films.

The tendons are anchored at one end with various devices. After the concrete hardens sufficiently and the forms have been removed, the tendons are prestressed by means of hydraulic jacks applied at the free end and the tendons anchored to maintain the prestress. The jacks are then removed.

The tendons of posttensioned members may be *bonded* or *unbonded*. If they are bonded, bond is established by forcing cement grout into the annular space surrounding the tendons after prestressing is completed.

Advantages and Disadvantages.   As stated, most prestressed members are precast and have the advantages of this procedure as compared with cast-in-place construction. In addition, there are advantages gained by prestressing when conditions are favorable, such as:

(1) Smaller dimensions of members for the same loading conditions, which may increase clearances or reduce story heights.

(2) Smaller deflections.

(3) Crack-free members.

(4) Smaller loads on supporting members because of the smaller dimensions required.

Among the disadvantages are the following:

(1) Higher unit cost of high-strength materials.

(2) Cost of prestressing equipment.

(3) Labor cost of prestressing.

(4) Not advantageous for short spans with low concrete stresses.

53.   PRECAST CONCRETE: FLOORS, ROOFS, AND WALLS

Floor and Roof Slabs.   Several types of precast slabs or panels are used extensively for floor and roof decks. The most common are the channel and double-T in Figure 8.4.

Channel slabs vary in depth from 9 to 12 in. Spans of 50 ft. or more have been used. The legs of the channels extend across the ends

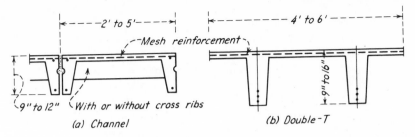

FIGURE 8.4   Channel and double-T decks.

if desired, and the legs and top slab may be stiffened with cross ribs. The tensile reinforcement is mild steel bars if the channels are not prestressed or high-strength steel tendons if they are prestressed. The top slab is reinforced with wire legs so that the joints between the slabs can be grouted to form keys between adjacent slabs. The long-line process is ordinarily employed.

Double-T's vary in width from 4 to 6 ft. and in depth from 9 to 16 in. Spans as long as 50 ft. have been cast. The top slab varies from 1½ to 2 in. thick and is reinforced with wire mesh. Except for short spans, the slabs are prestressed, usually by pretensioning using the long-line process. The tendons can be draped as shown in Figure 8.3(b) and (c).

Hollow or cored slabs or planks with various dimensions are manufactured. For the longer spans or heavy loads, the bottom reinforcement may be prestressed high-strength tendons; the top reinforcement may be mild steel bars designed to withstand prestressing and handling stresses.

Hollow slabs of special forms of concrete are assembled in rows and reinforced with bars grouted in grooves provided in the lower surfaces of the block. The vertical contact surfaces are grouted or ground smooth to obtain good contact.

Solid slabs are sometimes used for roof decks when the low cost of the forms offsets the saving in material cost that would result from the use of structurally more efficient slabs.

Floor and roof slabs are ordinarily connected to the supporting members by welded matching plates.

Walls and Partitions.   Bearing and nonbearing walls and partitions ordinarily are of panels precast in a horizontal position in a casting yard or on the floor of the building and placed in their vertical positions by cranes by the tilt-up procedure.

Usually, they are solid reinforced slabs from 5 to 8 in. thick, one or two stories high with lengths about equal to the clear distances between columns or other supporting members. Actual length depends upon the manner in which the slabs are fastened to the supporting members. Window and door openings are usually cast in the slabs. Steel window frames or pressed metal door frames are sometimes cast in exterior wall panels, but preferably openings are prepared for window and door frames. Extra reinforcement, both lateral and diagonal, should be provided around the openings.

The casting surface usually is a concrete floor slab with a smooth regular surface. Bond between this surface and the wall panels is prevented by coating the casting surface with some form of liquid bond breaker or covering with a plastic sheet material. The liquid bond breaker is more satisfactory. The upper surface of the panel may be finished in several ways such as troweling, floating, or brooming.

Precast channel and double-T panels, previously described, are used for exterior nonbearing walls. The ribs, placed vertically, are designed for adaptation to the different loading conditions. Prestressed double-T panels, with an intermediate support, have exceeded 60 ft. in height.

Sandwich panels, used for exterior walls, provide additional heat insulation. The face slabs, see Article 17, are tied together with wires, small rods, or in some other manner. Thicknesses vary from 5 to 8 in.

When precast concrete panels, described in Article 84, are used for curtain walls of multistory buildings, the panels are placed story by story. Wall panels are often connected to the building frame at their tops and bottoms and to each other by welded or bolted matching plates. Several other types of connections are used (Figure 8.1(f)). The joints between wall panels are made watertight by sealants on the outside and grouting on the inside.

Tilt-up construction is a construction procedure by which precast wall panels are rotated about their bottom edges to the vertical positions they are to occupy. The panels are designed to resist the lifting stresses. Small panels are sometimes hoisted by hand-operated cranes, but power equipment is usually employed. The simplest means of connecting the hoisting equipment to a panel consists of U-shaped bar loops cast in its top edge, which produce relatively high lifting stresses. A *strongback* made of steel I-beams or channels bolted to a panel reduces the lifting stresses, or a *vacuum pad* may be placed on the panel and held in position by creating a partial vacuum between the panel and the pad. This device is preferred to the strongback

FIGURE 8.5   Precast concrete tilt-up construction. Portland Cement Association.

because the panel is not disfigured by bolt holes which are cast in the panel for attachment of a strongback. See Figure 8.5.

Tilt-up construction is also used for placing rigid frames and arches.

54.   PRECAST CONCRETE: JOISTS, BEAMS, GIRDERS AND COLUMNS

Joists and Purlins.   Joists and purlins have T- or I-shaped cross sections (Figure 8.6(a) and (b)) and may be prestressed or of conventional design. Many manufacturers stock the common sizes for immediate delivery. Joists with inverted T-sections are available for composite construction where they support cast-in-place floor or roof slabs.

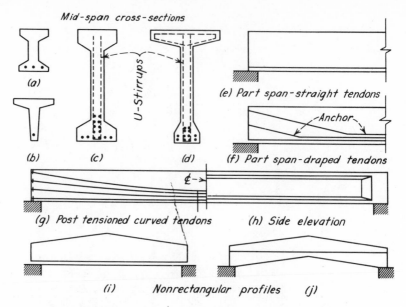

FIGURE 8.6   Precast and prestressed joints, beams, and girders.

Beams and Girders.   These terms are often synonymous; but when a distinction is made, those with the longer spans are called girders. They may be of conventional precast design or prestressed. The cross sections are usually I-shaped (*c*) except near the ends, where they are rectangular. T-shaped cross sections (*d*) are also used. The following discussion refers to prestressed members.

There are structural advantages in curving or inclining some of the tendons upward near the ends of a girder. In the earlier efforts of prestressing, it was considered necessary that pretensioned tendons be straight, as shown in Figure 8.6(*e*). It has been found, however, that tendons can be draped (*f*) by anchoring them at two or more points before or after prestressing, as long as the tendons are not overstressed.

In posttensioning girders, tendons are usually curved (*g*). Sloping tendons help carry shearing stresses. If additional shearing resistance is required, mild steel stirrups are used as in conventional reinforced concrete beams and as shown in (*c*) and (*d*).

Girders with clear spans exceeding 100 ft. may have nonrectangular profiles (*i*) and (*j*). The profile in (*j*) raises the center of prestress at the ends, which is advantageous, without curving the tendons as in (*g*).

The girders mentioned are simply supported; however, they may be continuous with single units or with several units, as shown in Figure 8.7(a). Overhanging girders with a *drop-in beam* are illustrated in (b). The ends of the drop-in beam rest on the ends of the overhangs and are stabilized by suitable anchors but do not develop continuity. These types may or may not be prestressed.

A prestressed *cap cable* develops continuity at a support of adjoining prestressed girders shown in (c). After the girders are positioned, cables are inserted in ducts cast in the concrete, anchored at one end, and prestressed. If the girders support a cast-in-place slab, another means of developing continuity is preferable, as explained in the paragraph on composite construction.

Types of connections between girders and columns are illustrated in Figure 8.1. In the simplest (a), matching plates are welded together after the members are in position. It develops no continuity. Others which develop partial or full continuity are shown in (d) and (e). When full continuity is developed, the combination becomes a rigid frame. Such action must be considered in the design of the girders and columns.

Composite Construction.   Composite construction consists of precast girders, beams, joists, or purlins on top of which a floor or roof slab is cast in place. The slab and precast units act together in resisting

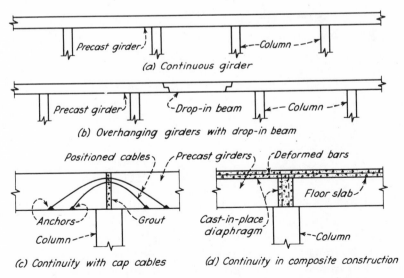

FIGURE 8.7   Overhanging girders and continuity connections.

the weight of the slab and the applied loads, provided movement is prevented between the slab and the precast units and if the slab anchors to these units preventing vertical separation. Horizontal movement of the slab along the beam is prevented by notched keys with vertical sides on the tops of the precast members or by special preparation of the contact surface by sand blasting or other means that insure adequate bond. Vertical separation is prevented by casting vertical steel dowels in the tops of the precast units or by projecting the ends of vertical stirrups in these units.

Beams are made continuous over supports by including longitudinal reinforcement over the supports in the slab (d). This procedure is preferable to using cable caps (c) if the slab is cast in place.

Columns. Precast concrete columns may be similar to cast-in-place columns, as described in Article 45, or they may be hollow. Prestressing the longitudinal reinforcement is advantageous only for columns subjected to relatively large flexural stresses. When prestressed, the longitudinal reinforcement is of high-strength tendons rather than mild steel bars. The tendons may be pretensioned; the long-line process may be used if many identical columns are to be cast. The lateral reinforcement may be small diameter mild steel rods or wires. Columns are cast in a horizontal position and erected by cranes.

A simple type of column base (Figure 8.1(g)) is suitable for square solid or hollow columns. The type shown in (h) may be used for hollow columns. A looped rod cast in the column footing projects upward into the hollow core when the column is set. A temporary opening in the side of the column permits the lower portion of the core to be filled with grout. The embedded loop forms an effective anchor. The opening is then dry-packed with mortar. Various other types are used.

55. PRECAST CONCRETE: LIFT-SLAB CONSTRUCTION

The lift-slab method was conceived by Philip N. Youtz and Tom Slick about 1950 and is usually known as the Youtz-Slick Method.

In lift-slab construction, all above-ground floor and roof slabs for an entire building are cast in stacks on a previously prepared ground floor. Cast around the columns that are to support them, they are lifted into their final positions by simultaneously operating hydraulic jacks mounted temporarily on the tops of the columns. The slabs are connected by *lift rods* to steel *lifting collars* cast in and anchored to the slabs and surrounding the columns.

Because long columns tend to buckle when not supported laterally at intervals, not more than three or four stories are usually constructed at one stage.

Sequence of Lifting Operations.   The sequence of operations for a 6-story building, including a basement, which requires two lifting stages is illustrated in Figure 8.8. It will be noted in (*f*) that the columns are extended to their final height at the beginning of the second stage. Taller buildings require more stages. These are carried out by extending the two-stage procedure.

Types of Columns.   The columns, usually of steel (Figure 8.9(*a*) to (*e*)) vary in external lateral dimension from 8 to 12 in. For a given building, column dimensions are usually constant; however, the steel areas are varied by specifying columns of different flange and web thicknesses. One dimension is limited by the spacing of the lifting rod attachments on the jacks. Occasionally, columns are made of precast reinforced concrete with longitudinal and lateral bar reinforcement (*f*). Prestressed columns with longitudinal wire tendons and mild steel lateral reinforcement are also used.

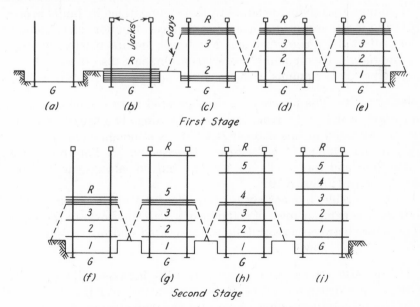

FIGURE 8.8   Sequence of lifting operations.

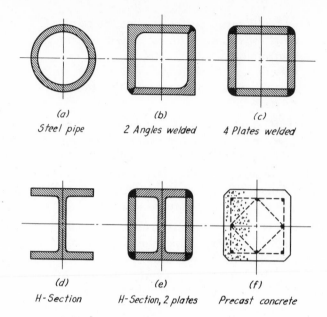

| (a) | (b) | (c) |
| Steel pipe | 2 Angles welded | 4 Plates welded |

| (d) | (e) | (f) |
| H-Section | H-Section, 2 plates | Precast concrete |

FIGURE 8.9   Types of columns.

Columns are rigidly anchored to their foundations by built-up steel bases to stand vertically. In fire-resistive construction, steel columns are protected by concrete, metal lath and perlite plaster, or other appropriate noncombustible materials after the slabs are placed. Filling hollow steel columns with concrete improves their fire resistance.

Size of Slab.   The hydraulic jacks, operated from a cont₁ol console, must act in unison to avoid local overstressing. The usual number of jacks controlled in this manner is 12 or 18, although equipment does permit the number of jacks to be increased to 36. For this reason, large floor and roof decks are usually, but not always, divided into several slabs (Figure 8.10).

It is desirable, but not always possible, for a slab to cantilever out beyond the exterior rows of columns of a slab, as shown in the figure. The joints between adjacent slabs are located midway between rows of columns, as shown.

Closure Strip.   In a space about 3 ft. wide between adjacent slabs, bars project from the edges of adjacent slabs into this space and overlap to permit welding. A *closure* or *pour strip* of cast-in-place concrete fills the space and provides continuity. By providing wider

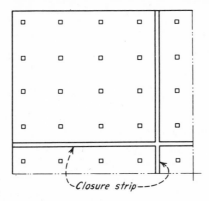

FIGURE 8.10   Subdivision of slabs.

spaces, stairways, elevator shafts, and other vertical shafts can be constructed. In addition, shear walls to provide lateral stability for the building can be built in these spaces as the construction proceeds.

Column Spacing and Slab Thickness and Types.   Common column spacings vary from 25 to 30 ft. or more and may differ in the two directions. The slabs, usually solid, vary from 5 to 10 in. thick. For the longer spans, *waffle slabs*, similar to the slabs shown in Figure 7.11, are used frequently. The areas adjacent to the columns must be solid. The top slabs are reinforced with wire mesh. Waffle slabs must be somewhat thicker than solid slabs of equal conditions.

Prestressing.   The slabs are usually prestressed by posttensioning in both directions (Article 52). Usually the tendons are inclined from the lower center portion of the slab to the upper portion at the columns and in the cantilevers to resist tensile stresses. Prestressing causes slabs to cant upward between columns. This is desirable because such slopes counteract most, if not all, of the deflection due to applied loads. In addition, prestressed slabs are crack-free, a quality which is always desirable.

Lifting Collars.   Before each column is erected, the lifting collar required for each slab is placed on the column with approximately ¼ in. clearance between the collar and the column. Collars are made of cast steel or are built up of steel sections.

Slab Casting.   Cast sequentially as they occupy the building from the ground up, the slabs are securely anchored to the lifting collars. These collars increase the resistance to shearing stresses near the columns which are caused by the vertical loads. To prevent bonding between slabs during casting, the top of each slab is coated with liquid wax or other liquid preparation or covered with a membrane of film before the next slab is cast.

Connecting Slabs to Columns.   After a slab has been lifted to its final position, the lifting collars are welded to steel columns. Provision may be made for temporary support before the final welding takes place. If reinforced concrete columns are used, this connection may be made by welding embedded matching plates. Slabs lifted in two or more stages require temporary positioning of some of the slabs (Figure 8.8(e) and (f)). Several slabs may be lifted simultaneously, the lowest one of which, when fixed in its final position, temporarily supports the slabs above. Slabs may also be parked in any predetermined position by provisionally supporting them on shear pins passing through holes in the columns. The rate of lifting varies from 3 to 5 ft. per hour.

Lateral Support.   During construction, special lateral support may not be required for low buildings or for the first stage of erection of higher buildings. It may be wise, however, to provide at least temporary horizontal braces between the tops of the columns in both directions. Anchored guy wires provide a simple method for temporary lateral support (Figure 8.8). If space is unavailable for this arrangement, crossed guy wires within the structure in the closure spaces, or around the outside, may be used. Of course, such guy wires can be placed only where slabs are in final position.

Permanent reinforced concrete shear walls constructed as stairway or elevator shafts provide lateral support during construction and for the completed building. Masonry end walls supported laterally at the floor and roof levels may provide permanent lateral support.

The joints between slabs and columns should be designed to resist lateral loads. The provisions for lateral support depend upon the conditions which prevail. Wind loads (and sometimes earthquake shocks) must be considered during the erection procedure.

Exterior Walls and Partitions.   Light exterior curtain walls may be located along the perimeters of the cantilever slabs. Projected slabs may effectively shade exterior enclosure walls located along the line of the exterior columns, although they are less effective on the east or west exposure. Interior partitions are constructed of the same materials as are used in other types of construction.

Advantages and Disadvantages. The primary objectives of lift-slab construction are to minimize form costs and to expedite erection. Some of the other advantages follow:

(1) Partition locations are unaffected by beams because they are eliminated, as in conventional flat-plate construction.

(2) The under surface of the slabs, as smooth as a well-troweled floor surface, does not require plastering to provide a finished ceiling.

(3) Floor construction requires less vertical space than some other types, so that the story height is reduced.

(4) If prestressed, deflections are minimal, and slabs are crack-free.

(5) Utility lines can be incorporated in slabs at the ground level.

(6) Slabs can be cast in cold weather within a minimum heated enclosure and can be lifted regardless of the weather.

Some of the disadvantages are the following:

(1) It is a type of construction with which architects, engineers and contractors are generally unfamiliar and costs may be affected adversely.

(2) Column arrangement and other planning features must be favorable.

(3) Lifting and prestressing costs are factors.

PRECAST CONCRETE: RIGID FRAMES

Rigid frames of various types are described in Article 21. Cast-in-place reinforced concrete rigid frames are considered in Article 49. Similar types can often be precast advantageously.

The joints between girders and columns and the members themselves can develop full continuity. Although the units function as rigid frames, they are not usually so termed.

The units included in precast rigid frames are cast in a horizontal position, usually on the finished concrete floor of the building. Single-span frames, cast in one piece, reduce the weight to be hoisted at one time. Multiple-span frames are cast in several appropriate units. They are raised into their vertical positions by cranes and supported temporarily until they receive support from other parts of the building. The units of the rigid frame are rectangular in cross section, with uniform thicknesses and varying depths. They are usually solid, but occasionally they have hollow cores made by heavy cardboard tubes cast in the concrete.

The foundations of the outer vertical members, called *legs*, may carry the outward horizontal thrusts at the bottom of each leg if soil conditions permit. A *tie beam*, consisting of steel bars surrounded with concrete cast within or under the floor, ties the tops of the foundations together to resist lateral thrusts. Buildings subject to seismic shocks may require tie beams.

The hinges in rigid frames may have horizontal pins which offer only frictional resistance to rotation. Instead, they may include bars or dowels which tie the abutting ends of the units together so that the joints offer a negligible resistance to rotation.

Single-span, single-story frame buildings may have adequate longitudinal stability after the members that enclose the building are in place. However, multiple-span frames usually require precast or cast-in-place horizontal or diagonal longitudinal struts between the frames and at the tops of the vertical supporting members of adjacent frames and tied to them. These struts are also called *tie beams*, although they do not function as beams.

A building using precast rigid frames for the main supporting members may be enclosed by precast roof and wall panels, as described in Article 53; or the exterior walls may be constructed of masonry units, as described in Chapter 4. Precast wall and roof panels attached to the frames in various ways are explained at the beginning of this chapter. Occasionally, the roof construction spanning between widely spaced rigid frames consists of a series of precast folded plates or of thin shell arches as shown in Figure 8.16.

One or both of the exterior legs of precast frames can be T-shaped in profile to provide cantilevers to support an outside projection of the roof called a *canopy* (Figure 3.10(*i*)). These may serve as shelters or sunshades.

Precast frames are sometimes prestressed, but usually they are not. Only frames which are not prestressed are considered in this discussion. They are designed as conventionally reinforced concrete members.

Single-Span Rigid Frames. Three types of precast single-span rigid frames are illustrated in Figure 8.11. The two-hinged frame in (*a*), if cast in two pieces, reduces the weight to be hoisted at one time. A construction joint, called a *wet joint*, joins the frame at the center of the top member. A detail of this joint is shown. The two-hinged frame in (*b*) is cast in three pieces by providing two joints in the top member. A precast beam, hoisted in between the upper ends of the legs, is called a *drop-in beam*. A detail of the joints at the end of this

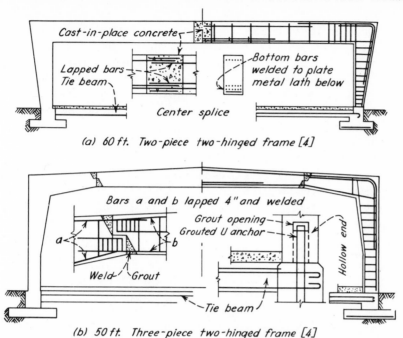

(a) 60 ft. Two-piece two-hinged frame [4]

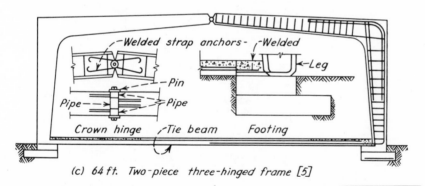

(b) 50 ft. Three-piece two-hinged frame [4]

(c) 64 ft. Two-piece three-hinged frame [5]

FIGURE 8.11   Single-span, precast rigid frames. U.S. Navy.

beam is shown in the figure. A seat or notch provides temporary vertical support for the ends of the beam. To establish continuity at each joint, short projecting ends of the reinforcing bars are overlapped and welded. The open spaces then are dry-packed with cement mortar.

The three-hinged frame in (c) is cast in two pieces. A detailed hinge at the crown is shown. The tie beams are usually not required.

Multiple-Span Rigid Frames. Three types of multiple-span frames are shown in Figure 8.12. A three-span continuous frame with three drop-in beams and hollow members is illustrated in (a). The joints at the ends of the drop-in beams are similar to those shown in Figure 8.11(b). A longitudinal tie beam is provided between frames at the top of each support. This type of frame can be designed for any number of spans.

A three-span frame consisting of two L-shaped exterior units and two T-shaped interior units is shown in (b). A joint functioning as a hinge is located at midspan of each top member. The frame functions as a series of three-hinged arches. This type of frame can be designed for any number of spans. Longitudinal tie beams extend between frames as shown.

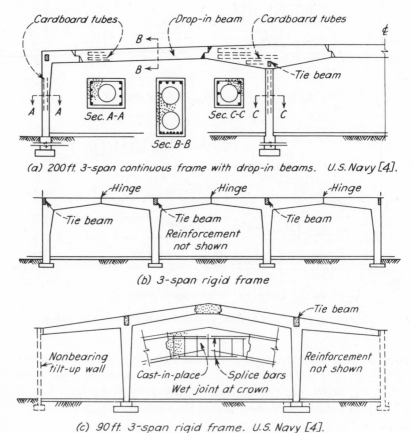

(a) 200 ft. 3-span continuous frame with drop-in beams. U.S. Navy [4].

(b) 3-span rigid frame

(c) 90 ft. 3-span rigid frame. U.S. Navy [4].

FIGURE 8.12   Three-span rigid frames. U.S. Navy.

A three-span frame, cast in two T-shaped pieces which are made continuous by a wet joint in the center of the top member, is shown in (c). The outer ends of the frame extend to, but are not supported by, solid slab tilt-up enclosure walls. Longitudinal tie beams extend between frames as shown.

Multistory Rigid Frames. The two-story, two-span frame in Figure 8.13 is similar to the one-story frames in Figures 8.11(b) and 8.12(a). The exterior and interior units include cantilever projections to receive the ends of the drop-in beams. Each of these two-story units is erected in one piece. The seated joints at the ends of the drop-in beams are similar to the type shown in detail in Figure 8.11(b).

Multistory, single-span frames are sometimes constructed and erected in one piece.

Reinforcement. As mentioned in Article 21, rigid frames are subjected to relatively large bending and shearing stresses. As shown in the frames in Figure 8.11, longitudinal reinforcement is provided near the outer and inner faces of the members. For clarity, layers of bars are shown to be more widely separated than they are actually separated. Reinforcement is provided normal to the members as required. This reinforcement corresponds to the stirrups used in beams. It is looped entirely around the longitudinal reinforcement. Additional to providing for shearing stresses, it retains reinforcement subjected to compressive stresses and thereby fulfills one of the functions of spirals or hoops in columns. It is essential in holding both tensile and compressive reinforcement in position in regions where the reinforcement is curved—for example, at the knees or haunches of the rigid frames in Figure 7.21, which is a cast-in-place frame but could be precast.

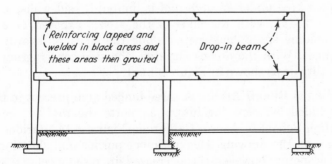

FIGURE 8.13 A 50-ft. two-span, two-story rigid frame with drop-in beams. Designed by Roberts and Schaefer for U.S. Air Force.

Lifting Hooks.   Appropriately located hooks cast in the members facilitate erection. Erection stresses produced by lifting must be provided for in the design of the members.

Acknowledgments.   The various offices of the U.S. Navy have done much to develop precast construction as described in publications of the Portland Cement Association. The frame illustrated in Figure 8.13(a) was patterned after a frame designed for the U.S. Air Force. It was designed by Roberts and Schaefer Company and described in the April, 1957, issue of *Civil Engineering*. The authors acknowledge the assistance from these sources in preparing this article.

57.   PRECAST CONCRETE: ARCHES AND PRESTRESSED DOMES

The general types of arches, irrespective of their materials, were described in Article 21. Cast-in-place reinforced concrete arches were considered in Article 49.

Both rigid frames and arches involve so-called arch action, but they vary in the relative amounts of flexure. Because of differing forms of their profiles, rigid frames are subjected to relatively large flexural stresses, although for arches the stresses are primarily compressive.

The horizontal thrusts on the foundations of both types of structures are comparable, and these thrusts are resisted by the foundations or abutments or by tie beams as described in the preceding article.

Most precast arches are three-hinged, but occasionally no hinges are provided and they are called *hingeless* or *fixed arches*. Precast arch ribs are usually solid with rectangular cross sections constant in width and depth. Precast arches may be prestressed, but this practice is uncommon and is not considered here.

The floor area under an arch roof is limited by sidewalls, as shown in Figure 3.8(a), with the ends of the arches exposed or enclosed within building space constructed along the sides, as shown by the dashed lines. With the arches elevated, the side construction carries the arch thrusts as shown in Figure 3.8(b).

Three-Hinged Ribbed Arch.   A three-hinged arch precast in two sections included between the hinges supports the roof of the Citrus Union High School in Azusa, California, Figure 8.14. Various details are shown on the drawing. Heavy timber purlins and sheathing were used for economy; however, this decreased fire resistance.

Other types of deck are used for precast ribbed arches. Precast con-

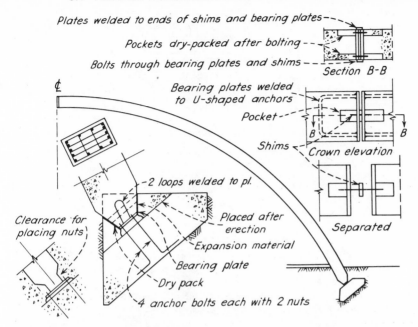

FIGURE 8.14   A 150-ft. span three-hinged arch. Architects: Austin, Field, and Fry. Structural Engineer: Ernest H. Lee.

crete purlins with dowel, matching plate, or bolted angle end connections to the arches anchor precast roof panels or a cast-in-place roof slab. The steep roof slope near the haunch of the arches makes cast-in-place concrete difficult. Precast concrete roof panels require secure anchorage to the purlins. These members are designed for the lateral load they carry. Securely anchored precast slabs, described in Article 53, can span between the arches and thereby eliminate the purlins.

Folded Plate Fixed Arch.   The arched roof for the Holy Trinity High School Gymnasium at Trinidad, Colorado, shown in Figure 8.15, is made up of 13 folded plate arch ribs about 10 ft. wide, as illustrated. Each rib consists of five identical segments about 31 ft. long. The arch segments, hoisted by cranes attached to strongbacks bolted to the segments, were supported on *falsework* during erection. Open spaces provided between the ends and edges of adjacent segments had overlapping bars projected from the ends and edges of the segments into the open spaces. These bars were welded and then grouted in pneumatically to establish continuity. The roof was waterproofed with a spray-on plastic [10].

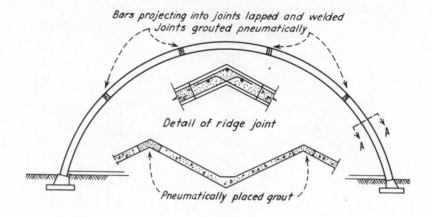

Bars projecting into joints lapped and welded
Joints grouted pneumatically

Detail of ridge joint

Pneumatically placed grout

Section A-A

Welded-wire fabric top and bottom except where grouted joint
bars added in ridges and valleys. Lapping bars welded in ridge joint

FIGURE 8.15   A 115-ft. span folded plate fixed arch. Architects: Toll and Milan. Consulting Engineer: Henry J. Boland.

Thin-Shell Barrel Arches.   A series of precast, thin-shell barrel roof arches supported by precast three-hinged rigid frames are shown in Figure 8.16.

Each arch unit was constructed of lightly reinforced, lightweight concrete 3½ in. thick, 15 ft. wide, and 61 ft. long, including an exterior cantilever projection of 9 ft. Spreading of the arches during erection was prevented by temporary bar ties located just above the springing lines.

The arches were anchored to the frames. Holes cast in the arches gloved over pipe dowels cast in the supports. After the arches were positioned, connections were welded to the tops of the dowels. The tie rods then were cut out. The arches spread slightly forming tight contacts. The edges of adjacent arches were tied together by welding matching plates. Built-up asphalt roofing covers the arches [11].

Thin-Shell Prestressed Concrete Domes.   The common procedure for constructing thin-shell dome forms is by conventional means, which require no explanation. This paragraph is devoted to two procedures which make use of compacted earth fills or natural mounds, appropriately shaped, instead of forms. A pierced thin-shell reinforced concrete dome roof, which was constructed in final position by a compacted man-made mounded earth fill as a form, was described in Article 49. It was not prestressed. Often the tension rings or ring

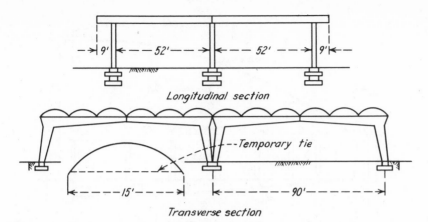

Longitudinal section

Transverse section

FIGURE 8.16   Thin-shell barrel arch roof deck. Architect: Mario Campi. Structural Engineer: Isadore Thompson.

girders of such domes are prestressed by posttensioning, as were those described in this paragraph.

A reinforced lightweight concrete thin-shell dome roof, constructed on a compacted earth fill and placed in final position by the lift-slab procedure, is illustrated in Figure 8.17. It has a diameter of 244 ft., a rise of 43 ft., and a uniform thickness of 4 in. The dome at the periphery is surrounded with a posttensioned concrete tension ring or ring girder 44 in. wide and 25 in. deep which is, in turn, surrounded with a cantilevered canopy 12 ft. wide.

The ring girder is supported by 36 equally spaced steel columns 26 ft. long, founded on spread footings and projecting 16 ft. above the ground surface.

The construction procedure was as follows:

(1) The concrete footings were constructed and the columns erected.

(2) The steel lifting collars were placed around the columns and positioned to become anchored into the ring girder.

(3) The ring girder or tension ring was formed and cast on the ground.

(4) The earth was shaped in a mound of compacted earth and covered with plastic slabs 1 in. thick to provide the form for the thin shell.

(5) The dome was reinforced on the mound.

(6) The concrete dome was poured.

(7) The tension ring, was prestressed, (after the concrete in the

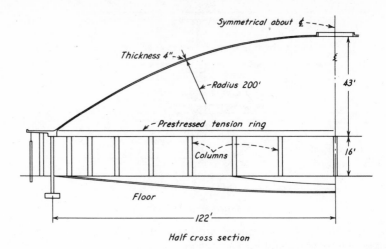

Half cross section

FIGURE 8.17 A 244-ft. diameter prestressed concrete dome erected by the lift-slab method. Warner Auditorium, Anderson, Indiana. Architects and Engineers: Johnson, Ritchart, and Associates. General Contractor: Lewis Construction Co.

ring and the dome had set) by wrapping around its outside edge with three rings of ¼ in. steel wire. Each ring consisted of 43 wires which were prestressed while being placed.

(8) A hydraulic lifting jack was placed on top of each column, connected to a single console for simultaneous control, and the lifting rods attached to the jacks and lifting collars.

(9) The dome was jacked into position and secured it by welding the lifting collars to the columns.

(10) The canopy was constructed.

(11) The earth fill was removed with the completion of the structure. The plastic slabs remaining on the underside of the dome serve as heat insulators.

This structure is the Warner Auditorium constructed in Anderson, Indiana, for the General Ministerial Assembly of the Church of God [16].

REFERENCES AND RECOMMENDED READING

1. *Minimum Standard Requirements for Precast Concrete Floor and Roof Units*, American Concrete Institute Standard ACI 711-58.
2. "Tentative Recommendations for Prestressed Concrete," ACI-

ASCE Joint Committee 323, *Journal American Concrete Institute*, 1958, p. 545.

3. Albert C. Smith, "Prestressed Concrete," *Construction Methods*, February 1957, p. 3.
4. *Modern Developments in Reinforced Concrete*, No. 29, Portland Cement Association, 1953.
5. *Modern Developments in Reinforced Concrete*, Portland Cement Association, 1957.
6. *Navy Builds All Precast Concrete Warehouses*, Portland Cement Association, 1955.
7. *Tilt-Up Construction*, Portland Cement Association, 1952.
8. "Lift Slab Goes Up Six Stories," *Engineering News-Record*, July 3, 1958, p. 37.
9. "Height Record Goes Up Again for Lift Slabs in the U. S.," *Engineering News-Record*, July 3, 1958, p. 37.
10. "Precast Sections Make a Corrugated Roof," *Engineering News-Record*, February 27, 1958, p. 40.
11. "Precast Arches for Gym Roof," *Engineering News-Record*, January 31, 1957.
12. T. Y. Lin, *Design of Prestressed Concrete Structures*, John Wiley and Sons, 1963.
13. Fred E. Koebel and Harvey R. Livesay, Jr., "Precast and Prestressed Concrete," Chapter 13 in *Handbook of Heavy Construction* by Frank W. Stubbs, Jr., editor-in-chief, McGraw-Hill, 1959.
14. "To Cover This Assembly Bowl: A 400-Ft. Prestressed Saucer," *Engineering News-Record*, June 1, 1961, p. 32.
15. "Designing, Bidding and Building World's Largest Edge-Supported Dome," *Building Construction*, February 1961, p. 28.
16. *Engineering News-Record*, December 14, 1961, p. 38.
17. *Engineering News-Record*, September 27, 1956, p. 36.

9

FLOOR CONSTRUCTION SYSTEMS

58. SELECTING TYPE OF FLOOR CONSTRUCTION

Factors Involved. Some of the factors involved in selecting the type of floor construction for a building are:

(*a*) general type of construction used in the building
(*b*) plan of the building
(*c*) floor loads
(*d*) lateral resistance
(*e*) floor thickness
(*f*) resistance to fire
(*g*) sound transmission
(*h*) weight

(*i*) ceiling, flat or exposed beams
(*j*) wearing surface
(*k*) position of floor, ground floor or above ground
(*l*) installation of utilities
(*m*) use of building
(*n*) cost

These factors overlap in many cases, and their relative importance varies with different classes of buildings. Local conditions including prevailing types of construction, availability of labor and materials, weather conditions, time available for construction, and other factors enter into the selection.

Unless precast concrete is specifically mentioned, the comments in this article refer to cast-in-place concrete construction. Special consideration is given to the use of precast concrete construction in Article 51, to the advantages and disadvantages of precast and prestressed concrete in Article 52, and to lift-slab construction in Article 55.

General Type of Construction. The floor system and the structural frame are so interrelated that selections of basic materials and types of construction are made for the building as a whole. For a given building, several alternatives may be investigated, the advantages and disadvantages of each being considered. The local building code requirements must always be satisfied. There are marked differences in the requirements of such codes.

498

Wood stud walls and partitions would naturally be associated with light wood joists and subfloors. Such floor systems might also be used with wood columns, beams, and girders; or a heavy timber floor system might be used.

If a steel frame is used, the floor system might be one of several types of reinforced concrete plain or ribbed floor systems, except the flat-slab, flat-plate, or slab-band. This type of frame is also suitable for use with thin concrete slabs on light steel or open-web joists or on light-gage steel cellular, ribbed, or corrugated panels. If the framing can be arranged so that there are many identical panels, some form of precast reinforced concrete or gypsum panel may prove advantageous.

If the beams and girders are to be of reinforced concrete, the floor system will be plain or ribbed reinforced concrete slabs and the columns usually will be reinforced concrete. Such columns will also be used with flat-slab, flat-plate, and slab-band floor systems. Precast concrete or gypsum panels may be advantageous for the same conditions given in the preceding paragraph.

Various types of floor systems can be used with masonry bearing walls, including light wood joist and subfloors, heavy timber floor systems, wood or concrete floors supported on steel I-beams or light steel or open web joists, reinforced concrete beams and slabs, reinforced concrete ribbed slabs, and precast concrete or gypsum slabs supported on steel joists.

Plan of Building.   If the floor is divided into panels which are square or very nearly square, reinforced concrete flat-slab or flat-plate construction may be desirable, or the two-way plain or ribbed reinforced concrete slab with beams on the four sides of each panel may be seriously considered. If the spans are short, the plain slab might be used, whereas for long spans the ribbed slabs will be more satisfactory. If the building is to be divided into rooms, flat-plate or slab-band construction may prove desirable, but flat-slab construction is not as satisfactory as it is for undivided areas because of the interference of the column capitals and drop panels with the partitions. Concrete ribbed slabs are particularly suitable for long spans.

The adaptability of a floor system to changes in the locations of partitions to suit changing occupancy is of importance in office and some other types of buildings.

Floor Loads.   Reinforced concrete flat-plate floors, light and open-web steel joists, and cellular, ribbed, and corrugated light-gage steel panels all with thin poured concrete slabs are suitable for light and medium floor loads, such as those of apartment houses, office buildings, hotels, and schools.

Reinforced concrete slabs supported on reinforced concrete or steel beams, girders, and columns, reinforced concrete flat-slab construction, and heavy timber matched or laminated decks supported on heavy timber or steel beams, girders, and columns are appropriate floor systems for heavy loads.

Even though the distributed floor load may not be heavy, consideration must be given to the possibility that a floor may be subjected to a heavy concentrated load. Such a load may be the determining factor in the thickness of the top slab in ribbed floors or floors with closely spaced joists.

Lateral Resistance.   Low tier buildings or even relatively high tier buildings with a low ratio of height to width will usually not require that special provisions be made to resist the lateral forces produced by wind. Girders framing into columns, either steel or concrete, offer the maximum opportunity for providing lateral resistance without the use of shear walls or other special arrangements. Reinforced concrete flat-plate floors and columns are not effective in this respect. Comments might be made on other systems, but these serve as extreme examples. In addition, if shear walls or wind bents are used, the floor system will be required to transfer the lateral load horizontally to the bents. This factor sometimes requires consideration.

Floor Thickness.   The minimum height of building for given number of stories and clear ceiling height is obtained with flat-plate, slab-band, or flat-slab construction.

Thin concrete slabs supported on light-gage cellular steel panels, wood subfloors, or thin concrete slabs supported on open-web steel joists, ribbed concrete slabs, and hollow precast panels form relatively thin floor systems if the girders that support them are located over partitions. Concrete slabs or wood decks supported on beams and girders take up the most room.

Any increase in height means an increase in the cost of walls, columns, elevators, stairways, and many other items, individually small, which may reach a total worth considering. By using thin floor systems, it may also be possible to construct one or more additional stories within a height limitation in a code.

Resistance to Fire.   When a building must be cheaply constructed and resistance to fire is not a decisive factor, ordinary wood joist construction with wood subfloor may be used. These conditions prevail in residence construction more than in any other class of building.

In warehouses and many types of industrial buildings, *heavy timber construction* may offer sufficient resistance to fire and provide a building at a lower cost than that of a more fire-resistive building. This is especially true if automatic sprinkler systems are installed.

In the congested downtown districts of many cities, building ordinances require *fire-resistive construction*, and therefore wood joist construction and *heavy timber construction* cannot be used. Floors constructed of open-web steel joists and concrete slabs protected with suspended ceilings of metal lath and vermiculite or perlite gypsum plaster are approved by some, but not all, cities as *fire-resistive construction*. It is always necessary to consult the local building code to determine the acceptability of the fire resistance of a floor system. The cost of fire insurance is an important factor entering into the choice of floor construction, the rates depending upon the fire resistance of the construction as well as the nature of the contents, the location, and many other factors. The rate is lowered in many classes of buildings by the installation of automatic sprinkler systems which come into action in any part of a building when the temperature in that part is raised by fire.

Sound Transmission. The various types of floor systems differ quite markedly in their effectiveness in resisting the transmission of sound. For some types of occupancy, such as warehouses, this factor is of minor importance, while for other types, such as hotels, schools, and apartment houses, it must be given serious consideration. Sound transmission of impact noises is influenced by the type of wearing surface as well as by the floor system. This subject is considered in Article 97.

Weight. The weight of a floor system to carry a given load is an important factor because it affects the weight and cost of the supporting members and the foundations. Wood construction is advantageous where its use is permissible. Systems with closely spaced joists which permit the use of thin lightweight slabs are lighter than those which require thicker slabs of reinforced concrete and may satisfy code requirements. Considerable weight reduction can be achieved by using lightweight aggregate in concrete. Various other obvious comparisons might be made, but enough comments have been made for illustrative purposes.

Ceiling. Ordinary wood joist construction, flat-plate construction, slab-band construction, and ribbed concrete slabs provide flat ceilings, but plain concrete slabs supported by beams and girders and some other types of construction require *suspended ceilings* if flat ceilings are desired. Girders running between columns interfere with flat ceilings whenever they are present. In addition to their better appearance, flat ceilings do not interfere with the locations of partitions. Partitions are often located under girders in which case girders are not objectionable.

Plastered ceilings are provided by applying two coats of plaster directly to the underside of ribbed slabs with clay, concrete, or gypsum tile fillers, flat slabs, or plain slabs supported by concrete or fireproofed steel beams. The ribbed slabs and the flat slabs provide a flat ceiling, but the slabs supported by beams necessitate breaks in the ceiling, and the cost of plastering is greater because of the increased area and the additional labor required in finishing around the edges of the beams. Wood joists, heavy timber construction, metal joists, and ribbed slabs with steel forms require lath and preferably three coats of plaster or plasterboard and two coats of plaster on the underside to provide a plastered ceiling. The ceiling supported by wood joists, ribbed slabs with steel forms, and metal joists will be flat, whereas the ceiling supported by heavy timber construction will probably follow around the beams. The cost of the lath and the additional cost of plaster should be considered in selecting the floor construction if plastered ceilings are required. Monolithic concrete ceiling surfaces are often finished by painting directly rather than on plaster.

Wearing Surface.   The type of wearing surface, as considered in the following articles, is a factor in selecting the type of floor system, or the reverse may be true.

If wood flooring is to be used for the wearing surface, light wood joist and heavy timber construction have an advantage in cost over other forms of floor construction where nailing strips with concrete fill between have to be provided. The closely spaced light steel joists may be covered with nailable concrete slabs and thereby avoid nailing strips, or nailing strips may be anchored to the tops of the joists with a light concrete slab between nailing strips. Some forms of wood floors may be cemented directly to carefully finished concrete surfaces with bituminous cement.

If the wearing surface is to be linoleum, cork, concrete, composition, asphalt tile, vinyl tile, cork carpet, rubber, ceramic tile, etc., any of the forms of floor construction which provide a concrete top surface are suitable, but the additional cost of providing a finished surface on the concrete must be considered.

With wood joist or heavy timber construction, a concrete surface required as a base for some wearing surfaces would add to the cost.

Asphalt tile, vinyl tile, and linoleum, etc., require a matched floor or plywood, but composition can be laid on the wood subfloor.

Ceramic tile, marble, slate, and terrazzo usually require a concrete foundation, and thus the various forms of floor construction which provide a concrete top surface are suitable for the installation of these materials without further expenditure, but wood subfloors require a 2 in. or 3 in. slab, which increases the cost considerably. The thin-setting bed for ceramic tile described in Article 63 may, however, be satisfactory.

For many purposes, a smooth troweled finish on a concrete slab may be satisfactory. It may be colored or, if subjected only to light traffic, a painted surface may be satisfactory. Such a finish can be provided readily by several floor systems.

Position of Floor.   Floors placed on the ground will normally be concrete slabs; and if not subjected to hydrostatic pressure, will have light wire mesh or no reinforcing.

Installation of Utilities.   The utilities which are installed in a building are not considered in this treatise, but the provisions which often must be made in floor systems to provide for conduits and pipes, wiring, plumbing, and heating, and ducts for ventilating and air conditioning require consideration. Open-web steel joists provide space through which pipes and conduits can be run to any position in a floor and ducts can be placed between joists. Light-gage cellular steel panels afford similar but more restricted opportunities for wiring conduits. The space above suspended ceilings is used extensively for such purposes. Wiring conduits are ordinarily located in concrete floor slabs or in concrete fills placed on top of the structural slabs, but water pipes should not be embedded in this manner because of the cost of repairing leaks. Wood joist floors have the same advantages as open-web steel joists but require that holes be bored for pipes and conduits running perpendicular to the joists.

Use of Building.   The use to be made of a building enters into the choice of the type of construction. It is a determining factor in the selection of a particular type of construction only to the extent that it influences the various factors that have been considered.

Cost.   Of course, one of the most important factors to be considered in selecting a specific floor system and wearing surface from several which satisfy all of the basic requirements is the cost. The total cost may be divided into the direct cost, the indirect cost, and the continuing annual cost.

The direct cost includes the cost of the floor system, including the wearing surface, the supporting beams, and ceiling surfaces, directly applied or suspended. The indirect costs are the costs of the girders, the columns, and their foundations. The relative effects of floor systems on the total height of a building may also be considered because differences in cost caused by this factor may be of some significance. Differences in height are reflected in differences in the costs of exterior walls, columns, elevators, stairways, vertical pipes, conduits, ducts, and other items. The different continuing annual costs include differences in costs of care and maintenance of the various wearing surfaces.

Finally, the selection of a floor system should take into account the differences in annual cost of fire insurance. For many classes of occupancy and types of construction, the installation of automatic sprinkler systems may markedly reduce the annual cost or make possible changes in the type of construction which reduce the overall cost.

Because of the many factors involved, including the effects of local customs and the relative availability and costs of different kinds of labor and materials, no specific comparisons can be made. When cost is an important factor in the selection of the floor system, and it usually is, comparative estimates are made of the costs of the types that warrant consideration after all other factors are considered.

59.   COMPOSITE STEEL AND CONCRETE FLOORS

Cellular Steel and Concrete Decks [1].   The floor decks of buildings with structural steel framing are often made of cellular panels consisting of light-gage galvanized or painted sheet steel with cross sections and interlocking edges, such as those illustrated in Figure 9.1(a). These panels span the distance between floor beams, which may be 20 ft. or more, and are covered with a poured lightweight concrete slab about 2 in. thick. No temporary intermediate supports are required. The panels are attached to the steel beams by welding.

The required fire resistance for the underside is provided by a suspended ceiling of metal lath and plaster. The thickness and composition of this ceiling is determined by the required fire-resistance rating. For example, a ceiling with $\frac{7}{8}$ in. of vermiculite plaster on metal lath, suspended to provide at least $2\frac{1}{4}$ in. of air space between the underside of the panel and the back of the lath, gives the construction a 4-hr. rating. If two coats of sanded gypsum plaster of specified mixes are applied on metal lath to give a total thickness of $\frac{3}{4}$ in. and no air space is provided, a $1\frac{1}{2}$-hr. rating is given.

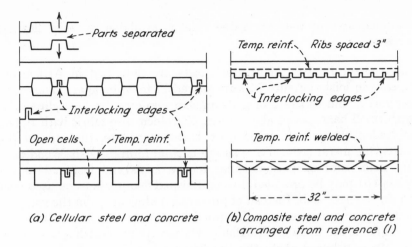

(a) Cellular steel and concrete   (b) Composite steel and concrete arranged from reference (1)

FIGURE 9.1   Types of lightweight steel and concrete decks.

This type of deck has light weight, provides a working platform before the concrete slab is placed, and makes possible the placing of present and future wiring in the open cells. It is suitable for use in office buildings, hotels, and apartment buildings because the floors do not carry heavy loads. Its use is permitted for *fire-resistive* buildings by many but not all building codes.

Composite Steel and Concrete Decks [1].   One type of light-gage galvanized or painted steel panel which is used in constructing short-span floor decks is illustrated in Figure 9.1(*b*). It is provided with longitudinal ribs to give rigidity during construction and to provide reinforcement for the concrete floor slab for which it serves as a form. Only the ribs, which are embedded in concrete, are considered effective in providing reinforcement. When necessary, additional reinforcement in the form of bars is included between the ribs, as shown. Transverse temperature reinforcement is provided. If the slab is continuous over the supports, reinforcement is placed near the top of the slab in this region, as described elsewhere. The panels are anchored to the steel beams which support them, ordinarily by welding. Temporary supports usually will be necessary between the beams while the lightweight concrete is being placed and hardening.

The light-gage corrugated steel panel (*b*) is used in a manner similar to the ribbed panel. However, since the temperature reinforcement is welded to the tops of the corrugations, as shown, it anchors or bonds the corrugations to the concrete so effectively that the full cross-sectional area of the corrugated steel is available to resist flexural stresses.

Light Steel Joists.   Light steel joists are described in Article 39. Open-web joists (Figure 9.2) are spaced 12 to 24 in. apart. They may support a thin concrete or gypsum slab reinforced in both directions with small bars spaced about 12 in. or welded-wire fabric with ribbed expanded metal lath placed over, and fastened to, the tops of the joists, the metal lath serving as the bottom form for slab. Wood nailing strips running perpendicular to the joists and fastened to the joists (*b*) may be provided if wood finished flooring is to be used over a thin concrete slab. Instead of providing nailing strips in the concrete slab, nailable concrete, as described in Article 43, may be used. If wood floors are to be used, nailer joists are available with a wood nailing strip anchored along the top flange of each joist to permit the nailing of the floor to the joists (*c*). Wood flooring may be fastened directly to the nailing strip shown in (*b*) and (*c*), or a wood subfloor may be used. If a subfloor is used, it should be laid diagonally, as has been described.

The light rolled-steel sections manufactured especially for use as floor and roof joists may be used in the same manner as the open-web joists, but nailer joists of this type are not available. The nailing strips are fastened to the joists on the job.

Horizontal or cross bridging is required for open-web joists to correspond to that used with wood joists. This bridging may consist of light steel rods or angles with their ends securely anchored to the flanges of the joists.

Ceilings may be provided on the underside of these types of floor

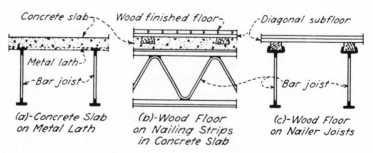

FIGURE 9.2   Concrete and wood floors on open-web joints. (*a*) Concrete slab on metal lath. (*b*) Wood floor on nailing strips in concrete slab. (*c*) Wood floor on nailer joists.

construction by plastering over metal lath, or other plaster base, fastened to the underside of the joists. Such ceilings improve the appearance and are usually required for that purpose as well as to increase the resistance to fire. A 4-hr. rating may be attained by using metal lath and vermiculite gypsum plaster and a noncombustible subfloor.

Ground Floors. Ground floors are usually constructed of concrete from 4 to 6 in. thick, which is either dampproofed or waterproofed as described in Article 15. Even after these precautions have been taken, it is quite probable that sufficient moisture will be present to cause some kinds of wearing surface to be short-lived. The various types of wearing surfaces are considered in subsequent articles.

Concrete slabs which serve as the ground floor of basementless houses should be protected against dampness as are other floors in contact with the ground; and in addition, in areas which experience low temperatures, the edges should be insulated with at least 2 in. of rigid waterproof insulation to avoid cold areas and condensation around the outside edges of the floors.

60. WOOD FLOOR SYSTEMS

Wood Floors on Wood Joists. The most common form of floor construction for *wood frame* and *ordinary construction* buildings consists of wood joists supporting a 1-in. wood subfloor and a matched-wood finished floor as described in Article 41, preferably with a layer of building paper or other material between the subfloor and the finished floor as shown in Figure 9.3. Magnesite composition as described in Article 62 may be used in place of matched flooring, or the construction may be changed to receive tile, terrazzo, or other material.

The joists are usually 2 in. wide and from 6 to 14 in. deep. For heavy loads, the joists may be 3 or 4 in. wide. Wider joists are used in *heavy timber construction,* which is considered under another heading. The usual spacings are 12 and 16 in., 24 in. being too great a space for good results with most lath. Light metal lath requires a spacing not greater than 12 in., but greater spacing may be used with heavier lath. In some cases, 1 × 2 in. furring strips, properly spaced for lath and running at right angles to the joists, are nailed to the underside of the joists to receive the lath. Then the spacing of the joists is independent of the lath. This construction enables fire to spread rapidly across the joists and is therefore objectionable. Another objection to

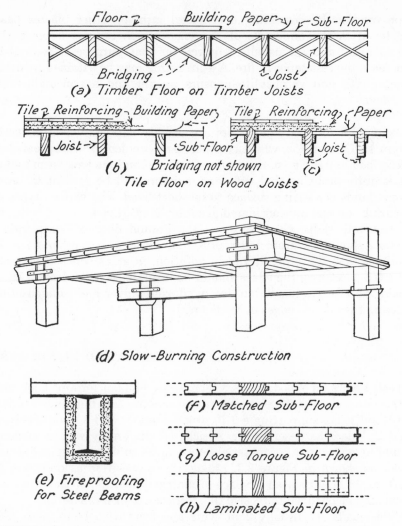

FIGURE 9.3 Wood floor construction. (a) Timber floor on timber joists. (b) and (c) Tile floor on wood joists. (d) Slow-burning construction. (e) Fireproofing for steel beams. (f) Matched subfloor. (g) Loose tongue subfloor. (h) Laminated subfloor.

the use of wood furring strips nailed to the underside of wood joists is the tendency for these strips to become loose and permit the ceiling to fall as the joists dry out and their grip on the nails weakens.

An important function of the subfloor is to provide a floor during the early stages of construction. A subfloor makes the floor more substantial, more fire-resistant, more soundproof, and warmer. When a subfloor is used, there is a tendency to nail the finished floor to the subfloor rather than to the joists. This practice is objectionable because the finished floor does not remain in place as well when nailed to the subfloor and tends to squeak. There is also a tendency to lay the finished floor parallel with the joists, which is objectionable. The subfloor should be laid diagonally, or if strips are placed on the subfloor and over the joists, the finished floor and subfloor may both be laid at right angles to the joists, the strips overcoming the uneven places in the subfloor. The subfloor may be of ordinary sheathing, matched boards, shiplap, or plywood. Sometimes the subfloor is omitted entirely to decrease the cost; and if so, the finished floor would probably have to be laid before plastering, during which operation it must be protected from dirt and water. It will absorb moisture given off by the plaster. This will cause swelling, which will be followed by shrinking and the opening of the joints.

The layer between the subfloor and the finished floor may be omitted entirely or may consist of building paper, asphaltic felt, asbestos paper, or gypsum board. Some form of layer should always be used; and if the cost will permit, the increase in fire-resistance and sound-deadening properties due to a layer of heavy asbestos paper or gypsum board will warrant the additional expenditure.

The joists are held in a vertical position by cross bridging as shown in Figure 9.3(a), one row being used where the span of the joists is over about 8 ft., and two rows where over 16 ft. Cross bridging also serves to distribute concentrated loads over several joists.

If a wood subfloor is laid before rain is excluded from the building, and it usually is, provisions should be made for the swelling of the subfloor, or the exterior walls may be cracked and pushed out and other damage may be caused. To prevent this, spaces $\frac{1}{4}$ to $\frac{1}{2}$ in. may be left between the boards, or every tenth or twelfth board may be omitted at first and placed after the building is under cover.

Where a tile floor is to be placed on wood joist construction, the floor is usually designed to carry a reinforced-concrete slab (Figure 9.3(b)) to act as a base for the tile. The construction shown in (c) is also used but is more likely to produce cracks than the construction in (b). A thin-setting bed for ceramic floor tile which does not require special construction is described in Article 63.

This type of floor construction is used in residences and other buildings of *wood frame* and *ordinary construction* and may be used in buildings with steel frames. It is inexpensive, light, and may be made sufficiently strong for heavy loads, but it is very combustible. If protected on the underside by plaster or metal lath, its resistance to fire is increased.

Heavy Wood Subfloor on Wood or Steel Beams. Wood subfloors varying in thickness from 3 to 10 in., depending upon the loading and span, may be supported directly by girders running between columns or by beams which are supported by the girders, as shown in Figure 5.14 and described in Article 29. In the first arrangement, the lateral spacing of columns commonly is not over 10 or 12 ft. because of the heavy subfloors required for longer spans. In the second arrangement, the beams are spaced 4 ft. or more apart, and the column spacing is not restricted by the strength of the subfloor.

To secure resistance to fire, wood beams and girders are made at least 6 in. wide and 10 in. deep, even though the loads may not require beams of this size. If steel beams are used, they may be protected against fire (Figure 9.3(e)), the beam first being covered with metal lath and then plastered.

Heavy wood subfloors may be of three types: matched (Figure 9.3(f)), loose-tongue (g), or laminated (h). The matched floor may be used for thicknesses of 3 or 4 in., but for greater thicknesses the waste in matching becomes so large that the hardwood *loose tongue* may be more economical. This loose tongue is also called a *slip tongue* or *spline*.

Floors 4 in. and more in thickness may be constructed by laying 2 in. lumber on edge and securing the adjacent pieces together with spikes spaced about 18 in., 2 × 4's being used for a 4 in. floor, 2 × 6's for a 6 in. floor, and so on. This type of floor is known as a *laminated floor*. A laminated floor is easier to lay than a heavy loose-tongue floor, for the pieces being smaller are more easily handled and drawn into position; however, more feet and board measure are required because 2 in. material is really $1\frac{5}{8}$ to $1\frac{3}{4}$ in. thick. The cost of the loose tongue is saved in the laminated floor.

When the details are properly worked out, this type of construction, using wood in large masses, is called *heavy timber construction*.

REFERENCES AND RECOMMENDED READING

1. Henry J. Stetina, "Steel Construction," included in *Floors, Ceilings and Service Systems*, Publication 441, Building Research Institute, National Research Council, 1956.
2. *Recommended Practice and Standard Specifications for Concrete and Reinforced Concrete*, report of Joint Committee representing affiliated committees of several societies, 1940. Seventh printing, American Concrete Institute, 1950.

10

FINISH FLOOR CONSTRUCTION

Types. Wood flooring is available in the following forms:

(*a*) *Strip flooring*, consisting of long narrow pieces or strips with tongued-and-grooved joints along the sides and often along the ends also for hardwoods. Generally called *matched flooring* or simply *flooring*.

(*b*) *Plank flooring*, consisting of wider boards than strip flooring with tongued-and-grooved joints along the sides and ends.

(*c*) *Parquet flooring*, consisting of short narrow boards cut to form patterns or mosaics.

(*d*) *Industrial wood block flooring*, consisting of heavy pieces cut in lengths of from 2 in. to 4 in. forming blocks which are set with the ends of the grain exposed to wear.

(*e*) *Fabricated wood block flooring*, consisting of small square or rectangular blocks formed by fastening short pieces of strip flooring together, tongued-and-grooved joints being provided on all sides.

Methods of Fastening. Wood flooring may be nailed to wood joists through a wood subfloor. *Nailing strips* are often provided on concrete floor slabs and other types of supporting floors. These strips may be beveled (Figure 10.1(*a*)) and embedded in concrete which holds them in place. Nails may be driven into the sides of the strips to grip the concrete. Expansion bolts or screws may be used as anchors. Special clips to receive the strips may be embedded in the top of the slab when the slab is poured, or the strips may be nailed to other strips at right angles to them, the whole grid being embedded in concrete (*b*).

Nailable concrete, described in Article 43, may be used, with more or less satisfactory results, instead of nailing strips. These concretes are of special composition which permits nails to penetrate. They are

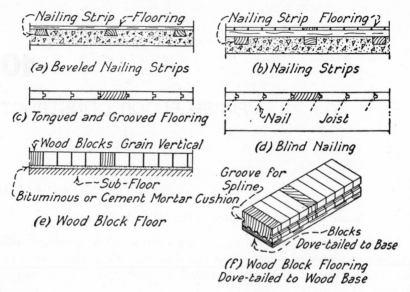

FIGURE 10.1   Wood floor surfaces. (*a*) Beveled nailing strips. (*b*) Nailing strips. (*c*) Tongued-and-grooved flooring. (*d*) Blind nailing. (*e*) Wood block floor. (*f*) Wood block flooring dove-tailed to wood base.

usually applied in a layer 2 in. thick with the top surface struck very accurately to receive the finished floor.

Industrial and fabricated wood blocks are usually cemented, and parquet, plank, and short pieces of strip flooring are often cemented to concrete floors, or to a concrete fill, with an asphalt mastic. These mastics are of two types, the *hot mastic*, which is made fluid by heating, and the *cold mastic*, which is made fluid by a solvent. The hot mastic is applied to the slab, and the flooring is bedded in it before it cools. The cold mastic gradually solidifies by the evaporation of the solvent after the flooring is placed. The layer of asphalt mastic is called an *underlayment*.

Flooring which is provided with tongue and groove is *blind-nailed* to joists or nailing strips (*d*) so that the nail does not show on the surface. To avoid injuring the floor by the last blow of the hammer which drives the nail home, a square bar, or the square end of a nail set, may be placed in the corner and against the nail head to receive the hammer blow and protect the flooring. Special flooring nails with small heads (Figure 5.4(*d*)) are used in blind nailing. The ends of plank flooring are commonly screwed to the subfloor, the heads of the screws being sunk in holes bored part way in the wood. These holes then are filled with wood plugs which are specially manufactured for that purpose. The plugs are quite visible, but this is the effect desired. To provide for expansion, plank flooring should be laid with temporary metal spacers $\frac{1}{32}$ in. thick placed in the cracks between the planks. Parquet floors are fastened to wood subfloors with brads driven through the face, the brads being set by driving their heads below the surface with a *nail set*. The holes above the heads are filled with putty.

Kinds of Wood. Strip flooring, plank flooring, parquet flooring, and fabricated wood blocks are available in the hardwoods: white and red oak, maple, beech, birch, and walnut. The oaks may be *plain-sawed* or *quarter-sawed*, often called *vertical-grain* or *edge-grain*. Strip flooring is also available in the softwoods: fir, yellow pine, white pine, cypress, and many others. Only the quarter-sawed or vertical-grain softwoods should be used, except in the cheapest construction or for attics or other floors which are little used. Industrial wood blocks are usually made of yellow pine, oak, gum, or Douglas fir.

Size, Grades, etc. *Strip flooring*, which is called flooring or matched flooring, is usually $\frac{25}{32}$ in. thick. The most common widths of hardwood flooring are $2\frac{1}{4}$, 3, and 4 in., which have exposed faces of $1\frac{1}{2}$, $2\frac{1}{4}$, and $3\frac{1}{4}$ in.; and the common widths of softwood flooring are 3 and 4 in., with exposed faces of $2\frac{3}{8}$ and $3\frac{1}{4}$ in. Oak flooring $\frac{3}{8}$ in. and $\frac{1}{2}$ in. thick and maple flooring $\frac{3}{8}$ in., 1 in., and $1\frac{1}{4}$ in. thick are available. Thicker maple flooring may be obtained on special order. The thinner flooring is for use over old floors and is not usually satisfactory for new buildings because of its tendency to squeak. Thick maple flooring is suitable for severe usage such as in warehouses, factories, and gymnasiums. Heavy trucking may break off the tongues or lower part of the grooves of ordinary strip flooring. Maple, beech, and birch flooring is furnished in three grades depending upon the quality, i.e., first, second, and third. Oak flooring is available in five grades, i.e., clear, sap clear, select, No. 1 common, and No. 2 common.

Plank flooring is usually $\frac{13}{16}$ in. thick and from 3 to 8 in. wide. It is commonly used in random widths and is known as *colonial plank flooring*. The exposed edges may be slightly beveled to form V joints.

Parquet flooring is usually $\frac{5}{16}$ in. thick and 1, $1\frac{1}{2}$, or 2 in. wide with square edges, but $\frac{13}{16}$ in. flooring, $2\frac{1}{4}$ in. wide, which is side- and end-matched is also available.

Industrial wood blocks vary in thickness from 2 to 4 in., depending upon the severity of the service. The width is from $2\frac{1}{4}$ to 4 in. and the length from 4 to 9 in. The blocks are often provided with wood or metal splines and may be assembled in strips, up to 8 ft. in length, whose width equals the length of a single block, as shown in Figure 10.1(f). The joints between the blocks are filled with bituminous filler. Expansion spaces about 1 in. wide should be provided against all walls and around all columns, thresholds, and permanent fixtures.

Fabricated wood blocks are made of flooring usually $13/16$ in. thick, but $1/2$ in. and 1 in. thick blocks are also available. The squares vary in size from $6 3/4$ in. to $11 1/4$ in., and the rectangles are about 6 in. by 12 in. The number of strips to a block may be 4, 5, or 6. Expansion spaces should be allowed around all walls, columns, thresholds, and permanent fixtures, the width being about $1/16$ in. for each foot of width or length. These spaces may be partly filled with asphalt mastic, may be completely filled with cork strips, or may contain springs to hold the edge blocks in place. If a room is much longer than it is wide, square blocks should be laid diagonally. The blocks are often sanded, filled, waxed, and polished at the factory.

Uses. Wood floors are warm and elastic and therefore are not tiresome to work on. They are clean and, if proper selection is made and they receive proper care, they are durable. They may be made attractive in appearance and are suitable, from that point of view, for the highest type of use. Wood floors supported by *fire-resistive construction* are used in many buildings classed as fire-resistive because they burn very slowly in case of fire. However, they are not used in the highest type of fire-resistive buildings. They are one of the most satisfactory floors for residences and are usually the cheapest.

Wood floors should not be placed until after all concrete work and plastering have been completed and the building has had a chance to dry out. The flooring should not be stored in the house during this period. The cracks in wood floors which are permitted to absorb moisture before they are laid will open up as the wood dries out. Special care should be used in wood floors laid over concrete to make sure that the concrete has dried out. An asphalt mastic dampproofing layer should preferably be placed over the concrete.

62. CONCRETE, TERRAZZO, AND MAGNESITE COMPOSITION FLOORING

Concrete [1]. Concrete wearing surfaces are used very widely where the structural part of the floor is of concrete. The wearing surface may be an integral part of the construction beneath, or it may be added as a separate layer. If the wearing surface is placed before the base has set, a thickness of $3/4$ in. to 1 in. is satisfactory; but if it is placed after the base has set, the thickness should be 1 in. to 1½ in., and the surface of the base should be roughened, thoroughly cleaned, and coated with cement grout just before the wearing surface is placed. This is done to secure a bond between the wearing surface and the base, but the results are uncertain.

The wearing surface which is an integral part of the floor may be considered as contributing to the strength of the structure; whereas a wearing surface which is added cannot be so considered and its weight increases the dead load that must be carried by all the beams, columns, and foundations. In spite of this disadvantage, a separate wearing surface may be cheaper than the integral wearing surface, for it is placed after the rough work on the building has been completed and does not need to be protected as carefully as the integral surface. The conditions which exist while the structural part of the floor is being placed may be such that the accuracy required of a finished floor is difficult and expensive to secure.

A truer surface can be secured with the separate wearing surface as the deflection due to the weight of the floor and the yielding of the forms has occurred before the surface is placed, and any discrepancies can be taken up in the wearing surface. The separate wearing surface is sometimes made thick enough to permit the placing of electrical conduits on top of the structural slab.

The coarse aggregate is excluded from the wearing surface, and a richer mixture than that used in the structural parts is used. The aggregate grains and not the cement resist the wear, and therefore the grains should have a high resistance to abrasion. Beyond a certain limit, an increase in the amount of cement reduces the wearing qualities of a surface and increases the tendency to crack. Usually the mixture or *topping* for the wearing surface consists of 1 part cement to not less than 2 or more than 3 parts fine aggregate.

For surfaces subjected to severe wear, the mixture should be 1 part cement, not more than 1 part fine aggregate, and not more than 2 parts coarse aggregate with a maximum size of ½ in. Various admixtures, or hardeners, are available for use in the topping to improve its wearing properties.

Concrete wearing surfaces or topping must be carefully laid using a minimum amount of water, troweling as little as possible, and protecting against drying out for at least ten days. Excessive troweling brings excess water and laitance to the surface, causing hair cracks to form and the floor to give off objectionable dust. This process is called *dusting.* Topping applied to a hardened base should be struck off and compacted by rolling or with tampers or vibrators and finished with a steel trowel. The use of dry cement or cement and fine aggregate sprinkled on the surface to stiffen the mix or absorb excess moisture is objectionable because it may cause hair-cracking, scaling, or dusting. Dusting may be prevented or remedied somewhat by the use of *floor hardeners* and other preparations, or by painting.

Painted surfaces are satisfactory when the amount of wear is small; but where subjected to severe use, frequent painting is required. Special paints are manufactured for use on cement floors. A cement-colored paint, of course, shows the wear less than paint of any other color. Paint should not be applied until after the floor has been in place three or four months. Before painting, the surface should be thoroughly scrubbed with a 10% solution of muriatic acid and washed so that the acid is completely removed. The floor should then be allowed to dry before the paint is applied. Paint for the first coat is thinned. Three coats are usually required.

Concrete wearing surfaces are sometimes colored and marked off to imitate tile. Colored concrete floors finished with wax may be attractive in appearance.

If artificial coloring matter is used, only those mineral colors should be employed which will not appreciably impair the strength of the concrete.

Mineral coloring material is preferred to organic coloring material because the latter fades more and may seriously reduce the strength of concrete. Mineral coloring may reduce the strength of concrete somewhat, but where the quantities used are less than 5%, this is not serious. The use of colored aggregates is preferable in obtaining color effects, the surface of the floor being brushed or ground to expose the aggregate.

Concrete floors are inelastic and cold and tiresome to work on, but they are durable if well constructed, are easily cleaned, and are relatively inexpensive. Concrete wearing surfaces should not be used over wood subfloors without taking special precautions to prevent cracking. The construction should be similar to that used for tile floors (Figure 9.3(*b*).

A cement wall base is often used with concrete floors. If a separate wearing surface is used, the wall base may be placed at the same time as the wearing surface with a curving fillet at the junction. This facilitates cleaning. A cement base will not adhere to lime plaster, hard wall plaster, gypsum block, Keene's cement, or plasterboard. The backing should be brick, stone, hollow tile, concrete, or metal lath.

A concrete floor is sometimes called upon to serve the double purpose of a floor and a roof. Under these conditions, a built-up roofing, as described in Article 78, is placed on the structural slab. A wearing surface, consisting of a concrete slab about 3 in. thick, is placed over this roofing and is divided into sections not over 16 ft. square by expansion joints about ¾ in. wide with a bituminous filler.

Terrazzo [1]. Terrazzo wearing surfaces are constructed in a manner similar to concrete wearing surfaces, but a special aggregate of marble chips or other decorative material is always used, and this aggregate is exposed by grinding the surface.

The mortar-base course should be at least 1¼ in. thick and should be composed of 1 part portland cement and 4 parts sand with only enough water to produce a mortar of the stiffest consistency that can be struck off accurately with a straightedge.

The *mortar base* can be placed directly on the concrete slab and bonded to it by first cleaning this slab, thoroughly wetting it, and applying a thin coat of neat cement broomed into the surface for a short distance ahead of placing the mortar base; or the surface of the slab can first be covered with a thin smooth layer of fine dry sand about ¼ in. thick, on which is placed a layer of waterproof paper with end and side laps of at least 1 in., on which the mortar base is placed.

The mortar base is struck off at least ¾ in. below the finished floor level. Metal or plastic *dividing strips* are inserted in the mortar base before it hardens, in positions which will control the cracking and conform to the design or pattern desired. The tops of dividing strips should extend at least $\frac{1}{32}$ in. above the finished floor level so that they can be ground down flush with the floor surface when the terrazzo is being ground.

The *terrazzo mixture* should consist of 1 part of gray, white, or colored portland cement, according to the decorative effect desired, to not more than 2 parts by weight of marble chips, other decorative aggregate or abrasive aggregate, or such a mixture of any of these as is desired. The amount of water used should be such as to produce a workable plastic mix. Wet mixtures do not produce good results. Any special coloring agents should be mineral pigments.

After the mortar base has hardened enough to stand rolling, the terrazzo mixture should be placed to the level of the tops of the dividing strips and struck off. It should then be rolled in both directions to secure a thorough compacting. Additional aggregates of the desired color should be spread over the surface during the rolling process until at least 70% of the finished surface is composed of aggregate. As soon as the rolling is completed, the surface should be floated and troweled once without attempting to remove trowel marks. Further troweling is objectionable.

After the terrazzo has hardened sufficiently to hold the aggregate firmly, it should be ground by hand or with a grinding machine, the floor being kept wet during the process. The material ground off should be removed by flushing with water. Any air holes or other defects should be filled with thin cement paste spread over the surface and worked in. After the paste has hardened for at least 72 hours, the floor surface should receive its final grinding. It should be kept continuously wet for at least ten days, scrubbed clean with warm water and soft soap, and mopped dry.

This type of floor is more expensive than concrete and less expensive than tile or marble. It is used for floors of buildings where an attractive and durable floor is desired, but is inelastic and cold. The greatest objection to terrazzo floors is their tendency to crack. Dividing strips greatly reduce this objection.

Terrazzo is not used directly over a wood subfloor. If placed over wooden construction it should have a base similar to that used for tile floors, as shown in Figure 9.3 (b).

A terrazzo wall base is commonly used with terrazzo floors. It is usually made in the form of a sanitary cove base, the angle between the floor and the wall having a fillet to facilitate cleaning. A terrazzo base will not adhere to lime plaster, hardwall plaster, gypsum block, or plasterboard. The backing should be brick, stone, concrete, hollow clay tile, or metal lath.

Magnesite Composition. Several basic types of magnesite composition floors are on the market. In general, magnesite composition floors consist of a dry mixture of magnesium oxide, asbestos or other inert material, fine aggregate such as crushed stone or sand, and a pigment to which liquid magnesium chloride is added on the job to form a plastic material which is troweled to a smooth finish and sets hard in a few hours.

Magnesium oxide is obtained by calcining magnesite, which is magnesium carbonate, the carbon dioxide being driven off in the process. When magnesium chloride is added to magnesium oxide, a cementing material, known as magnesium oxychloride, is formed. This is the cementing material in magnesite composition floorings, the asbestos being the inert aggregate. Asbestos is chosen because of its toughness and cushioning effect.

The finished surface is usually $\frac{1}{2}$ or $\frac{5}{8}$ in. thick, but floors as thick as $1\frac{1}{2}$ in. are used. The $1\frac{1}{2}$ in. flooring is usually applied in two layers of about equal thickness. The lower layer is fibrous and serves as a cushion for the upper layer which is harder and forms the wearing surface. Magnesite flooring may be applied to a subfloor of wood, concrete, or steel plates. If the subfloor is wood, a base course or foundation is required in which metal lath or wire mesh is placed to prevent cracking.

A wall base of the same material may be placed at the same time as the floor, and it may be made monolithic with it, with a rounding corner between the two forming a sanitary base which is easily cleaned. This base should not be applied over hardwall plaster, Keene's cement, gypsum blocks, or plasterboard since it will not adhere to these surfaces. It will adhere to hollow clay tile, brick, stone, or concrete masonry and to metal lath. Metal lath should preferably be galvanized.

This type of floor is less attractive and less durable than ceramic tile, terrazzo, and marble but is more comfortable to work on and less noisy than these floors. It is dustproof, easily cleaned, fire-resistant, and oil and grease resistant.

Magnesite composition, of some type, is appropriately used on the floors of schools, office buildings, industrial buildings, and many other types of buildings.

63.   CERAMIC TILE, BRICK, QUARRY TILE, AND STONE

General Comments.   Hard materials of various kinds, including ceramic tile, brick, stone, and glass, are used in tile or slab form for wall surfaces and, with the exception of glass, for floor surfaces also.

Ceramic tile of various shapes, sizes, thicknesses, colors, and surface finishes are manufactured for use as a surfacing material for interior and exterior floors and walls where a quality surface is desired.

Structural clay facing tile, as described in Article 26, are manufactured for use as structural units in constructing partitions and interior and exterior wall surfaces. They are manufactured with dull and glazed exposed surfaces to provide a finished wall surface. Extruded wall ashlar, which is similar to structural clay facing tile but is more accurately finished, is available for use as a combined structural and facing wall material and for use as a surfacing material only.

Clay brick, as described in Article 23, and special paving brick are used for floor surfaces under conditions of heavy wear. Facing brick and brick with glazed and enamel exposed surfaces are used as facing material as well as a structural material for interior and exterior surfaces of brick walls and partitions.

Various forms of natural stone, including marble, travertine, granite, sandstone, limestone, and slate, are used for interior and exterior floor and wall surfaces in the form of tile or large slabs. These materials are discussed in Articles 24 and 25.

Structural glass is available in the form of tile or slabs in thicknesses from $\frac{1}{4}$ to $1\frac{1}{4}$ in., in various opaque colors, and with polished or honed finish, for use as a finish on exterior and interior wall surfaces. It is known by various trade names such as *Carrara Glass*, *Vitrolite*, and *Opalite*.

These materials may cover the entire wall surface, or they may extend upward only a few feet to form a wainscot and thereby protect the portion of the wall which receives the most severe use. They are all easily cleaned.

All the more important materials, except ceramic tile, are described in other articles and therefore will not be considered further here.

Ceramic Tile. Ceramic tile are usually set in portland-cement mortar when used on the interior of buildings. Quarry tile and promenade tile are set in cement mortar over membrane waterproofing on roof gardens or in similar positions where a watertight floor is required. Floor tile are usually set on a concrete slab foundation or base; so wood floors must be specially constructed to receive this base, as described in Article 60 and illustrated in Figure 9.3(b). Wall tile are set in a bed of portland-cement mortar applied to masonry walls or over portland-cement plaster on metal lath if applied to walls or partitions with wood studs. A thin-setting bed for floor tile has also been developed. It consists of emulsified asphalt and portland cement and is designed for use over wood, steel, and concrete surfaces. Its small thickness of $\frac{1}{16}$ to $\frac{3}{8}$ in. makes possible the use of tile floors in locations which formerly would not have been suitable.

Ceramic tile are divided into many classes depending upon the processes of manufacture, the degrees of vitreousness, and other properties. These classes are considered more in detail in subsequent paragraphs. The Tile Manufacturers Association divides tile into *exterior* and *interior tile*, according to their exposure, and into *wall* and *floor tile*, according to their position. *Trimmers* are available for angles, corners, recesses, and special uses.

Ceramic tile are made by burning special clays or mixtures of clays which have been pressed into the desired shape. Two processes are used: the plastic process and the dust-pressed process.

In the *plastic process*, the clays are mixed with water and run through pugging machines until a uniform plastic consistency is secured. They are then pressed by hand or machine in dies or molds and, after drying, are burned in kilns. The plastic nature of the materials has a tendency to produce tiles which are slightly irregular in shape.

In the *dust-pressed process*, the clays, after being finely ground and mixed with water, are passed into filter presses where the excess water is pressed out. The resulting mass is dried, pulverized, pressed into shape in metal dies, and burned in kilns.

The production of special sizes and shapes in the dust-pressed process involves special dies and handling and is a deviation from the regular routine of manufacture. In the plastic process, special sizes and shapes may be produced without distinct departure from the methods of production common to the regular tile.

Ceramic tile may be glazed or unglazed. For use in residences, all types of floor glazes are sufficiently durable for floors; but for public buildings subjected to severe traffic, a special high-fire type of glaze is available.

The colors in unglazed tiles are produced either by the selection of clays which will burn to the desired colors or by the addition of certain materials, such as the oxides of cobalt and chromium. Some clays and color ingredients can be fired to complete vitrification, producing vitreous tile, while others will not stand this high temperature and produce semivitreous tile. A great variety of colors and textures are available.

Unless otherwise noted, the various types of vitreous tile are obtainable in the following colors: white, celadon, silver gray, green, blue-green, light blue, dark blue, pink, cream, and granites of these colors. The semivitreous tiles are available in buff, salmon, light gray, dark gray, red, chocolate, black, and the granites of these colors. The term granite means a mottled color resembling granite.

The most common shapes of tile are square, rectangular, hexagonal, octagonal, triangular, or round; the sizes vary from $\frac{1}{2}$ in. to 12 in., and the thickness from $\frac{1}{4}$ in. to $1\frac{1}{2}$ in. Trim tile are available for use as wall base or to meet any other decorative or utilitarian demands.

Various names are given to the tiles of different sizes and shapes. The more common types used for floor surfaces are the following. *Ceramic mosaic* include unglazed dust-pressed tile $\frac{1}{4}$ in. thick with an area of less than $2\frac{1}{4}$ sq. in. They are vitreous or semivitreous, depending on the color, and may be square, oblong, hexagonal, or round. These tile usually are mounted with exposed face stuck to paper in sheets about 2 ft. by 1 ft., the paper being removed after the tile are set. If desired, the tile can be obtained loose.

Plastic mosaic include the same size and shape tile as ceramic mosaic, mounted or loose, but these tile are made by the plastic process, and the colors are those that result from the firings of natural clays.

Cut mosaic floors are made from unglazed, dust-pressed, vitreous or semivitreous strips, ¼ in. thick and ½ and ⅝ in. wide, which are cut into the irregular pieces necessary in the production of ungeometric designs and pictorial work. These tile are furnished in loose strips or are assembled in designs mounted with exposed face on paper which is removed after the tile are set.

Vitreous tile and *semivitreous tile* are names applied to unglazed, dust-pressed tile ½ in. thick. These tile are vitreous or semivitreous, depending on the color. They are furnished in the same shapes as ceramic mosaic, except round, but are larger, having an area of 2¼ sq. in. or greater, the largest vitreous tile being 3 in. square; and the largest semivitreous tile, 6 in. square.

Paving tile are unglazed, dust-pressed tile ¾ in. thick. *Flint tile* are vitreous paving tile, and the semivitreous are called *hydraulic tile*. These tile may be square, oblong, hexagonal, or octagonal. With the exception of the oblong tile, the smallest size is 4¼ in. and the largest 6 in. Oblong tile vary in size from 6 in. by 3 in. to 10 in. by 5 in.

Corrugated paving tile are semivitreous, unglazed, dust-pressed paving tile 1³⁄₁₆ in. thick, and 6 in. square with corrugated face.

Rough red paving tile are semivitreous, unglazed, dust-pressed tile, ½ in. or ⅝ in. thick, depending on the size, and 6 in. or 9 in. square with the corresponding oblong half-tile.

Inlaid or *encaustic tile* are unglazed dust-pressed decorative tile, ½ in. thick, produced by inlaying a figure or ornament of one or more colors into a body of a contrasting or harmonizing color before firing. They are viterous or semivitreous according to colors.

Quarry tile are machine-made unglazed tile, ¾ in. to 1½ in. thick, made from common clays. They are always square, the usual sizes being 6 in., 9 in., and 12 in. The colors may be various shades, plain red, or the following granites: red, light gray, dark gray, black, chocolate, light brown, dark brown, or green.

Promenade tile are machine-made, unglazed tile 1 in. thick, made from common clays. The size is always 6 in. × 9 in., and the color some shade of red.

Plastic tile are unglazed tile made by the plastic process from natural clays. Any size or shape can be obtained. The thickness is ½ in. or more, depending upon the size.

64.   RESILIENT FLOORING

General Comments. Elastic or *resilient* materials of various kinds, including linoleum, cork, rubber, asphalt, and vinyl plastic, are used for floor and wall surfaces. These materials are usually cemented to wood, concrete, or plaster surfaces with special cements.

Linoleum. Linoleum is used as a covering for wood and concrete floors. In making linoleum, linseed oil is oxidized by exposure to the air into a tough, rubber-like substance which is mixed with ground cork, wood flour, coloring matter, and other ingredients, and the resultant plastic substance is pressed upon a backing of burlap. It is then passed into drying ovens where it is thoroughly cured and seasoned. There are three common types of linoleum: plain, printed or stamped, and inlaid. Linoleum is furnished in thicknesses varying from $\frac{1}{20}$ in. to $\frac{1}{4}$ in. in rolls usually 2 yd. wide but in some cases 4 yd. wide. It is also available in tile form.

Plain linoleum is a solid color throughout its entire thickness. It is furnished in several thicknesses, varying from $\frac{1}{12}$ in. to $\frac{1}{4}$ in., and in many colors. The thicker grades are known as *battleship linoleum* and are the most satisfactory grades for heavy traffic.

Stamped or *printed linoleum* has a pattern printed on the surface with oil paint. It varies in thickness from $\frac{1}{20}$ to $\frac{1}{12}$ in. and is satisfactory for light service only because the pattern wears off in time. After the pattern shows wear, the linoleum is still serviceable, but is unattractive. Occasional varnishing will preserve the pattern.

Inlaid linoleum consists of small units of linoleum of various colors and shapes arranged in patterns and pressed on a burlap back. The color of each unit is constant throughout the entire thickness; so the pattern remains as long as the linoleum lasts. Inlaid linoleums are furnished in thicknesses varying from $\frac{1}{12}$ in. to $\frac{1}{8}$ in. They are used extensively and will give satisfactory service wherever their use is appropriate.

The best method of laying linoleum on matched wood flooring is to paste a layer of heavy unsaturated felt paper to the wood floor and then to cement the linoleum to the felt. Linoleum is usually cemented directly to smooth concrete, plywood, or hardboard subfloors. Linoleum is sometimes tacked to wood floors, but this method is unsatisfactory.

Linoleum which is cemented directly to a matched wood floor tends to split as the boards shrink. If it is tacked around the edges and not cemented, it buckles because the traffic on linoleum tends to make it spread slightly. Matched floors should be sanded before linoleum is laid. Even if sanded, the outline of matched wood flooring eventually tends to show through linoleum, especially if the boards cup somewhat, as they often do. For this reason, $\frac{5}{8}$-in. plywood is an excellent subfloor for linoleum because the joints will be 4 ft. apart and may not show at all if securely nailed. Hardboard ¼ in. thick, closely nailed to a wood subfloor, forms a good base for linoleum.

When a suitable linoleum is properly laid, it will last for many years. It is sanitary, easily cleaned, resilient, warm, and attractive. Considering the length of life and satisfactory service, it may be classed as an inexpensive floor covering. Linoleum should not be used in basements. A special cove wall base is available for use with linoleum, and special linoleums are available for wall coverings.

Linoleum floors are often used without any surface treatment, but if plain and inlaid linoleums are waxed and stamped linoleums are varnished as often as the use requires, the floors will be more easily cleaned and will last longer.

Linoleum tile are of the same composition as linoleum and have the same properties and uses, but may be arranged in patterns to form a floor which may be more attractive than linoleum. They are cemented to a wood or concrete floor in the same manner as linoleum.

Cork.   Cork flooring is available in two forms, cork carpet and cork tile. Floors of wood or concrete may be covered with *cork carpet*, which is a covering similar to linoleum and is laid in the same way. It is composed of the same material: oxidized linseed oil, ground cork, and wood flour, pressed to a burlap back, but it is not subjected to as great pressure as linoleum and so is more resilient and porous and less durable. Cork carpet is ¼ in. thick and is furnished in rolls 2 yd. wide. Many colors are available.

To secure the best service, cork carpet should be laid by first cementing a layer of unsaturated felt to the wood or concrete floor, and then cementing the cork carpet to the felt. Cork carpet is often cemented directly to wood or concrete, and is sometimes tacked to wood floors. The latter method is particularly unsatisfactory.

Cork carpet makes a very quiet floor covering, and is more elastic, more absorbent, and less durable than linoleum. It is particularly suitable for use in churches, theaters, public libraries, and other places where a noiseless floor covering is essential. Cork carpet should not be used in basements.

Cork tile are made from pure cork shavings compressed in molds to a thickness of ½ in. and baked. They are used over wood or cement floors, to which they are cemented. Cork tile are elastic, noiseless, fairly durable, and quite absorbent. They are available in various shades of brown. The rosin in the cork liquefies during the baking process and cements the shavings together when the tile cools. The brown color of cork tile is largely caused by the baking it receives, the darker browns having been baked longer than the lighter browns. Cork tile are used for wall coverings as well as for flooring.

Rubber. Rubber flooring is made of synthetic rubber combined at high temperatures with fillers, such as cotton fiber, various minerals, and with the desired color pigments. It is made in form of sheet rubber and rubber tile. The thickness varies from $\frac{1}{8}$ to $\frac{1}{4}$ in. Many colors and patterns are available. Rubber flooring is cemented to concrete or wood in the same manner as linoleum. It is attractive in appearance, elastic, noiseless, durable, easily cleaned, but quite expensive. Rubber floors are not resistant to oil, grease, and gasoline and should not be used on floors in contact with the ground. Rubber is used as a wall covering as well as a flooring material.

Asphalt Tile. Asphalt tile are made from asphalt or resinous binders, asbestos fiber, mineral pigments, and inert fillers by amalgamating under heat and pressure. The sheets produced by this process are cut into tile of various sizes usually between 9 in. squares and 18 × 24 in. rectangles, but smaller and larger sizes are available. The thicknesses are $\frac{1}{8}$, $\frac{3}{16}$, and $\frac{1}{4}$ in. The thinnest tile are suitable for placing on concrete when the traffic is not heavy. The most commonly used thickness for placing on wood and concrete is $\frac{3}{16}$ in., and for severe service, the $\frac{1}{4}$ in. thickness is used. Asphalt tile are available in a great variety of colors varying from light to dark. Asphalt tile for industrial use are available in $\frac{1}{4}$, $\frac{3}{8}$, and $\frac{1}{2}$ in. thicknesses. They are available in various colors. For the lighter colors, the binder is chiefly or entirely resinous.

Asphalt tile floors are cemented to matched wood flooring, plywood, hardboard, or concrete in the same manner as linoleum.

Asphalt tile are resilient, nonabsorbent, reasonably stainproof and acidproof, relatively inexpensive, attractive, and moistureproof so that they can be used on concrete floors below the ground level. They are not resistant to grease and oil, but special "greaseproof" tile are available. Their resistance to indentation is relatively low. They are suitable for use in schools, apartment and office buildings, hospitals, laboratories, residences, and many other kinds of structures. Asphalt tile are used as a wall covering as well as a flooring material.

Vinyl Plastic. Vinyl plastic floor coverings have thermoplastic binders, as described in Article 93, together with plasticizers and granular mineral fillers and pigments selected and proportioned to give the desired properties and colors. They are manufactured in rolls 6 ft. wide and in tile of many sizes. The thicknesses available are $\frac{5}{64}$,

$\frac{3}{32}$, and $\frac{1}{8}$ in. A large assortment of solid colors, as well as marbelized and other patterns, are available. It may be unbacked or backed with various materials.

Vinyl plastics are wear- and indentation-resistant, impervious, resistant to oil and grease, resilient, easily maintained, and attractive.

Vinyl asbestos tile include asbestos fibers in their composition. They are less flexible than the other types, are less resistant to indentation, and are available only in tiles 9 in. square.

Enameled Felt Base. The lowest-cost floor covering consists of an asphalt-saturated felt with a baked enameled design on the top surface and a painted bottom surface. The covering is not usually cemented or fastened in any way to the base, except possibly at door openings, because it is torn easily. Attractive designs are available. It is easily cleaned, but it is not resistant to oil, grease, and alkaline cleaners. It is suitable for use where a low-cost floor covering is desired and it is subjected only to light foot traffic. Widths of 9 ft. are available.

Seamless Flooring. A resilient flooring material that features a unified and unjointed surface is called *seamless flooring*. These types of flooring are frequently applied integrally, forming a wall base; sometimes the material is extended up the wall, forming a wainscot. Both risers and treads of stairways may be surfaced with this material. Most seamless flooring materials may be classified as plastics. Some have a polyester binder, or vinyl resin base, or a two-component epoxy resin base; some have a urethane finish coat. Most seamless flooring systems offer a decorative floor surface in a variety of colors, with embedded marble chips. These chips, cast in the unset base material, are intended to give the appearance of traditional terrazzo. The finish texture of seamless floors is frequently a pebbled or "orange peel" surface.

Depending upon the manufacturer, seamless floors are applied by brush, roller, or trowel in thicknesses that are measured in fractions of an inch—about $\frac{1}{8}$ to $\frac{1}{2}$. These floor surfaces have a combination of features offered by few other resilient floor finishes. Seamless floors are resistant to most common chemicals. They are a resilient, comfortable, easy-to-maintain, water-resistant, nonslip, and durable surface. Some are sparkproof and are most suitable for hospital operating rooms and other areas where spark resistance and sanitation are mandatory. These floors are used in public, institutional, and commercial buildings.

65. SELECTION OF FINISH FLOORING

The selection of a proper floor surface is one of the most important and, at the same time, one of the most difficult problems in the construction of a building. The appearance, usefulness, and cost of upkeep of a building are greatly affected by the type of floor installed. Considering the importance of the subject, it is unfortunate that it is not possible to devise a satisfactory basis of selection.

The report of a Committee on Floors of the American Hospital Association has been of considerable value in preparing this article. The definitions of the properties of floor surfaces follow those of the committee quite closely.

The following discussion of the relative merits of the various floor surfaces is prepared while realizing that the value of such discussion is limited because of the great variations in the materials and workmanship and in the kind of usage and care a floor will receive.

Floors which are manufactured complete and ready for installation, such as tile, linoleum, and rubber, will show less variation in quality than such floors as terrazzo, concrete, and magnesite composition which are manufactured on the job, and are not subject to as rigid control as factory-made products. Natural flooring materials such as marble and slate are also quite variable in quality.

Each property of a wearing surface will be discussed, and an attempt will be made to classify roughly the various materials according to the degree to which they possess that quality.

Appearance. Appearance is the attractiveness of the material, its color range, texture, and its decorative value in an architectural sense.

There are many floor surfaces which are attractive when suitably used, such as hardwood when properly finished, terrazzo, ceramic tile, marble, and, to a somewhat less degree, vinyl plastic, rubber tile, cork tile, linoleum tile, asphalt tile, linoleum, cork carpet, sheet rubber, slate, and magnesite composition. Concrete without special treatment, and industrial wood blocks are not suitable for use where appearance is a factor. Concrete floors may be painted or waxed and are not unattractive as long as the surface is maintained, but this is difficult to do where the traffic is at all heavy.

Durability. Durability may be defined as the resistance to wear, temperature, humidity changes, decay, and disintegration. The adhesion of a material to its base is also a factor in durability.

The most durable floor surfaces for foot traffic are ceramic tile, terrazzo, slate, and concrete, but terrazzo floors are likely to crack if not divided into blocks by dividing strips or laid in the form of tile. Marble is widely used in floors subject to severe wear, but it does not stand up as well as the materials just mentioned. Some marbles are much more resistant to wear than others. Concrete surfaces to be durable must have durable aggregates.

Hardwood, linoleum, linoleum tile, vinyl plastic, and rubber tile give very satisfactory service, whereas cork carpet, cork tile, asphalt tile, and magnesite composition are fairly satisfactory.

With the exception of concrete, none of the materials mentioned so far is suitable or satisfactory for heavy traffic, such as trucking. Brick and wood block may be used under these conditions. Heavy maple flooring may also be satisfactory.

Comfort.   Comfort under foot is determined by the shock-absorbing qualities, sure-footedness, evenness of surface, and conductivity. A floor which is a good heat conductor will always feel cold.

The most comfortable floors to work on are cork tile, cork carpet, and rubber. Wood, linoleum, vinyl plastic, magnesite composition, and asphalt tile, are very satisfactory; but concrete, terrazzo, ceramic tile, marble, slate, and brick are tiresome and cold.

Noiselessness.   Cork tile, cork carpet, and rubber are practically noiseless; wood, linoleum, vinyl plastic, magnesite composition, and asphalt tile are slightly less satisfactory but still very good; but concrete, ceramic tile, marble, slate, and brick are the noisiest of flooring materials.

Fire Resistance.   Materials may be noncombustible but still suffer severely in case of fire. Concrete, ceramic tile, and brick are probably the most fire-resistant floor surfaces, but terrazzo, marble, and slate are very satisfactory. Magnesite composition, or asphalt tile, will not burn but may suffer seriously in a fire. Linoleum, cork carpet, rubber, vinyl plastic, and wood are combustible; but if laid on a fire-resistant base, they are not considered a serious defect in a fire-resistant building.

Sanitation.   To be sanitary, a floor surface must be nonabsorbent and easily cleaned. Joints which are not watertight are an unsanitary feature.

The most sanitary floor surfaces are terrazzo, ceramic tile, marble, and slate. Magnesite composition, asphalt tile, rubber, vinyl plastic and linoleum are quite satisfactory. Cork carpet is unsatisfactory because of its porosity, concrete because of the difficulty in cleaning, and wood because of its porosity and the presence of open joints.

Acid and Alkali Resistance. The factors that should be considered under this heading are immunity from damage by occasional spillings of strong acid solutions and resistance to the continuous use of soap, lye, cleaning and scouring compounds, and disinfectants.

Ceramic tile is the most satisfactory floor surface in this respect; asphalt tile and vinyl plastic are quite resistant; rubber, terrazzo, marble, concrete, and magnesite composition are sufficiently resistant for ordinary purposes; but linoleum, cork carpet, and cork tile should not be subjected to the action of acids and alkalies.

Grease and Oil Resistance. Grease and oil are not absorbed by ceramic tile and by vinyl coverings and do not affect these materials. They are absorbed by wood, brick, concrete, terrazzo, linoleum, cork carpet, and cork tile and therefore detract from their appearance, but they do not seriously affect their durability. Asphalt and rubber floors, except greaseproof asphalt tile, are seriously affected by grease, oil, and gasoline.

Dampness. Ceramic tile, brick, concrete, terrazzo, and asphalt tile are not affected by dampness and are suitable for use on floors located on the ground such as basement floors; but wood, rubber, linoleum, cork carpet, and cork tile are not suitable for use in such locations.

Indentation. The hard flooring materials, such as ceramic tile, concrete, terrazzo, and brick, do not suffer indentation from chair legs, heels of shoes, and other objects which rest on them or strike them. Marple and oak flooring yield very little and do not retain imprints. Other materials such as linoleum and rubber yield considerably under such loads but recover quite well when the load is removed. Asphalt and vinyl asbestos tile may become permanently indented.

Trucking. Three factors are pertinent when considering the suitability of a floor for trucking. It must stand the abrasive action of the truck wheels, the tractive effort required to pull the truck must not be excessive, and the flooring must have structural resistance sufficient to carry the load transmitted to it by the truck wheels. Concrete, heavy maple flooring, and industrial wood blocks are satisfactory in all three respects if the materials are of high quality and if the trucking is parallel to the length of the maple flooring rather than crosswise. The maple flooring must be heavy enough so that the weight of the trucks will not break the tongued-and-grooved joint. The aggregate in the concrete may have to be specially selected so as to have a high resistance to abrasion. Rubber-tired wheels are much easier on all types of flooring than wheels which are steel-tired.

Maintenance. This heading includes such items as the ease with which a flooring is cleaned, the necessity for care and surface treatment, such as waxing and painting, the necessity for repairs, and the cost of such operations.

Ceramic tile, marble, terrazzo, slate, vinyl plastic, and rubber tile floors are easily cleaned and require very little care. Linoleum, asphalt tile, and magnesite composition are easily cleaned but should receive surface treatment occasionally. Cork carpet is not easy to clean and requires surface treatment. Hardwood floors are fairly easy to clean if in good condition, but require frequent surface treatment. Concrete is not as easy to clean as ceramic tile, linoleum, etc. if it is not painted or waxed. Painting makes cleaning easier but requires frequent renewal.

The monolithic floors such as terrazzo, magnesite composition, and concrete are difficult to repair satisfactorily. Floors composed of separate units of tile, slate, or marble are more easily repaired, but require skilled mechanics. Linoleum, cork tile, and cork carpet may be easily repaired by replacing the damaged parts.

The maintenance costs of wood block, heavy asphalt mastic, brick, and concrete are relatively low except under extremely severe traffic. With the exception of concrete, these materials are easily repaired. They receive no surface treatment.

Initial Cost. One of the first factors in selecting a floor surface is the initial cost, but even the most expensive materials do not possess all the desirable features.

Flooring materials may be roughly divided into classes according to their cost in place. In the following list, the most expensive materials are given first.

(a) Ceramic tile, marble, vinyl tile, and rubber tile.
(b) Terrazzo, magnesite composition, cork tile, and hardwood.
(c) Cork carpet, linoleum, and asphalt tile.
(d) Concrete.

This list assumes that a concrete base is available to receive the wearing surface. Brick and wood block floors are not included in these lists, for they are used for a different class of traffic from that to which the materials included, except concrete, are subjected, and a comparison would be of no value.

Weight. The heavier floor surfaces add indirectly to the cost by requiring stronger floor construction, beams, girders, columns, and foundations. Ceramic tile, marble, and slate are bedded on ½ in. or more of cement mortar which has no structural value; terrazzo requires 1¼ in. of material which is simply a dead weight; and hardwood flooring usually requires about 2 in. of filling between the nailing strips if placed over concrete. Rubber tile, magnesite composition, cork tile, cork carpet, linoleum, asphalt tile, and concrete do not require this additional material. This additional weight exists when the structural floor is some form of concrete slab.

REFERENCES AND RECOMMENDED READING

1. *Recommended Practice and Standard Specifications for Concrete and Reinforced Concrete*, report of Joint Committee representing affiliated committees of several societies, 1940. Seventh printing, American Concrete Institute, 1950.
2. Ben John Small, *Flooring Materials*. Circular Series Index F4.6, Small Homes Council, University of Illinois, 1955.
3. *Sweet's Architectural Catalog File*, F. W. Dodge Corp.

11

FINISH WALL CONSTRUCTION

66. PLYWOOD AND WALL BOARD

Wall boards of various materials are usually manufactured in sheets 4 × 8 ft. They are used mainly for interior wall surfaces attached to a supporting framework. These frameworks most frequently consist of wood studs (2 × 4 or 2 × 6), a framework of pressed, stamped metal, or cold rolled steel channels. These studs are commonly spaced at 12, 16, or 24 in. centers—spacing depending upon the height of the stud wall and its function as either a nonbearing or load-bearing partition.

The space between studs is most often braced with a horizontal member located at midheight which acts as fire blocking and an insulation serving thermal and sound deadening purposes. The finished sheet materials are then placed on the studs and secured with nails, staples, screws, or special clips.

Plywood.   Plywood pertaining to structural framing and formwork has been discussed in Chapters 5 and 7. Our discussion here concerns plywood manufacture and plywood used as a finished surface for walls and ceilings. Generally, this type of construction is called *paneling*; and the plywood sheets, *panels*.

Plywood consists of thin layers or *plys* of wood glued or pressed together. The plys are arranged with the grain direction of adjacent plys at right angles to each other. Plywood ranges in thickness from ⅛ in. to 1 in., the actual thickness depending upon the number of plys and the core material. The most commonly used thickness for interior paneling is ¼ in. or ⅜ in. Widths are usually 4 ft. and lengths are usually 8 ft., although they may be as long as 16 ft. in certain species. The number of plys, always an odd number to prevent warping, is 3, 5, or 7. The outside plys are called the *faces* or *face* and *back*. The intermediate plys with grain perpendicular to the face grain are called *crossbands*, while those in the middle are called the *core*. The core of plywood may be another ply, solid strips of wood, or a sheet of particle board.

On the face that is to be exposed, panels may have a special finish, texture, or design. Plywood panels that have a hardwood veneer have the finished hardwood surface on one side only. The core and the back side are made of less expensive wood.

The grain configuration of plywood panels is determined by the manner of cutting the veneer ply from the log as well as the species of wood. The wood ply that is cut from the log is called a *flitch*. Plywood panels, as they are designed for paneling a room, may be arranged in several ways for layout patterns. Layout patterns are based upon the grain configuration of the panels. If panels are to follow a determined order continuously, such as bookmatched pattern, they are referred to as bookmatched *in sequence*.

The texture of plywood panels may be smooth or *sanded; rough sawn*, which shows saw marks; or *weathered*. The panels are sometimes *grooved* in several widths—even or random. Panels are also made to give the appearance that the panel is made up of several individual boards.

Plywood panels are made with finished surfaces in more than 20 hardwoods and in about 5 softwoods. Among the hardwood veneers are black walnut, birch, red oak, white oak, ash, pecan, and cherry. The plywoods that have an attractive softwood finish are redwood, several of the pines, and Engelmann spruce.

Particleboard.   A panel which consists of compressed wood chips, slivers, sawdust, and shavings combined with an adhesive is called *particleboard*. As stated by ASTM D 1554, particleboard is compressed by two methods. In the *flat-platen pressed* method, the sheets are formed by compressing perpendicular to the faces; whereas by the *extruded* method, the binder and wood materials are extruded through a heated die, and the pressure is applied parallel to the face of the sheet and in the direction of the extruding. This is known as *extruded particleboard*. Particleboards are further classified by their density as *low density* of less than 37 lbs. per sq. ft.; *medium density* between 37 and 50 lbs. per sq. ft.; and *high density* of more than 50 lbs. per sq. ft. Particleboard is made in sheets 4 ft. wide and 8 ft. long in thicknesses from ⅜ in. to 1¾ in. The surface finish of these panels may be smooth or grooved.

These panels are dimensionally stable and nonwarping due to their lack of grain. They are ideally suited as an underlayment or core for hardwood veneer and as a backing for the laminated plastics mentioned below. Particleboard is also used in doors and cabinet work, as backing panel to receive fabric wall surfaces, and directly as a finished wall surface to receive paint.

67.   PLASTICS AND FABRICS

Laminated Plastic Veneers.   Laminated plastic consists of sheets of plastic impregnated papers. Sometimes called *melamine sheets*, these high pressure laminates are made by assembling several sheets of phenolic-impregnated kraft paper into which are impressed a melamine-impregnated printed pattern sheet and a translucent melamine overlay. These materials are assembled and compressed at pressures of about 1400 lbs. per sq. in. at temperatures exceeding 300°F. The sheets are marketed in several widths: 2, 2½, 3, 4, and 5 ft., with lengths of 6, 7, 8, 10, and 12 ft.

Three grades are commonly made with respect to use and the thickness of the finished sheet. *General purpose grade* is ¹⁄₁₆ (0.062) in. thick and is used primarily for horizontal applications, and for some vertical applications. This grade is the most durable and is intended for the most severe use. *Postforming grade* is made with stretchable papers and a thermosetting laminate that will soften upon reheating and reset when cooled. It is used mostly for kitchen counter tops and splashbacks where a small radius curved surface of coves and bullnoses must be formed. This grade, 0.050 in. thick, is thick enough for self edging. *Self edging* means that the laminate is beveled or chamfered at the square edges. The durability of this grade compares to that of the general purpose grade. *Vertical grade* is thinner, approximately ¹⁄₃₂ (0.035) in. thick. It is used primarily for veneering vertical surfaces that take little abuse. It is made in a postforming or a nonpostforming material that may be self edged.

Laminated plastics are impervious, hard, nonbreakable, durable, fire-resistant, and attractive. They are cemented as veneers to a backing of plywood, hardwood, particleboard, or asbestos cement. A great variety of plain colors and wood and marble imitations are available. This easily cleaner veneer is very resistant to acids, alkalies, and alcohol. They are used for table and counter tops and as finish wall panels. Panels with laminated plastic veneers are fastened by special nails, clips, or adhesives.

High pressure laminates are frequently identified by their trade names: *Formica, Nevamar,* and *Texolite.*

Fabric Wall Covering. A plastic covered, fabric backed wall covering that is applied to a wall surface is called *fabric wall covering*. These vinyl coated coverings are made in rolls 48 in. to 54 in. wide and in lengths up to 36 yds. They may be installed over almost any surface that is smooth, dry, clean, and free of loose material. They should not be applied over surfaces that have been covered with oil base paints or stain.

Wall fabrics are frequently made in colors and patterns intended to resemble other materials such as brick, wood, linen, and burlap. The fabric is generally attached with a wheat paste.

Fabric covered wall surfaces are easily cleaned and are resistant to most agents that soil walls.

68. WALL TILE

Ceramic tile is probably the most frequently used wall tile, although several other materials are used. For all tile—copper, stainless steel, aluminum, or plastic—installation and construction criteria are similar. As discussed in Article 63, ceramic wall tile is installed similarly to floor tile. Differences, if any, are in the backing materials, the method of set, adhesive, and the use of more sculptured tile in wall installations. Sculptured tile or tile with a patterned relief are seldom used for floors. Masonry, a common wall backing material, is not used as a backing surface for floor tile work, although application to masonry walls is similar to a concrete surface.

When wall tile are set, foreign matter, loose material, and oiled or soiled material are removed. Each of the two setting methods, cement mortar setting bed or thin set method, has advantages. The traditional mortar setting bed method permits a wall backing surface to be horizontally evened and plumbed before installation commences. Thin set methods permit little correcting of a backing surface.

When wall tile are to be set, the lateral center of the wall surface is established and marked. This center line will be a joint line, or the center of a tile will be placed on this line. The number of full size and fractional part tile for half the lateral distance of the wall are then determined. The desired result is to have each end tile the same size, either a full tile or a fractional part tile preferably larger than one-half tile.

Adhesives and mortars are spread on the back up surface with a *serrated trowel*. The *saw toothed* or *notched* edge of the trowel assures an even depth of spread of the mortar or adhesive.

Mortars and Adhesives. There are several adhesives that are used to secure ceramic tile. The use of traditional *mortar cement* is suitable for most backing surfaces. The *thin-set* methods are newer methods and are suitable for certain applications.

Cement mortar methods for wall installations require a mixture of portland cement, sand, and hydrated lime, their proportions varying with job conditions and backing materials. The cement mortar scratch coat is trowled onto a supporting or backing surface in a thickness of approximately ¾ in. A leveling coat of about ¼ in. to ½ in. removes the imperfections that may exist from an uneven supporting surface. The tile may then be placed on a neat cement skim coat, or cement adhesive may be buttered to the back of larger individual tiles before placing. Tile intended for mortar beds should be presoaked with water to prevent their drawing water from the cement mortar. The cement mortar method provides a more water-resistant and durable installation and is recommended where water penetration may be a problem, such as in showers or where severe exposure is to be encountered. This method is exceptionally well suited where masonry or concrete backing is the wall material. It may, however, be used with wood framed structures; moistureproof kraft paper as well as reinforcing with metal lath or mesh on solid wood sheathing should be used.

There are several mortars referred to as *thin-set* mortars because the actual thickness of the mortar or adhesive bed is approximately ¼ in. One of these, *dry-set mortar*, is a premixed mortar to which a small quantity of water is added. Some include sand; others are unsanded. Suitable for most surfaces, this adhesive has excellent holding qualities.

Closely resembling dry-set mortar is *latex-portland cement mortar*, a mixture of portland cement, sand, and a latex solution. Types of tile adhesives called *two-part mortars* or adhesives are *epoxy mortar*, *epoxy adhesive*, and *furan mortar*. These materials are used in thin-set methods. The epoxy is mixed with a catalyst which causes a rapid set of the epoxy resin. Furan is likewise mixed with a catalyst. The two-part adhesives are made for application to a solid surface that requires excellent bonding. Another thin-set adhesive is called *organic adhesive*. This material, which is like a mastic, requires no catalyst and is ready for immediate use. Its adhering qualities are not as reliable as those of other thin-set adhesives.

After tile are set, the joints are grouted. Various types of grout, similar to the adhesives, should be water-resistant and compatible with the mortar or adhesive used. The grout is worked into the tile joints diagonally over the tile, filling all interstices.

Mosaics. Mosaics are pictorial (or abstract) designs made of ceramic tile. These designs, created by the architect, are executed by artisans. The tiles of proper color are mounted on paper backing (on the face of the tile) in approximate 12 in. squares and numbered according to the design. The mounting papers are removed after the tile have been cemented in place; the tiles are then grouted.

REFERENCES AND RECOMMENDED READING

1. *Sweet's Catalog File*, F. W. Dodge Corp., New York, New York.
2. *ASTM 1971 Annual Book of Standards, Part 16*, American Society for Testing and Materials, Philadelphia, Pennsylvania.
3. *U.S. Product Standard PS 1–66 for Softwood/Plywood.* Supplied by American Plywood Association, Tacoma, Washington.
4. *Guide to Plywood Sheathing for Floors, Walls, and Roofs*, American Plywood Association, Tacoma, Washington, 1970.

12

PLASTER AND STUCCO

69. PLASTERING MATERIALS

The ceiling and wall surfaces of many buildings are finished by covering them with two or three coats of a mixture that consists of a cementing material, an inert fine aggregate, and water. This mixture, known as *plaster*, is mixed to a plastic state and troweled on.

The cementing material for interior surfaces is usually gypsum plaster or hydrated lime, but surfaces which are to be subjected to extreme moisture conditions or hard usage are plastered with portland cement. For exterior surfaces, mixtures of portland cement and hydrated lime are used. When used on the exterior, this mixture is frequently called *stucco*.

Usually the terms plaster or stucco are used irrespective of their composition but denote only their use and locations. The term *mortar*, however, frequently implies a plastic mixture of a cementing material, fine aggregates, and water regardless of its use.

Quicklime. Quicklime is frequently called *lime*. Commercial quicklime, which is pulverized, contains varying amounts of magnesium oxide (MgO) resulting from the presence of magnesium carbonate ($MgCO_3$) in the raw material. Silica, alumina, iron, and other impurities are always present.

Lime comes from the kiln in lumps or in unpulverized form called *lump lime*—it is not sold in this form.

Manufacture of Quicklime. Quicklime is the product derived from the burning of pure limestone ($CaCO_3$) at a temperature sufficiently high to drive off the carbon dioxide, leaving calcium oxide (CaO), a pure quicklime.

If pure limestone is used, the process may be shown by the following formula:

$$CaCO_3 + Heat = CaO + CO_2$$

Calcium oxide (CaO) is a white solid, and carbon dioxide (CO_2) is a gas. If magnesium carbonate is present, a corresponding reaction occurs, leaving magnesium oxide and driving off carbon dioxide gas. The fuel used in the kiln is coal or coke.

Classification and Grades. Quicklime is divided into four classes depending on the relative amounts of calcium oxide and magnesium oxide present: *high-calcium* contains 95% of calcium oxide; *magnesian* has between 25 and 40% of magnesium oxide; *high-magnesian* or *dolomitic* contains a high percentage of magnesium oxide. However, common practice recognizes the division of quicklime into only two classes—calcium limes and magnesian limes.

Quicklime is divided into two grades: *selected,* a well-burned lime free from ashes, core, clinker, and other foreign material; and *run-of-kiln,* a well-burned lime without selection.

Hydrated lime is the dry product produced by treating quicklime with water in sufficient quantity to complete its hydration. Hydrated lime is furnished in two classes according to plasticity. *Masons' hydrated* lime, of lower plasticity than finishing hydrated lime, is used for mortar and for the scratch and brown coats of plaster. It is further classified as *Type N* (Normal) and *S* (Special) according to its ability to develop high early plasticity [ASTM C 207]. *Finishing hydrated lime,* with higher plasticity, is used for the finish coat of plaster as well as the uses made of masons' hydrated. It, too, is further defined as *Type S* [ASTM C 206].

Slaking. In preparing lime mortar, quicklime is mixed with water, forming, calcium hydroxide, $Ca(OH)_2$. This is a fine white powder, but an excess of water is always used, forming a paste called *lime paste* or *lime putty.* The form of lime known as *hydrated lime* is simply the hydroxide formed by adding water to quicklime at the manufacturing plant rather than at the jobsite. During slaking, considerable heat is generated, and a marked increase in volume occurs.

The calcium limes slake more rapidly than the magnesium limes and give off a greater amount of heat. *Quick-slaking limes* begin slaking in less than 5 minutes, while *medium-slaking limes* start to react in 5 to 30 minutes. Limes that do not start slaking for about one half hour are termed *slow-slaking limes.* For quick-slaking limes, *Type S* (Special) [ASTM C 206], the lime should be added to the water; and when steam appears, the lime should be hoed and enough water added to reduce the steaming. For medium-slaking and slow-slaking limes, *Type N* (Normal) [ASTM C 6], add the water to the lime. To avoid slow-slaking lime in cold weather, it may be necessary to heat the water. There is little danger that too much water will be added to quick-slaking lime, but an excess of water may cause magnesium lime to be "drowned." If too little water is added to either calcium or magnesium limes, they may be "burned." In either burning or drowning, a part of the lime is spoiled, for it will not harden, and the paste will not be as viscous and plastic as it should be.

Slaked lime should age two weeks before it is used for plastering, but 24 hours is sufficient if the lime is for masons' mortar.

In making putty or paste from hydrated lime, the lime is sifted slowly into the water and the mixture stirred constantly. The putty is aged or soaked for at least 24 hours. The aging process increases the workability and sand-carrying capacity of the putty. Because hydrated lime has been slaked before shipping, the increased volume of the putty is small. A sack of hydrated lime which weighs 50 lbs. and contains 1 cu. ft. will make about 1.1 cu. ft. of lime putty.

Setting. In setting, the excess water evaporates and the calcium hydroxide combines with carbon dioxide from the air, forming calcium carbonate, as shown by the formula

$$Ca(OH)_2 + CO_2 = CaCO_3 + H_2O$$

A corresponding reaction occurs when magnesium hydroxide is present. The absorption of carbon dioxide occurs very slowly, and in heavy masonry walls the setting may never occur.

The term *air-slaked* is often applied to quicklime that has become slaked by absorbing moisture from the atmosphere; but the process does not stop when this change has occurred, for the hydroxide thus formed absorbs carbon dioxide to form calcium carbonate, which has lost its cementing properties.

Besides plastering, quicklime is used in a variety of manufacturing processes and in several construction procedures. In road building and foundations, it is used for the treatment of soils having a high content of expansive clay. As previously mentioned, slaked lime is used in masonry mortars. Lime is used in the manufacturing processes of paper, bleaches, brick, and steel.

Gypsum. The basic material in all gypsum plaster, the mineral gypsum, occurs in three forms: *gypsum rock*, *gypsum sand*, and *gypsum earth* or *gypsite*. Gypsum rarely is found in a pure state but contains clay, limestone, iron oxide, and other impurities. Pure gypsum is a soft, white mineral, a form of which is alabaster.

Manufacture of Gypsum. Heated to a temperature between 212°F and 400°F, gypsum loses three-fourths of its combined water, i.e., some of the water is driven off as steam. The remaining product is *calcined gypsum*, or *plaster of Paris* if pure gypsum is processed. Plaster of Paris and cement plaster are made by *calcining* or burning gypsum in large kettles or in rotary kilns. If kettles are used, the gypsum is finely ground before burning; but in rotary kilns, the gypsum is crushed to a size of about 1 in. The final pulverizing is accomplished after burning.

Cement or *hard-wall plaster* results if certain impurities are present or are added in the manufacturing. These impurities cause the product to set more slowly than the rapid setting plaster of Paris. If gypsum is heated to about 1400°F, practically all the combined water is driven off, forming a *dead-burned*, or *anhydrous*, plaster. If certain substances such as alum or borax are added, *hard-finish plaster* is produced.

The time of set of hard-wall plasters is regulated to the convenience of the craftsmen. Ordinarily the setting time must be delayed; therefore, a retarder such as glue, sawdust, or blood is added. The working quality and sand-carrying capacity of plasters are improved by adding clay or hydrated lime by the manufacturers; and the cohesiveness is increased by the addition of cattle hair, wood fiber, or glass fiber. In localities where good sand is not available, plaster may be purchased with the proper amount of sand premixed.

Generally, gypsum plaster is grouped into two classifications— base plasters and finish plaster. *Base plasters* are troweled on a suitable firm base and followed with a *finish plaster*, which forms the finished surface.

Base Plasters. ASTM C 28 identifies four types of base coat plasters. *Gypsum Ready-Mixed Plaster* is a calcined gypsum plaster that has a mineral aggregate added to the plaster during milling. This type of plaster may be mixed with a vermiculite, perlite, or sand aggregate. This plaster, which serves as a base coat plaster, requires only the addition of mixing water at the jobsite to prepare it for application. These plasters may have setting times ranging from 1½ hours to 8 hours.

Another calcined gypsum plaster is *Gypsum Neat Plaster*, which is premixed with materials that control setting time and working quality. This plaster permits an extended setting time ranging from 2 to 16 hours. Sold in *unfibered*, *fibered*, or *extra fibered* form, this plaster is intended for jobsite mixing that requires the addition of water and aggregates. The term *neat* means that the plaster (though it may be fibered) is sold without aggregates.

Similar is *Gypsum Neat Plaster, Type R*, which is intended for mixing with sand only. It has a shorter setting time of not more than 3 hours.

Gypsum Wood-Fibered Plaster requires the addition of a sand aggregate. This material, which serves as an excellent base coat plaster, has a normal setting time of $1\frac{1}{2}$ to 8 hours. The nonstaining, mill-mixed, shredded wood fibers give this plaster superior binding qualities. *Glass fibered* plaster is also available and exhibits excellent binding qualities.

Gypsum Bond Plaster contains mill-mixed ingredients that permit a bond to untreated concrete and masonry surfaces. This plaster serves as a base coat under finished gypsum plaster surfaces. It has normal setting times.

Finish Plasters. There are several plasters classed as finishing plasters. *Gypsum Ready-Mixed Finish Plaster* is a mill-mixed neat finish material that requires only the addition of water for jobsite mixing.

Gypsum Gauging Plaster, a finely ground gypsum finishing plaster, is mixed with slaked lime putty for improved workability. This material, which is troweled on in a coat about $\frac{3}{32}$ in. thick, results in a hard, smooth, white finished surface. It is provided in *slow set* or *quick set*. When setting time is not retarded, it should set up in about 20 to 40 minutes. Otherwise, slow set requires 40 minutes or more [9].

The most common and best known form of hard finish plaster is *Keene's Cement Plaster* [ASTM C 61], first processed in 1838 by R. W. Keene of England [12]. This material forms by calcining lump gypsum, immersing it in a 10% alum solution, recalcining it, and finally pulverizing it. This gypsum plaster, which is referred to as *"dead burnt"* or *"hard burnt,"* offers a hard, blanched, impervious finish. It is characterized by high resistance to abrasion and water penetration. It is ideally suited for kitchens, bathrooms, and other locations requiring a finished wall material that will not absorb water. It is available in *quick setting* or *standard setting*—20 minutes to 6 hours.

Gypsum Molding Plaster, a very white gypsum plaster, is employed primarily for molding cornices, column capitals, seals, and other decorative features. It is used to form a suitable mold for precast concrete where special forming demands a design of unusual requirements.

Plaster of Paris is used for ornamental castings, but the rapidity with which it sets makes it unsuitable as a wall plaster or for mortar.

Portland cement is used in *portland cement plaster*, which is both a finish plaster and base plaster. Type I, Type II, and Type III portland cements are used; and where the advantages of air entrainment are required, Type IA, Type IIA, and Type IIIA are used. Portland cement plaster is applied where exposure to moisture is a problem—especially at exterior applications.

Portland cement-lime plaster, which contains Type S finishing hydrated lime, is applied in similar locations as portland cement plaster. The lime serves as a plasticizing agent, resulting in better workability.

Plastic cement is a portland cement, Type I or Type II, containing a mill-mixed plasticizing agent—usually a quantity not in excess of 12% by weight [13]. Plastic cement improves workability, increases water retentivity, and reduces shrinkage in the drying plaster. The addition of plasticizing agents is prohibited during mixing.

Bonding Agents.   A *bonding agent* is a liquid resinous water emulsion applied to base surfaces requiring adhesion between a base surface and a base plaster. The colored liquid is evenly sprayed, rolled, or brushed. The water soluble, dried bonding agent is re-emulsified by the water in the plaster. As the plaster dries, the bond film dries again, thus forming a bond between the plaster and the base material.

Bonding agents are effective between most plasters (gypsum, cement) and several backing materials, namely: concrete, cinder block, brick, ceramic tile, gypsum lath, gypsum wallboard, glass blocks, gypsum block, wood, and steel. Backing materials may be new or used, unfinished or painted with an oil base, rubber base, or vinyl base paint. Bonding agents should not be used over water soluble paints, glues, or wallpaper [14].

Aggregates.   Plaster aggregates are identified as sand aggregates and lightweight aggregates. Aggregates should be well graded, clean, inert material, free from excessive amounts of organic impurities and water-soluble impurities. Sieve gradations should conform to standard grading criteria as defined by the American Society for Testing and Materials.

Sand.   Sand is classified as natural or as manufactured sand by ASTM C 35. This standard defines *natural sand* as "the fine granular material resulting from the natural disintegration of rock or from the crushing of friable sandstone"; and *manufactured sand* as "the fine material resulting from the crushing and classification by screening, or otherwise, of rock, gravel, or blast furnace slag." Sand for gypsum base plaster, gypsum-lime plaster, and portland cement plaster must pass a #4 sieve, with 90 to 100% retained on a #100 sieve. Up to 5% of the sand should be retained on the #8 sieve, while proportional amounts of graded sand must be retained on the #16, #30, and #50 sieves.

Lightweight Aggregates.   Perlite or vermiculite aggregates improve the plaster's fire-resistive and heat-insulating properties.

Natural *perlite* is a siliceous rock of volcanic origin ranging from light gray to glossy black. When heated, its volume increases up to twentyfold, resulting in a snow-white product. Heated above 1600°F, combined water and occluded gases in the crude crushed perlite explode, forming numerous minute glassy air bubbles. This cellular structure gives the processed perlite its extraordinary insulating properties, unusual light weight, and exceptional fire resistivity.

Perlite, a form of natural glass, is chemically inert; thus, it does not combine chemically with plasters or cements. This inert aggregate does not accelerate or retard the plaster setting.

In addition to its high insulation qualities, perlite or vermiculite as an ingredient of plaster results in a plaster with marked acoustic absorptive ability. As an aggregate for plaster, these lightweight materials must meet grading requirements—that is, all perlite particles must pass a #8 sieve with only 5% of the aggregate retained thereon. At least 88% of the perlite must pass the #100 sieve [ASTM C 35].

High insulating values and a low unit weight of about 7½ to 15 lbs. per cu. ft. make perlite ideally suited as an aggregate for plasters that are used for fire protection.

Vermiculite, which is a micaceous mineral, is greatly exploded into a fan-shapped aggregate when subjected to temperatures of about 2400°F.

Vermiculite, which weighs from 6 to 10 lbs. per cu. ft., exhibits the same above mentioned physical properties of perlite. Grading requirements are similar, although slightly coarser grading is permitted.

Water.   Although potable water is usually satisfactory, mixing water should not contain oils, acids, minerals, or organic substances in amounts sufficient to accelerate or retard plaster setting. Moreover, water in which equipment has been cleaned should not be used for mixing.

70. PLASTER BASES

Classification.

Plaster is applied to properly prepared bases such as masonry, monolithic concrete, or lath bases. The *masonry bases* include brick, stone, hollow clay tile, concrete block, and gypsum block. The *lath bases* which are designed to receive plaster include metal lath and gypsum lath. Gypsum plaster can be applied to any of these, but lime or portland cement plaster should not be applied to untreated gypsum block or gypsum lath.

Masonry. A base coat on masonry is held in place by *bond*. To insure adequate bond, the surface of the masonry must be rough and free from glaze, oil, dirt, efflorescence, or any other foreign material that would prevent the development of a strong bond. Wire brushing or chipping may be necessary to roughen monolithic concrete surfaces because dressed lumber, metal, or plywood oiled forms and well-vibrated concrete tend to produce smooth surfaces that inhibit adequate bond.

The first coat of gypsum plaster applied to untreated monolithic concrete or concrete block should consist of gypsum bond plaster. If a gypsum bond plaster is not used, a liquid bonding agent should be employed. Several proprietary solutions are available that will prepare the monolithic concrete or concrete block surface. Lime plaster or stucco should not be applied to untreated monolithic concrete or concrete block.

Except in arid climates, plaster should not be applied directly to the interior surfaces of exterior masonry walls but should be furred as described below.

Metal Lath. *Metal lath* is an expanded or perforated sheet metal material into which a base coat of plaster is "keyed." Metal lath, which is attached to a supporting structure of solid material or a frame of studs, is primarily intended for interior plaster applications free from excessive moisture. Various types of metal lath are manufactured from zinc alloy or galvanized strip steel or from copper bearing strip steel that has been coated with a rust-inhibiting paint after fabrication. The basic types ordinarily used in plastering are *diamond* mesh (small) or flat expanded metal lath, $\frac{1}{8}$ in. *flat rib* lath, $\frac{3}{8}$ in. *rib* lath, and $\frac{3}{4}$ in. *rib* lath, *sheet* lath, and *large diamond mesh*. Metal lath sheets are generally 27 in. wide and 8 ft. long, with the ribs, if any, running longitudinally; however, $\frac{3}{4}$ in. rib lath is made in 24 and 29 in. widths and in 8, 10, and 12 ft. lengths.

In manufacturing *expanded metal lath* types, 8 in. wide steel strip is cut with sized and spaced longitudinal slits. The strip is then expanded laterally by pulling, thus forming sheets 27 in. wide. Sheet lath is stamped and perforated as full-sized sheets.

Expanded lath is either flat or self-furring. Rib lath does not require a furring system. *Self-furring* expanded lath has spaced depressions or crimps that resemble a dimple. This presents raised points on the lath side that is placed adjacent to the supporting surface. This results in an airspace or furred space of about ¼ in.

Metal lath is usually packaged in bundles of 10 sheets to each package.

Wire Lath.   *Wire lath* is a fabrication of wire woven or electrically welded into a fabric, screen, or mesh. Wire lath generally backs exterior plaster or interior applications exposed to excessive moisture. Usually the wire is copper-bearing, cold drawn steel that has been galvanized or painted with a rust-inhibiting paint. Wire lath is made flat or self-furring; or it may be unbacked or backed with a water absorbent kraft building paper. If paperbacked, wire lath is usually self-furring.

Woven wire fabric lath is a galvanized, woven, hexagonal mesh. It resembles "chicken wire," and is sometimes so called. It is produced in 1 in. hexes with 18 gage wire, 1½ in. holes of 17 gage wire, and in 2¼ in. hexagonals of 16 gage wire. It is marketed in rolls of 36, 48, and 60 in. widths. This wire lath is frequently used as a stucco mesh.

Electrically *welded wire fabric* (plain), which is composed of galvanized wire of not less than 16 gage, is made in rectangular or square mesh of not less than 2 in. All wires of this mesh must be welded at each intersection. This lath may be flat or self-furring.

Paperbacked welded wire fabric lath, an excellent stucco lath, is a self-furring type similar to plain welded wire fabric. However, the paper backing is sandwiched between the vertical wires and the horizontal strands. Horizontal slots in the paper permit the vertical wires to be welded to the horizontal wire. Frequently, additional horizontal reinforcement rods give added support and rigidity. The paper backing, a waterproof, absorbent kraft paper, is held securely in this arrangement. This type of lath is made into sheets 28 in. wide and 50 and 98 in. long.

Wire cloth lath consists of a mesh of about ⅜ in. square made of not less than 19 gage wire and galvanized or finished with a rust-inhibiting paint.

Metal Lath Installation.   The maximum permissible spacings of vertical and horizontal supports for metal lath and wire lath vary from

12 to 24 in., depending upon the type and weight of the lath. For wire fabric, this spacing is always 16 in.

The various types of metal plaster bases are anchored to wood supports with roofing nails, common nails, or staples—the choice of fastener varying with the type and the direction of the supports. They are fastened by wire, not smaller than 18 gage, to open-web steel joists, to steel studs, and to concrete ceilings with wires partially embedded. The anchors should be not more than 6 in. apart and side laps not less than 1 in. between supports for metal lath, wire lath, or wire fabric. They are stitch-wired together at intervals not more than 9 in. The long dimension of metal lath sheets is oriented perpendicular to the supports; however, wire lath and fabric may be placed with the long dimension parallel to supports.

Metal lath are commonly nailed to both sides of wood studs to form partitions (Figure 12.1(a)). Special prefabricated metal studs 2 to 6 in. deep are used in a similar manner for nonbearing partitions from 4 to 7½ in. thick (Figure 12.1(b)). Lath are wired to the studs that are firmly attached to steel *runners*, which in turn are anchored to the floor and ceiling or end-anchored. The studs, spaced from 16 to 24 in. as determined by the type and weight, can be used for partitions 26 ft. high or more. Two ¾-in. channels braced together may be used instead of a prefabricated stud as shown in Figure 12.1(b).

Nonbearing solid plaster partitions are constructed by anchoring light hot- or cold-rolled steel studs to the floor and ceiling, placing wiring metal lath to one side of the studs, and plastering both sides of the metal lath to embed completely the studs by backplastering (Figure 12.1(c)). The studs vary in depth from ¾ to 1½ in., depending upon the partition height; and spacing varies from 12 to 24 in., depending upon the type and weight of lath used. The thicknesses vary from 2 to 3 in., depending upon the height and the ratio of the height to the unsupported length of the wall. For heights exceeding 20 ft., horizontal channel or ¼ in. *rod stiffeners*, spaced not more than 6 ft. vertically, are wired to the channel side of the lath.

Solid *studless nonbearing* partitions are constructed of diamond mesh or of ⅜-in. rib metal lath with the long dimension vertical. They are secured at the ends to anchored floor and ceiling horizontal channels called *runners* (Figure 12.1(d)). The sheets are edge-lapped 1 in. and wired together at least every 9 in. of height. Temporary bracing is required until the scratch and brown coats are applied to the unbraced side. The total thickness is usually 2 in., the optimum height 8½ ft. or less, and the maximum permissible height 10 ft.

Furred and suspended ceilings are constructed with metal lath, supported so that the permissible spacing of supports is not exceeded

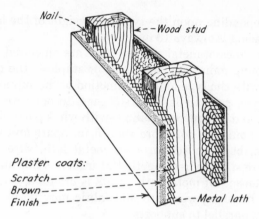

(a) Metal lath on wood studs

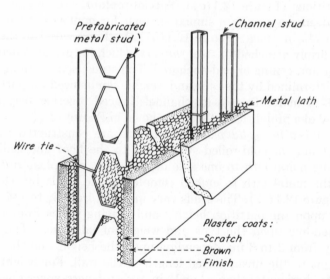

(b) Metal lath and plaster hollow partition with metal studs

FIGURE 12.1   Types of metal lath partitions. Metal Lath Manufacturers Association.

for a given type and weight of lath. If available structural members are not spaced to exceed the permissible spacing or permissible span of the lath, no further furring is required.

Sometimes *cross furring* with the appropriate spacing is provided perpendicular to its supports. Rods ¼ to ¾ in. or 1½-in. channels are used; the choice depends upon the distance between supports.

When the cross furring is supported directly by attaching to the underside of the structural members, such as joists, the ceiling is a *furred ceiling*.

When the supports described are unfeasible, when a flat ceiling surface is desired, or when this method of support would not provide such a surface, a suspended ceiling may be used. For a *suspended ceiling*, the cross furring is attached to main runners, which are usually ¾ in. or 1½ in. cold-rolled steel channels, the size depending upon the spacing of their support. The *main runners* are perpendicular to the furring and suspended from the floor system above by heavy galvanized wires, light steel rods, or light strap iron hangers. The sizes and spacings of the hangers depend upon the ceiling area each supports. All attachments for metal lath, metal furring, and metal runners are made with annealed galvanized steel wire or special metal clips.

Furred or suspended ceilings conceal the structural members and provide flat unobstructed ceilings, or they are used to increase the fire resistance of floor and ceiling construction. For the latter purpose, plaster with vermiculite or perlite aggregate is more effective than plaster with sand aggregate:

A stucco surface reinforced with expanded metal, sheet-metal lath, or wire fabric and applied over waterproof building paper or felt serves as a thin concrete slab. This is especially the case when stucco cannot be bonded adequately to masonry, when the walls are wood sheathed, or when the structure is of steel frame. The paper serves as a backing against which the stucco is forced to provide full embedment of the metal reinforcement. Supports to which the metal reinforcement is attached must not be spaced more than 16 in. The reinforcement is furred out from the backing at least ¼ in. by a spacing device so that the lath will be completely surrounded with stucco.

Gypsum Lath. Gypsum lath, sometimes referred to as *gyplath* or *plasterboard*, consists of a sheet of gypsum plaster with the surfaces covered with heavy paper. This is called *plain gypsum lath*. The sheets are ⅜ in. or ½ in. thick, 16 or 24 in. wide, and 48 in. long, although sheets are made as long as 12 ft. and are so identified as *long length gypsum lath*.

Similar to plain gypsum lath is *perforated gypsum lath*, which has holes at least ¾ in. in diameter and spaced about 4 in. apart both ways. Plain gypsum lath that has a sheet of aluminum foil on the back is called *insulating gypsum lath*. This foil serves as a vapor barrier and as a reflective heat insulator. Gypsum lath that is classified as fire retardant by ASTM C 37 is identified as *Type "X" gypsum lath.*

In combination, an assembly of ½ in. finish plaster over Type "X" gypsum lath will have a fire rating. The plaster is held to the gypsum lath by adhesive bond, and supplemented by keys if perforated lath is used.

Generally, gypsum lath is attached to supports spaced not more than 16 in.; however, ½ in. plain lath may be used on supports spaced at 24 in. centers. The long dimension is usually placed horizontally. Usually the vertical joints are staggered, but they may be continuous if covered with 3 in. strips of expanded metal lath. They are fastened to horizontal or vertical metal supports with special metal clips.

Gypsum lath is used for hollow partitions with metal studs and for solid partitions with or without metal studs for constructing non-bearing partitions similar to those constructed with metal lath. It is also used for furred and suspended ceilings in a method similar to that described for metal lath.

For solid studless partitions, ½ in. thick gypsum lath 16 to 24 in. wide, with a length equal to the ceiling height, is placed vertically. Fastened top and bottom similar to the manner described for the corresponding metal lath partition, it is braced until the brown coat has been applied to one side. The usual finished thickness is 2 in.

Grounds and Other Accessories. The required uniform thickness of a plaster coat is of paramount importance in attaining first-class workmanship. The thickness is controlled by installing various guides called *grounds*, corner beads, and casing beads on the perimeters of openings and at the tops, corners, and bases of walls. Cornices and baseboards conceal grounds at the top and bottom. The accessories are usually made from solid sheet metal that has been stamped and formed, from expanded metal, or from welded or woven wire. Several of the beads or screeds that are made from sheet metal have attached reinforcement strips of expanded diamond mesh. These reinforcements are called *expanded flanges* or *expanded wings*. Accessories are galvanized or painted with a rust-inhibiting finish.

In locations where guides cannot be installed, narrow strips of plaster are accurately built up to a thickness that will be flush with the brown coat. These plaster screeds become integral with the brown coat. The screeds, which are often placed horizontally across walls and across and around ceilings, assist in forming true surfaces for the brown coat.

Grounds, screeds, and other accessories can be set to provide a minimum thickness of ⅝ in. over metal lath, wire lath, and masonry and concrete walls; ¼ in. over gypsum lath; and from ⅛ to ⅜ in. over monolithic concrete ceilings.

Stucco with metal reinforcement should be at least ⅞ in. thick measured from the face of the reinforcement. Stucco on unit masonry should not be less than ¾ in. thick.

Walls that are backed with gypsum lath require reinforcement for the plaster at all vertical and horizontal interior corners and on perimeters of openings. Ell-shaped reinforcement used at interior corners is known as *cornerite*. There are two types—a smooth-edge expanded diamond mesh and a woven or welded wire fabric.

Strip reinforcing, which is nailed to the perimeters of openings, is a flat, smooth-edged, expanded diamond mesh 4 or 6 in. wide.

At projecting (exterior) corners, plaster is reinforced and protected with *corner beads*. The two types which also serve as plastering guides are *small nose* corner beads and *bull nose* corner beads. Both types are made of solid sheet metal or of solid sheet metal with attached expanded flanges.

At doors and other edges where the plaster is terminated, grounds or *metal plaster stops* that are also called *casing beads* are installed. They are sheet metal, or sheet metal with expanded flanges. Several profiles are made—square, ¼ round, and beveled.

Metal plaster stops that are called *base screeds* serve as dividers between plaster and a dissimilar material such as stucco or terrazzo.

Drip screeds are used at plaster eaves or rakes or at water tables. They provide an even termination of the plaster with provisions for a water drip.

There are several other sheet metal plastering accessories such as *picture moldings*, *partition terminals*, and *control expansion joints* that minimize plaster cracking.

Furring. In general, *furring* is a light, nonstructural framework of strips or metal channels that is applied to a supporting surface such as walls, studs, columns, ceilings, joists, or beams. *Wood furring* of 1 × and 2 × size is usually nailed to the supporting surface. *Cold rolled steel channels*, ¾, 1½, and 2 in. deep, are supplied in 16 and 20 ft. lengths.

Furring must provide a true, level, and even horizontal and plumb surface. It enables closer appropriately spaced supports for lath that will not span between widely spaced structural supports and encloses areas called *furred down* spaces that form a ceiling to cover unsightly pipes, conduits, or duct work. Furring provides a nonstructural frame necessary for framing various built-up shapes such as arcades, vaults, domes, and decorative features. It also creates a dead air space between the structural support and the base plaster. The air space formed by the wall furring has four functions:

(1) It intercepts moisture that penetrates an exterior wall.

(2) It acts as an insulator and prevents condensation from forming on a cold interior wall surface.

(3) It reduces heat transmission through the wall. The insulating properties of the air space realize a conservation of fuel in cold weather and keep the building cooler in hot weather.

(4) It permits a space for the base plaster to penetrate a perforated or expanded lath. This results in a more positive bond or "keying."

There are several methods of furring. One consists of a direct application of furring strips vertically to a wall surface or to the bottom of beams and joists. Another employs specially designed blocks of structural clay or gypsum tile that are veneered to the masonry wall. They are held to the masonry walls by any one of a variety of masonry ties, or by hooking the tiles over nail heads of nails that have been driven into the mortar joints. The blocks are pointed with mortar, then coated with a bonding agent if necessary, then plastered. Still another method of furring is used for ceilings. The furring strips may be suspended by soft, annealed wires that are called *hangers* or *tie wires*.

71. PLASTER MIXING AND APPLICATION

Coats.   Plaster and stucco applications are classified, according to the number of coats, into *two-coat* work and *three-coat* work.

The last or final coat is called the *finish coat*. The coat, or combination of coats, applied before the finish coat is termed the *base coat*. The coat directly beneath the finish is known as the *brown coat*. In two-coat work, the brown coat is also the base coat and is applied directly to the plaster base. In three-coat work, a *scratch coat* is applied to the plaster base before the brown coat is applied. Its outer surface is scratched to improve the bond to the coat that follows. This operation gives the coat its name. Three-coat work is required over metal lath.

In two-coat work, the base coat may consist of a scratch coat, which is not scratched, and a brown coat placed before the scratch coat has set. This is called the *double-up method*, or *double-up work*, and is widely used in applying plaster to unit masonry and gypsum lath. It is also called *laid-off* and *laid-on work*.

Application.   Plaster is applied by hand or machine. If applied by hand, the mortar is dumped on a mortar board. The hand tools used by the plasterer are shown in Figure 12.2(a). Mortar is shoved from the board onto the wood or metal hawk with a metal trowel, transferred from the hawk to the wall, and spread with the trowel.

In applying the scratch coat, exerted pressure forces some of the mortar through the open spaces of the lath, forming *keys* which clinch the mortar to the lath. The hair or vegetable fiber included in the scratch coat permits only enough mortar to enmesh onto the lath surface, forming good keys. Fiber also adds strength to the key during the early stages of setting. This coat is about ¼ in. thick and is not accurately finished. Its surface is *scratched* with appropriate tools while it is still soft to provide a better bond for the coats that follow.

Machine application is by pneumatic plastering machines, which blow the plaster into position.

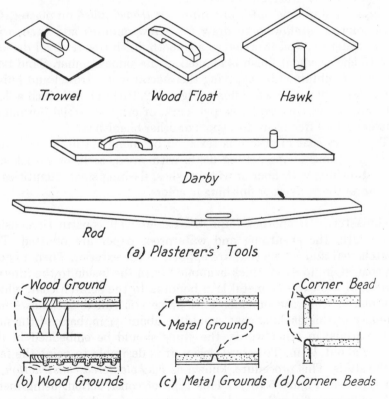

Trowel   *Wood Float*   *Hawk*

Darby

Rod

(a) Plasterers' Tools

Wood Ground   *Corner Bead*

Metal Ground

(b) Wood Grounds   *(c) Metal Grounds (d) Corner Beads*

FIGURE 12.2   Plasterers' tools, grounds, and corner beads.

The brown coat is applied directly to bases of unit masonry or gypsum lath or to a scratch coat over metal lath after the scratch coat has set firm and hard. It is troweled on and brought to a true surface with a *darby* or *rod* (Figure 12.2(*a*)), the ends of which bear on grounds, screed, or other guides. Given a partial rotary motion, the tools leave the brown coat surface rough enough to receive the finish coat.

Gypsum lath are not wet down. Unit masonry surfaces on which suction must be reduced before applying the brown coat directly to them should be properly wet down. Irregularities in stone walls should be filled with mortar before applying the brown coat. Hopefully, one would not find it necessary or be so insensitive to plaster over stonework that has an architectural quality; but if it must be done, that is the way to do it. The *finish coat* is applied to a partially dry brown coat or to a thoroughly dry one which has been evenly but not excessively wetted by brushing or spraying.

For a gypsum-lime putty or *white-coat trowel finish* on plaster, the finish mortar is allowed to draw for a few minutes and then, with ample watering, well troweled to a smooth finish free from all defects. The Keene's cement finish is applied in the same manner. Sand float finishes are obtained by applying the mortar wih a trowel and finishing by floating with a wood float (Figure 12.2(*a*)) or with such a float with its surface covered by carpet, cork, or other material, forming a plaster surface free from slick spots or other blemishes.

The finish coat of stucco is textured by manipulating trowels and tools of various shapes, using floats with wood, cork, or carpet surfaces, brushing with fiber or wire brushes, dashing small quantities of mortar on the surface, or finishing in colors.

Backplaster. In another form of exterior wall construction, using metal lath, the sheathing and waterproof paper are omitted. The scratch coat and brown coat are applied to the exterior. Then a backing coat $\frac{5}{8}$ in. to $\frac{3}{4}$ in. thick is applied from the inside to the interior surface of the exterior metal lath bonding to the scratch coat, which was placed from the outside. This process is called *backplastering*. The face of this backing coat should be about $\frac{1}{4}$ in. back of the face of the studs; in other words, the studs should be embedded in the plaster about $\frac{1}{4}$ in. To retard decay, it is desirable to paint the face of the studs. This procedure, known as *backplastered construction*, is illustrated in Figure 5.18(*f*). This type of construction, using metal lath, will be sufficiently rigid if the corners of each wall are braced diagonally by 1 × 4 in. boards let-in to the studs on the inner side and securely nailed.

Ornamental Plastering. Ornamental interior plastering includes moldings, cornices, panels, decorative ceilings, or rosettes made of plaster or similar material.

The base for moldings or small projections is built up solid with the same material as the brown coat into approximately the finished shape, assuring that there is adequate clearance for the finish coat. The finish coat, which is composed of lime putty and gauging plaster, is applied to this base and is cut to the desired profile by means of a sheet metal template operating as a guide. The finish coat cannot be cut with sharp outlines in one operation but must be gone over several times. The low places are filled in with mortar. For larger moldings, the base is built up of metal lath and braced.

Moldings and similar ornamental plastering are placed before the finish coat on the remainder of the walls and ceiling because the guides cannot be placed on the finish coat without marring. Parts of moldings, such as internal and external miters which cannot be run with templates, are hand-formed or cast and placed in position before the moldings are run. Dentils and brackets may be cast separately and placed after moldings are run. Ornaments that cannot be run are cast in gelatin molds, or they may be purchased from firms that specialize in this kind of work.

Acoustic Plaster. Acoustic plasters are special plasters with porous aggregates troweled on as a final coat to improve the sound-absorbing properties of walls and ceilings. See Article 97.

Fireproofing. Fire-resistive plaster applied on metal lath over structural steel members to prolong their resistance to high temperatures is made with perlite or vermiculite instead of sand.

Proportions. Usually, the proportions for plaster materials are measured by parts of volume—the cubic foot. However, fiber, if added, is measured by weight. For the base coats of gypsum plaster, the proportions of sand, vermiculite, or perlite to a 100-lb. bag of gypsum neat plaster should not exceed those given in Table 12-1 [2].

Fibered plaster is sometimes used for the scratch coat over metal lath. The proportions for lime-cement stucco on metal lath are:

Scratch coat. Stiff lime putty 1 part, portland cement 1 part, aggregate 6 parts, with 6 lbs. of fiber per cu. yd.;

TABLE 12-1
Proportions for Plaster

	Sand, lb. Damp, Loose	Vermiculite or Perlite, cu. ft.
Two-coat work (double-up method)		
Over gypsum lath	150	2½
Over unit masonry	300	3
Three-coat work		
Scratch coat over lath	200	2
Over masonry	300	3
All brown coats	300	3

Brown coat. Stiff lime putty 2 parts, portland cement 1 part, aggregate 9 parts, with 3 lbs. of hair per. cu yd.

The proportions for lime-cement stucco on unit masonry except gypsum block which do not provide a suitable base for stucco are: base coat of two-coat double-up work, stiff lime putty 2 parts, portland cement 1 part, aggregate 9 parts, with 6 lbs. of fiber per cu. yd.

For portland-cement plaster and stucco, each coat should consist of 1 part of portland cement to not less than 3 but not more than 5 parts of damp, loose aggregate.

Lime, finely divided clays, and asbestos flour are added to increase the workability of portland-cement plaster and stucco. White portland cement is used in the finish coat for white and the lighter-colored stucco. Colors are obtained by using mineral pigment. .

Gypsum ready-mixed plaster is furnished with the sand aggregate added at the mill. No further additions, except water, are made on the job. *Gypsum bond* plaster is applied as the first coat over monolithic concrete surfaces without the addition of aggregate on the job. *Gypsum wood-fibered* plaster, also used in this same manner, is applied over all types of lath. When applied to unit masonry, equal parts by weight of damp sand and wood-fibered plaster may be used.

The proportions of lime plaster are as follows:

(1) For the scratch coat required only on metal lath—1 cu. ft. lime putty, 2 cu. ft. sand, and 7½ lbs. hair or vegetable fiber per cu. yd. of mortar.

(2) For all brown coats—1 cu. ft. lime putty to 3 cu. ft. sand and 3½ lbs. hair or fiber per cu. yd. or mortar.

Finish Coats. The final or finish coat of plaster is extremely important. It must be attractive, the desired texture, hard, and durable.
Many types of finish can be used; the most common is known as the
white-coat finish. The *sand float* or *sand finish* is also widely used,
and the lime-*Keene's cement finish* provides an especially hard surface that is impervious to water. The proportions for these finishes
follow:

(1) For application over either a gypsum or lime plaster brown
coat, a *white-coat finish* of 1 volume of gypsum gauging plaster, which
is calcined gypsum, is mixed with not more than 3 volumes of lime
putty.

(2) *Gypsum-sand float finish*—1 part gypsum neat, unfibered
plaster to not more than 2 parts of fine sand by weight.

(3) *Lime-sand float finish*—3 parts lime putty, 3 parts fine sand,
and 1 part gauging plaster by weight.

(4) *Keene's cement finish*—on 100 lb. bag of Keene's cement to
$1\frac{1}{2}$ cu. ft. lime putty, for medium hard finish, or $\frac{5}{8}$ cu. ft. of lime
putty for hard finish.

(5) *Lime-cement stucco*—the proportions for a smooth trowel finish
are: stiff putty 5 cu. ft. and portland cement 1 cu. ft.; for sand finish
—stiff lime putty 5 cu. ft., portland cement 1 cu. ft., and silica plastering sand or marble dust 500 lbs.

Prepared Finishes. Several finishes are prepared at the mill and
made ready for use on the job by adding water.

Water Proportions. Water is added to the specified mixtures in the
amounts required to make them workable. Too little water makes
the plaster or stucco difficult to spread. If too much water is added,
the mixture slides off the trowel or off the wall surface. If mixtures
are over-sanded and the excess water is added to make them workable, the strength, hardness, and durability are reduced; and the
service they will render is impaired. By avoiding over-sanding, the
need to add excess water is eliminated.

Mixing. The ingredients of which the mortar for plaster and stucco
is made are mixed in batches, or *batched,* by a mortar mixer. For
lime plaster, hydrated lime, which is a dry powder, is usually used.
Water is mixed with the lime on the job to form lime putty. The
manufacturer's instructions should be followed in preparing this
putty. Sometimes it may be used immediately, and sometimes a
soaking period of approximately 24 hours is required.

When gypsum is present in the mortar, all tools and equipment should be thoroughly cleaned after the mixing of each batch. Cleaning is important because the presence of partially set gypsum may cause plaster to set too rapidly.

REFERENCES AND RECOMMENDED READING

1. *Standard Definitions of Terms Relating to Gypsum*, ASTM Designation C11-58.
2. *American Standard Specifications for Gypsum Plastering and Interior Lathing and Framing*, American Standards Association, 1955.
3. *Specifications for Lime and Its Uses in Plastering, Stucco, Unit Masonry and Concrete*, National Lime Association, 1945.
4. *Plasterers Manual for Applying Portland Cement, Stucco and Plaster*, Portland Cement Association.
5. *Gypsum Lathing and Plastering*, Gypsum Association.
6. *Metal Lath Technical Bulletins*, Metal Lath Manufacturers Association.
7. *Standard Specifications for Portland Cement Stucco and Portland Cement Plastering*, American Standards Association, 1946.
8. *Standard Specifications for Lime-Cement Stucco*, American Standards Association, 1960.
9. *ASTM 1971 Annual Book of Standards, Part 9*, American Society for Testing and Materials, Philadelphia, Pennsylvania.
10. *Plasterer's Manual*, Portland Cement Association, Skokie, Illinois, 1964.
11. *Sweet's Catalog*, 1971 Edition, F. W. Dodge Corp., New York, New York.
12. John R. Diehl, *Manual of Lathing and Plastering*, National Bureau of Lathing and Plastering, MAC Publishing Association, New York, New York, 1960.
13. *Lathing and Plastering Reference Specifications*, Third Edition, California Lathing and Plastering Contractors Association, Inc., Los Angeles, California, 1965.
14. *Sweet's Catalog*, 1968 Edition, F. W. Dodge Corp., New York, New York.
15. *Uniform Building Code*, 1970 Edition, Vol. I, International Conference of Building Officials, Whittier, California.
16. Charles G. Ramsey and Harold R. Sleeper, *Architectural Graphics Standards*, Sixth Edition, John Wiley and Sons, 1970.

13

ROOF DECKS, ROOFING, AND SIDING

Roofs of buildings are divided into various types, depending upon the shape.

Flat roofs (Figure 13.1(*a*)) are extensively used on all kinds of buildings. They are sloped from ⅛ in. to 2 in. vertical to 12 in. horizontal to insure proper drainage, although some roofs are designed without any slope whatever. Even though watertight roofs can be secure without any slope, a *dead-level* roof is not recommended.

Roofs which slope in one direction only (*b*) are called *shed roofs*. This type of roof is used on entire buildings or in connection with other types shown in (*c*) to form a lean-to. A *butterfly roof* slopes in two directions: two shed roofs meet at their low *eaves*. *Gable roofs* also slope in two directions (*d*); two shed roofs meet to form a *ridge*. This type of roof, widely used on residences, slopes often as low as 4 in. vertical to 12 in. horizontal and is occasionally as steep as 20 in. vertical to 12 in. horizontal. The most common slopes are between 4 in. vertical to 12 in. horizontal and 12 in. vertical to 12 in. horizontal.

Hip roofs slope in four directions (*e*). This widely used roof has the same slopes as a gable roof.

Gambrel roofs slope in two directions, but there is a break in the slope on each side (*f*). The gambrel roof is used for residences because the space under the roof can be used efficiently, especially when a long shed dormer (Figure 13.2(*a*)) is used. *Mansard roofs* slope in four directions but have a deck at the top Figure 13.1 (*h*).

The various types of roof *dormers* are shown in Figure 13.2(*a*).

The *saw-tooth roofs* shown in Figure 13.2(*b*) and (*c*) are used extensively on industrial buildings because of the opportunities they offer in light and ventilation. The steep face of (*b*) and the vertical face of (*c*) are mostly glass and usually faced north. Northern light (in the northern hemisphere) is more constant throughout the day

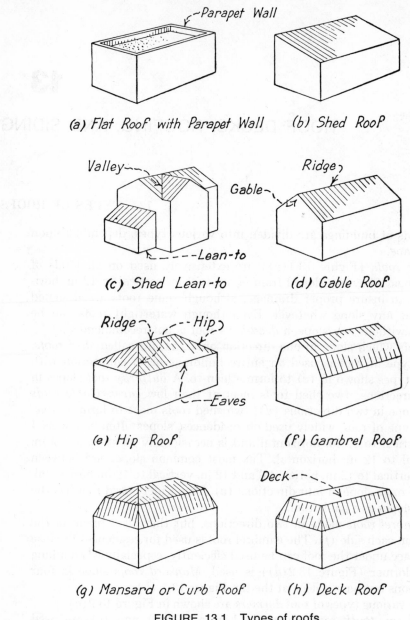

(a) Flat Roof with Parapet Wall   *(b) Shed Roof*

(c) Shed Lean-to   *(d) Gable Roof*

(e) Hip Roof   *(f) Gambrel Roof*

(g) Mansard or Curb Roof   *(h) Deck Roof*

FIGURE 13.1 Types of roofs.

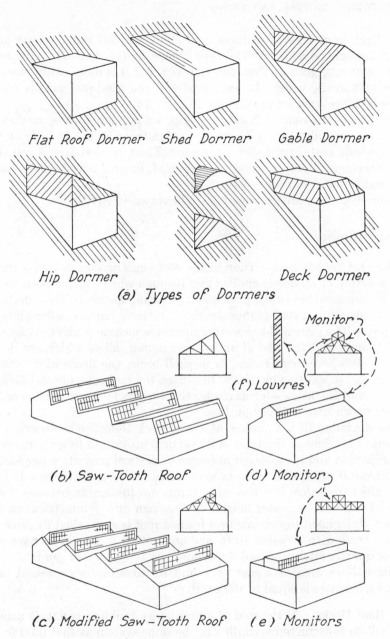

Flat Roof Dormer   Shed Dormer   Gable Dormer

Hip Dormer   Deck Dormer

(a) Types of Dormers

Monitor

(f) Louvres

(b) Saw-Tooth Roof   (d) Monitor

(c) Modified Saw-Tooth Roof   (e) Monitors

FIGURE 13.2 Types of dormers and roofs.

561

than light from other directions, and the glare and radiant heat of direct sunlight is avoided. The type shown in (c) does not have as large an area of glass as that shown in (b), but it is more easily made watertight in the valley. At least a part of the windows, usually the upper half, are open for ventilation.

The parts of a roof such as ridge, hip, valley, gable, rake, cricket, and eaves are indicated in Figure 13.1. Monitors (Figure 13.2(d) and (e)) provide better light and ventilation. Their vertical face is called the *clerestory*. If only ventilation is desired, louvres may be used for ventilation in the clerestory (d).

Thin-shell roofs are described and illustrated in Article 56.

73. ROOF DECKS

The *roof deck* is that portion of the roof construction to which the roof covering or roofing is applied and through which the loads on the roof are transmitted to the principal supporting members. Roof decks include sheathing, roof planks or slabs, rafters, purlins, subpurlins, and joists. The principal supporting members include girders, trusses, rigid frames, and the ribs of arches and domes, all of which are described elsewhere. Sometimes, as in shell roofs, the decks also serve as principal supporting members. In others, the roof covering and deck are one and the same—for example, the various forms of corrugated roofing which span between purlins.

The deviation of the surface of a roof deck from the horizontal is measured in terms of the ratio of its vertical projection to its horizontal projection and is expressed in inches of vertical projection per foot of horizontal projection. This ratio is called the *slope* or *incline*. It is also called the *pitch*, but the other terms are preferable because the ratio of the *rise*, or center height, to the span of a symmetrical roof truss with inclined top chords or a framed roof is also called its *pitch*. Such a truss with a rise of 10 ft. and span of 40 ft. is said to have a $^{10}\!/_{40}$ or quarter pitch; the roof slope, however, is $^{10}\!/_{20}$ or 6 in. per ft.

The surface area of a roof deck and its covering is expressed in squares; a square is equal to 100 sq. ft.

Flat Roof Decks.   If the roof of a multistory building is flat, it usually will be constructed according to the same system as that used for the floors. When all of the factors involved are considered, such a procedure is usually cheaper and more satisfactory than any other system of equal quality. Various types of floor construction are discussed in Article 58. Some systems will result in level decks. Roof decks which do not slope to the drains are objectionable because shallow pools of water, called *birdbaths*, remain in slight depressions and deflections of the deck surface. A further objection to dead-level roof decks is that the roofing surface will have early deterioration due to standing water. However, slopes of at least ⅛ in. to the foot are provided by using shaped wood members or, more often, fills of lean, lightweight concrete placed on top of a flat concrete deck.

For single-story buildings, flat roof systems may be of the same types as some of the floor systems described in Article 33. Types which are especially suitable for heavy loads, such as the flat-slab, would not be appropriate for such buildings because roof loads are relatively light.

Decking especially designed for roofs include precast gypsum and concrete planks and wood-fiber cement planks in various forms installed on steel rafters or purlins. Gypsum is nailable; and by using appropriate aggregates, concrete can be made nailable. Cast-in-place or poured gypsum slabs are also used for roof decks but are not extensively used in floor constructions. Lightweight, noncombustible termite-proof *wood-fiber-cement* planks 2, 2½, and 3 in. thick, made of a compressed chemically treated wood fiber and portland cement, are manufactured for roof decking. The standard width is 32 in., with lengths up to 9 ft. Usually supported on bulb-tee subpurlins spaced 2 ft. 9 in., the edges are rabbeted to fit under the bulb projection so as to be anchored when joints are filled with mortar. These planks have good heat-insulation and sound-absorbing properties.

Also designed especially for roof decks are light gage steel-ribbed panels of various types similar to the floor panels illustrated in Figure 9.1 but installed with the flat surface on top and without concrete. The flat surface permits a rigid insulation or roofing with less chances of a surface rupture.

The only types of roofing suitable for flat or gently sloping roofs are the built-up roof and the sheet-metal roof with flat soldered seams. The built-up roofs can be used on all types of flat decks, but sheet metal roofing involves nailing and therefore cannot be used directly on nonnailable concrete.

Sloping Roofs. Three sloping roof decks and their parts are shown in Figure 13.3. The principal supporting members shown in (*b*) and (*c*) are trusses, but the same arrangements can be used with other types. A typical wood frame construction is shown in (*a*). Usually there are no main supports for the rafters between the plate and the hip. In (*b*), timber or steel purlins support the rafters, with 1 in. wood

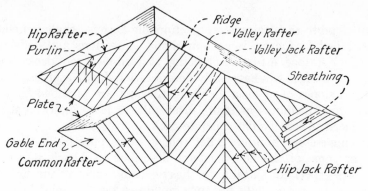

(a) Frame Construction Roof

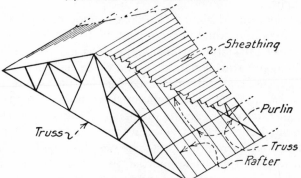

(b) Roof with Trusses, Purlins, Rafters, and Sheathing

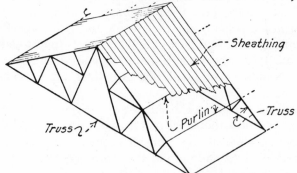

(c) Roof with Trusses, Purlins, and Sheathing

FIGURE 13.3   Roof framing.

sheathing, on light wood rafters. The assembly in (c) does not include rafters, but the sheathing spans between purlins. The construction might consist of wood sheathing, usually 2 in. thick, and timber or steel purlins; precast gypsum; wood-fiber cement or nailable concrete planks and steel or precast concrete purlins; or some form of corrugated sheets spanning between timber or steel purlins while serving as a roof covering. The methods shown in (b) and (c) are suitable for a flat roof, although corrugated sheets would not serve as a roof covering.

The coverings for sloping roofs must be fastened to the decks with mechanical fasteners, a process which usually employs nails if the deck is continuous. For this reason, a type of nailable deck must be selected or wood nailing strips can be installed which will have adequate nail holding ability. Included in such decks are those constructed of wood, gypsum, and nailable concrete. The gypsum and nailable concrete are usually in various forms of precast planks because this procedure is more economical and the installation is faster than casting-in-place.

Cast-in-place decks of ordinary concrete are not nailable; thus, for this form of deck, provision must be made for fastening the roofing to the deck. This objective is accomplished by providing wood nailing strips, called *sleepers* or *nailers*, that are treated with preservatives. For built-up roofs, these nailers are embedded in the concrete at intervals of 3 ft., running parallel with the slope and with their top surfaces flush with the top surface of the concrete. If rigid insulation with a thickness at least equal to that of 1 in. nominal thickness lumber is used, the nailing strips are placed between the sheets of insulation with their tops flush. The usual method for fastening built-up roofing to poured concrete decks with slopes of less than 1 in. per foot is to cement the roofing to the deck with roofing pitch, but this procedure is not suitable for greater slopes because of the tendency of the roofing to creep down the slope.

Heat Insulation. Preferably, the roof or the ceiling of the top story of buildings should be insulated. This reduces the heat losses and condensation on the ceilings and reduces temperatures. Various insulating materials are described in Article 95.

Roof Drainage. Roofs may range from flat with no slope whatever to roofs with a considerable slope. Some engineers and architects advocate the absolutely flat or dead-level roof because of its simplicity of construction, maintaining that standing water for short periods on a roof, owing to slight irregularities, does not harm but may actually protect the roof from the effects of the sun. The type of roofing used on dead-level roofs is a built-up roofing consisting of coal-tar pitch and tarred felt with a wearing surface of gravel or slag.

Gutters and Downspouts.   The slopes required for various types of roofing are discussed in this chapter. The rainwater that falls on a roof may run off and drip from projecting eaves, but usually it is necessary or desirable to collect the water in *gutters* placed along the lower eaves. The water is carried in the gutters to vertical pipes called *downspouts, conductors,* or *leaders.* Flat roofs or other roofs not having projecting eaves are drained by downspouts or conductors placed at joints where the water is carried by the slight slope provided in the roof. The size of the gutters and conductors is determined by the contributing area and by the intensity of rainfall.

Several types of gutters for sloping roofs are shown in Figure 13.4. The *hanging gutter* is the simplest form, although not as attractive as the *crown-mold gutter* or the *wood gutter,* which fit into the design of the cornice. An *O.G. gutter* has an exposed side shaped like an Ogee curve.

Gutters made of metal, copper, aluminum, or galvanized steel are rust resistant. Redwood is used for constructing gutters which have a simple Ogee curve face.

The standing gutter is inconspicuous and easily constructed, but the concealed is quite expensive. Gutters are sometimes called *eave troughs.*

Conductors or downspouts should be provided with strainers at their upper ends (Figure 13.4) so that debris cannot clog them. *Conductor* or *leader heads* are installed as shown in Figure 13.4. It may be desirable to enclose conductors on the interior of a building rather than to expose them on the outside walls because the heat of the building keeps the pipes from freezing. They may be placed in *chases* on the inside of exterior walls, along columns, or in partitions. If placed on the outside of exterior walls, it is desirable to keep them off north walls if at all possible. Steam outlets are sometimes provided in exposed conductors so that, by discharging steam into them, they can be kept from freezing. Cleanouts should be provided so that clogged conductors and the connecting drains can be easily cleaned. Exposed conductors are commonly made of copper and galvanized steel. Copper, cast-iron, and steel pipe are used for concealed or interior conductors or where appearance is not a factor.

Scuppers.   Frequently, flat roofs that have a parapet wall have a sheet metal spillway or channel which penetrates through the parapet

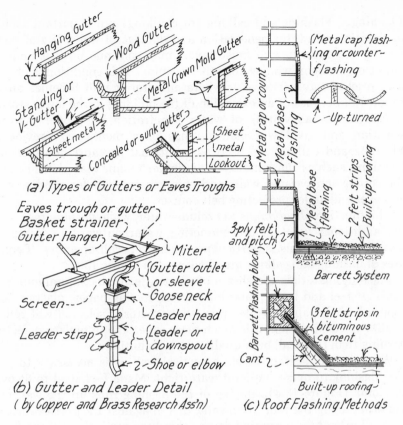

(a) Types of Gutters or Eaves Troughs

(b) Gutter and Leader Detail
(by Copper and Brass Research Ass'n)

(c) Roof Flashing Methods

FIGURE 13.4   Roof drainage.

wall. Known as *scuppers*, these devices are designed to conduct rain-water runoff from the roof through the parapet and discharge into a gutter, a leader, or directly in a free fall. Scuppers should be made of noncorrodible (copper or galvanized) metal of leak-proof construction. Scuppers should be located at the low points of the roof slopes with bottom or *invert* of the scupper at the low point of the roof line. Usually they have a 1 in. vertical projection called a *gravel stop*. All roofs that are equipped with scuppers and roof drains should have overflow drains that are located remote from the scuppers and drains. Because scuppers and drains are subject to clogging by leaves and other floating debris, overflows should be located with their inverts slightly above the inverts of the scuppers. This will deter the clogging of the overflows.

Flashing.   Flashing and calking (or caulking) are important factors in reducing rainwater penetration and damage of walls. Water penetrating the outer face of finish layers may continue until it finds an outlet over a door, window, or other opening. This condition is critical over oriel or bay windows or where an exterior wall becomes an interior wall because of an extension of the building in the lower stories. Subsidence or alleviating of water penetration by proper material selection and good workmanship has been mentioned previously. Flashing and calking are used as additional precautions at critical locations, such as the junction of dissimilar roofing materials or under the copings on parapet walls and chimney tops, under window sills, and over and under projecting belt courses.

The function of *calking* is to exclude water, while *flashing* may have the additional function of conducting water outward to lessen its harmful effects. This is particularly true over wall openings. Typical uses of flashing and calking are illustrated in Figure 13.4.

Flashing materials are sheets of copper, lead, zinc, aluminum, galvanized steel and terne plate; bituminous roofing paper; plastics; and copper-backed paper. While appropriate material selected will reflect quality of construction, it is much more expensive to replace defective flashing than to install high-quality flashing initially.

Where a roof surface meets a vertical wall, it is necessary to provide flashing to make the joint watertight. Made L-shaped to fit over the joint (Figure 13.4), one leg of the flashing runs up the wall and the other along the roof. Rainwater driven against the vertical face of the wall is kept from running down behind the vertical leg of the flashing by *counterflashing* or *cap flashing*, also made in the form of an L. In counterflashing, the L is inverted; the horizontal leg is built or fitted into a joint or reglet (Figure 13.4), and the vertical leg fits over the flashing. A *reglet* is a continuous formed joint in masonry or concrete. Reglets may be formed of sheet metal and filled with caulking. Counterflashing inserted in the reglet is secured by friction. Counterflashing allows expansion of the metal or differential settlement and movement of the structure without rupturing the flashing from its anchorage. Built-up roofing is often flashed, as shown in Figure 13.4, without the use of sheet metal. To avoid the sharp corner between the wall and the roof, cant strips or boards or concrete cants are often used. Flashing block, as shown in the figure, are frequently used. The acute angle between the back side chimney or other projection and a sloping roof is protected with a *saddle* or *cricket*, which is two sloping surfaces meeting in a horizontal ridge perpendicular to the chimney. The valleys on sloping roofs are made watertight by sheet-metal strips, preferably of copper, bent to fit the two intersecting roof surfaces. Roll roofing is used for valley flashing on asphalt-shingle roofs, but it is short-lived.

74. SHINGLES

Wood Shingles. The best wood shingles are sawn or hand-split cypress, red cedar, or redwood. Standard lengths are 16, 18, 24, and 32 in. Shingles are tapered in thickness; the thickness at the *butt* or thick end is expressed in terms of the number of shingles required to produce a total butt thickness of a designated number of inches. Sixteen inch shingles are usually 5 butts in 2 in.; 18 in. shingles, 5 butts in 2¼ in.; and 24 in. shingles, 4 butts in 2 in.; they are called 5/2, 5/2¼, and 4/2, respectively. Although regular shingles are variable in width, they must be uniform in their measurements; the maximum width permitted by grading rules is 14 in., and the minimum width 3 in. for 16 and 18 in. shingles, and 4 in. for 24 in. shingles. Shingles are termed *random shingles* if the widths in a bundle vary or *dimension shingles* if their widths are cut to specified uniform widths.

Not more than one-third of the length of a shingle should be exposed to the weather on roofs, nor more than one-half on side walls. Heartwood is more decay-resistant than sapwood; edge-grain shingles are less likely to warp or cup than flat-grain shingles. Thick shingles warp less than thin shingles, and narrow shingles warp less than wide shingles. Shingles should be clear, entirely heartwood, and entirely edge-grain. It is not economical to use any but the best grade of shingle except for temporary construction.

Thick heavy shingles are called *shakes*. They are usually hand-split, and sometimes they are then sawn. They are called *resawn shakes*. In their manufacture, a mallet is used to strike a sharp bladed steel cleaving tool called a *froe*. This tool is placed on the end of a block of wood parallel to the grain of the wood. Hand-split shingles, used when their rough-textured surface is desired, are nonuniform and considerably thicker than sawn shingles.

Redwood Shingles and Shakes. Redwood shingle sizes are indicated as 18 in. 5/2¼ in. random shingles; 16 in. 5/2 in. random shingles; 16 in. × 4 in.—5/2 Dimension Shingles; 16 in. × 5 in.—5/2 Dimension Shingles; and 16 in. × 6 in.—5/2 Dimension Shingles. These shingles are graded as *No. 1 Grade, No. 2 Vertical (VG) Grade,* and *No. 2 Mixed Grain (MG)*. Redwood shingles are not less than "5 butts to 2 inches" [9].

There are five redwood shake types; all are made from clear heartwood. They are: *Handsplit-and-Resawn Shakes* that have split faces and sawn backs, *Taper Sawn Redwood Shakes* that have split or rough edges, *Machine Grooved Shakes* which are the same as Handsplit-and-Resawn except that they may contain flat grain material, and *Tapersplit Shakes* and *Straight-Split Shakes*, which are handmade. Redwood shakes are not less than ¾ in. thickness nor less than 4 in. wide. Lengths are 18, 24, and 32 in.

Western Red Cedar Shingles and Shakes.   These widely used shingles are made in five distinct grades. The finest grade is No. 1 (Blue Label), which is made of clear, edge-grain, all heartwood. Grade No. 2 (Red Label), a flat grain, contains some sapwood. Where economy is a factor, No. 3 (Black Label) is used. Grade No. 4 (Under Coursing) is used as the under course in side wall applications. Another grade, No. 1 or No. 2 Rebutted-Rejointed, consists of material that conforms to the above grades, although they are machined to assure parallel sides and right angle butts.

Western red cedar shakes are made in three grades: No. 1 Handsplit and Resawn, No. 1 Tapersplit, and No. 1 Straight Split. Cedar shakes are 18 and 24 inches in length. The thickness of the butts is up to 1¼ in.

Machine grooved *sidewall shakes* are made in No. 1 grade (Blue Label). These shakes have a combed or striated face and parallel edges. They are made in the standard shingle sizes.

Fancy-Butt Shingles.   A type of shingle that has specially shaped butts is called a *fancy-butt shingle*. These specially designed shingles (Figure 13.5) are used where decorative effects are desired. Their use is especially effective where used in limited applications on side walls as accents to contrast with the standard butt shingles. These shingles, cut to order, are made in lengths of 16 and 18 in. and 5 in. wide. They are usually laid with a 6 in. weather exposure.

Shingle Roof Applications.   Wood shingles are nailed to wood sheathing or slats; each row of shingles is lapped over the row below to give an exposed surface varying from 4 to 7 in.; the smaller exposure to the weather is for short shingles on flat slopes, and the greater exposure is for long shingles on steep slopes. Shingles are laid in *courses.*

Shingles should be fastened with two aluminum or two steel hot-dipped, zinc-coated, steel-wire nails, 3d or 4d, depending on the thickness of the shingle.

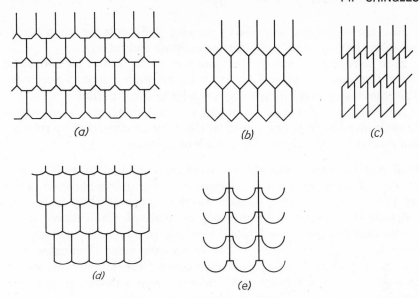

FIGURE 13.5 Fancy-butt shingles.

The sheathing may be square edged (Figure 13.6(a)), matched (b), or shiplapped (c). The use of wood slats, or strips, illustrated in (d), is sometimes called *open* or *spaced sheathing*. If spaced sheathing is used, the open space should not exceed the width of the sheathing board or exceed 4 in. The portion of the shingle that is exposed is referred to as the *weather exposure*. Weather exposure varies from $3\frac{3}{4}$ in. to 13 in., depending upon the type of shake, the location (wall or roof), and the length of the shake.

When the sheathing is not spaced, or when matched sheathing or shiplap is used, a layer of waterproof paper is sometimes "shingled" between the sheathing and the shingles. This paper makes the roof more air- and watertight; but where there is heavy rainfall, it may

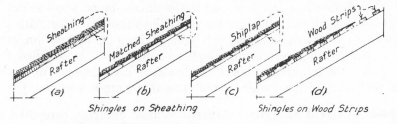

FIGURE 13.6 Types of sheathing for wood shingles.

reduce the life of the shingles by making it difficult for them to dry, out from the underside, thus causing cupping and rotting.

Shingle roofs with a slope less than 4 in. vertical to 12 in. horizontal are not recommended; a 6 in. to 12 in. minimum slope is preferred. The chief use of wood shingles is in residence construction.

An ever-present danger attendant to wood shingles or shakes may be their propensity to fire; however, this may be alleviated by treating shingles with a proprietary fire-retardant chemical.

Wall Applications.   Shingles are frequently applied to walls as well as roofs. Wall application is referred to as *sidewall application.* They are usually applied with greater weather exposure than in roofing. Sidewall shingles are made by two methods of applications, single course and double course. *Single coursing* is similar to roof applications, whereas *double coursing* has an underlayment of shingles. This underlayment, made of grade under course shingles, permits an increased weather exposure and dramatic deep shadow lines at the course lines.

Sidewall applications are frequently accented with fancy-butt shingles mentioned above.

Asphalt Shingles.   Asphalt shingles, also called *composition shingles,* are made of heavy felt composed of rag, paper, or wool fiber, saturated and coated with asphalt. Crushed slate or other aggregate embedded in the top surface coat forms a weather-resistant and colored exposed surface. They are made in various shapes, sizes, and colors. A common shape is a strip 12 × 36 in. with the exposed surface cut to resemble three 9 × 12 in. shingles. These are called *strip shingles.* To resist the tendency of winds to raise the exposed butts, some of the types are tapered, provided with mechanical fasteners or tabs, and made in interlocking shapes. Some are provided with a bank or area of self-sealing adhesive along the underside of the lower edge of the butts.

Asphalt shingles are laid over a layer of roofer's felt and nailed to solid wood sheathing, as described in the discussion on wood shingles, with hot-dipped, zinc-coated steel nails. The exposure varies with the dimensions of the shingles and will usually give a lap of 2 or 3 in. over the upper end of the second shingle underneath. This is called the *head lap.* The minimum slope that should be used depends upon the characteristics of the shingle; usually the minimum is 4 in. per ft.

Asbestos-Cement Shingles.   Asbestos-cement shingles, composed of asbestos fiber and portland cement under pressure, are made in various shapes, sizes, and colors in a thickness of about $5/32$ in. and $1/4$ in. They are laid on a wood deck over a layer of roofer's felt, with a head lap of 2 in., using copper or hot-dipped, zinc-coated steel nails, on a minimum slope of 4 in. per ft.

Asbestos shingles are much stiffer, more fire-resistive, and more durable than asphalt shingles. They are cheaper than clay tile and slate and more expensive than asphalt shingles.

75. CLAY TILE AND SLATE

Clay Tile. Clay tile, made by shaping moist clay in molds and burning in kilns, are made in several patterns, colors, and textures. The most common shapes are *French, Spanish, English, Greek, Roman,* and *mission* or *pan* tile (Figure 13.7(*a*) to (*d*)). *Shingle* tile are rec-

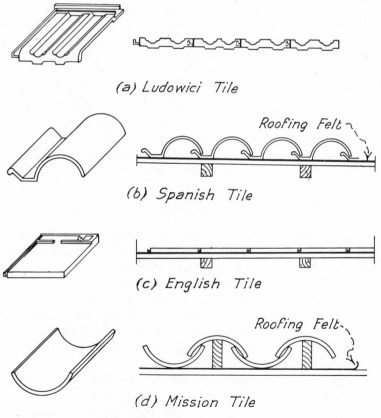

(a) Ludowici Tile

(b) Spanish Tile

(c) English Tile

(d) Mission Tile

FIGURE 13.7 Clay roofing tile.

tangular slabs about ½ in. thick in widths up to 9 in. and lengths up to 14 in. Tile of constant width are always used. They are laid in rows to break joints with the length exposed not more than one-half the length of the tile, minus 2 in. The rows may be regular or irregular.

All forms of tile may be nailed to wood sheathing. Clay tiles may be nailed, through prepared holes, to wood nailers or solid blocking. They are laid over a layer of asphalt-saturated rag felt, gypsum, or nailable concrete. Nails should preferably be copper, but hot-dipped galvanized steel nails are often used. Some forms of tile may be wired to closely spaced steel-angle subpurlins or special anchorage devices. They may be wired (through prepared holes) with annealed wire that is attached to a strip of noncorrodible metal. The metal strips, which have perforations or hooks to receive the wire, saddle across the ridge and run down the slope of the roof. These strips are nailed securely to the roof.

Special tile are made for the ridges and hips. Preferably, valleys should be made with sheet copper, but in most localities galvanized steel is satisfactory. One or two layers of roofers' felt placed under the tile serves as a cushion, keeps air currents from lifting the tile from beneath, and sheds water while damages or defects are remedied. This is an *underlayment*.

The usual forms of clay roofing tile should not be used on slopes less than 4 in. vertical to 12 in. horizontal. A special form of tiles called *pavers'* or *promenade tile*, laid on a waterproof base, forms the floor of a roof an inch thick. A waterproof base is constructed in the same manner as a built-up roof; but the surfacing is omitted and the tile are bedded in a 1 in. layer of cement mortar.

Clay tile are fireproof, durable, and attractive, but they are expensive and require strong supporting roof members.

Slate.   Slate roofing is quarried and split from the natural rock by shaping it into rectangular pieces of the desired dimensions. Roofing slate should be hard, tough, and have a bright metallic luster when freshly split. It should ring clear when supported horizontally on three fingers and snapped with the thumb of the other hand.

Slate has a great variety of colors such as gray, green, dark blue, purple, and red. Furnished in almost any size from 6 to 14 in. wide, 12 to 24 in. long, and ⅛ to 2 in. thick, the most common sizes are 12 × 16 in. and 14 × 20 in., ³⁄₁₆ and ¼ in. thick.

Roofs may be made of pieces of uniform size, thickness, and color, but random sizes, thicknesses, and colors are also used.

Slate may be laid like shingles, each course lapping 3 in. over the second course below, or they may be laid at random as long as care is taken to give sufficient sidelap. They are nailed through prepunched holes to matched wood sheathing, nailable concrete, or gypsum slabs. A layer of underlayment of 30 lb. asphalt-saturated felt is used between the slate and the deck. The nails should preferably be copper or yellow-metal slater's nails, although redipped galvanized nails and copper-coated nails are often used.

Slate is joined at the ridges by a *saddle ridge* or a *combed ridge.* Hips of slate roofs are built by one of four methods: a saddle hip, Boston hip, mitered hip, or fantail hip.

In dryer regions, slate may be nailed to properly spaced wood strips without an underlay of felt. They are sometimes supported directly on steel subpurlins to which they are wired. Slate roofs should not be used on slopes less than 4 in. vertical to 12 in. horizontal.

Slate flagging is used for roof gardens by omitting the surfacing on the ordinary built-up roof and by bedding the slate in a 1 in. layer of cement mortar.

Slate roofs are fireproof, durable, and attractive. All the slate used in the U.S. is quarried in the Eastern states, mostly in Vermont and Pennsylvania. Its cost increases in relation to the distance from these sources. The first or initial costs of slate roofs are relatively high. The first cost may eventually be offset by slate's exceptional durability—100 years or more.

Cement Tile.   Roofing tile made of Portland cement and a fine aggregate are known as *concrete tile* or *cement tile.* These tile are flat and resemble various clay tiles, such as English tile or French tile. These tile are manufactured in several colors and in various sizes that compare to the clay tiles that they imitate. They are supplied with various trim pieces that are designed for rake closure, ridges, and hips. Their installation is similar to that of the clay tiles that they resemble. Their physical properties include a high resistance to fire and deterioration by weather.

Miscellaneous Shingle Materials.   Several other materials have been processed into roofing and siding materials. Porcelainized aluminum and mineral based materials have been shaped into shingle material. These materials are usually shaped or pressed into forms that are intended to resemble other materials, such as wood shakes or shingles.

76.   SHEET-METAL ROOFING AND SIDING

Method of Laying.   Sheet-metal roofing of galvanized iron or steel, terne plate, copper, zinc, aluminum, or lead is widely used. To prevent leakage, it is necessary in the installation of sheet-metal roofs to provide for the large amount of thermal expansion which takes place. The expansion of various metals in inches per 10 ft. of length for a temperature rise of 100°F is as follows: steel and iron, 0.08; copper, 0.11; aluminum, 0.16; lead, 0.19; and zinc, 0.21. Expansion is the chief factor controlling the design of the seams and the methods used in fastening the sheet metal to the roof. Laid with a *flat seam* (Figure 13.8(*a*)) or with a *standing seam* (*b*), seams on steeper slopes run parallel or perpendicular with the slope of the roof. The seams are clinched tight; the separation shown in the figures is shown for clarity. *Cross seams*, always flat, connect two sheets and run between two standing seams for two flat seams. Flat seams are soldered, but standing seams are not. The sheet metal is fastened by metal cleats nailed 8 to 12 in. apart on the deck and locked into the seams.

If possible, the nails should be of the same material as the roofing to avoid galvanic action between dissimilar metals: tinned nails with terne plate roofs, copper nails with copper roofs, galvanized nails with galvanized iron or steel and zinc roofs, and aluminum nails with aluminum roofs. Nails should not be driven through the sheets. *Batten* or *ribbed roofs* are formed by using wood battens or ribs. Special extruded aluminum battens are available for aluminum roofs. Sheet-metal *troughs* are fitted between these battens, and caps are placed over the battens (Figure 13.8(*c*)).

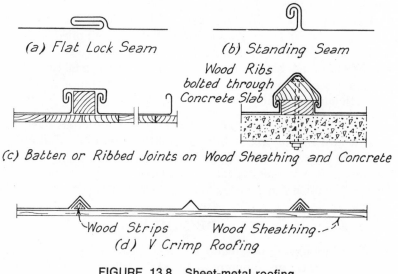

(a) Flat Lock Seam   (b) Standing Seam

(c) Batten or Ribbed Joints on Wood Sheathing and Concrete

(d) V Crimp Roofing

FIGURE 13.8   Sheet-metal roofing.

Before sheet metal is placed over matched wood sheathing, gypsum, or nailable concrete slabs, a layer of *roofing felt* is placed between the metal and the deck. An important function of this felt is to prevent quick ignition of the wooden decking when the roof is exposed to burning brands or radiated heat. It should be at least $\frac{1}{16}$ in. thick. Tar paper should not be used under tin or terne roofing on account of its possible deleterious effect on the metal, but a rosin-sized paper is satisfactory.

The processes used in manufacturing the various metals used for roofing are described in Articles 35, 92, and 93.

Longitudinal *crimps*, shaped like an inverted V, formed in sheet-metal roofing (Figure 13.8(*d*)) increase its stiffness. Such crimps may be only at the edges, but a crimp may also be in the middle, as shown in the figure. For further stiffness, double crimps are formed along the edges and the middle. Triangular wood strips are provided under the outside crimps for nailing. Such roofing is unsuitable for flat roofs.

Galvanic Action. As reported by the Bureau of Standards [4],

"*Galvanic action* or *electrolysis* is defined as the deterioration by corrosion of one metal while protecting a dissimilar metal in metallic contact with it. This action will occur only when metals occupying different positions in the electro-potential series (aluminum, zinc, iron and steel, tin, lead, and copper, arranged in order) are in intimate metallic contact in the presence of a suitable electrolyte, such as moisture."

Terne Plate. Tin plates are made by dipping plates of sheet-metal or iron in a molten bath of tin to form bright tin plates. If dipped in a molten bath of tin and lead, they form *terne plates*. When taken from the molten bath, the plates are usually passed through rolls; the pressure on the rolls determining the thickness of the plate. Bright tin plates are superior to terne plates but are so expensive that terne plates are used for roofing. The common sizes of sheets are 14 × 20 in. Several sheets may be assembled at the factory by jointing their ends together to form long sheets which are shipped in rolls, but sheets 50 ft. long can be seamless.

The sheets are usually painted on both sides at the factory; but if they are not, the bottom side must be painted before laying. Terne plate roofs will last for many years if kept properly painted.

A recent development is *terne-coated stainless steel*. This roofing sheet is a superior terne plate that provides a maintenance-free and durable sheet-metal roof.

Titanium, Copper, Zinc Alloy.   Sheets made of copper, zinc, and titanium alloy are used for a durable, lightweight roofing. Noncorrosive, these sheets weather to an attractive blue-gray patina. They may be easily installed using batten seams or standing seams supplemented with soldering if desired. This alloy sheet may be used in combination with several other metallic materials without fear of galvanic action. These sheets may be used with galvanized steel, aluminum, lead, or tin. If used in contact with other more electrolytically active materials, they must be protected with a coat of asphaltum paint. Like several other sheet-metal roofing materials, this alloy is used widely for flashing, facias, gutters, and other roofing accessories.

Lead.   Lead sheets are used for roofing to a limited extent. Lead is particularly suitable for curved or irregular surfaces, for it can be stretched easily and worked to fit such surfaces without cutting. It has a high coefficient of expansion and is difficult to hold in place, particularly on pitched roofs. It has a long life and need not be painted.

A roofing known as *hard lead* is composed chiefly of lead, but 4% to 6% of antimony is added to increase the elastic limit and decrease the coefficient of expansion. This material has the advantages of ordinary lead without its disadvantages. It may be used on any slope.

Lead is also used for flashing difficult places and as a waterproof barrier for garden roofs that are to be finished with a tile wearing surface.

Copper.   Sheets of soft copper make an excellent roof covering which has a high initial cost but is durable and requires no painting. Exposure to the weather causes a green copper carbonate to form on the surface. It protects the remainder of the metal and is attractive. Sheet copper is available in sheets 24, 30, and 36 in. wide and up to 10 ft. long.

Copper roofs are made in batten seams, standing seams, or flat seams. The joints of these seams are frequently accentuated. Heavy shadow lines created by the seams emphasize either a horizontal, diagonal, or vertical design. A chevron or horizontal pattern which is sometimes called a *Bermuda roof* is quite popular.

Copper sheets lend themselves to application of nearly all parts of a roof covering and the necessary accessories. Flashing, reglets, facias, gutters, and downspouts are some of the parts of roofs that are made of copper.

Galvanized Iron and Steel.   Iron and steel sheets are galvanized by dipping clean sheets in a bath of molten zinc. The zinc protects the sheet from corrosion in proportion to the thickness of the coating. Sheets which are to be sharply bent in forming the joints or for other reasons are given thinner coats to decrease their tendency to flake while being bent. For use in laying galvanized sheet roofing with standing seams, several sheets are assembled at the factory to form continuous sheets, usually 50 ft. long, which can be rolled for shipping. The ends of the sheets are joined by double cross-seam locks. Other forming necessary for the standing seams is done on the job. U-shaped caps are placed over the standing seam with the sides squeezed together with tongs to form roll and cap roofing. V-crimp roofing (Figure 13.8) and sheets with U-shaped edges, which are overlapped and squeezed to form standing seams, are also available. Galvanized sheet roofing is used extensively, is relatively low in cost, and lasts for many years if kept painted.

Stainless Steel.   Stainless steel sheets are occasionally used for roofing if the importance of long life, because of inaccessibility or for other reasons, justifies their high cost. They are highly resistant to corrosion and do not require painting.

Aluminum.   Aluminum sheets are available for laying on sloping roofs with standing or batten seams to provide for the relatively high coefficient of expansion. The battens may be of wood or specially designed extruded aluminum. Aluminum is highly resistant to corrosion and has a relatively long life as a roofing material. It does not require painting.

Aluminum sheets are finished in various colors. The colors are processed permanently by a baked enamel process, by an electrochemical process that is identified as a duranodic finish, and by fluorocarbon finish.

77.   CORRUGATED SHEET ROOFING AND SIDING

Metal Siding.   Corrugated steel sheets are widely used for roofing and siding of industrial buildings. Sheeting, which is a relatively thin material, forms the outside surface of exterior walls and roofs to exclude weather. In many respects, the siding and roofing materials are similar and often identical, or the same types of materials are used for both purposes. Siding seldom serves a structural purpose, although siding may be anchored and able to serve as a shear wall to resist horizontal forces.

The size of corrugated sheets is indicated by their corrugations, thickness, and sheet size. The nominal size of the corrugations of sheets is measured from crown to crown of the corrugation. This measurement is called the *pitch*, whereas the measurement of the height from crown to valley is the *depth*. Corrugation sizes range from 1¼ in. to 3 in. pitch and have depths from ¼ in. to 1 in. The most widely used sheet is the nominal 2½ × ½ corrugation. The actual pitch of the sheet is 2⅔ in. The thickness of sheets is indicated by the gage, which ranges from 29 gage, the thinnest, to 12 gage, the thickest. The usual width of sheets is 26 in. for siding sheets and 27½ in. for roofing sheets. Lengths are up to 10 ft.

Steel sheet siding and roofing is further identified by that which is *uncoated, galvanized,* painted, ceramic coated, asbestos protected, or black.

Corrugated sheets may be nailed to wood sheathing or may be supported directly by wood or steel purlins spaced 2 to 9 ft. with spacing dependent upon the gage of the corrugated sheets, which depends upon the imposed loads. The sheets are lapped 1½ to 2 corrugations on the sides and 6 or 8 in. on the ends, depending on the slope of the roof. On slopes as flat as 2 in. vertical to 12 in. horizontal, standing seams should be used instead of side laps.

Where steel purlins are used, the corrugated steel may be fastened to nailing strips (Figure 13.9(*b*)), or the sheets may be held in place by straps passing around the purlins and riveted to the corrugated steel on each of the purlins (Figure 13.9(*c*)). Long malleable nails called *clinch nails* or *J-bolts* may be driven through the corrugated steel and clinched around the purlins (Figure 13.9(*d*)). The side laps are held together by galvanized-iron rivets spaced about a foot apart. All rivets and nails should be driven in the crown or peak of the corrugations to prevent leakage. The fasteners should be heavily zinc coated. Lead or neoprene washers are used under the heads of all fasteners.

Unfinished sheets must be protected against corrosion by frequent painting. Some sheets are coated with colored plastics.

Because galvanized sheets are not subjected to severe bending when forming the corrugations, they may have a heavy coating of zinc. Because of their relatively low cost and fair length of life, they are extensively used on industrial buildings. They are not suitable where appearance is paramount.

Standing-seam corrugated steel roofing is shown in Figure 13.9(*e*). Thicknesses range from 16 to 18 gage, painted or galvanized. Cleats fasten the sheets to wood or steel purlins without the use of rivets or nails. The use of this roofing is the same as that of ordinary corrugated steel.

Ribbed or crimped sheets of steel, aluminum, and stainless steel are

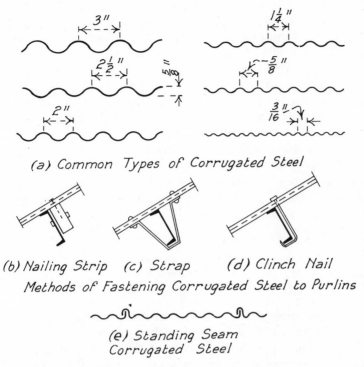

(a) Common Types of Corrugated Steel

(b) Nailing Strip   (c) Strap   (d) Clinch Nail

Methods of Fastening Corrugated Steel to Purlins

(e) Standing Seam
Corrugated Steel

FIGURE 13.9   Corrugated steel roofing.

used for roofing and siding. There are various patterns or profiles of these sheets which suit jobsite functions and designs. There are those identified as 3-V-Crimp and 5-V-Crimp. The sheets, 24 in. wide, are made in various trapezoidal patterns. These are used for siding and for roofing on both sloping and flat roof decks.

Many of these materials have porcelain baked enamel coatings, vinyl plastic coatings, and asbestos protected finishes in various colors. These are frequently made into *sandwich panels* that have a rigid heat insulation core between the metal sheets.

These materials are attached to a rigid structural frame which may be of wood or timber, concrete, or structural steel. They are usually attached with neoprene washered, sheet metal screws.

At the eaves and the sills of these structures where the metal corrugated and ribbed sheets terminate, a filler strip of neoprene material serves as a gasket or filler strip between the sheet metal and the supporting frame. These filler strips, which are identical in profile to the sheet metal, serve to close the gap from the weather.

Aluminum. Corrugated aluminum sheets are illustrated in **Figure** 13.10, and one form of ribbed sheet in (*b*). Sheets are 35 and 48 in. wide, 0.024 and 0.032 in. thick, and up to 12 ft. long. Their use is similar to the use of corrugated steel. Aluminum should not be used in direct contact with steel, however, because of possible galvanic action. Therefore, asphaltum paint or felt strips should be placed between aluminum sheets and steel purlins; aluminum fasteners, such as nails, should be used.

Uncoated aluminum sheets are usually employed, but sheets coated with porcelain or baked enamel provide a variety of colors. In addition, several colors are obtained by *anodizing*, which consists of applying a natural oxide and inorganic dye to the surface by an electrochemical process.

Stainless Steel. Corrugated stainless-steel sheets are specified for roofs subjected to the highly corrosive effects of some industrial atmospheres, where their superior resistance to corrosion justifies their higher initial cost.

Asbestos-Protected Metal. Asbestos-protected metal consists of sheet steel covered first with a layer of asphalt, then a layer of asbestos, and finally a heavy waterproof coating. It may be obtained in flat sheets or corrugated sheets of the same size and shape as the plain corrugated sheets just described. The method of application is the

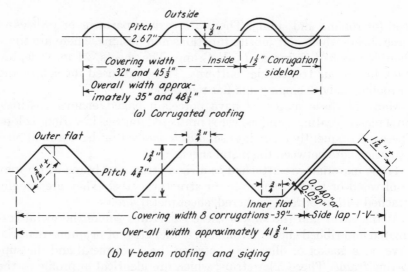

FIGURE 13.10   Corrugated and V-beam aluminum roofing.

same as for plain corrugated steel sheets, but it does not require painting for it is well protected from corrosion. It is a poor conductor of heat and therefore may be used in places where the uncoated sheets are not suitable.

Because of their relatively light weight and fairly good insulating properties, corrugated sheets are extensively used on long-span roofs, such as those of hangars and field houses. They are fastened directly to the purlins. The development of plastic coatings has made it possible to obtain sheets of many colors.

Asbestos-Cement Board. Corrugated sheets made of asbestos fiber and portland cement under pressure are used in the same manner as corrugated steel. This material is a good nonconductor of heat, and no trouble is experienced with condensation. It is durable and does not require painting. Sheets are also made with plastic coatings in a great variety of colors.

Wire Glass. Corrugated wire-glass sheets similar to corrugated steel sheets can be used alone or in connection with corrugated roofing to assist in lighting the interior of buildings. These sheets are $\frac{1}{4}$ in. thick and have wire netting embedded in the glass to strengthen it and to hold it in place when a break occurs. The glass sheets are not laid with side laps, but the joint is covered with metal caps held in place by bolts passing between the sheets. The sheets are held to steel purlins by clips bolted to the metal cap covering the joint between sheets. Strips of asphaltic felt are placed over the purlins to cushion the glass.

Where a fire resistance rating is required, codes usually require wire-glass where glass is permitted.

Asbestos-Cement Materials. Asbestos-cement shingles as described in Article 74 and plain or corrugated sheets are manufactured in many colors and used for siding materials. Sandwich panels with a core of heat-insulating material and outer surfaces of asbestos-cement boards are manufactured.

Asphalt Materials. Asphalt materials are used for siding in two forms. They may be either in the form of shingles as described in Article 74 or in rolls as described in Article 78. Roll sidings, applied over wood siding, are made to resemble brick or stone masonry.

Plastics. Transparent and translucent acrylic plastic sheets are used for roofing, siding, or skylight installations. These flat sheets, which resemble glass materials, are usually $\frac{1}{8}$ or $\frac{1}{4}$ in. thick and are either clear, bronzed, or gray. Several colors are available and are used where

effects are desirable. Thinner acrylic sheets that have a fiberglass reinforcement are manufactured in profiles, usually corrugated or flat,
that are identical to many of the metal roofing and siding sheets.
These plastic sheets are used in roofing and siding installations which
require light as well as weather protection, such as in greenhouses
and warehouses.

The fire resistivity or flamespread resistivity of most plastics is
somewhat limited; therefore, most codes do not permit more than a
specified area on a roof.

78.   BUILT-UP AND ROLL ROOFINGS AND PLASTIC COATINGS

Built-Up Roofing.   Built-up roofings consist of several overlapping
layers of bituminous-saturated roofing felt cemented together with
bituminous roofing cement, usually applied so that in no place does
felt touch felt. The bituminous material may be tar or asphalt. Tar is
a by-product driven off in the process of converting coal to coke and
is called *coal tar pitch*. Asphalt is a by-product driven off in the
process of refining petroleum. Roofing felt is made from organic or
asbestos fibers and is furnished in rolls 3 ft. wide. The organic fibers
are primarily graded rags, and for that reason, the felts from which
they are made are called *rag felts*, but other organic fibrous materials
are used along with rag fibers. *Asbestos felt* is composed of at least
85% asbestos fibers, with other materials used to provide desirable
properties.

Both asphalt and coal-tar roofing cements are solid at ordinary
temperatures. Before applying, they are heated until they become
liquid. Tar has a lower melting point than asphalt, a quality which
is desirable because it tends to be self-healing if punctured or ruptured
but undesirable because tar is less stable on sloping roofs when it is
heated by the sun. Tar is also less affected by standing water than
asphalts. Asphalt roofing cement is used with asphalt-saturated felts
and coal-tar roofing pitch with felts saturated with coal tar. Asphalt
is usually preferred for sloping roofs and coal tar for flat or nearly flat
roofs.

In addition to the roofing felts and cements, built-up roofs include
various surfacing materials, which are embedded in the top layer of
roofing cement to protect them from the action of the elements. These
are usually fine gravel and crushed slag, but felts with smooth top
surfaces designed to be exposed are available. Flat roofs that serve as
promenades carrying foot traffic may be covered with built-up roofs

surfaced with floor tile or flagstones bedded in a 1 in. layer of portland-cement mortar.

Built-up roofs are adaptable to the various types of *nailable* roof decks, such as those of wood, poured or precast gypsum, precast nailable concrete, and other less common materials, and to those which are *nonnailable*, such as cast-in-place concrete. They are also adaptable to flat and sloping roofs. They can be applied over decks that are not insulated and to decks covered with rigid insulation. A bituminous vapor seal is applied to the deck before the insulation is installed. The seal prevents the accumulation of moisture from the air on the interior of the building in the insulation. The top surface of the insulation is mopped with roofing cement before the first ply of felt is applied. Sheet-steel decks must always be insulated. Wherever insulation is used, it must be adequately secured to the deck.

The simplest form of the best type of built-up roof usually installed is illustrated in Figure 13.11, in which the roof deck is made of cast-in-place concrete and its slope does not exceed 1 in. per ft. The plies are lapped the distance required to produce a cover whose thickness at every point is equal to the number of plies desired. The roof in the illustration is a 4-ply roof. The plies are laid with their lengths perpendicular to the slope. The laps are made in the direction required to shed water, the direction of flow in the illustration being downward.

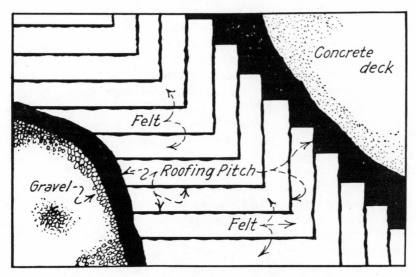

FIGURE 13.11   Four-ply built-up roofing on concrete deck.

The surface of the deck is first covered with a mopping of hot roofing cement, the portion of the surface of each ply which is to be overlapped is similarly covered before placing the next ply, so that in no place does felt touch felt, and finally a uniform coating of hot roofing cement is mopped or poured over the top surface of the top ply, and in this coating, clean fine gravel or crushed slag is embedded.

For nonnailable decks with slopes greater than 1 in. to the foot, nailing strips are required, to which each ply can be nailed in a specified manner using tin-plate disks 1 in. in diameter under the nailheads. The nailing strips are made of wood which is creosoted to preserve it. They are usually placed 3 ft. apart, parallel with the slope, with their top surface flush with the surface of the deck and with sides beveled for anchorage or anchored in some other manner.

If a wood or other nailable deck is used, a layer of sheathing paper or unsaturated felt is nailed to the deck. The top surface of this layer is mopped with hot roofing cement, and then the desired number of plies of saturated felt are placed and surfaced as for the concrete deck illustrated in Figure 13.11. A 5-ply roofing on a nailable deck, placed as described, is considered equivalent to a 4-ply roofing on a concrete deck, also placed as described.

Various manufacturers' specifications for laying built-up roofing differ in some respects. They are essentially in accord with the descriptions already given but are more detailed.

Built-up roofs are widely used and durable. Roofs such as those described are guaranteed for 20 years against repair and maintenance expense by the manufacturers if their specifications are followed, but the expected life is much longer. If the number of plies is reduced by one, the guaranty period is reduced five years.

Built-up roofs are sometimes placed by cold processes; in one process, the bituminous cementing material is thinned with kerosense to produce what is called a *cutback*. The cementing action occurs as the solvent evaporates. The cold process is used primarily for maintenance rather than new construction.

Roll Roofing.   Roofing consisting of sheets composed of a felt base impregnated and coated with asphalt is furnished in rolls and therefore is called *rool roofing*. Because it is prepared ready for application, it is called *prepared* or *ready roofing*; and it is also called *composition roofing*. To prevent sticking when rolled, both surfaces are dusted with mineral matter, such as mica, talc, or fine sand, or the top surface is coated with crushed slate, quartz, burned clay, or other minerals with or without coloring agents embedded in the asphalt coating.

Roofings with smooth surface coatings are lowest in cost, but they are less desirable because coarse mineral surfacing gives considerable protection from the weather and increases the fire resistance. The surfacing also makes possible the manufacture of roofing in a variety of colors.

Roll roofing, available in various weights or thicknesses, is usually furnished in rolls 3 ft. wide and 36 ft. long. It is also furnished in rolled sheets 18 in. wide and 76 ft. long. The most extensive use of roll roofing is over wood decks to which it is nailed with galvanized roofing nails which have large heads. In addition, it is cemented in various ways, using roofing cement which can be applied cold and which hardens as the thinner evaporates.

One type of roof covering composed of roll roofing, when laid on a wood deck, is placed over a layer of lightweight roofing felt laid dry. The roofing is 18 in. wide and is laid with 8 in. exposed, leaving an unexposed *selvage* of 10 in. It is cemented to the ply below over the selvage area. A row of nails is driven along the upper edge of each ply and another above the lower edge of the selvage where it will be covered by the next ply. This arrangement provides 2 plies of protection.

A similar type of construction makes use of roll roofing 36 in. wide, laid with an 11½ in. exposure and a 24½ in. selvage. This arrangement provides 3 plies of protection.

The least expensive and shortest-lived roof covering composed of roll roofing on a wood deck makes use of 36 in. roofing with all its surface except about 2 in. exposed to the weather. This 2 in. edge is side- and end-lapped, cemented, and nailed with the heads of the nails exposed. Because of the distance between rows of nails and its light weight, this type of installation can be blown off by air currents filtering through the deck during windstorms and, being only 1 ply thick, is easily punctured by hail. It is suitable only for temporary structures.

Roll roofings can be used for slopes as flat as 2 in. to the foot.

Plastic Coatings.   During recent years, several liquid coatings which can be sprayed or rolled on thin-shell concrete roof decks to make them waterproof have been developed. One of them is polyethylene rubber in combination with pigments and fillers. It is resistant to the temperatures to which a roof may be subjected and to atmospheric gases and weathering. The coating forms a membrane which expands and contracts with the base to which it is applied and maintains a continuous covering over hairline cracks which may develop in concrete surfaces. It will not fill visible surface imperfections. The membrane should be reinforced over cracks other than hairline cracks. The coating is available in several colors.

After the surface to be coated has been prepared and cleaned, a prime coat is applied and permitted to dry for a few hours. A minimum of four topcoats is then applied. The coats may be applied in alternating colors to insure that each coat covers the entire surface. This plastic coating is suitable for application to plywood and metallic surfaces as well as to concrete.

A relatively new product, this plastic coating has been successfully used on many thin-shell roofs, and accelerated weathering tests indicate that it will have a life of at least 15 or possibly 20 years.

79. SELECTION OF ROOF COVERINGS

Factors to Be Considered.   Many factors enter into the selection of a roofing material for a given building. Some are quite definite, but others vary with local conditions, the class of occupancy, and the preferences of the architect and owner.

The most pertinent factors (not listed in order of their importance) are: slope of roof deck, type of roof deck, weight of roofing, climatic conditions, fire exposure, durability, cost, appearance, and maintenance.

Slope of Roof.   All forms of shingles, tile, and slate are suitable for use on roofs with a slope of 6 in. or more to the foot and are often used with slopes of 4 in. to the foot. The flatter the slope, the greater the tendency for driving rains to cause leakage. The various kinds of corrugated roofs require a slope of 4 in. to the foot. Special attention should be given to the tightness of the end laps. Sheet-metal roofs can be used on any slope from flat to steep by employing appropriate joints or seams. Built-up roofings can be used on any slope from flat to steep. Roll roofings require a slope of at least 2 in. to the foot.

Type of Roof Deck.   All forms of roof covering must be fastened to the decks that support them. The methods of fastening the various types of roof coverings to various types of roof decks have been explained. Because of these relationships, it is seen that the type of roof covering and the type of deck should be selected concurrently.

Weight of Roofing.   The weight of the roofing affects the design, weight, and cost of the roof deck and the supporting members. The weight of a given roofing depends upon the specifications being used and range as shown in Table 13-1.

TABLE 13-1
Weights of Roofing Materials in Place

Material	Weight*	Material	Weight*
Terne plate	60–75	Built-up roofing	400–675
Copper	80–120	Roll roofing	130–325
Corrugated galvanized steel	120–220	Shingles, wood	200–300
Clay tile	850–1800	asphalt	130–325
Slate	550–1000	asbestos-cement	275–600

* Values in pounds per square of 100 sq. ft.

Climatic Conditions.   The durability of roofing materials is affected by the climatic conditions to which they are subjected. *Strong winds* may damage slate, tile, and asbestos-cement shingles, and asphalt shingles and roll roofing are especially vulnerable unless special provisions are made. *Hail* punctures asphalt shingles and roll roofing and may break tile and slate, the thicker types being less vulnerable than the thinner.

The expansion and contraction of sheet-metal roofs from *extreme temperature changes* may cause the sheets to crack and the joints to leak if adequate provision is not made for this action. Asphalt shingles and roll roofing are also adversely affected by extreme temperatures, especially *high temperatures*, and *sunlight*. Those with mineral surface coatings are more resistant to these actions than smooth surfaces.

Fog, salt air, smoke, and other gases in industrial areas tend to corrode metal roofings. Copper, lead, aluminum, and zinc are highly resistant to *atmospheric corrosion* and do not require protection by painting. Terne plate is resistant to corrosion if protected by painting. Although zinc is resistant to atmospheric corrosion, galvanized steel and iron may corrode because of defects in the zinc coating. Corrugated sheets can be more heavily coated than sheets which are bent sharply to form seams and can be therefore more durable. Galvanized sheet metal should be painted to prolong its life.

Clay tile, slate, asbestos-cement, and built-up roofings are unaffected by the composition of the atmosphere.

Fire Exposure.   Building code requirements for roof coverings are based upon their resistance to fire. The classification of roof coverings included in most codes is based on the classifications by the Underwriters Laboratories. These laboratories arrange the coverings into three classes.

Classes, A, B, and C include roof coverings which are effective against severe, moderate, and light fire exposures, respectively, and which are resistant to flying brands. The *National Building Code* of the National Board of Fire Underwriters requires class A or B coverings on all buildings, although class C coverings are accepted on dwellings, buildings of *wood frame construction*, and buildings outside the fire limits which, on the basis of height and area, could be of *wood frame construction.*

The Code Manual of the State Building Construction Code of the State of New York follows the classification of the Underwriters Laboratories but is more complete and includes classes 1, 2, 3, and 4 as specified below. According to that code, "Roof coverings labeled by the Underwriters Laboratories as Class A, B and C, respectively, are acceptable under Code classifications as Class 1, 2 and 3." Some noncombustible roof coverings such as slate, tile, and concrete are not classified by the Underwriters Laboratories. Certain other materials, such as wood shingles and lightweight felt roll roofings, which do not meet the requirements for class 4 roof coverings, are not classified under the Code.

The following classifications are abstracted from the Code Manual, each roof covering to be laid as specified in the manual. All weights are expressed in pounds per 100 sq. ft.

Class 1.   Clay roof tile with underlay, slate not less than ³⁄₁₆ in. thick, asbestos-protected sheet metal.

Asphalt-saturated asbestos felt, smooth-surfaced, 4-ply sheet roofing, laid in a single thickness and with a total weight of not less than 80 lb.

Five layers of 15-lb. asphalt-saturated asbestos felt or equivalent, cemented together with asphalt and surfaced with asphalt paint.

Four layers of 15-lb. asphalt- or tar-saturated asbestos or rag felt or equivalent cemented with asphalt or tar and finished with gravel, stone, or slag in asphalt or tar.

Three layers of 15-lb. asphalt-saturated rag felt or equivalent, cemented with asphalt and finished with asphalt roof tile or ½ in. asphalt-impregnated fibrous board applied with mastic asphalt.

Class 2.   Asbestos-cement shingles not less than ³⁄₁₆ in. thick, laid to provide one or more thickness over underlay.

Asphalt-asbestos felt, smooth-surfaced, 3-ply sheet roofing laid in single thickness and weighing not less than 60 lbs.

Asphalt-asbestos felt shingles, surfaced with granular materials laid to have a total weight of not less than 180 lb.

Asphalt mastic shingles surfaced with granular materials.

Sheet roofing of copper, galvanized iron, or tin (terne)-coated iron, with an underlay.

Tile or shingle pattern roofing of copper, galvanized iron, or tin (terne)-coated iron, with an underlay.

Four layers of 15-lb. asphalt- or tar-saturated asbestos or rag felt or the equivalent, cemented with asphalt and finished with asphalt cement.

Three layers of 15-lb. asphalt- or tar-saturated rag felt or the equivalent, cemented with asphalt and finished with gravel, slag, or stone, in asphalt or tar cement.

Class 3.   Asphalt-asbestos felt, surfaced with sheet or roll roofing, laid in a single thickness with laps, to have a total weight of not less than 48 lb.

Asphalt-saturated asbestos felt, granular-surface sheet or roll roofing laid in a single thickness, to have a total weight of not less than 85 lb.

Asphalt-saturated rag felt, granular-surfaced sheet or roll roofing, laid in a double thickness with laps, to have a total weight of not less than 80 lb.

Asphalt-saturated rag felt individual or strip shingles surfaced with granules, laid with lap, to have a total weight of not less than 80 lb.

Sheet roofing of copper, galvanized iron, or tin (terne)-coated iron, without an underlay or with an underlay of rosin-sized paper.

Tile or shingle pattern roofing of copper, galvanized iron, or tin (terne)-coated iron, without an underlay or with an underlay of rosin-sized paper.

Three layers of 15-lb. asphalt-saturated rag felt or the equivalent, cemented with asphalt and finished with asphalt cement.

Class 4.   Asphalt-saturated rag felt, smooth-surfaced rool roofing, laid in a single thickness with laps, to have a total weight of not less than 45 lb.

Asphalt-saturated rag felt, granulated-surfaced roofing, laid in single thickness with laps, to have a total weight of not less than 80 lb.

Two layers of 15 lb. asphalt- or tar-saturated rag felt, cemented with asphalt or tar or such other combinations or roofing felt that do not meet the requirements of class 3 built-up roofing.

Wood Shingles.   Wood shingles are not classified according to fire exposure. For a discussion of the resistance of wood shingles to fire exposure, see Article 74.

Limitations in Use.   The requirements of the *National Building Code* for roof coverings are given earlier in this discussion. According to the New York State Code,

"Within the fire limits, roof coverings with or without insulation, shall be Class 1 or 2 except that where the distance separation between buildings is more than 20 ft. and the horizontal projected area of the roof does not exceed 2500 sq. ft., Class 3 roof coverings may be used.

Outside the fire limits, roof coverings, with or without insulation, shall be Class 1, 2 or 3, except that where the distance separation between buildings is more than 20 ft. and the horizontal projected area of the roof does not exceed 2500 sq. ft. and the building does not exceed two stories in height, Class 4 roof coverings or wood shingles may be used."

Specific requirements for the application of roof coverings to satisfy the State Building Construction Code classifications are given in the code manual.

These include requirements concerning lap, sealing for water tightness, nailing shingles or tile, minimum and maximum slopes, and weights of felt and applying built-up roofs. The underlay for tile, shingle or sheet metal roofs, when required, is one or two layers of 15 lb. asphalt-saturated asbestos felt or two layers of 15 lb. asphalt-saturated rag felt. Other building codes include requirements for roof coverings. Those for the New York Code are given as examples.

Durability.   Durability of a given roofing material is affected by its quality, its suitability for the purpose used, climatic conditions, quality of workmanship in laying, effectiveness of maintenance, and many other factors. Thus, no definite comparisons of different materials may be made.

Considering the best quality of each type, proper installation, and appropriate maintenance, roofing materials may be arranged in groups in decreasing order of length of life somewhat as follows.

(1) Sheet copper, lead, stainless steel, aluminum, clay tile, slate, asbestos-protected metal.

(2) Terne plate, galvanized iron, asbestos-cement tile, built-up roofing.

(3) Wood shingles.

(4) Asphalt shingles.

(5) Roll roofing.

Cost.   The cost of roof coverings in place, exclusive of the cost of the deck, vary with some exceptions in about the same order listed under durability. Special comment should be made about slate because of the transportation costs involved in shipping from the limited sources of supply.

Appearance.   Appearance is an important factor in selecting a roof covering for some buildings and is of little consequence in other cases, such as many industrial buildings. Furthermore, some roof coverings are installed on flat or nearly flat roofs where they will usually not be seen. The architectural style of a building and the class of occupancy may be determining factors.

Clay tile, slate, and wood shingles are considered to be attractive roof coverings. Asphalt shingles are surfaced with granular material in a great variety of colors to suit individual preferences. Unless thick butt shingles are used, such coverings have a flat appearance. Copper weathers to a pleasing soft blue-green color. Aluminum weathers to a dull gray. Terne plate and galvanized sheets are painted to protect them from corrosion, and any color can be applied. Corrugated sheets of various materials are extensively used but are not considered to be attractive unless colored sheets are used.

Maintenance.   The initial costs of roof coverings are naturally considered in selecting a roof covering, but the cost of maintenance must also be given consideration.

Copper, aluminum, lead, and stainless-steel roof coverings are noncorrosive and do not require painting. Built-up roofs require very little maintenance. Terne plate and galvanized sheet steel require periodic painting because of the tendency of the base metal to corrode when defects develop in the coatings. Asbestos-protected metal requires little maintenance.

Individual slate, clay tile, and asbestos-cement tile may be broken by large hailstones or other objects and require replacement; asphalt shingles and roll roofing may be damaged by hail or wind and require repairs.

The life of wood shingles can be prolonged by coating them occasionally with creosote and, less effectively, with shingle stain. Oil paint should not be used because it causes them to warp and curl.

The life of smooth-surfaced asphalt roll roofings can be prolonged by frequent recoating with asphalt roof coatings.

REFERENCES AND RECOMMENDED READING

1. *Recommended Minimum Requirements for Small Dwelling Construction*, Building Code Committee, Department of Commerce, 1932.
2. Hubert R. Snoke, *Asphalt-Prepared Roll Roofings and Shingles*, Report BMS 70, National Bureau of Standards, 1941.
3. *Slate Roofs*, National Slate Association, 1953.
4. Leo J. Waldron, *Metallic Roofing for Low-Cost House Construction*, Report BMS 49, National Bureau of Standards, 1940.
5. *Manufacture, Selection, and Application of Asphalt Roofing and Siding Products*, Asphalt Roofing Industry Bureau, 1959.

6. *Test Methods for Fire Resistance of Roof Covering Materials*, UL 790, Underwriters' Laboratories Inc., National Board of Fire Underwriters, 1958.
7. *Code Manual*, generally accepted standards applicable to State Building Construction Code, State of New York, 1959.
8. *Architectural File*, Sweet's Catalog Service.
9. *Standard Specifications for Grades of California Redwood Lumber*, Redwood Inspection Service, San Francisco, California, 1970.
10. Charles G. Ramsey and Harold R. Sleeper, *Architectural Graphic Standards*, Sixth Edition, John Wiley and Sons, 1970.
11. *Sweet's Catalog*, 1971 Edition, F. W. Dodge Corp., New York, New York.
12. Certi-Split Manual of Handsplit Red Cedar Shakes, Donald H. Clark, Red Cedar Shingle & Handsplit Shake Bureau, Seattle, Washington.

14

WINDOWS AND CURTAIN WALLS

80. DEFINITIONS AND GENERAL DISCUSSIONS

In general, a *window* is an opening in a wall of a building to provide any or all of the following: natural light, natural ventilation, and vision. The term also refers to the construction installed in the opening to provide protection against entry and the weather. To be satisfactory, windows must be durable, weathertight, reasonable in cost, readily installed, and, for many uses, attractive in appearance. Windows are also used to a limited extent in partitions for vision from room to room, for *borrowed* light, or for other reasons.

Only exterior windows are considered in this chapter. In addition to the glass, they are made of wood, steel, aluminum, stainless steel, and bronze. They are available in a great variety of types to suit many requirements and individual preferences.

Much has been done by the associations representing the manufacturers of windows of each material to standardize the types, sizes, and construction details of their windows. Lower costs and prompter deliveries are possible if *stock*, rather than *custom-made*, windows can be used.

Parts of a Window. A window frame includes the members which form the perimeters of a window; it is fixed to the surrounding wall or to other supports. The horizontal top member of a frame is the *head*; the vertical side members, the *jambs*; and the horizontal bottom member, the *sill*. A vertical member that subdivides a frame or a vertical member placed between and attached to adjacent frames is called a *mullion*. Similar horizontal members are also called mullions.

A *sash* is a framed unit which may be included within a window frame and may be fixed in position or arranged to open for natural ventilation or cleaning. The top member of a sash is the *top rail*; the bottom member, the *bottom rail*; and the side members, the *stiles*. If one sash is placed above another, the adjacent rails are called *meeting* or *check rails*. If one sash is placed beside another, the adjacent stiles are called *meeting stiles*. If a separate member is placed between adjacent sash, as described, it is called a meeting rail or meeting stile.

The units of glass included within a window frame or sash area are called *panes* or *lights*. If more than one light is included in this area, the vertical and horizontal members between panes and those that directly support them are called *muntins*. The panes of glass are held in position by putty, glazing compounds, metal clips, wood or metal moldings, or beads. The last one mentioned is called a *glass bead* or *stop*. Placing the glass in position is called *glazing*. Windows may be designed for glazing from the outside or the inside.

Narrow strips of sheet material in various forms made of metal, felt, rubber, or plastic, called *weatherstrips* or *weatherstripping*, may be placed around the edges of operating sash to exclude moisture and air infiltration.

A *screen* consists of wire mesh surrounded by a rigid frame which is placed in a window or sash opening to exclude insects. A *storm window* or sash is an additional glazed unit placed in a window opening to reduce the heat loss during winter. It is usually placed on the outside of the sash but may be inside. It is often interchangeable with a screen. A *shutter* is a panel provided to cover a window to exclude light, obscure vision, or to provide protection or decoration. Usually one edge is hinged to a jamb and shutters are usually provided in pairs. They are rarely used, and then only for decorative purposes, on residences, in which case they are often fixed in the open position. They are also called *blinds*. A *fire shutter*, also called a *fire window*, is fire-resistant and provides protection against outside exposure fires. It usually operates automatically.

Window Types.   The usual types of windows, regardless of the materials of which they are constructed, are shown diagrammatically in Figure 14.1. The glass areas in these examples are subdivided in various ways which may have no significance to the type of window illustrated. Some windows may include only a single pane of glass. Subdivisions may be made to permit the inclusion of ventilating units, for convenience in cleaning, to limit glass size for safety, for architectural effect, or for other reasons. Various types of windows or sash are often combined in a single window opening.

A *fixed window* is illustrated in (*a*). It makes no provision for natural ventilation.

A *double-hung window* is shown in (*b*). Both sash slide vertically, with the weight of each counterbalanced by sash weights, spiral spring

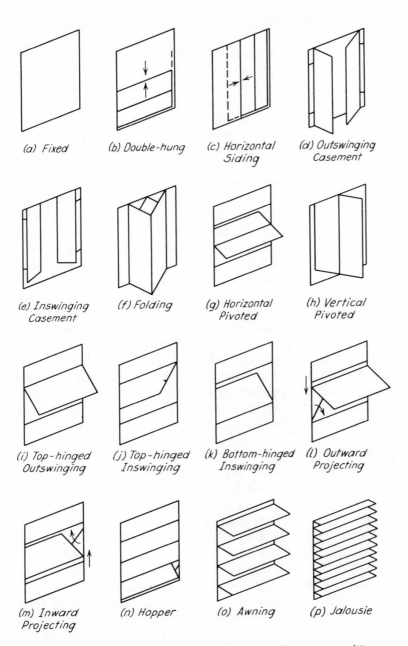

(a) Fixed

(b) Double-hung

(c) Horizontal Sliding

(d) Outswinging Casement

(e) Inswinging Casement

(f) Folding

(g) Horizontal Pivoted

(h) Vertical Pivoted

(i) Top-hinged Outswinging

(j) Top-hinged Inswinging

(k) Bottom-hinged Inswinging

(l) Outward Projecting

(m) Inward Projecting

(n) Hopper

(o) Awning

(p) Jalousie

FIGURE 14.1 Window types—outside views for open positions.

balances, or tape spring balances similar to clock springs; the sash is easily operated and edge friction will hold it in any set position. For a *single-hung window*, only the lower sash operates. Some types are arranged so that the sash can be removed from the inside.

A *horizontal sliding window* is shown in (*c*). One or both sash may be arranged to slide. Some types are arranged so that the sash may be removed from the inside. Heavy sash are often provided with nylon rollers for ease in operation. Sash are sometimes suspended from rollers operating on overhead tracks.

In general, any hinged window is a *casement window*. It may swing out or in and may be hinged at either side, the top, or the bottom, but the term is usually applied only to side-hinged windows. An *outswinging casement window* with two sash is shown in (*d*). Each sash swings on extension hinges attached to the hanging stile of the sash and the jamb of the frame. The extension provides an open space between the hanging stile and the jamb to facilitate cleaning the outside. An *inswinging casement window* is shown in (*e*). Extension hinges are used to make the sash swing clear of the inside surface of the wall. One or more casements of either type may be included in a single opening. For example, three sash could be included by providing a mullion between a single sash and a pair of sash. Outswinging casements are much more widely used than inswinging. The *folding window* is illustrated in (*f*). It is a form of outswinging casement window with the two sash hinged together on their meeting stiles rather than each to its outside stile. Projection arms, which are shown in the figure, are so arranged that the sash operate symmetrically.

A *horizontal pivoted sash*, pivoted at the center, is shown in (*g*). Such sash are often arranged in a row to form a continuous or *ribbon window* located in a sawtooth roof or monitor and are operated in unison from the floor by a mechanical operator.

A *vertical pivoted sash* is shown in (*h*). Such sash are often arranged to swing in a full circle.

Sash may be *top-hinged* and swing out (*i*) or top-hinged and swing in (*j*). They may also be *bottom-hinged* and swing in (*k*).

Windows with ventilator sash which operate like those in (*l*) are called *projected windows*. The ends of the arms are pivoted to the stile of the sash and to the frame. Shoes are attached to the top rail of the sash and move vertically along the stiles. They are guided by tracks attached to the vertical members at the sides of the opening. The window shown in (*l*) is an *outward-projecting window*, and that shown in (*m*) is an *inward-projecting window*. If the latter is located at or near the bottom of a window (*n*), it is called a *hopper ventilator*.

If several outward-projecting ventilators are located vertically adjacent to each other (*o*) and are arranged to be operated simultaneously by a single operator, the window is called an *awning window*. Fixed meeting rails sometimes are provided between adjacent sash. The downward movement of the top rail of projected sash provides an opening through which the outside of outward-projecting sash can be cleaned.

A *jalousie window* is shown in (*p*). It is similar to an awning window except the ventilating units are heavy glass slats from 3 to 8 in. wide with metal end supports to which the operator is attached. Adjacent edges of the slats overlap ½ in. or more to exclude rain and reduce air infiltration. There is considerable air leakage when closed, and they are usually used only for enclosed porches or where air leakage is not objectionable.

A *wicket panel* is a small sliding or hinged panel in an inside screen provided so the operating mechanism of an outswing sash can be reached from inside.

Designations. A procedure for indicating the type of rotating operating units included in a window is illustrated in Figure 14.2(*a*). The meeting point or intersection of the two diagonal lines on a ventilator is located on the axis about which the ventilator rotates. The procedure does not indicate whether the movement is outward or inward. There is no distinction between ventilators rotating about horizontal or vertical axes through their centers, but the axis is usually horizontal.

A system has been devised for designating windows according to the arrangement of lights and ventilators. This designation consists of digits arranged in the following order: Lights wide; lights high; number of ventilators; lights in lower, or only, ventilator; lights in upper ventilator, if any; lights between lower or only ventilator and bottom; lights between upper ventilator, if any, and bottom. Items not present are omitted in designation. The use of this legend is illustrated in Figure 14.2(*b*). The legend was devised for steel windows and is used occasionally for aluminum windows.

Factors in Selection of Type. Several of the factors considered in the selection of the type of window for a specific use are given in the following paragraphs.

Ventilation. The following comments refer primarily to buildings that are not air-conditioned but depend on natural ventilation. A vertical or horizontal sliding sash has little effect on the direction of flow of outside air into a room. A rotating ventilator tends to guide

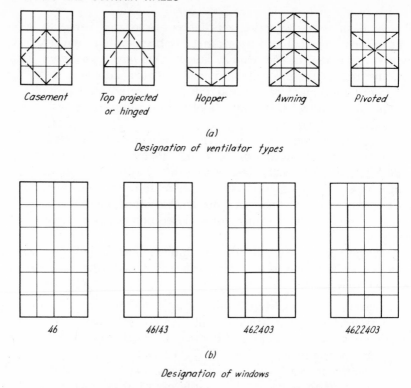

Casement   Top projected   Hopper   Awning   Pivoted
             or hinged

(a)

Designation of ventilator types

46   46/43   462403   4622403

(b)

Designation of windows

FIGURE 14.2   Designations of ventilators and windows.

the flow in the direction at which the sash is set. For this reason, projected hopper, awning, jalousie, and horizontal pivoted sash, as usually arranged, direct the inflowing air upward toward the ceiling, an effect which is usually, but not always, desirable. Similar but horizontal effects result from the directions at which casement and vertical pivoted sash are set. Outswinging casements can be set to direct air currents flowing parallel to the wall surface into a room. Outward projecting sash, awning windows, or jalousie windows serve as canopies in deflecting rain while the sash are open for ventilation. Folding windows are said to have a chimney effect which tends to draw air out of a room. Ribbon or continuous horizontal pivoted or projected ventilators operated mechanically in groups and located high up on the walls of industrial buildings or in roof monitors or sawteeth are especially effective in ventilating industrial buildings.

Screening. Any operating unit can be screened, but sometimes screening is difficult. The screen wire mesh is installed in frames and is renewable. The wire mesh may be painted or galvanized steel, plastic-coated glass fiber, aluminum, or bronze, the last two being the most durable. Screens are arranged to be installed from the inside.

Double-hung, single-hung, horizontal sliding, inswinging casements, and ventilators which open inward (Figure 14.1) may be provided with outside screens. Ventilators which open out, also shown in this figure, may be provided with inside screens. The inside screens for outswinging casement, awning, and jalousie windows do not interfere with the operators, but special provision must be made for operating outward-projecting ventilators. Horizontal and vertical centrally pivoted ventilators are difficult to screen.

A special type of screen is available which keeps all or part of the direct sunlight from entering a window, depending upon the exposure. Instead of horizontal wires, the screen has flat miniature metal slats corresponding to the slats of a venetian blind. The slats are set at the angle for greatest effectiveness. The vertical wires are more widely spaced than the usual screen wires. This type of screen, if placed outside, is quite effective in reducing temperatures during the summer, especially for rooms with south exposure.

Cleaning and Repairing. The possibilities for cleaning, replacing broken lights, or eliminating water leakage are important factors to consider in selecting the type of window. Double-hung, single-hung, and horizontal sliding sash are often removable for cleaning or repair from the inside. Outswinging casements are fitted with extension hinges which provide an open arm space between the hanging stile and the jamb. Sometimes the entire outside surface of a window can be reached from the inside for cleaning by adjusting the position of the ventilator. In others, the ventilator or movable sash provides for access to the outside for the window cleaner. Except for windows close to the ground or an adjacent roof surface, anchors for the safety belts of window cleaners are often attached to the jambs of windows. Various types of installations have been devised to provide for window washing to be done entirely from the outside even if all the windows are fixed.

Vertical pivoted sash which can rotate through a full circle and top-hinged inswinging sash are designed especially for use in air-conditioned buildings where the sash are only opened while their outsides are being cleaned. Off-season and emergency ventilation may be provided by including a hopper ventilator in the window with one of the types of sash mentioned.

The suitability of the various types of windows for cleaning can be judged by examining Figure 14.1. If the windows are screened, it will be necessary to remove the screens on some types.

Other Factors.   Many other factors must be considered in selecting the type of window, regardless of the material of which it is constructed. A ventilator which projects inward may interfere with the use of space adjacent to a window and with the operation of shading equipment such as window shades and venetian blinds; vertical and especially horizontal members in a window may interfere with desired vision through a window; large glass areas may result in personal injury from glass breakage when someone falls against a window; need for access to the controls of operating units may restrict the placing of furniture or equipment against walls; and the type of occupancy and the appearance may have a significant effect on the selection. Such factors as durability and cost are more closely related to the designs and the materials used in construction than to the type of window.

Minimum Area.   Every habitable room, that is, every room arranged for living, eating, or sleeping purposes, must be provided with natural light and ventilation by one or more windows opening on a street, alley, or court. The total glass area in the windows must be not less than $\frac{1}{10}$ of the floor area, with a minimum total glass area of 10 sq. ft., except in bathrooms where it is 3 sq. ft.

Windows or other openings required for ventilation must have a total openable area of at least 50% of the glazed area required for lighting.

This requirement does not apply to rooms provided with artificial ventilation or air conditioning. Windowless rooms are common, and even windowless buildings are frequently constructed because of the better control they afford for illumination, ventilation, temperature, and humidity than is possible with windows.

Special requirements are included which apply to other types of occupancy and to artificial lighting and ventilation.

81.   WOOD WINDOWS

General Comments. Wood windows are used extensively in many types of buildings when not excluded by the fire-resistive requirements of building codes.

Wood windows have been used for many centuries. The large-scale production of such windows by machines in factories started in this country about 1840 [1]. They are now used more than windows of any other material for residences. The most widely used type is the double-hung window, as illustrated in Figure 14.1(b), although out-swinging casement in (d) is quite common, and horizontal sliding in (c), the outward projecting in (l), the hopper in (n), the awning in (o), and the jalousie in (p) are also used.

Double-Hung Window. A detailed drawing illustrating the various parts of a wood double-hung counterweighted window for placing in a masonry wall is given in Figure 14.3. It consists of three principal parts: the frame, the sash, and the interior trim. The principal parts of the frame, including the head, jambs, and sill and those of a sash, including the top rail, bottom rail, and stiles, are defined in Article 80. These and the other parts of a double-hung window are shown in Figure 14.3, and the other parts may be defined as follows:

Sash weights. Metal weights to counterbalance the weight of a sash and facilitate its operation and to hold it in a set position when raised.
Weight box. The box in which the weights operate.
Sash cord or chain. Cord or chain which connects a sash to the weight.
Sash pulley. A metal unit consisting of a box containing a sheave or special type of wheel over which the sash cord extends between the sash and a sash weight. Also called a *frame pulley*.
Pulley stile. The part of a weight box adjacent to the sash and along which sash slides.
Back lining. The part of a weight box next to the masonry.
Box casing. The side of a weight box.
Pocket. The removable section of a pulley stile which gives access to the interior of the weight box.
Pendulum. A thin partition of wood or sheet metal in the weight box to keep the weights from interfering with each other. It is fastened at the top only, so that it can be pushed aside to reach both sides of the weight box through the pocket for repairs.
Parting strip. The guide between the two sash.
Stop or stop bead. The outer and inner guides for sash.
Yoke. The top member of the frame, also called the head.
Sill. The bottom member of the frame.
Stool. The sill of the interior finish.
Apron. The part of the finish below the stool.
Casing. The interior trim at the sides of the opening.
Head casing. The interior trim at the top of the opening.
Grounds. The strips under the interior trim arranged to serve as guides to fix the thickness of plaster and as nailing strips for the interior finish.
Brick mold. The mold in the outside corner between frame and brickwork.
Staff bead. The general term for brick mold to apply to other materials.
Reveal. The exposed masonry on the jamb between the frame and the outside face of the wall. Also, the corresponding surface on the inside.

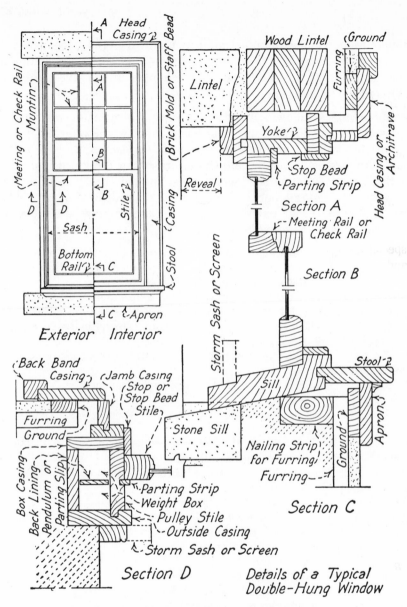

FIGURE 14.3 Details of wood double-hung window.

Jamb casing. The interior trim on jambs of the opening where the frame is set to give a reveal on the inside of the opening.

To permit the use of narrower interior trim on window openings, narrow rectangular sash weights may be used instead of the common cylindrical sash weights. The same result is accomplished by spiral-coiled springs instead of sash weights. These are placed in recesses in the exposed faces of the frame stiles and the sliding edges of the sash stiles. Sash balances consisting of flat coiled-steel springs, like clock springs, with metal housings are mounted at the tops of the frame stiles with their ends attached to the edges of the sash stiles. They balance the weights of the sash by coiling and uncoiling as the sash are operated.

The sash of all types of counterbalanced double-hung windows are held in set positions by friction between the stiles of the frame and the sash.

Casement Windows. An outswinging casement window with two sash for a masonry wall is illustrated in Figure 14.4(a); and an inswinging casement window, in (b). The principal parts of each window are the frame, the sash, and the trim. Many of the parts of these windows have corresponding parts in the double-hung window, which were defined. They are illustrated in the figures. A part peculiar to this type of window is the *astragal*, which may be provided to cover the joint between meeting stiles and make it more weathertight.

Casement windows may include one or more sash, the two-sash casement being the most common. All types can be controlled by operators located on the inside.

Species of Wood. The wood used for the exposed parts of wood windows should be highly weather-resistant and should not warp or shrink. The most commonly used wood is ponderosa pine, but white pine, sugar pine, fir, redwood, cypress, and cedar are also used extensively, the selection being governed largely by relative availability at a given location. The jambs of double-hung windows are subjected to considerable wear, and therefore they should be made of some relatively hard wood such as yellow pine. They are sometimes protected with aluminum. The parts which are exposed only on the inside are made of the same wood as the interior trim.

Frames and sash should be treated with water-repellent wood preservative. All exposed exterior surfaces of wood windows should be protected by periodic painting and interior surfaces by appropriate finishes, except the contact surfaces of double-hung and sliding windows, which should be protected by oiling.

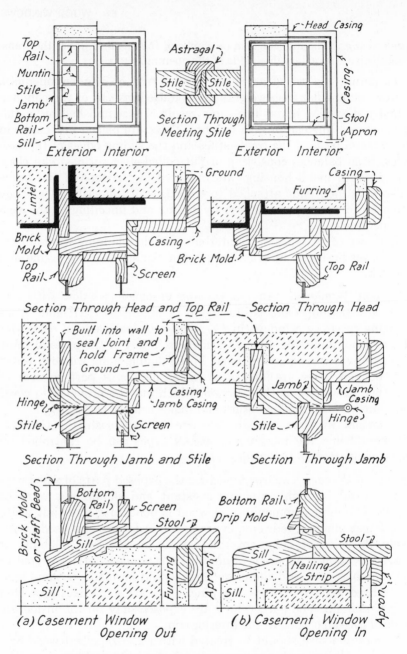

FIGURE 14.4 Details of wood casement windows.

Thickness of Sash. The actual thickness of sash most often used is $1\frac{3}{8}$ in., but $1\frac{3}{4}$ in. material is used for the better class of construction and $2\frac{1}{4}$ in. for the best large windows. Small basement sash may be $1\frac{1}{8}$ in. thick. Screens are normally $\frac{3}{4}$ or $1\frac{1}{8}$ in. thick and storm sash $1\frac{1}{8}$ in. thick.

82. METAL WINDOWS

Metals used for window construction are aluminum, steel, bronze, and stainless steel (arranged in order of the quantity used). Steel windows were introduced in the U. S. early in the twentieth century. The use of aluminum windows on a commercial scale started about 1930.

Aluminum Windows. All types of aluminum windows are made from sections or shapes made by extruding an aluminum alloy through dies which produce cross sections of the desired size and shape. The thickness of the metal used for windows in residences and low-cost housing projects varies from about $\frac{1}{16}$ to $\frac{1}{8}$ in.; and for those in industrial, commercial, and monumental buildings such as stores, factories, schools, and office buildings, the thickness varies from about $\frac{1}{16}$ to $\frac{3}{16}$ in. All the window types illustrated in Figure 14.1 are available. The most widely used type is probably the projected window.

The cross sections of the members of aluminum windows are similar to those of corresponding solid-section steel windows shown in Figure 14.5.

Aluminum windows will not corrode, except under extreme conditions, and thus do not require painting. The usual types of finish are the mill or natural finish; the satin finish produced by etching, belt polishing, or rubbing with abrasives; the bright finish produced by buffing; and the anodized finish, which is an electrolytic finish that provides a much thicker and more protective oxide coating than the one that forms naturally. The anodized finish or coating can be given any one of the three finishes first mentioned. Aluminum windows should always be coated with a special temporary clear lacquer applied by the manufacturer to protect against the action of mortar and plaster during construction [1].

Any possible galvanic action between aluminum and other metals, as explained in Article 76, should be avoided. This possibility should be considered in the selection of hardware and weatherstripping. The preferable nonaluminum metals for use with aluminum are nonmagnetic stainless steel, heavily galvanized steel, and zinc. Copper and

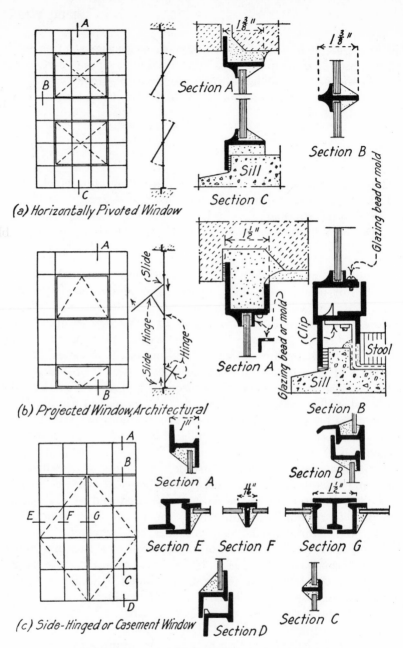

(a) Horizontally Pivoted Window

Section A

Section B

Section C

Sill

(b) Projected Window, Architectural

Section A

Section B

Glazing bead or mold

Glazing bead or mold

Slide Hinge

Slide

Hinge

Hinge

Clip

Stool

Sill

(c) Side-Hinged or Casement Window

Section A

Section B

Section E   Section F   Section G

Section D

Section C

FIGURE 14.5   Solid-section steel windows.

nickel alloys should be avoided. Since moisture must be present for galvanic action to occur, watertightness and drainage are important precautions against galvanic action, as are protective coatings, suitable gaskets, and seam compounds. The drainage of salts from non-aluminum metals over aluminum windows should also be avoided. Window details which require one aluminum surface to slide over another such surface should not be used because of the excessive wear which may occur [1].

Steel Windows. Steel windows are constructed of hot-rolled solid sections and of cold-roll-formed strip steel. The solid sections are used for nearly all industrial types of window and for the casement and projected types for nonindustrial buildings. Cold-formed strip steel is used principally for double-hung windows in nonindustrial buildings. It has not proved to be economically feasible to shape steel for window construction by extrusion through dies. Almost any form, type, or size of steel window can be manufactured if the cost can be disregarded. Because of the high cost of equipment used for forming solid sections or cold-formed sections and for assembling the shapes into windows, the shapes of sections and types and sizes of windows have been standardized to a considerable extent, but substantial savings could be made by further standardization [1]. All the types of window illustrated in Figure 14.1 are available.

Solid hot-rolled section steel windows are classified according to the kinds of building for which they are best suited: *Architectural* for buildings in which appearance and quality are of major importance, such as residences, apartments, hotels, office buildings, school buildings, and hospitals; and *commercial* and *industrial* for which a plain substantial design is suitable, such as stores, factories, warehouses, storage buildings, and power houses.

Cross sections of some of the members of three types of steel windows are shown in Figure 14.5. All the members of these windows are made of hot-rolled sections.

Windows made of hot-rolled sections are usually protected against corrosion by a special factory treatment of the surface, called *bonderizing*, to improve the adhesion of paint, and then by painting periodically as required. Windows made of hot-rolled steel sections are sometimes galvanized by hot-dipping the completed window, but this practice is not common.

The use of steel windows has been somewhat curtailed. They have been largely replaced by aluminum windows. Their use is now confined mostly to security installations such as psychiatric hospitals and prison works. There is, however, a plastic-covered steel window that is used for public buildings.

Bronze and Stainless Steel.   Bronze and stainless steel are attractive materials which do not corrode and are highly weather-resistant. The shapes used in constructing bronze windows are extruded in the same manner as those for aluminum windows. For this reason, any of the window types available in aluminum can also be constructed of bronze.

Stainless-steel windows are made from cold-roll-formed strip metal following details similar to those used for double-hung steel windows. The vertical pivoted type is also manufactured using similar details.

Windows constructed of these materials are expensive and therefore are used only on the highest type of buildings. Because of the small demand, they are custom-made.

83.   GLAZING MATERIALS

Composition of Glass.   Glass is composed of about six parts white sand, one part lime, and one part soda, with small amounts of alumina and other materials.

Classification.   Glass for glazing purposes is classified as follows by Specification 123 of the Federal Specification Board.

Polished plate glass	{ Second silvering quality { Glazing quality	
Clear window glass	Single-strength	{ A quality { B quality
	Double-strength	{ A quality { B quality
	Heavy sheet	{ Glazing quality { Factory-run quality
Processed glass	Chipped	{ No. 1 processed { No. 2 processed
	Ground	{ Acid-ground { Sand-blasted
Rolled figured sheet	{ Figured sheet { Colored figured sheet	Large variety of patterns

	⎧ Polished wire
	⎪ Polished (one side)
Wire glass	⎨ Figured
	⎪ Corrugated
	⎩ Colored

| Ornamental | Figured plate (polished one side) |

	⎧ Pressed tile
Prism glass	⎨ Rolled sheet
	⎩ Rolled and pressed sheet

Definitions and Manufacture.   The definitions and brief statements of the methods of manufacture of the various kinds of glass, as given by Specification 123 of the Federal Specification Board, are as follows.

"Plate glass.   Transparent, flat, relatively thin glass having plane polished surfaces and showing no distortion of vision when viewing objects through it at any angle. Plate glass is made at present by casting and rolling large sheets periodically or by rolling a continuous sheet. The sheets are then ground and polished.

Clear window glass.   Transparent, relatively thin, flat glass having glossy, fire finished, apparently plane and smooth surfaces, but having a characteristic waviness of surface which is visible when viewed at an acute angle or in reflected light. Clear window glass is made at present by hand blowing or by machine blowing and drawing into cylinders and flattening, or by drawing directly into a sheet, the surface finish being that obtained during the drawing process.

Processed glass.   There are three kinds of processed glass either in plate or window glass, viz., ground glass, chipped one process, and chipped two processes. The ground glass is made by either sandblasting or acid etching one surface. The chipped glass is made by applying either one or two coatings of glue to the ground surface. The glue is applied hot and in cooling and drying shrinks and pulls small chips off of the surface of the glass.

Rolled figured glass.   A flat glass in which the vision is more or less obscured either by the roughened surface produced in rolling or by the impression of a large variety of decorative designs in one surface of the sheet.

Wire glass.   Rolled flat glass having a layer of meshed wire incorporated approximately in the center of the sheet. This glass is produced with polished or figured surfaces.

Ornamental plate.   A figured plate glass made by rolling or rolling and pressing and having the plane surface ground and polished.

Prism glass.   A flat glass having prism-shaped parallel ribs designed for deflecting light. This is made as a rolled plate or as a pressed plate, of which one side may be ground and polished, or as a pressed tile."

Thicknesses and Grades.   The most common thickness of polished plate glass is $\frac{1}{4}$ to $\frac{5}{16}$ in., but $\frac{1}{8}$ in. and $\frac{3}{16}$ in. glass is fairly common. Other thicknesses up to $1\frac{1}{4}$ in. are obtainable by special order. *Second silvering quality* is used where the highest quality is desired, but most of the glass used is known as *glazing quality*.

The most common thicknesses of clear window glass are *single strength*, varying in thickness from $\frac{1}{12}$ to $\frac{1}{10}$ in., and *double strength*, varying in thickness from $\frac{1}{9}$ to $\frac{1}{8}$ in. Other thicknesses up to $\frac{1}{5}$ in. are obtainable. All clear window glass should be relatively flat, but a slight irregular curvature is not objectionable if it does not exceed 0.5% of the length of the sheet. Glass with a reverse curve or which is crooked should be rejected. A small amount of AA quality window glass is sometimes selected for special purposes and may be obtained at a high price. The grade most often used in windows where appearance is an important factor is A quality, but B quality is used quite extensively. Two grades inferior to B quality are on the market. They are Fourth quality and C quality. Fourth-quality glass contains many defects and distorts the objects viewed through it. It should never be used where vision is important. C quality is too poor for use in buildings.

Rolled figured sheet glass is made in thicknesses from $\frac{1}{8}$ to $\frac{3}{8}$ in. It is made in a great variety of surface finishes which obscure the vision, diffuse the light, and give decorative effects.

Wire glass is made in thicknesses from $\frac{1}{8}$ to $\frac{3}{4}$ in., the standard thickness being $\frac{1}{4}$ in. Only one quality is manufactured for glazing purposes. It is made with polished surfaces and with a great variety of surface finishes which obscure the vision, diffuse the light, and give decorative effects. The wire is in the form of a wire mesh, the standard size of mesh being $1\frac{1}{4}$ by $\frac{7}{8}$ in. and the weight of wire No. 24 B & S gauge.

The wire is placed in the glass by any of three methods: (1) by rolling a sheet of glass, placing the mesh on it while the glass is still plastic, pressing the mesh into the glass, and finishing the surface; (2) by rolling a thin sheet of glass, placing the mesh, and rolling another sheet of glass on the first sheet; (3) by placing the wire on the casting table and holding it in position while the glass is poured around it. Polished wire glass is not of the same quality as polished plate glass.

Safety and Bulletproof Glass. Laminated glass, built up of layers of glass between which are cemented layers of a colorless transparent plastic resembling celluloid, is called *shatterproof* or *safety glass*. The chief use of a glass of this type is in automobiles, but it is also used for skylights and in the windows of asylums. *Bulletproof glass* or *bullet-resisting glass* is a thick safety glass used in banks. Ordinary safety glass is ⅛ to ¼ in. thick, but bulletproof glass has several laminations built up to thicknesses of ½ to 2 in. The thickness most commonly used is 1⅛ in. This glass will not be penetrated by bullets from most firearms, although 2 in. glass is recommended to resist shots from a 30-30 rifle. These types of glass will crack under impact, but the plastic layers hold the various pieces of glass together so that it does not shatter.

Tempered Plate Glass. Tempered glass (see Article 88) is used where safety precautions are required. Store fronts (and other areas accessible to the public where glass breakage could injure) should be glazed with tempered glass. Glazing installation may be from ¼ in. to 1¼ in. in thickness. Tempered glass may frequently be identified by two tong marks on one edge of the glass, although these tong marks are usually covered by glazing materials. Tempered glass is obtainable without tong marks.

Heat-Absorbing Glass. *Heat-absorbing* or *actinic glass* is a kind of glass which, because of its special composition, excludes a high percentage of the ultraviolet rays to which it is exposed, transmits less solar heat into a building than other kinds of glass, and has less bleaching effect on colored fabrics.

This type of glass has the advantage of reducing air-conditioning loads and costs. It should not be set in frames where the glass will be partially shaded. For if a portion of the glass is shaded and a portion is in direct sunlight, a differential in expansion and contraction will likely result. If differential expansion does result, cracking of the glass is likely.

Quartz glass transmits a larger percentage of ultraviolet rays than ordinary window glass. Since these rays are beneficial to the health, this kind of glass may be used where this quality is desired.

Multi-Glazing. Windows are often glazed with factory-produced panes consisting of two or three parallel sheets of glass separated by a thin dehydrated air space maintained by a tight seal between the sheets and located around their edges. Because of the seal and the dehydrated air, there is no condensation or accumulation of dirt within the air space. The heat transmission through the window is substantially reduced during the winter, and the heat gain during the summer is correspondingly reduced, especially if the outer sheet is heat-absorbing glass. Various other combinations, which may include tempered plate, are used. Multi-glazing also reduces sound transmission. This type of insulation is called *insulating glass*.

Reflective Glass.   A glass that is of laminated construction with a reflective film in the middle is called *reflective glass*. This glass behaves as a mirror reflecting heat and glare. Thus, it is also used as a heat absorbing glass.

Low-Transmission Glass.   Tinted or coated glass is designed to reduce light transmission. Called *low-transmission glass*, it is made in several shades of gray and bronze. These glasses are identified by their transmission capabilities, which are expressed as a percentage of the light that passes through.

Psychiatric Glass.   This glass, which features a mirror on one side and on the other side appears as transparent glass, is called *psychiatric glass*. It is popularly known as *one way glass*. To function properly, the side of the glass that is viewed as transparent must be on the side in which the light intensity is less than the side that appears as a mirror. This glass is used for security installations or where private controlled observations are desired.

Selection of Glass.   Polished plate glass is used for exposed windows in the better grades of buildings. It is much superior to window glass in appearance and in the clearness of vision through it but is much more expensive. Large windows such as show windows are always made of polished plate glass.

Clear window glass is extensively used in all classes of buildings.

Chipped and ground glass are used to a limited extent where light is to be admitted but vision is to be obscured. Chipped glass is more attractive than ground glass and is used in interior partitions. Rolled figured glass serves the same purposes as chipped and ground glass, is usually cheaper, and is available in more attractive designs.

Rolled figured glass is extensively used on the exterior and interior of buildings to obscure the vision or diffuse the light. Many attractive designs are available, and the cost is less than that for clear window glass.

Wire glass may be used in outside windows because of its resistance to fire. When heated and drenched with water it will crack but, owing to the action of the wire mesh, it will remain in position and protect the interior of a building from fires originating outside except when exposed to extremely high temperatures. Wire glass is also used in doors and in other positions where breakage is likely to occur. It will continue to give service even though badly fractured.

Prism glass is used to light areas remote from a window. Light striking the glass is deflected so that it is effective for a considerable distance away from the glass. A common use for prism glass is in the upper part of store windows where the stores are deep and windows are in the front only.

It is desirable to use some type of figured glass in basement windows below grade because such windows cannot be kept clean and dirt is conspicuous on clear glass. The diffusing effect of figured glass is also usually desirable.

Glazing.   The process of placing glass in windows is called *glazing*. Rebates or rabbets are provided in the edges of the members that support the panes of glass. They must be deep enough to receive the glass and provide the space required by the medium which holds the glass in place. This medium may be putty, glazing compounds, and wood or metal moldings called *glass beads* or *stops*. Before putty is placed, the panes of glass in wood sash are held in place by small triangular or diamond-shaped pieces of zinc called *glazier's points*, driven into the rabbet. For metal windows, glass is held in a similar manner, before the glazing compound is placed, by spring wire *glazing clips* or other devices.

Wood windows are usually glazed from the outside. Some types of metal windows are glazed on the outside and others on the inside. If a pane of glass is set in position and the putty applied with a putty knife or by other means as shown in Figure 14.6(*a*), the operation is called *face puttying. Back puttying* consists of forcing putty into any spaces that may be left between the edges of the sash and the back of the glass. Before the glass is placed in position, a layer of putty may be spread on the surfaces of the sash to receive the glass. This operation, called *bedding* the glass in putty, is illustrated in (*b*), where the final step consists of face puttying. In (*c*), a pane which is bedded in putty in a wood frame is held in position by a wood *glass bead*. In (*d*), a similar pane in a steel frame is held in position by a steel *glazing angle*. Such continuous metal moldings are required on industrial windows. The objectives of bedding are to provide a cushion

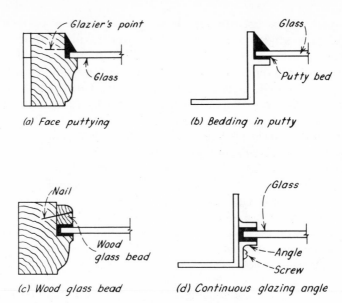

FIGURE 14.6   Methods of glazing.

for uniform support of the glass in order to avoid cracking and to make a water- and airtight seal. Other means are used for holding panes of glass in position.

The selection of glass and the shading of windows to reduce the solar heat transmitted through them are considered in Articles 92 and 97.

Structural Glazing Gaskets.   Neoprene gaskets that are designed to receive and hold glass are called *structural glazing gaskets*. These gaskets, placed in a structural supporting frame, are capable of holding large and thick sheets of glass. The glass is placed in the gasket and a neoprene strip insert is placed in a prepared slot. The gasket then firmly holds the glass in place. See Figure 14.6.

84.   CURTAIN WALLS

Metal Curtain Walls

by W. S. Kinne, Jr., Director, University Facilities Research Center, Madison, Wisconsin

A metal curtain wall is an exterior building wall which carries no roof or floor loads; it is fastened to the structural frame and acts solely as an enclosing envelope. It usually consists of metals in combination with glass, plastics, and other surfacing materials, supported on metal unit frameworks. There are three basic types of curtain walls now in general use. *Custom* walls are designed specifically for one major project, using parts and details made specially for this purpose. *Commercial* walls are made up principally of parts, components, and details standardized by a manufacturer and assembled either in the fabricator's stock patterns or in accord with the architect's general design. *Industrial* walls are those in which ribbed, fluted, or otherwise preformed metal sheets in stock sizes are used for major surface coverage, along with standard metal windows, as the principal elements. The commercial and industrial types are sometimes referred to as *window walls*.

Custom curtain walls are most often used for high-rise buildings or multibuilding complexes—office buildings; large-scale housing and apartment developments; corporate building programs; large municipal service expansions such as airport terminals; and major new campus expansions of high school, junior college, college, and university organizations. Commercial curtain walls have had their most frequent use for local educational buildings at the primary, elementary, and secondary school levels, for commercial and industrial office accommodations, and for semispeculative buildings built primarily for rentals to other than the original owners. Industrial curtain walls have been used for large-scale, essentially low-cost wall enclosures for factory and loft types of occupancies, as well as for low-budget applications sometimes considered to be adequately serviced by the commercial curtain wall types.

In metal curtain walls, major emphasis is placed on repetitive factory parts production and prefabricated assembly of reasonably thin, large-area wall surface units, with the objective of keeping assembly and erection labor on the building site at a minimum.

The chief functions of a metal curtain wall are to *withstand* the actions of the natural environment—wind, rain, snow, hail, sleet, and sunlight; to *prevent* access by intruders, the absorption or retention of water vapor, and damage by fire; and to *control* the flow of heat and the passage of air, light, and sound. It must be aesthetically appropriate and have permanence consistent with the design life expectancy of the building enclosed by the wall. The possibility of curtain wall replacement, or substitution, within the useful life of the basic structural frame is always present.

The principal elements of a metal curtain wall are the supporting curtain wall frameworks, usually vertical but sometimes horizontal *mullions*; the *glazing units*, operating windows (see Article 80) or fixed glass panels; the opaque enclosing *panels*; the *anchors* for the attachment of the curtain wall system units to the structural frame of the building; and the *joints* and *sealants* at the junctions between wall units, which compensate for movements within the curtain wall caused by external forces of nature, by movements of the building frame, and by thermal changes. *Solar screens* are often made an integral part of the curtain wall to be a means of controlling the thermal effect of sunlight both on the wall itself and on the internal building spaces enclosed by the wall.

The *mullions* are usually manufactured to a structurally stable profile made from extruded aluminum or aluminum sheet and solid stock, often protected with electrochemical anodic coatings; or they are made of rolled stainless or carbon steel or extruded bronze. Color can be applied to the mullions as part of the basic manufacturing process by the use of coatings of porcelain, baked enamel, vinyl, and other plastics or specialized anodic treatments. Vertical mullions usually extend through one or two stories of the building's height, with slip joints to provide watertight junctions while allowing linear expansion movement.

Sash frames for *glazing units* are usually attached to, or are an integral part of, the mullion system. Operating window sash are often chosen primarily for the ease of washing glass surfaces, since many buildings of the kind that use metal curtain wall have year-round air conditioning, and window ventilation is not an important factor except in between the seasons. Where fixed glass is used, exterior glass washing is often accomplished by power-driven vertical stagings suspended from the roof and restrained from side sway by movable attachments to the mullion system portions of the curtain wall.

Panels are usually composite fabrications of two or more materials into a single assembly. Two principal types are produced, classified by the method of assembly. *Mechanical panels* are held together by screws, rivets, or other mechanical fasteners, and *laminated panels* are held together with adhesives. Panels are designed to provide adequate strength, dimensional stability, thermal insulation, resistance to weather and corrosion, control of condensation, sufficient sound isolation, the necessary degrees of fire resistance, and desired appearance. Structural glass is sometimes used as a panel, backed up with an air space and a secondary panel to provide the desired physical qualities. Composite panels are covered with a wide variety of exterior facings in a wide variety of color and texture ranges. Porcelain on steel sheet or aluminum sheet, thin-set ceramic tile, thin precast concrete or terrazzo, preformed chemically treated metal sheet, and plastic-coated inorganic sheets are frequently employed. Among the more common core materials for laminated panels are treated paper expanded honey comb, foamed closed-cell insulation units, and compressed glass fiber insulation boards. Vapor barriers on the warm side of the assembly are advisable for all conditions needing condensation control.

Several notable custom curtain wall systems have been designed with the idea of having a preformed metal panel of large size become the entire curtain wall unit assembly. Here the large single-sheet metal units have the structural capability of acting as substructural mullions, present an arrangement possible for the introduction of glazing elements within their unit area, and make possible logical jointing and sealing junctures with adjacent similar units. These units can be laminated to, or backed up with, other materials to make overall physical performance acceptable.

Anchors which provide for the support of curtain wall units at the points of contact with the building frame are usually factory-fabricated from structural steel shapes which are protectively coated by being galvanized or bonderized for corrosion resistance; these frequently employ stainless steel bolts and point-of-contact fasteners. Anchor assemblies must be designed to allow rather sizable dimensional tolerances, in the range of ½ in. to ¾ in. plus or minus, adjustable during the field erection process in three directions: (1) *vertical;* (2) *horizontal, parallel* to the curtain wall surface; and (3) *horizontal, perpendicular* to the wall.

Curtain wall units and assemblies are manufactured and fabricated to closely predictable sizes, but the basic building structural elements to which the wall will be attached are much less precisely dimensioned at the erection site. Thus, flexibility of anchorages is very important in assuring a logical and sure junction of wall and structure.

Joints and *sealants* for metal curtain walls are perhaps more critical to the satisfactory performance of the wall than they have been to conventional stone and brick masonry exterior walls. For the metal curtain wall, materials are essentially nonabsorbing. Infiltrated water or internal condensation in an essentially metal assembly will normally collect in substantial quantities. Then it may be forced through the wall as a leak or remain entrapped, causing most undesirable effects on insulation materials and metal fittings subject to corrosion. In contrast, moisture in more absorbent masonry and its joints can be evicted by flashings and removed by slow evaporation without serious local effects.

Metal curtain walls are essentially *multimaterial* assemblies. The variety of materials used have varying coefficients of thermal expansion, as shown in Table 14-1.

Thus, it may be seen that large metal curtain wall units or their

TABLE 14-1

Dimensional Change, in Inches, Caused by 100°F Temperature Difference for 10 Lineal Feet of Material

Aluminum	0.156	Glass	0.060
Stainless steel	0.115	Concrete	0.096
Porcelain on steel	0.084	Stone	0.048
Bronze	0.113	Reinforced plastic	0.194
Carbon steel	0.084		

components will grow and shrink in different magnitudes within ranges of temperature encountered on the exterior of a building. The *joints* between the units, and within them, are designed to provide for these movements mechanically. Then they are sealed to control air and water migrations within the wall, with the idea of repelling and expelling water to the outside while preventing inward or outward air movements which might be disturbing to control of interior air conditioning and ventilation. The *sealants* or gaskets must have qualities of long life, adequate elongation without failure, and good adhesion. They may be considered to be of two types, *liquid sealants* and *preformed gaskets*, both manufactured as synthetic plastics. The usual sealants are polysulfide polymers, known as Thiokol (a trade name). These are applied in liquid form, but at room temperature they convert to flexible synthetic rubber. The preformed gaskets are factory-made as extrusions in profiles appropriate for their use. They are most often polyvinyl chloride plastics or synthetic neoprene rubbers, chemically designed to provide desirable properties of hardness, durability, elongation, and adhesion.

Solar screens should be considered an integral part of metal curtain walls or as a potential adjunct. The idea of controlling the effect of solar energy on the interior of a building has long been understood; but the metal curtain wall assembly, using materials with relatively high coefficients of heat transmission and relatively large glass areas, emphasizes the desirability of diverting or reflecting solar heat before it touches the wall surface. This solar energy can then be dissipated in the atmosphere before it enters the interior of the building envelope.

Glass normally covers a rather substantial surface area of metal curtain walls, ranging from less than 20% to about 80% of the total exposed surface of the wall. The general subject of glass is covered in Article 83. It might be well to describe here the variations of heat transmission expected from the kinds of glass that are usually employed in metal curtain wall assemblies.

Heat transmission, sunlight through glass

¼-in. plate glass—88% heat transmitted through.

¼-in. heat-absorbing glass—66% heat transmitted through.

¼-in. heat-absorbing glass, air space, ¼-in. plate glass—50% heat transmitted through.

These data represent the sum of reradiation and transmission. For further information, see Article 94.

There are three types of exterior solar controls now in general use: *horizontal overhangs*, canopies, or eyebrows. These may be solid or louvered and are usually fixed but sometimes movable. They are most effective for southern exposures in the Northern Hemisphere but can be quite effective on northern exposures in the tropics. Perforated *vertical screens*, or preassembled grid systems placed parallel to the wall and window system, can be effective for all orientations. They are usually fixed installations and often become a rather dominant aesthetic factor, for they are literally a screen wall outside of the curtain wall. Outside *vertical fins* or *louvers* are most effective for eastern and western exposures in the middle and high latitudes of the Northern Hemisphere, especially when the fins are movable to compensate for varying early-morning and late-afternoon sun angles through the year.

Interior solar screens, although somewhat less effective than the exterior type because the diverted solar energy is trapped within the building, are still important components of metal curtain walls with large glass areas. With their use, trapped heat can be isolated on the inside of glazed areas and diverted by air-conditioning installations designed for the purpose. Horizontal or vertical retractable venetian blinds placed in pockets or tracks built into the curtain wall system, often in conjunction with heating or cooling elements of mechanical systems, are common. One prime advantage of the interior solar screen idea is that it allows greater flexibility of use under the direct control of the individual building occupant—dictated by considerations of sun position, cloud cover, and glare implications.

Concrete Curtain Walls

Precast concrete curtain walls are considered in Article 53. They are appropriate for use when windows occupy a relatively small portion of the wall area. The following discussion is concerned with such walls when appearance is an important factor. It is based on a bulletin entitled *Concrete Curtain Walls* published by the Portland Cement Association [8].

Concrete curtain walls are made up of precast reinforced concrete slabs whose size and shape can be selected to meet specific requirements. A common size of slab is 8 × 14 ft. The thicknesses range from 4 to 6 in., depending upon code requirements and upon the degree of fire-resistance and heat insulation required.

Composition.   The slabs are constructed of lightweight concrete not only to reduce their weight but also to improve their heat insulation properties. They may be of the same composition throughout, or a layer about 1 in. thick, adjacent to the exposed face, may have a special composition for decorative effect.

Color.   Various colors are provided for the facing layer by selecting appropriate aggregates or mixtures of aggregates and mortars. Some of the aggregates used are quartz, marble, granite, gravel, tile and ceramic, and other vitreous materials. Each of these is available in a wide range of colors. The ceramic and vitreous materials are manufactured and thus can be produced to match samples submitted by the architect. Ceramic facing tile are available in a great variety of colors and patterns. The tile are laid in cement mortar applied to the face of the concrete slab and are usually laid in regular rows which occupy up to 90% of the exposed surface. They are often 1 in. square or 1 × 1½ in. rectangular tile.

The mortar or matrix in which the aggregate is embedded has a marked effect on the color. Curtain walls whose color depends primarily on the color of the matrix can be made. White portland cement is usually employed in the facing mix to insure purity of color even for the darker shades. The matrix is colored by using mineral oxide pigments. Practically all colors can be obtained in this manner.

Texture.   The textures of the exposed surfaces of curtain walls are important factors in their appearance. They vary from glossy to rough or rugged textures, depending upon the kind of surfaces against which they are cast, upon the treatment the surfaces receive after removal from the forms, and upon the size and shape of the aggregates. An exposed aggregate finish is often used. It is produced by using various techniques in casting and finally brushing the surface to expose the aggregate. Exposed aggregate facings may be ground smooth to resemble terrazzo. Desired effects are sometimes secured by mechanical means such as bush-hammering, tooling, and sandblasting. Finally, the characteristics of the surface against which the facing is cast largely determines its texture unless the surface is treated after removal from forms.

Patterns.   Patterns of various designs are created on the exposed faces of panels by high and low relief in adjacent areas, by differences in the aggregate colors, and by contrasting textures. In addition, any desired form can be produced on the exposed face by casting against an appropriate mold.

Panel Shapes.   Panels of almost any desired shape can be manufactured. The most common shapes are rectangular and square, but diamond-shaped panels have been used.

Panels are cast with open grillwork over a part or all of their surface to serve as a solar screen on the outside of glazed areas. The same effect is achieved by building a wall of small perforated units located just outside the wall containing the windows.

Erection.   The procedures followed in placing panels in position and anchoring them in place correspond to those described in Article 53.

Heat Insulation.   The effectiveness of lightweight concrete aggregate curtain walls can be increased by plastering the inside surfaces with perlite or vermiculite plaster. Sandwich panels, described in Article 53, are also used to achieve a high degree of insulation.

Fire Resistance.   Concrete curtain walls can be constructed with the thickness and quality of concrete necessary to conform to any building code requirements.

Other Materials

Curtain walls are also constructed of brick masonry, stone masonry, structural clay tile, concrete block, and glass block, as described.

Fire-Resistance Requirements

Curtain walls as an important feature in the architectural treatment of buildings have developed markedly in recent years. Few building codes have been modernized to recognize this development adequately. There are also major differences in those that have been modernized. As one example, the requirements of the National Building Code for the highest-quality *fire-resistive building* are quoted in part. The requirements for lower types of buildings are less severe. According to this code, with minor editorial changes:

Horizontal separation means a permanent open space between the building wall under consideration and the lot line or the center line of a facing street, alley, or publicway. Where two or more buildings are on a lot, the horizontal separation of the wall under consideration shall be measured from an imaginary line drawn at a distance from the facing wall equal to the horizontal separation applicable for that wall.

(a) Bearing and nonbearing walls shall be of approved noncombustible material. Interior bearing walls shall have a fire-resistance rating of not less than 4 hours. The fire-resistance rating of each exterior wall and the maximum allowable percentage of windows to wall area in any story, except in the first story facing a street or public place, shall be as shown in Table 14-2 for the horizontal separations indicated.

(b) Where there are 2 or more buildings on the same lot and the total area of the buildings does not exceed 1½ times the allowable area of any one of the buildings, no fire-resistance rating is required for the nonbearing portions of exterior walls of these buildings that face each other.

(c) Openings in exterior walls shall be protected and vertical separation between openings in exterior walls shall be provided.

(d) Lintels over openings in walls shall have a fire-resistance rating not less than required by this section for the wall in which the lintel is placed; however, no fire-resistance rating shall be required when the opening is spanned by a masonry arch designed to carry all imposed loads, or the opening is spanned by a beam above the lintel which has a fire-resistance rating not less than required by this section for the wall in which the beam is placed, or the span does not exceed 4 ft.

TABLE 14-2

Horizontal Separation	Exterior Bearing Walls		Exterior Nonbearing Walls	
	Minimum Fire-Resistance Rating (hours)	Maximum Window-to-Wall Area (%)	Minimum Fire-Resistance Rating (hours)	Maximum Window-to-Wall Area (%)
0–3 ft.	4	0	3	0
Over 3–20 ft.	4	20	2	30
Over 20–30 ft.	4	30	1	40
Over 30 ft.	4	40	None required	No limit

Stone lintels shall not be used unless supplemented with steel lintels, reinforced concrete or masonry arches designed to support the imposed loads.

Except as listed below, every opening in an exterior wall of a building shall be protected by an approved fire window, fire door or other approved protective when such opening: Faces on a street and is less than 30 ft. from the opposite building line; is less than 30 ft. distant in a direct unobstructed line from an opening in another building or from a wood frame building; is above and less than 30 ft. from any part of a neighboring roof of combustible materials or any roof having openings within this distance; faces on and is located less than 15 ft. from an adjacent lot line.

Except as listed below, the exterior openings located vertically above one another shall have not less than 3 ft. vertical separation provided by an assembly of noncombustible material having a fire-resistance rating of not less than 2 hours between the top of one opening and the bottom of the one next above, or the exterior openings shall be separated by such an assembly extending outwardly from the building wall a horizontal distance of not less than 3 ft. No vertical separation is required between exterior openings under certain specified conditions.

REFERENCES AND RECOMMENDED READING

1. James Arkin, "Wood Windows"; William Guillett, "Steel Windows"; and John P. Jansson, "Aluminum Windows," *Windows and Glass in the Exterior of Buildings*, Publication 478, Building Research Institute, National Research Council, 1957.
2. *Selecting Windows*, Circular Series Index Number F11.1, Small Homes Council, University of Illinois.
3. *Sweet's Catalog Architectural File*, F. W. Dodge Corp., issued annually.
4. *Curtain Walls of Stainless Steel*, 1955; *Data on Stainless Steel Curtain Walls*, 1957; *Joints in Metal Curtain Walls*, 1957; *Thermal Behavior of Curtain Walls*, 1957; American Iron and Steel Institute, Princeton University School of Architecture (pamphlets).
5. *Metal Curtain Walls*, Publication 378, 1955; *Sealants for Curtain Walls*, Publication 715, 1959; Building Research Institute, National Research Council (pamphlets).
6. William Dudley Hunt, Sr., *The Contemporary Curtain Wall*, F. W. Dodge Corp., 1958.

7. *Metal Curtain Wall Manual*, National Association of Architectural Metal Manufacturers, 1960 (looseleaf binder).
8. *Concrete Curtain Walls*, Portland Cement Association.
9. Peter P. F. Dejongh, "The New Look in Buildings," *Civil Engineering*, August 1958, p. 52.
10. Marianne Stern, "What You Should Know About Thin-Wall Buildings," *Engineering News-Record*, April 14, 1958, p. 38.
11. *Architectural Data Handbook*, Fourth Edition, Pittsburgh Plate Glass Company, 1961.
12. *Sweet's Catalog*, 1971 Edition, F. W. Dodge Corp., New York, New York.

15

DOORS AND DOOR HARDWARE

85. DOOR NOMENCLATURE

A *doorway* is an opening through a building wall or partition, providing passage for persons or vehicles. The opening may be closed by a movable barrier, a *door*, held in position by a *door frame*, the members of which are the *sides* or *jambs* and *top* or *head* of the opening. A doorway is often considered to be the clear opening within the frame. Saddles or thresholds are provided at the bottom or sill of exterior doorways.

Classes of Doors. Ordinary doors, whose primary function is to permit the passage of persons, are classed as *exterior* and *interior doors*. Interior doors between rooms are known as *communicating doors*. Doors at the principal entrances are commonly called *entrance doors*. Light *screen doors*, much of whose area is covered with insect screen, are mounted on the outside of the frames of exterior doors. During the winter months screen doors are often replaced with *storm doors* which have glass panels. Doors with interchangeable screened and glazed panels which can be changed seasonally are called *combination doors*. Doors designed to resist the passage of fire are called *fire doors*. A *wicket door* is a small door within a large door, provided to permit passage without opening the large door.

In addition to the classes of doors mentioned in the preceding paragraph, many others, such as elevator doors, garage doors, hangar doors, industrial doors, and scuttle doors, are given little, if any, consideration herein. See reference 1.

Operation of Doors. Doors usually open by swinging about a vertical axis or by sliding horizontally, but sometimes they may swing about a horizontal axis or slide vertically. These axes are provided by means of *hinges* or *pivots* fastened to the door and the door frame. Horizontal sliding doors are suspended from *hangers* containing wheels which operate on tracks placed at the top of the door openings. Vertical sliding doors move between guides provided at the sides and are operated by cables or chains passing over pulleys in much the same way that double-hung windows are operated. They may be either counterweighted or counterbalanced, as will be explained later.

The most common type of door is the swinging door shown in Figure 15.1(*a*).

Two doors hinged at opposite sides of an opening (*b*) are referred to as *double doors* or a *pair of doors*. Such doors are used at the entrances of buildings, of large rooms, and even of small rooms in residences to give a more spacious effect and to provide more clear passageway.

The *double-acting door* (*c*) is provided with special hinges which keep the door closed when it is not held open. The door can easily be pushed open in either direction.

The *folding* or *accordion doors* shown in (*d*) and (*e*) are used singly or as folding partitions so that two rooms may be used as a single room or separately. They are made for very wide openings. Doors are also hinged together (*e*), forming a folding partition.

The sliding doors shown in (*f*) and (*g*) were once used extensively in residences. Swinging doors or *cased openings* without doors have replaced them. The slide-into pockets provided in the partitions are concealed when not in use. Sliding doors are used in many other locations, particularly residential closets.

Doors sliding on one side of a wall or partition (*h*) and (*i*) are extensively used for fire doors which normally stand open, but they close (released by a *fusible link*) in case of fire. They may be made self-closing by sloping the tracks from which they are suspended or by properly arranging weights and pulleys.

Sliding doors (*i*) and (*j*) are often used for elevator doors. They are so aranged that they will all open when one is pulled back. The inner doors in (*j*) are arranged to move faster than the other two so that they will all be completely open at the same time. Two doors opening to the same side are more extensively used than the four-door unit shown.

The right-angle door (*k*) is used to a limited extent on garages. The doors are suspended from an overhead track.

Vertical sliding doors in (*l*) and (*m*) are counterweighted and may be operated electrically. They are pulled up by cables or chains which are placed at each side of the opening and which operate over pulleys in the same manner as for double-hung window sash. Such doors are

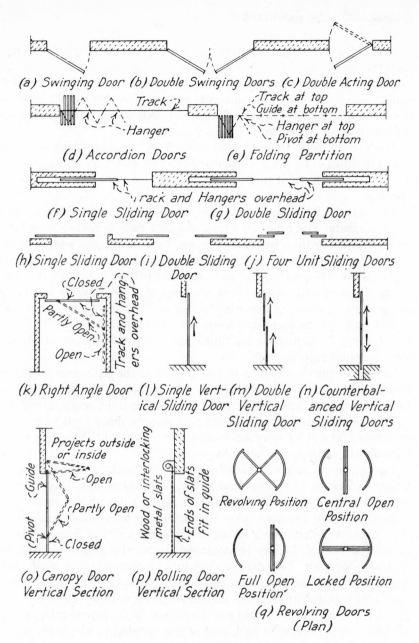

(a) Swinging Door (b) Double Swinging Doors (c) Double Acting Door

Track
Hanger

Track at top
Guide at bottom
Hanger at top
Pivot at bottom

(d) Accordion Doors   (e) Folding Partition

Track and Hangers overhead

(f) Single Sliding Door   (g) Double Sliding Door

(h) Single Sliding Door (i) Double Sliding (j) Four Unit Sliding Doors
Door

Closed
Partly Open
Open
Track and hangers overhead

(k) Right Angle Door (l) Single Vert- (m) Double (n) Counterbal-
ical Sliding Door Vertical anced Vertical
Sliding Door Sliding Doors

Projects outside
or inside
Guide
Open
Partly Open
Pivot
Closed

Wood or interlocking metal slats
Ends of slats fit in guide

Revolving Position Central Open
Position

Full Open Locked Position
Position

(o) Canopy Door   (p) Rolling Door
Vertical Section   Vertical Section

(q) Revolving Doors
(Plan)

FIGURE 15.1   Operation of doors.

used for large openings in industrial buildings, particularly for freight elevator doors.

Where conditions permit their use, the counterbalanced vertical sliding doors showns in (*n*) are convenient. Used for freight elevator doors, they are easily operated by hand. When one panel moves up, the other panel moves down an equal distance.

The canopy door shown in (*o*) is used for large openings in industrial buildings. The door is counterweighted and may be electrically operated. If desired, it may open outward to form a canopy over the opening.

The *rolling door* shown in (*p*) operates in the same manner as a window shade. The roller on which the door rolls is operated by hand or electrically or by a spring which counterbalances the weight of the door. The door is made flexible by using interlocking slats of wood or sheet metal. The ends of the slats are held behind guides at the sides of the doors. Wood rolling doors are used to form movable partitions in the same manner as accordion doors and folding partitions shown in (*d*) and (*e*). Steel rolling doors are used extensively for large exterior doors of industrial buildings and for fire shutters which will close automatically in case of fire.

The revolving door shown in (*q*) is used at the entrances of certain buildings, such as banks and stores. It does not permit much cold air to enter from outside when it is in use. During mild or warm weather when it is not in use, the revolving part may be moved out of the way as shown. The door may be locked when desired. These doors are allowable but highly restricted by codes.

Two forms of garage doors which move vertically to open are illustrated in Figure 15.2. The door in (*a*) consists of four leaves which are hinged together as shown. Wheels in the sides of the doors move in guides so that the door takes the position shown in the figure when it is open. A coiled spring balances the weight of the door so that it is easily operated. It may be operated electrically and is often called an *overhead door*. The door shown in (*b*) consists of a single leaf, which is pivoted on each side as shown. It is opened by lifting vertically and rotating around the pivots. A spring is arranged to assist in opening the door and to hold the door open. This door requires a wall or other support close to each side of the opening to which the pivots can be attached.

Door Materials. Wood, aluminum, carbon steel, and glass are the materials most extensively used in the manufacture of doors, but

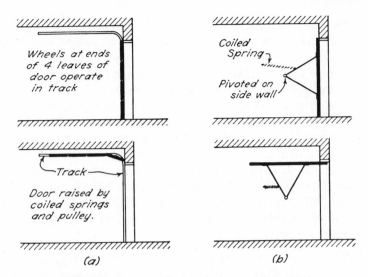

FIGURE 15.2   Vertical-opening garage doors.

stainless steel, bronze, and copper are also used, as are facing materials such as hardboard, fiberboard, asbestos, and plastics.

Door Sizes.   The sizes of ordinary stock doors whose primary function is to permit the passage of persons vary with the types and the materials used in their construction.

The common widths of single wood doors are 2 ft. 6 in., 2 ft. 8 in., and 3 ft. for interiors, and 2 ft. 8 in., 3 ft., and 3 ft. 4 in. for exterior doors. Heights are 6 ft. 6 in., 6 ft. 8 in., and 7 ft. for exterior doors. Thicknesses are $1\frac{3}{8}$ in. and $1\frac{3}{4}$ in. for interior and exterior doors, but doors 3 ft. 4 in. wide are a minimum thickness of $1\frac{3}{4}$ in. Exterior doors with a thickness of $2\frac{1}{4}$ in. are not unusual, and such doors with heights of 7 ft. 6 in. and 8 ft. are sometimes used.

When double doors are used (Figure 15.1(b)), each door may be somewhat narrower than a single door. Storm doors, screen doors, and combination doors are usually $1\frac{1}{8}$ in. thick.

The stock sizes of metal doors are more limited than those for wood doors but correspond quite closely with the sizes mentioned.

Panel and Flush Doors.   The most common type of door is the *panel* or *framed door*, usually consisting of vertical and horizontal members which frame rectangular areas in which opaque panels of panes of glass or louvers are located. Such doors with glass panels are called *sash doors*.

Another common type of door is the *flush door*, which usually consists of flat, relatively thin, face panels the full height and width of the door; the door is rigidly bonded or otherwise attached to a solid or hollow core with finished edges. Openings may be provided in flush doors for the insertion of glass vision panels or of louvers for ventilation. A *vision panel* is a small glazed panel at eye level enabling a person approaching a door to see a person approaching the door from the opposite side. Doors with panels that are of approximately the same thickness as the stiles and rails are called *flush panel doors*.

Panel and flush doors of various materials correspond in appearance with the wood doors illustrated in Figure 15.3. A great variety of other types with standard sizes are manufactured and carried in stock as *stock doors*. Other sizes and types are manufactured on order to the architect's specifications.

Parts of Panel Doors.   The parts of panel doors, regardless of the material of which they are constructed, have the same designations as those of wood doors illustrated in Figure 15.3. These parts may be defined as follows:

Stiles.   Vertical members at each side of the door which extend its full height. For doors swinging about a vertical axis located on one edge, the stile to which the hinge or pivots are attached is called the *hinge* or *hanging stile*. The edge of the hinge stile is called the *butt*. The stile to which the lock, latch, push, or pull is attached is called the *closing* or *lock stile*.

Rails.   Horizontal members extending the full width between stiles and framing into them. The rails at the top and bottom of the doors are called the *top rail* and the *bottom rail*. Rails located between these two rails are called *intermediate* or *crossrails*. The crossrail at lock height is called the *lock rail*.

Mullions.   Relatively heavy vertical members which subdivide the areas between stiles and rails. *Muntins* are light vertical or horizontal members corresponding in position to mullions and crossrails.

Panels.   Members which fill the areas between, and which are surrounded by, the stiles, rails, mullions, and muntins. Wood panels may be *flat panels* of plywood or *raised panels* with a border thinner than the remainder of the panel. One or more panels of a door may be replaced by glass to form a *sash door* for vision, light, or architectural effect or by louvers for ventilation.

Parts of Flush Doors.   Flush doors vary widely in their construction, but all types include *face panels* which form the faces; a *core* which is the construction between the face panels to which they are attached, which gives them support, and which may be solid or hollow; and

Ledged and Braced · Flush · One Panel · Two Panels · Four Panels

Five Panels · Six Panels · Louvered · French · Dutch

Division

(a) Types of Doors

Glass Bead

Solid Door with Raised Panels · Solid Door with Raised Panels, Flush Molding · French Door Glass set with Stops

Raised Panel · Flat Panel

Veneered Doors

(b) Types of Door Construction

FIGURE 15.3 Types and construction of wood doors.

edge strips which surround the rim of the core to provide finished edges to match the material in the face panels. Hollow cores may include vertical or horizontal *ribs*, rectangular *grids*, or small *cells*. Some cores include *stiles* and *rails*. Cores usually include special provisions such as solid blocking for attachment of locks and other hardware. Openings may be provided through flush doors for the insertion of panes of glass for light or vision or of louvers for ventilation.

Astragals. An *astragal* is an overlapping strip of material usually on the exterior surface of a door. The strips are made of metal, wood, or rubber. They may be attached or an integral part of the door. Astragals are located on the lock stile of the active leaf of a pair of swing doors. They lap over to engage the inactive leaf, thus covering the space between the closed doors. The function of an astragal is to protect the building interior against driving rain and tampering with locks. Astragals may also be used to cover the crack created by the meeting of sliding doors. Pivoted hinge entrance doors that do not have latch sets may have rubber astragals that act as closure strips to keep out rain and cold.

Labeled Doors. Building codes require that doors provide not only safe egress but also restrict the spread of fires. Doors that meet these conditions are known as *fire doors*. Fire doors are *labeled* and known also as *labeled doors*, or *rated doors*. A labeled door is classified according to its ability to resist fire penetration or destruction by fire. This rating is established by conducting standard fire tests, which simulate actual fire conditions, against doors and door assemblies. These tests indicate the ability of certain doors to resist fire and destruction by a fire hose. Accordingly, the doors of a specified construction are *fire rated* as capable of resisting fire for a specified time in certain locations.

Fire ratings are identified according to time in hours. Hourly ratings are 3, 1½, and ¾ hours. A letter designation of A, B, C, D, or E, following the hourly rating, designates the *class location* for which the door is intended. The actual label is an embossed metal plate attached to the butt side of the door.

Fire doors are produced under the factory inspection and label service program of the Underwriters Laboratory and are identified by a label designating hourly rating, class location letter, and the UL stamp. Factory Mutual Research Corporation also rates doors. Their approval is indicated by the initials FM within a diamond.

Most importantly, fire doors should be installed only in frames, walls, and assemblies that are equal to the fire door rating. Therefore, walls that require fire resistivity, such as fire towers, division walls, and corridors, should have door assemblies that are compatible with the wall construction.

Other factors that determine door ratings are the location usage; the minimum door thickness, which is $1\frac{3}{4}$ in.; the maximum glass size if permitted; the door opening dimensions; and hardware requirements.

86. WOOD DOORS

Material.   The most commonly used wood for door construction is ponderosa pine. A considerable portion of the wood resembles white pine in appearance and texture [6] and is usually accepted when specifications call for white pine unless true white pine is specified. It is used for constructing panel doors, as the base to which veneers are applied, and sometimes for the construction of cores for flush doors. Among the other woods used for constructing panel doors are Douglas fir, western red cedar, and sitka spruce.

Veneers of various decorative woods are used for facing panel and flush doors. These include ponderosa and white pine, red and white oak, walnut, hard maple, gum, red and white birch, and African and Philippine mahogany. These face veneers are bonded directly to the solid cores of the stiles and rails of panel doors. They are also used for exposed outer plys called *face veneers*, for plywood used for the panels of panel doors, and for the face panels of flush doors. For a description of plywood, see Article 30. The required width of face veneers is obtained by assembling separate strips matched for grain and color. The selection of the species of wood for face veneers is based on the appearance and suitability to receive the surface coating to be applied. The cross bands of plywood used for face panels of flush doors are usually hardwood.

The face veneers of flush doors are often made of thermosetting laminated plastics. Such veneers are available in a great variety of colors, are not painted or surface treated in any way, and are durable and easily cleaned. Patterns can be obtained if desired. The veneer thickness is usually $\frac{1}{16}$ in. Edge strips which match the face veneer are used.

Manufacture of Veneer.   Veneers are thin sheets of wood. They are made by three processes: the sawing process, the slicing process, and the rotary-cut process. In the *sawing process*, the thin sheets of wood are cut from large blocks of wood by a circular saw. This process is wasteful because of the wood consumed in the saw-kerf, but it produces the best grade of veneer because it injures the wood less than the other processes. In the *slicing process*, the thin sheets of wood are cut from large blocks with a cutting knife. The wood is softened by steaming to make cutting without splitting possible. In the *rotary-cut process*, a log which has been softened by steaming or boiling is placed in a lathe and revolved against a wide stationary blade which gradually moves toward the center of the log and cuts off a continuous slice. The entire log cannot be cut up in this manner; there remains a core which cannot be cut for veneer.

Before the exposed surface is finished by sanding, sawed veneers are from $\frac{1}{8}$ to $\frac{1}{4}$ in. thick. They are more natural in appearance and more durable than the veneers produced by the slicing and rotary-cut processes. Rotary-cut veneers are from $\frac{1}{28}$ to $\frac{1}{16}$ in. thick and sliced veneer $\frac{1}{20}$ in. and less thick.

Panel Doors.   Many designs and layouts are used for wood panel doors, some of which are illustrated in Figure 15.3. The *French door* shown is also called a *casement door*. The ends of the rails are usually fastened to the edges of the stiles, against which they abut by wood dowels held in position by hidden water-resistant joints. Panels are fastened to the stiles and rails in grooves provided in these members (Figure 15.3); or a separate continuous mold, called *sticking*, may be used, usually on one side only.

The panels of doors with solid stiles and rails may be solid raised panels not less than $\frac{7}{16}$ in. thick or flat plywood with a minimum thickness of $\frac{1}{4}$ in. but they may be raised.

The stiles and rails of veneered doors have cores built up of low-density wood blocks not more than $2\frac{1}{2}$ in. wide (Figure 15.3(*b*)) and of varying lengths with end joints in adjacent rows staggered. These core blocks are bonded with water-resistant adhesive. Side edge strips of the same species of wood as the veneer are provided. Top and bottom edge strips may be of any suitable species of wood. The face veneers of stiles and rails should be not less than $\frac{1}{8}$ in. thick before sanding.

The panes of glass in sash doors are held in position by *glass beads* of the same species of wood as a solid door or as the veneer of a veneered door. The panes of outside doors are bedded in putty for watertightness.

Flush Doors.   Various types of flush doors are illustrated in Figure 15.4. Any of these may have openings to provide for panes of glass or louvers. Stile and rail frames may be used with or without inter-

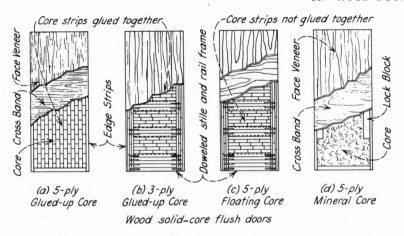

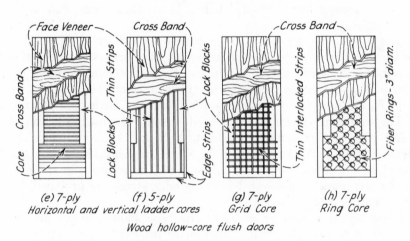

FIGURE 15.4 Wood flush doors.

mediate rails. The rails are connected to the stiles by glued dowels.

There are two general types of flush doors: the *solid-core door* and the *hollow-core door* (Figure 15.4). The cores of both types are usually made entirely of low-density kiln-dried wood, such as ponderosa pine, but other materials are used, especially in special-purpose doors.

The face panels are usually constructed of plywood, which may have 1, 2, or 3 plys. The face panels should include at least 1 ply whose grain runs in the direction normal to the predominant direction of the grain of the units that make up the core. The face panels are bonded to the core under pressure with water-resistant adhesives and surfaced by machine sanding. The exposed edges of the core are covered with matching edge strips about $\frac{3}{4}$ in. thick, of the same species of wood as the face veneer. The face veneer is always placed with the grain of the wood vertical. Face panels are also made of laminated plastic. They are especially suitable for exterior doors that have hard usage.

The core of the most common type of solid-core door is built up of vertical strips made of wood blocks about 2 in. wide and several inches long, with the ends of blocks in adjacent strips staggered (Figure 15.4(a)). The strips are glued together to form a solid panel. No frame is required, but matching edge strips are provided. Each face panel is plywood with 2 plies for a 5-ply door and 3 plies for a 7-ply door, the core being counted as 1 ply. Particle board is used for solid cores also. These doors have solid wood stiles and rails and a 3-ply plywood veneer finish. Particleboard is an ideal core that resists warping.

The solid-core door shown in (b) consists of a doweled stile and rail frame with one or more intermediate rails and horizontal blocks glued together and assembled in the frame. Each face panel usually consists of 1 ply of sawed veneer $\frac{1}{4}$ in. thick for doors $1\frac{3}{4}$ in. thick and 1 ply $\frac{1}{8}$ in. thick for $1\frac{3}{8}$ in. doors. Matching edge strips are provided.

The solid-core shown in (c) is made up of a doweled stile and rail frame with one or more intermediate rails and horizontal blocks or strips interspaced between the rails. Since these blocks are not glued together, the core is called a *floating core*. The entire core is bonded to the face panels, however. Matching edge strips are provided. The doors may be 5-ply or 7-ply.

Fire-resistive solid-core doors of the type shown in (d) are constructed with noncombustible mineral or asbestos-type cores or impregnated wood chip cores, plywood face panels, each with 2 or 3 plies, and impregnated matching edge strips.

A solid-core door similar in construction to the door in (a) has the core, cross bands, and edge strips chemically impregnated to improve its fire resistance.

Solid core doors that serve specialized functions may have an interior shield. Doors that house X-ray equipment or other sources of radiation require *shielded doors* containing a sheet of lead. The sheet of lead, which may be from $\frac{1}{16}$ in. to $\frac{1}{2}$ in. thick, is located on the interior and covered with a veneer of finished material—usually wood. Doors to areas that house sensitive electronic equipment usually require shielding. These shielded doors have an interior sheet or mesh of copper. The shield is provided with grounding wires, which must be grounded at the jobsite.

Several general types of *hollow cores* are used in the construction of flush doors. Various kinds of spacers are located between the face panels to which they are glued and assembled within stile and rail frames with matching edge strips. The bond to the face panels holds the spacers in position, and the spacers provide lateral support for the panels. A door with a *ladder core* which includes closely spaced horizontal fiberboard strips is shown in Figure 15.4. One with wood or fiberboard strips placed vertically is shown in (*f*).

Another type (*g*) includes thin vertical and horizontal wood strips located and interlocked to form a *mesh, lattice,* or *grid,* with cells or spaces between the strips. A core that includes closely spaced fiber rings is shown in (*h*).

To provide for the face attachment or mortising of locks and other hardware to hollow-core doors, *lock blocks* are provided on each stile, as shown in the figures, so that the door can be hung at either edge.

Solid-core doors are heavier, more fire-resistant, and better insulated against the transmission of heat and sound than wood panel and wood hollow-core doors. They are more suitable for curved heads than either of these types and can be more easily adapted to receive glassed and louvered openings than hollow-core doors.

87. METAL DOORS

General Discussion.   The metals most commonly used in the manufacture of doors are aluminum and carbon steel; but bronze, stainless steel, and copper are also used.

The metal members for door construction are usually hollow, with or without some form of nonmetallic filling; and therefore such doors might be classed as hollow metal doors. The usual practice, however, is to class aluminum and bronze doors separately and restrict the use of the term *hollow metal doors* to doors constructed of hollow members of steel with fillings or core materials other than wood, except plywood cores in the panels. Aluminum and bronze doors have hollow metal members, but they are usually classed separately. In addition, doors with metal facings fitting tightly over wood cores are called *metal-covered doors.* Finally, crudely formed solid-wood fire doors covered with terne plate, although covered with metal, are called *tin-clad doors.* There are many other special types.

Aluminum Doors.   Both panel and flush doors are constructed of aluminum. The most common type is the panel door with tubular extruded aluminum stiles and rails approximately rectangular in cross section (Figure 15.5(*a*)), with a minimum thickness of ⅛ in. The

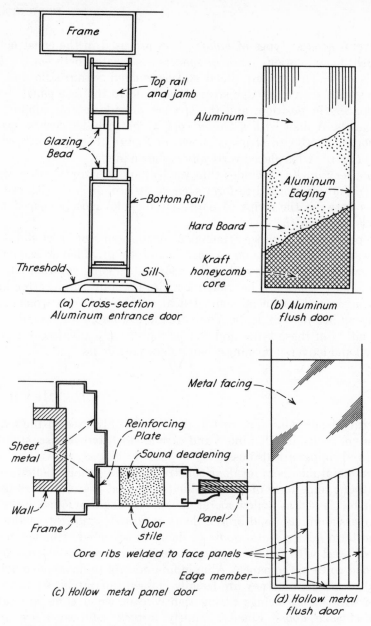

(a) Cross-section
Aluminum entrance door

Frame

Top rail
and jamb

Glazing
Bead

Bottom Rail

Threshold

Sill

(b) Aluminum
flush door

Aluminum

Aluminum
Edging

Hard Board

Kraft
honeycomb
core

(c) Hollow metal panel door

Reinforcing
Plate

Sound deadening

Sheet
metal

Wall

Frame

Door
stile

Panel

(d) Hollow metal
flush door

Metal facing

Core ribs welded to face panels

Edge member

FIGURE 15.5   Metal and metal-covered doors and frames.

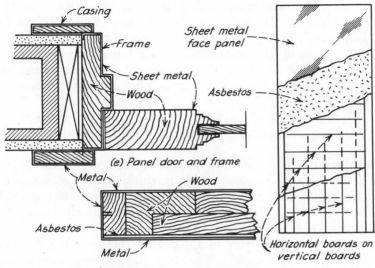

(e) Panel door and frame

(f) Metal covered flush door

members are joined by electric-arc welding with the joints located and finished to be invisible. The frames are constructed of the same material and are also tubular and rectangular in cross section. The doors are usually 1¾ in. thick, with stiles and top rails varying in width from 2⅛ to 5 in. and bottom rails, from 4 to 10 in. The frames vary in cross section upward from a minimum width of 1¾ in. and a depth of 4 in. The finish of the doors and frames is usually a special polished or satin finish but may be of various architectural colors.

For entrance doors, the space between the stiles and rails is usually occupied by a plate-glass panel but may be divided by a narrow aluminum center cross panel. The glass is held in position by continuous aluminum glass stops with vinyl glazing beads, or in other ways to make putty unnecessary.

Instead of using two glass panels, the lower panel is often faced with aluminum sheets. It is of sandwich construction with a core ¾ in. thick of resin-impregnated honeycomb, with ¾ in. hexagonal cells, on each face of which is bonded a hardboard layer ⅛ in. thick, to each of which is bonded an outer layer of sheet aluminum about 1/32 in. thick. The bonds are formed under heat and pressure.

Aluminum *flush doors* have face panels of fine vertical ribbed or fluted aluminum sheets. For one type of door (Figure 15.5(*b*)), these sheets are about 1/32 in. thick and are bonded to ⅛ in. hardboard which are in turn bonded, under heat and pressure, to a resin-impregnated honeycomb core with ¾ in. hexagonal cells. The edges of the door may be extruded aluminum sections to which the surface sheets are securely attached. Various types of flush doors are available. They may be pierced for one or more lights of glass or for louvers.

Another form of flush door consists of a framework with extruded aluminum stiles, a top and a bottom rail, and intermediate rails, with the space between rails filled with a solid fibrous core and faced with aluminum sheets.

The *flush panel door* has a stile and rail frame with panels of about the same thickness as that of the stiles and rails.

Hollow Metal Doors.   Hollow metal doors are either the panel or flush types. They are usually 1¾ in. thick. They are constructed primarily of cold-rolled sheet steel especially processed to give smooth flat surfaces. They ordinarily have a factory finish consisting of one or more coats of baked enamel.

The *panel doors* resemble in appearance the wood panel doors of various designs (Figure 15.3). The stiles and rails are made of sheets about $\frac{1}{20}$ in. thick and are accurately formed as shown in Figure 15.5(c), with the joints welded and ground smooth. The edges of the doors are usually reinforced internally with channels welded in position. Reinforcement is provided where hardware is to be attached. The stiles and rails are filled with mineral wool, asbestos, or other material for sound deadening so that the doors will not have a metallic sound when closing.

The panels are of sandwich construction with sheet-metal faces about $\frac{1}{20}$ in. thick and cores of asbestos, composition board, sheetrock, hardboard, or other material.

Flush doors (Figure 15.5(d)) consist of sheet-metal face panels about $\frac{1}{20}$ in. thick and the full size of the door. They are separated by vertical sheet-metal ribs of various forms, spaced 6 or 8 in., to which they are spot-welded at frequent intervals.

The rim of the door is closed and strengthened by channels or in some other manner, and the face panels are firmly secured to the rim. Sound-deadening material may be bonded between the stiffening ribs on the inside of the face panels or the entire space between ribs may be filled with asbestos or mineral wool insulation or other sound-deadening material. Provisions are made for the attachment of hardware. Other types are available.

The frames for hollow metal doors are made with various profiles which are break-pressed from steel sheets welded and ground smooth at the corners (Figure 15.5(c)).

Metal-Covered Doors. Doors of this general type, which are commonly called *kalamein doors*, consist of sheet-metal facings securely bonded or attached to nonresinous kiln-dried wood interior construction (Figure 15.5(*e*)). All joints in the metal covering are recessed locked joints finished with solder and ground smooth. The metal is usually galvanized steel or furniture steel, which is a sheet steel specially processed to give a smooth flat surface. Copper, bronze, aluminum, and stainless steel are also used. The usual thickness is 1 ¾ in.

Panel doors are constructed of doweled or mortised and tenoned stile and rail frames with wood cores and sheet-metal coverings tightly filled by drawing through dies. The panels consist of cores of asbestos board, for the maximum fire resistance, or of composition, plywood, or other board to which metal facings are glued under heat and pressure. The types of panel doors available are similar to those for wood panel doors, some of which are shown in Figure 15.3.

A common type of *flush door* (Figure 15.5(f)) consists of a wood core with a horizontal and a vertical ply, which may have a thin covering of sheet asbestos glued to one or both sides, both sides covered with sheet metal without seams pressure-glued to the core. The edges are reinforced with wood strips covered with sheet metal. The frames and trim may be sheet metal break-pressed to the desired profile, or wood with metal-covered exposed surfaces.

Kalamein doors are obsolete, having been largely replaced by the flush, solid-core, fire-rated doors.

Tin-Clad Doors. Tin-clad doors consist of a wood core covered with terne plate. The cores are made of either two or three layers of 1 in. boards, preferably tongued-and-grooved, and not more than 8 in. wide. If two layers are used, one is vertical and the other horizontal. If three layers are used, the outer layers are vertical and the inner layer horizontal. The layers are securely fastened together by clinched nails or in some other manner to give smooth surfaces. The covering of terne plate is made of 14 × 20 in. sheets, preferably with double-lock joints. Solder, if used, must serve only to improve the appearance. The terne plate is held flat against the core by nails.

The 3-ply doors are required where the most effective resistance is desired, and the 2-ply where only a moderate degree of protection is necessary.

Steel-Plate Doors. Steel-plate doors consist of steel plates fastened to one side of an angle-iron frame or both sides of a channel frame, the frames being braced by intermediate members.

Corrugated-Steel Doors.   This type of door is constructed of heavy corrugated-steel sheets supported by a structural steel frame. The better doors consist of two thicknesses of corrugated sheets, one with corrugations vertical and the other with corrugations horizontal, with an asbestos lining between the sheets. This lining is from $\frac{1}{8}$ in. to 1 in. thick. Sometimes the structural steel frame is covered with a $\frac{1}{8}$ in. layer of asbestos. A cheaper and less fire-resistant door is made of one thickness of corrugated steel with corrugations vertical, riveted to an angle-iron frame, intermediate braces being provided where necessary.

Steel Rolling Doors.   This type of door consists of a curtain of interlocking corrugated steel slats which rolls up on a roller or drum in much the same way as a window shade (Figure 15.1(p)). The edges of the curtain operate in vertical guides, and the roller is housed in a steel hood. The curtain may be counterbalanced by springs so that it can be easily raised or lowered by hand; it may be operated by an endless chain with sprocket and gear, by a crank, or by electric motor. Devices for closing the door automatically in case of fire are available.

88.   GLASS DOORS

This article is concerned with the common type of *tempered-plate-glass* single- or double-acting swinging door.

In the tempering process, plate glass is heated to a high temperature and then cooled suddenly. The sudden cooling has an immediate effect close to the surfaces only. As the interior of the glass slowly cools to a normal temperature, it contracts. This action causes compressive stresses to develop near the surface and balancing tensile stresses to develop in the interior, an equilibrium which remains permanently.

Tempered plate glass is much more resistant to pressure and impact than normal plate glass and is more flexible. It cannot be cut, drilled, or altered in any way after tempering. The glass thicknesses used are $\frac{1}{2}$ and $\frac{3}{4}$ in. The surfaces of the glass are usually smooth and polished, but rough-textured glass is available when it is desired to obscure vision. The edges of the glass are finished smooth.

All fittings to receive hardware to be attached to the door are cemented in position to the glass at the factory.

Aluminum, bronze, or stainless steel hardware is used. Top and bottom pivots are used rather than hinges to provide for the swing of the doors. Pivot door checks or closers are recessed in the floor.

Push bars and plates, door pulls, and other hardware required are adapted when necessary for use on this type of door, with appropriate connections to the door.

Cross sections of one type of glass door, with its frame and metal fittings designed to receive the glass and to provide for hardware connections at the top and bottom of a door are illustrated in Figure 15.6(a) and (b). They may extend entirely across a door or may be located in any of the positions near the edges shown in (c), or in others as required. Holes to receive bolts for attaching certain hardware may be drilled in the glass before tempering but not after.

The frames for glass doors are tubular metal approximately rectangular in cross section, reinforced with steel channels, with the rods located inside the members (Figure 15.6(a) and (b)). The metals used in their construction are extruded aluminum or bronze, or stainless-steel sheets pressed into shapes similar to the extruded shapes.

Sliding doors are also available.

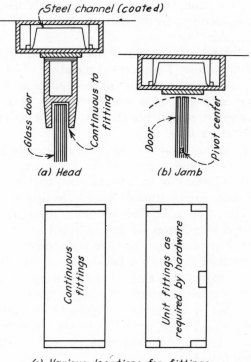

(a) Head   (b) Jamb

(c) Various locations for fittings

FIGURE 15.6   Tempered-plate-glass doors and their frames.

The steel channel that reinforces the finished aluminum frame should be coated to prevent electrolysis.

89. DOOR FRAMES

Door frames retain the door, anchor door hardware, and present a finished appearance by forming a transition between the door and the wall. Commonly used frame materials are wood, solid steel sections, hollow aluminum, stainless steel, and hollow cold-rolled pressed steel.

The side members of a door frame are called the *jambs;* and the top member, the *head. Door stops* are projections or strips on the faces of the jambs and head against which one-way swinging doors close (Figure 15.7). The term is also applied to projections or strips between which doors slide. A small sash, called a *transom*, which may be fixed or operated for ventilation, may be located above a door and included in the same frame. A horizontal bar, called a *transom bar*, is used to separate the door and the transom. Formerly, operable transoms were very common, but now they are rarely used. Fixed transoms of material identical to the door are now quite popular. These do not have a transom bar.

Door frames intended for a pair of doors frequently have a framed member called a *mullion*. This vertical member, which has the same profile as the door frame, is located in the middle of the double door frame. It serves as a lock base and door stop. These mullions are sometimes removable, permitting the full double opening to be utilized as a wide open passage.

Frames have various profiles. Some have a simple lip and plaster stop, whereas others have fluted moldings. Profiles or *plaster keys* intended to meet plaster and wall board are sometimes called *plaster stops* or *backbends*. Frame profiles usually have a ½ in. protrusion that stops the closed door. To dampen noise, these stops may have a rubber lining or rubber *mutes* set in predrilled holes.

To eliminate dust pockets, stops are sometimes cut off about six inches above the floor line with a beveled or square cut. These cuts are called *cut-off stops*, sometimes *hospital stops* or *sanitary stops*. This cut permits the base material to wrap around the door frame or a stainless steel *spat* to be installed (Figure 15.7). Frames frequently have a *subframe, rough buck*, or *subbuck*. The subframe provides the heavy anchorage that the door requires, whereas the door frame is a *casing* or *finished frame*.

The members ordinarily used to provide a finish around door open-

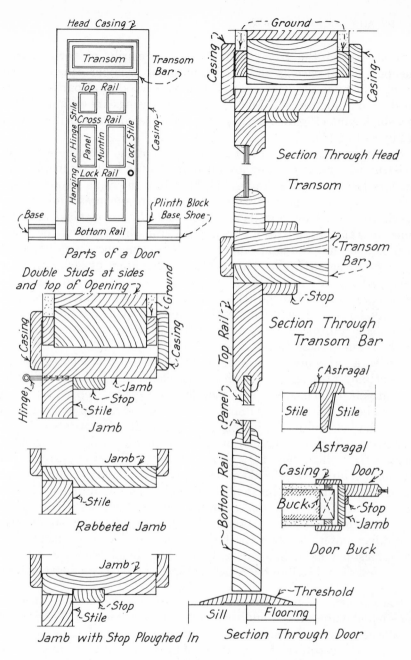

Head Casing

Transom

Transom Bar

Top Rail

Hanging or Hinge Stile

Cross Rail

Panel

Muntin

Lock Stile

Casing

Lock Rail

Plinth Block

Base Shoe

Base

Bottom Rail

Parts of a Door

Double Studs at sides and top of Opening

Ground

Casing

Casing

Hinge

Jamb

Stop

Stile

Jamb

Jamb

Stile

Rabbeted Jamb

Jamb

Stop

Stile

Jamb with Stop Ploughed In

Ground

Casing

Casing

Section Through Head

Transom

Transom Bar

Stop

Section Through Transom Bar

Top Rail

Panel

Bottom Rail

Astragal

Stile

Stile

Astragal

Casing

Door

Buck

Stop

Jamb

Door Buck

Threshold

Sill

Flooring

Section Through Door

FIGURE 15.7   Parts and details of a framed wood panel door.

647

ings are called the *trim*. Trim consists primarily of vertical side members called *casings* and a horizontal top member called a *head casing* which, as used for a wood door in a wood stud partition, are illustrated in Figure 15.7. Metal frames and casings are often included in a single unit (Figure 15.5(*c*)). A metal-covered frame is illustrated in Figure 15.5 (*e*).

Door frames may be set in masonry walls at the time the walls are constructed; and if so, they are anchored to the walls by metal *Tee-anchors* fastened to the frames and built into the walls. Wood blocks may be built into masonry walls to form an anchorage for frames which are set after the walls are built. Door frame openings are provided in walls and partitions with wood studs by doubling the wood members forming the jambs and heads. These members may be called the *subbuck* or *subframe*. Masonry partitions are provided with *door bucks*, which are rough frames (Figure 15.7) set at the time the partitions are built and anchored to the masonry by metal anchors, such as bolts, lag screws, and nails. The openings provided are larger than the outside dimension of the finished door frames, thus permitting the finished frames to be plumbed and trued.

An exterior doorway is also provided with a *sill* located across the bottom of the opening. A *threshold*, sometimes called a *saddle*, is a relatively thin member with beveled or rounded top edges. It is located below the bottom of an exterior door and on top of the sill. It raises the bottom of a swinging door to provide clearance so that it will swing free above the floor or floor covering but will be reasonably weathertight when closed. It also provides a finished appearance for the floor material at the door. Hardwood thresholds are used with wood doors and metal thresholds with wood and metal doors. Thresholds are used with exterior doors of various materials. Thresholds are sometimes used with interior doors, providing a finish for two dissimilar materials when the flooring on two sides of a door differ.

Frames may be factory preassembled or shipped "knock-down."

Frames should be mitered for appearance as well as for construction.

Wood Frames.   Wood frames of various species complement the door design and materials with finishes to match or to contrast. The hidden side of the casing or frame is *kerfed* full length with shallow saw cuts that prevent warping. Wood exterior frames should be treated to resist termites and decay. They are finished or painted in accordance with wood painting and stains.

Wood frames are usually anchored with nails in stud walls or with anchor bolts in masonry and concrete walls. These members may also be *rabbeted* or *rebated* to provide the stops, or the stops may be *ploughed* or *plowed* in.

Pressed Metal Frames. Pressed metal cold-rolled steel frames are made in 12, 14, 16, or 18 gages. They are used in most wall type construction. Pressed metal frames provide a variety of profiles in single rabbeted or double rabbeted shapes. Hinge points are predrilled and mortised and reinforced for the reception of mortised, template butt hinges. The strike plate is likewise reinforced.

Frames set in cast-in-place concrete walls are frequently set and plumbed in place in the formwork. The concrete is then poured with the frames integral with the wall. Obviously, no other anchorage is needed. Frames anchored to precast concrete are secured by installing flat head machine screws into a steel expansion shield. The screw is threaded through a pipe sleeve that prevents frame collapse or rupture. Each frame should have a minimum of three anchors for each jamb; heavier doors and doors wider than three feet should have at least four anchors for each jamb, otherwise cracking or frame tear-out from the wall is likely to happen. Masonry anchors are loose tees, galvanized. They may be corrugated and are adjustable so that they may be placed in a mortar joint. Pressed metal frames in masonry walls or precast concrete are usually grout filled to obtain added strength or for noise dampening. In stud walls, these frames are secured with adjustable sheet metal anchors that are nailed to a wood subframe. *Floor knees* or *clip angles* provide anchorage to floors. Hollow frames in stud walls are usually filled with insulation for noise abatement.

Frequently, it is necessary to have a pressed metal subframe. These subframes, called *cabinet frames*, are installed in the rough door opening, and the finished pressed metal casing is screwed to the cabinet frame.

Pressed metal frames may be treated by a Phosteem process, consisting of an iron phosphatizing coat which increases the resistance to rust. Steel frames should then be primed with red oxide-zinc chromate primer and baked.

Aluminum Frames. Aluminum frame extrusions, usually rectangular in profile, are used extensively as door frame assemblies for glass entrance doors. They usually have the anodized aluminum finish.

Aluminum frames that are used where high winds prevail may have steel subframes. Precautions must be taken to prevent electrolysis between the two metals by coating them at their points of contact.

Steel Frames.   Steel channels and angle sections are used for door frames on sliding and roll-up type industrial doors. Some sliding fire resistant doors require steel channel frames approximately 4 to 8 in. deep. Steel bar or channel stops of approximately ¾ in. are welded to the channel jamb and head.

Labeled Frames.   A *labeled frame* has met the requirements of thickness and construction specified for fire rated door frames. They must be at least 14 gage steel for a three hour (A) rating and of 16 gage steel for a lesser rating. These ratings are established by the Fire Underwriters Laboratories (UL) or the Factory Mutual Research Corporation (FM). Special frames of bronze or stainless steel, labeled or nonlabeled, are for special purposes, such as matching a door design.

90. DOOR HARDWARE

Basically, door hardware requirements are fulfilled with the installation of a hinge set and a latch system. Specialized devices such as automatic closers and miscellaneous hardware items meet specific needs. Code requirements may demand a panic device.

Traditionally and by definition, hardware is metallic, although ceramics and plastics are now used. In general, function decides the most appropriate material. Stainless steel is best suited for the corrosive atmospheres of marine and industrial areas. Wrought iron and steel are used where heavy industry demands ruggedness. The softer materials (brass, bronze, and aluminum) in either cast or wrought forms may be more aesthetically appealing for commercial, public, and residential uses.

Handing of Doors.   The standard procedure for indicating door swings is called *handing* or *handing of the door.* Many hardware items must be oriented or handed to the door swing.

Handing is determined thusly:

(1) For exterior doors, handing is from the outside.

(2) For interior doors with locksets, handing is from the key side.

(3) For communicating doors without locksets, handing is determined from the side from which the hinges are not visible.

(4) For closet and cabinet doors, the handing is from the room side.

With hinges on the left, doors are *left hand* (*LH*); hinges on the right are *right hand* doors (*RH*); strike bolt bevels that face the ob-

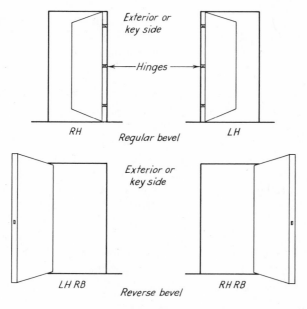

Exterior or
key side

—Hinges—

RH   Regular bevel   LH

Exterior or
key side

LH RB   Reverse bevel   RH RB

FIGURE 15.8   [11]

server are *regular bevel*; strike bolts with the bevel facing away are *reverse bevel* (*RB*). Figure 15.8 will clarify handing.

Hinges.   Through usage, the term *butt* has become synonymous with *hinge*, a usage that refers to the part of the door where hinges are frequently applied—the *butt edge*. The term *butt hinge* refers to hinges applied to the door surface as well as those mortised into the butt edge. Usually the hinge is comprised of three basic parts: a *male leaf* with an even number of knuckles and normally attached to the door; a *female leaf* with an odd number of knuckles designed to receive the male leaf, and a *pin* inserted into the cylindrical hole formed by the joined knuckles which compose the *barrel*.

Pins.   On exterior doors where the barrel is exposed, unauthorized pin removal is deterred by using nonrising loose pins, sometimes called *nonlifting pins*. They are retained by an inaccessible set screw on the inner top knuckle and are adjustable only when the hinge is fully open. *Loose* or *removable pins* are used on interiors where pin removal may be desired. *Fast pins* or *rivet pins*, which may have both ends machine-spun, are oval shaped and factory sealed. They are nonremovable and are intended for industrial and institutional buildings. *Hospital hinges* require a type of nonremovable pin. *Friction pins* on friction hinges are special pins with an adjustment enabling the pin to operate as a hold-open device or to close the door gently. Stainless steel pins are used on bronze, aluminum, and stainless steel butt hinges.

Tips.   The most common tip design is the convex top *button tip,* replacing the once very popular *ball tips. Oval head* or *oval tips* form an integral surface with the barrel and are used where simplicity is desired. *Hospital tips* have a curved tip formed as a part of the top knuckle. The bottom knuckle may be shaped like the top. They improve sanitary conditions; but more importantly, these tips do not provide a surface or catch to which cords or clothing may be attached. *Steeple tips* are cone shaped and are used where appearance is paramount. The bottom tips of certain mortise hinges have a small hole used to facilitate the removal of the loose pin. These holes are called *plugs. Flat tips* are flush with the top and bottom knuckles of the barrel. *Spun tips* have a convex profile and are used on fast pins. Figure 15.9 illustrates several tip designs.

Hinge Classification.   There are four general types of butt hinge classifications with respect to application. *Full mortise hinges* (Figure 15.10 (*a*)) have both leaves embedded flush with the door and door frame. Conversely, *full surface hinges* (*b*) have leaves applied to both the door surface and the front face of the door frame. Hinges that have one leaf mounted on the front face of the door frame and the other leaf mortised into the door butt edge are *half-mortise hinges* (*c*). *Half-surface hinges* (*d*) are those that have one leaf surface mounted on the door and the other mortised to the frame. The exposed faces of surface hinges are usually beveled. See Figure 15.10.

Hinges are further classified by the screw hole pattern and by function. Those having a standardized screw hole size and pattern are *template hinges.* Their primary use is for metal doors and pressed metal frames. The countersunk screw holes are in several template patterns. *Nontemplate hinges* are used principally on wood doors and wood frames; their screw hole pattern is usually staggared. *Blank*

FIGURE 15.9   Hinge tip designs.

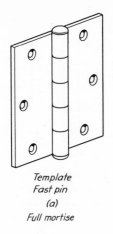

Template
Fast pin
(a)
Full mortise

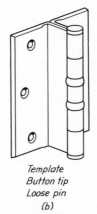

Template
Button tip
Loose pin
(b)

Ball bearing hinge:
Full surface, regular weight,
five knuckles, jambs

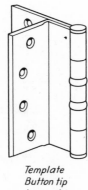

Template
Button tip
Loose pin
(c)

Ball bearing hinge:
Half mortise, regular weight,
five knuckles, iron jambs

FIGURE 15.10

face, also called *plain hinges* (*e*), have no predrilled holes. Their chief use is on welded full surface or mortise installations as needed for industries and institutions having maximum security requirements. Also, where field-drilled holes are desired for bolted construction, blank face hinges are used.

Several hinge types are identified by function. *Clear swing hinges* (*f*), (*g*) feature an offset pivot located to allow the door to swing

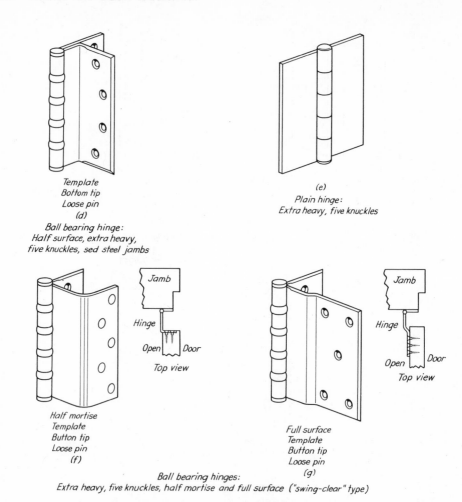

Template
Bottom tip
Loose pin
(d)
Ball bearing hinge:
Half surface, extra heavy,
five knuckles, sed steel jambs

(e)
Plain hinge:
Extra heavy, five knuckles

Jamb

Hinge

Open   Door

Top view

Half mortise
Template
Button tip
Loose pin
(f)

Jamb

Hinge

Open   Door

Top view

Full surface
Template
Button tip
Loose pin
(g)

Ball bearing hinges:
Extra heavy, five knuckles, half mortise and full surface ("swing-clear" type)

free of the passageway, permitting full use of the opening. Such hinges are used in hospitals or other institutions where the frequent movement of wide furniture and equipment requires the extra space. These may be installed by any of the four methods of application. Mortise butts called *wide throw hinges* are used where a wall reveal or return or applied trim would interfere with the door swing. These hinges have extra wide leaves, permitting the pivot point to extend well beyond the door face. *Anchor hinges* are a special full mortise hinge featuring an extension of one or both leaves bent at a right angle and mortised into the top and bottom edges of the door. They are handed. *Pivot hinges* (*h*), usually used on exterior doors, may be applied full mortise. The double knuckle features an extended pivot

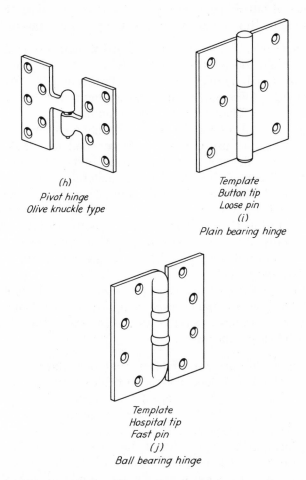

(h)
Pivot hinge
Olive knuckle type

Template
Button tip
Loose pin
(i)
Plain bearing hinge

Template
Hospital tip
Fast pin
(j)
Ball bearing hinge

located beyond the face of the door which swings fully open. *Olive knuckle types* (*h*), so named because of their shape, are a pivot hinge or *paumelle* type usually full mortise. These handed hinges meet aesthetic criteria. *Slip-in hinges* are designed to slip into a slot in doors and frames. The screw holes are not countersunk.

Hinge Selection.   Proper hinge selection in respect to application is determined by the door or door frame materials. All wood doors and wood frames should have mortised butt hinges. Mortised applications are mandatory for hollow metal doors and pressed metal frames. These various combinations of door and door frame materials will determine if a surface or mortise application is appropriate.

The choice of butt hinge bearing types is determined by traffic loads and door weight. Butt hinges feature three general types of bearings:

(1) *Plain bearing hinges* (*i*) of standard weight are used on doors with light traffic loads or on lightweight doors of expected low-frequency use (such as doors to church confessionals). Plain bearing hinges are made of wrought bronze and wrought steel in polished and unpolished finishes. They are not made of aluminum.

(2) *Ball bearings hinges* (*j*) are used on doors having moderate weight and moderate use and are recommended for all doors using automatic closing devices. Communicating doors on public institutional and commercial buildings have moderate use requirements, making the use of ball bearing hinges imperative. They generally have five knuckles with two ball bearing raceways and beveled leaves. For a smooth operation that will result in longevity to hinges, the ball bearings of polished hardened steel are enclosed in a *bearing raceway* containing a nonfluid lubricant. These hinges are made of wrought bronze and wrought steel.

(3) *Extra-heavyweight ball bearings hinges* (*d*) are employed on heavy doors of high frequency use, such as entrance doors and restroom doors on buildings accessible to the general public. The barrels have four ball bearing raceways.

Hinge Sizing.   The principle factors that determine the hinge dimensions are the door height, width, and thickness and the necessary clearances. When sizing butt hinges, the length, which is the length of the barrel, is stated first. The width, if indicated at all, is stated secondly and is the measurement of the unfolded hinge. The hinge length is determined by the width of the door specified. The width of the hinge is determined by the thickness of the door plus the offset necessary to assure a clear swing of the jamb and the trim. Table 15-1 will clarify the requirements for sizing.

The minimum number of hinges recommended for passage doors is three; however, doors 90 in. or more in height should have a minimum of four butts. Doors 60 in. or less require two hinges, although doors of this size seldom occupy the entire door frame opening, i.e., toilet stall doors. Hinges are billed in pairs; most applications require 1½ pair.

Swaging.   *Swaging* is the bending of the hinge leaf intended to bring the leaves into closer contour with the barrel. Fully swaged hinges have parallel leaves when closed or folded. All full mortised hinges should be swaged. Hinges may be swaged flat on one leaf. Template

TABLE 15-1
Rules for Sizing Hinges [22]

Door Thickness (in inches)	Door Width (in inches)	Height of Hinges (in inches)
Cabinet doors	To 24	2½
Screen doors or combination doors	To 36	3
Doors 1⅜ in. thick	To 32	3½
	Over 32	4
	To 36	4½
Doors 1¾ in. thick	36 to 48	5
	Over 48	6
Doors, 2, 2¼, 2½ in. thick	To 42	5 extra heavy
	Over 42	6 extra heavy

hinges intended for application to metal doors and metal frames must be swaged to produce equal width leaves. Nonswaged hinges called *flat back hinges* are used for full surface applications such as blank face hinges. On quality hinges, the leaves (in addition to swaging) have *beveled joints*, which means the leaves are bent near the barrel, improving appearance.

Surface Closers and Door Checks. *Door closers* and *surface closers* are mechanical devices designed for closing and controlling hinges on pivoted doors. The closer, while permitting an easy opening, protects against door damage. The closing is the checking action, thus closers are called *checks, door checks,* or *checking floor hinges* (see Figure 15.11). The checking action should be smooth throughout the closing. Floor checks and surface closers have two closing speeds. The *main*

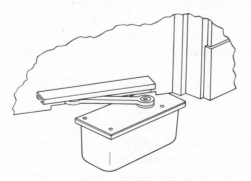

FIGURE 15.11 Checking floor hinges.

closing speed, or *sweep speed*, is the closing speed up to the last five degrees of closure. The *final closing speed*, a deceleration, is called the *latch speed* and may be adjusted independently, bringing the door to a gentle, soft closing. An adjustment made on closers to stop the opening swing action at a determined point is called the *back-checking* and protects the door, the closer, and the surrounding construction from damage. All doors should have stationary *door stops* that will assure positive stopping, notwithstanding back-checking. Adjustments and control on accessible regulating valves of the closers and checks are made by special key operated plugs that deter tampering. Casings are made from cast iron, aluminum alloy, or cast bronze. They are of one-piece construction to prevent excessive leakage of the hydraulic fluids, which should be a glycerine-based mixture, such as glycerine alcohol, petroleum lubrication oil, or a silicone fluid capable of operating in freezing weather.

Surface Closers.   Surface closers may be mounted on either side of the door and concealed by mortising into the door or by encasement within the transom bar. The three general types of surface closers are the *rack and pinion* type (Figure 15.12 (*a*)), the *crank shaft* type (*b*), and the *rotary vane* (*c*). Closers may be handed or nonhanded to permit interchangeability.

Arms.   The arms forming the linkage between the door and the closing mechanism are made of steel, cast iron, or malleable iron. There are two operating type arms: *regular arms* perpendicular to the door and *parallel arm types* with arms parallel to the face of the door and mounted to the side opposite the hinges. Various special features are associated with arm design. A *hold-open feature* allows the door to be held open, able to be released with only a slight push or pull. *Hospital hold-open arms* are adjustable to hold the door in any of several positions. *Fusible links* installed on hold-open arms are set to hold a door open. These arms have a small piece of low melting-point (eutectic) metal. They are used on fire doors kept open for normal use but which must be closed when excessive heat fuses the link. Delayed action may be incorporated into the arm, permitting cart and service trucking through an open door that will close of its own accord.

Brackets.   *Brackets* hold the exposed surface closer in proper position, permitting it to function efficiently. They are made of malleable iron, cast iron, or steel suitable for mounting on the door frame. *Corner brackets* (Figure 15.13 (*a*)), when properly located, provide leverage to the closer, enabling it to operate more efficiently. *Soffit brackets* (*b*) are used where no other adequate mounting surface is

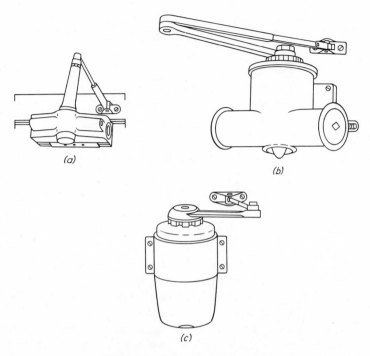

(a)

(b)

(c)

FIGURE 15.12 Door closers.

available or where space allowances are insufficient. *Adjustable brackets* (*e*) are for rounded or arched in-swinging doors.

Floor Checks. Floor checking devices, which are set flush with the floor, may be single acting or double acting. Some types have an adjustment enabling conversion from one action to the other. They may be centrally pivoted with pivots integral with the checking device. Otherwise, the checking device is independent of the hinge action as with butt hinge installations and offset pivots. Helical springs with a preset or adjustable tension are used on helical torsion and helical compression types of checks. The closing force is adjusted by the spring tension on the check or by positioning the check on its mounting plate. Floor plates are usually used for covering the embedded floor checking types. Hydraulic-electric actuated door checks are used on more expensive public buildings where a high frequency use warrants the added cost for installation.

Closer Selection. A number of factors determine door closer choice. The choice of surface applied closer or concealed closer depends upon maintenance, weather conditions, frequency of use, appearance con-

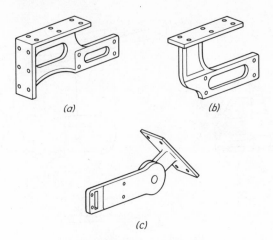

FIGURE 15.13   Brackets.

siderations, and, most importantly, the type of closer that would function with the selected door. Surface closers may be unsightly but present fewer maintenance problems than concealed floor checks. Checking floor closers are subject to water damage and dust penetration. On the other hand, concealed overhead closers, located in the transom bar of the door frame, are more accessible for maintenance though initial cost may be higher. Exterior doors are subject to more traffic, weather conditions, drafts, and severe winds that would require a closer meeting these conditions. The door swing will be a factor if center-hung by pivots, or by offset pivots, or if hung by butt hinges. Most closers handle a variety of closing conditions; however, certain concealed closers are designed for specified hanging methods. Surface closers may be mounted on the frame or on a bracket that is mounted on the frame, or the closer may be mounted on the door itself. *Bracket mounted closers* are located on the interior side of the door, providing weather protection in addition to improving the appearance of an otherwise cluttered exterior elevation.

Locking Devices.   The latch system should have smoothly coordinated durable moving parts devoid of looseness. The use of plastics is confined to nonmoving parts not subject to excessive wear.

There are four types of lock sets with respect to installation:

Rim locks are mounted on the interior surface of the door and have square or rectangular boxes with matching strikes. They are little used now except where traditional residential doors complement the architecture. They frequently have lever type handles.

Cylinder locks provide better lock concealment. They are installed by inserting the lock into a hole at the lock stile. The lock is located laterally by measuring from the beveled edge of the stile to the keyhole centerline, a distance called the *backset*. A door or a wall return makes a standard backset awkward; therefore, an *extended backset* positions the lock a few inches toward the middle of the door. This is used in conjunction with wide throw hinges mentioned above. The core is replaceable when rekeying.

Mortise locks are set flush to the lock stile edge. The locking portion may be similar to the cylinder lock or rim lock in other respects.

Unit locks are applied to the door surface and seated into a square cut made into the locking stile. These one-piece units are installed quickly and efficiently, requiring little adjustment.

Lock Fronts and Face Plates. Lock fronts and face plates are used on the lock stile edge. They are a part of the lock assembly acting as a guide sleeve to the latch bolt while sealing the lock. Their shape conforms to the door edge—flat where required, or beveled $\frac{1}{16}$ in. per inch on beveled edges. Rounded lock fronts with a 4 in. radius are used on double acting doors.

Panic Hardware. Building codes usually require that exit doors on public buildings shall have hardware easily openable. The code specifies that panic hardware shall have a *push bar* to release the locking mechanism and, moreover, that the bar must extend across at least three-fourths of the door width. Push bars may be adjusted, or *dogged*, in an open or down position that will effect a quick release. There are three types of push bar panic devices:

(1) *Rim types* (Figure 15.4(a)) have the locking mechanism mounted directly on the interior surface of the door. A strike plate is required for the inactive leaf. When paired doors are used, the inactive leaf must employ a *vertical rod type*, protected by an overlapping astragal. Rim devices may be used singly or in pairs; but if paired, they require a fixed mullion between them to hold the strike box. They are handed.

(2) *Mortise type devices* (b) have the concealed lock mortised into the door edge. They are used singly; if used in pairs, they must have a mullion between the leaves. Mortised locks may be used in combination with the vertical rod type when the mortised device is the active leaf.

(3) *Vertical rod types* (c) are surface mounted with a surface mounted top and bottom vertical rod threaded through several guides

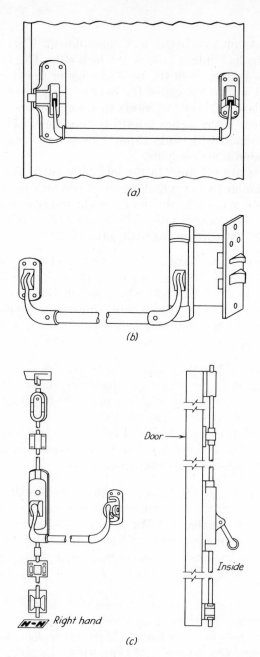

FIGURE 15.14 (a) Rim type. (b) Mortise type. (c) Vertical rod type.

mounted on the locking stile. The vertical strike bolts enter a soffit mounted strike box and a sill strike mortised flush into the threshold. For appearances, a variation has concealed rods within the lock stile. These installations have mortised strike plates in the head and sill. They are installed singly or in pairs or as the inactive leaf in pairs combined with other types. The vertical rod devices may be used on both leaves of the doors providing a *split* or *meeting astragal* is used. Where an overlapping astragal is used on the active door, the inactive leaf may be a vertical rod type; however, the active leaf must be a rim type or mortise type. Restated, vertical rods cannot be used in pairs on double doors with an overlapping astragal.

Miscellaneous Hardware.   A *coordinator* is used on double doors that have a rabbeted face or an astragal on the active leaf. This device, designed to coordinate the closing sequence of the doors, is centrally mounted on the top of the door frame. See Figure 15.15.

Strike Plates.   Flat *strike plates* mortised into the jamb are slightly curved and are designed to engage smoothly and hold the beveled strike bolt. *Extended lip strike plates* are used on wide jambs or deep-set reveals. *Box strikes*, with the interior side of the strike box deleted, allow independent door operation on double doors. They are located on the inactive leaf. *Dust proof strikes* are floor mounted and mortised. When not engaged, they present a flush surface that will prevent clogging.

Stops and Holders.   *Stops* and *holders* protect and retain the door in an open position. *Plunger stops* and *lever stops* suited for interiors have rubber tipped arms mounted on the door. *Two-piece holders* for exterior doors have a friction roller mounted on the door that engages a wall or floor mounted catch. Automatic smoke detector *hold-open*

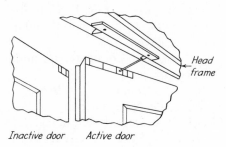

Inactive door   Active door

FIGURE 15.15 Door coordinator: for double doors with overlapping astragals, or with rabbeted stiles.

devices connected to a fire detection system hold the door open for normal routine but release it when energized.

Plates.   *Flat plates* of various sizes protect the door from damage. They are made from stainless steel, wrought brass, bronze, aluminum, or plastic. Surface mounted plates are usually beveled; if flush mounted, they are square edged. *Kick plates* are installed on doors with a high frequency use. *Mop plates* are similar but smaller. Full width *stretcher plates* with wraparound edges protect doors where service carts are used. *Armor plates* cover the entire lower portion of a door. *Push plates* are simple, flat, light material installed on the lock stile. They may be made of plastic.

Escutcheons are plates that cover the entire locking mechanism, including keyhole and knob spindle. Simple or elaborate and decorative, escutcheons may portray some motif. *Roses* are similar to escutcheons but smaller.

Finishes.   Economic factors aside, hardware finishes are selected for appearances and the ability to withstand corrosion and exposure. Brushed and sprayed finishes are applied to aluminum, stainless steel, and brass. Dull bronze, gold, statuary bronze, and jappaned are examples of sprayed surface applied finishes; whereas chrome, galvanized, and anodized finishes are plated. Bonderized and lacquered finishes serve as priming for painting.

When specified, the finish is applied to hardware on both sides of the door. However, if the sides require different finishes, a *split finish* is stated—the exterior is named first, the interior second.

Surface finishes have been standardized by the United States Department of Commerce.

REFERENCES AND RECOMMENDED READING

1. *Sweet's Architectural Catalog File*, F. W. Dodge Corp.
2. *Pondersoa Pine Doors*, Commercial Standard CS120-58, U. S. Department of Commerce.
3. *Hardwood Veneered Doors*, Commercial Standard CS171-58, U. S. Department of Commerce.
4. *State Building Construction Code*, State of New York.
5. *National Building Code*, National Board of Fire Underwriters.
6. *Wood Hand Book*, U. S. Department of Agriculture.
7. *Exterior and Interior Solid Core Flush Doors*, Architectural Woodwork No. 5, Architectural Woodwork Institute.

8. *Standard of National Board of Fire Underwriters for the Installation of Fire Doors and Windows*, No. 80, 1961.
9. *Building Materials List*, Underwriters Laboratories, Inc., January 1961, p. 137.
10. *Stanley Hardware*, Catalog #120, The Stanley Works, New Britain, Connecticut.
11. Charles G. Ramsey and Harold R. Sleeper, *Architectural Graphic Standards*, Sixth Edition, John Wiley and Sons, 1970.
12. *Von Duprin Panic Exit Devices*, Catalog 59V, Vonnegut Hardware Co., Indianapolis, Indiana, 1959.
13. *Builders' Hardware Specialties*, Builders Brass Works Corporation, Los Angeles, California.
14. *Locks and Builders' Hardware*, Catalog No. 24, Lockwood Hardware Manufacturing Co., Fitchburg, Massachusetts, 1957.
15. *Builders Hardware*, Catalogs, Schlage Lock Company, San Francisco, California.
16. *Builders Hardware*, Catalog No. 12, Quality Hardware Manufacturing Co., Inc., Hawthorne, California.
17. Catalog 1964, Sargent & Greenleaf, Inc., Rochester, New York, 1964.
18. Catalog: Falcon Lock Co., South Gate, California.
19. Federal Specification, *Hardware, Builders'; Locks and Door— Trim*; FF–H–106a—Amendment—1, 23; November 1948, U. S. Government Printing Office.
20. Supplement to the above. December 10, 1952.
21. Federal Specification, *Hardware, Builders'; Door—Closing Devices*; FF–H–121c; June 18, 1954, U. S. Government Printing Office.
22. Federal Specification, *Hinges, Hardware, Builders'*; FF–H–116c; July 3, 1957, U. S. Government Printing Office.
23. *Recommended Standard Steel Door Frame Details*, S.D.I. 111–A, Technical Data Series, Steel Door Institute, Cleveland, Ohio.
24. Brochure: *How to Select the Proper Hinge*, AIA File No. 27–B, Hager Hinge Company, St. Louis, Missouri, March 1970.

16

MISCELLANEOUS METALS

91. FERROUS METALS

Pig Iron

Definition, Composition, and Uses. *Pig iron* may be defined as the product obtained by the reduction of iron ores in the blast furnace. It contains 91 to 94% iron, 3.75 to 4.50% carbon, 0.25 to 3.50% silicon, 0.03 to 1.00% phosphorus, and less than 0.10% sulfur. A vertical section through a blast furnace is shown in Figure 16.1.

Pig iron may be used directly in making castings, but its most important use is in the manufacture of cast iron, wrought iron, and steel, where it may be used in a molten state direct from the blast furnace or in the form of *pigs*, which are castings made from the pig iron as it is drawn from the furnace.

Raw Materials. The raw materials used in the manufacture of pig iron are the iron ores which furnish the iron, the fuel which furnishes the heat and the *reducing agent* to reduce the carbon content, the *flux* which provides a fusible *slag* that carries off ash of the fuel and some of the impurities of the ore, and the air which supplies the oxygen for the combustion of the fuel.

The principal iron ores of commercial importance may be divided into four classes: iron oxides, iron carbonates, iron silicates, and iron sulfides. Only the oxides are of importance in this country. The oxides of iron used in the manufacture of pig iron are:

Hematite, Fe_2O_3, containing about 70% iron when pure;
Limonite, Fe_2O_3, n.H_2O, which is hydrated hematite containing about 60% iron when pure;
Magnetite, F_3O_4, containing about 72% iron when pure.

Impurities in the ore, such as sand and clay, are called the *gangue*.

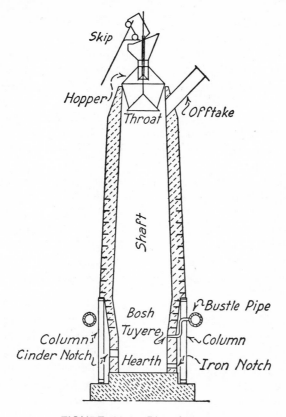

FIGURE 16.1   Blast furnace.

The fuel has two functions in the manufacture of pig iron: to furnish the necessary heat and to supply the reducing agent to combine with the oxygen of the ore. The fuel used is coke because of its porosity and resistance to crushing.

The functions of a *flux* are to make the impurities in the ore and fuel, such as silica and alumina, more easily fusible and to provide a fusible slag in which these and other impurities may be carried off. The flux used in the blast furnace is usually limestone, which, when pure, is calcium carbonate and has the chemical formula $CaCO_3$, although a considerable amount of magnesium carbonate ($MgCO_3$) is usually present.

The Blast Furnace.   The blast furnace is a structure 100 ft. or more in height, approximately cylindrical in shape, built of fire brick, enclosed more or less completely in a steel shell. The furnace is divided into three main parts: the *hearth*, or *crucible*, the *bosh*, and the *stack*, as shown by the diagram in Figure 16.1.

The hearth is provided with an *iron notch* and a *cinder notch* through which the iron and slag are periodically removed. Just below the bosh, a ring of 10 to 16 *tuyeres* penetrates the lining. The air necessary for the process is blown through these tuyeres, which are all connected to the *bustle pipe* encircling the furnace. This pipe is in turn connected to the hot-blast stoves, which heat the air, and finally to the blowing engines which provide the air at a pressure of from 15 to 30 lb. per sq. in.

At the top of the furnace is a device that permits the raw material to be charged into the furnace but prevents the escape of the gases. These gases are heated to a very high temperature and contain a large amount of carbon monoxide (CO), which is combustible. Formerly, these gases were permitted to burn as they left the furnace, but now they are used in the hot-blast stoves, under steam boilers, and in internal-combustion engines. This practice results in a considerable saving in fuel.

Operation.   A blast furnace is operated continuously by charging the *burden*, which consists of ore, fuel, and flux in proper proportions, in the top of the furnace and drawing off the slag and iron at the bottom, the slag being tapped at intervals of about 2 hours and the iron at intervals of about 5 hours. The iron may be used in the molten state in the manufacture of wrought iron or steel or may be cast into bars called *pigs* for subsequent use.

Chemical Changes.   Air contains about one part of oxygen to four parts of nitrogen and other inert gases. The air enters the blast furnace through the tuyeres at a temperature of about 1000°F and a pressure of from 15 to 30 lbs. per sq. in. It immediately comes into contact with hot coke, and the oxygen of the air combines with the carbon of the coke forming carbon dioxide (CO_2):

$$C + O_2 = CO_2 \tag{1}$$

Because of the excess of carbon, the carbon dioxide is reduced to carbon monoxide (CO):

$$CO_2 + C = 2CO \tag{2}$$

The ore as it moves down through the furnace encounters this carbon monoxide, which is a powerful reducing agent, and the following reactions result, the first occurring near the top of the furnace at a temperature of approximately 600°F, and the last occurring in the lower part of the stack at a temperature of about 1400°F:

$$2F_2O_3 + 8CO = 4Fe + 7CO_2 + C \qquad (3)$$
$$3Fe_2O_3 + CO = 2Fe_3O_4 + CO_2 \qquad (4)$$
$$Fe_3O_4 + CO = 3FeO + CO_2 \qquad (5)$$
$$FeO + CO = Fe + CO_2 \qquad (6)$$

Reactions 1 and 3 produce most of the heat required by the other reactions and heat to dry the raw materials, to decompose the limestone, to flux the impurities, and to melt the iron and slag. The temperatures in the furnace increase from about 400°F at the top to about 3000°F in the crucible.

Other changes occur, but they are too confusing to consider here. However, it may be well to mention that the heated coke acts as a reducing agent in much the same manner as the carbon monoxide already considered.

The limestone in the charge breaks up into calcium oxide and carbon dioxide at about 1600°F:

$$CaCO_3 = CaO + CO_2 \qquad (7)$$

The CO_2 is reduced to CO when it encounters the hot coke. At about this same temperature, the iron, which is in a spongy form, absorbs carbon from the coke. This lowers its melting point and it becomes fluid.

At the top of the bosh, the lime (CaO) combines with some of the gangue, a little unreduced iron oxide, and manganese oxide from the ore and forms slag which, with the molten iron, runs down through the coke to the hearth. There the slag and the molten iron separate into two layers because of the difference in their densities; the slag, being the lighter, remains on top.

In the hearth, part of the oxides of manganese (Mn_3O_4), silicon (SiO_2), and phosphorus (P_2O_5) are reduced by the carbon of the coke, which extends through to the bottom of the furnace, and join the iron. The remainder of the oxides are not acted upon by the carbon and are found in the slag. All the phosporus present in the charge is found in the iron, but the amount of silicon and manganese in the pig iron depends upon furnace conditions. Sulfur is introduced into the blast furnace mainly as an impurity in the coke in the form of iron sulfide. Some of this reacts with the lime to form calcium sulfide and joins the slag, but the remainder is found in the iron, for iron sulfide is soluble in iron. Conditions that tend to decrease the amount of sulfur present in the iron also tend to increase the amount of silicon.

Cast Iron and Malleable Cast Iron

Production.   Cast iron is manufactured by remelting pig iron in a *cupalo*, which is similar to a blast furnace but much smaller, and pouring it into molds to form castings of the desired shape when the metal solidifies in cooling. Scrap iron, consisting chiefly of discarded castings, is used with the pig iron because it is less expensve than pig iron. No chemical changes of importance take place during the process of remelting. Cast iron contains from 2.5 to 4.0% carbon.

Molds and Patterns.   The process of making castings is called *iron founding*. The molds are made of sand; the impressions in the sand are usually made by *wood patterns*. In *loam molding*, no pattern is used, the impression being formed in the sand by hand or machine. Patterns must be slightly larger than the objects to be cast to allow for the shrinkage of the metal in cooling. Vertical surfaces must be slightly tapered to facilitate the withdrawal of the pattern from the mold. This taper is called *draft*.

Gray and White Cast Iron.   When molten cast iron solidifies, the carbon which is present remains combined with the iron as *carbide of iron* (Fe_3C) or may separate from the iron as *graphite*. White cast iron contains carbon chiefly in the combined state. It has a white metallic fracture and is very hard and brittle. Gray cast iron contains carbon chiefly in the form of graphite mechanically mixed with the iron, but some carbon is present in the combined state. It has a gray, crystalline fracture and is not as hard and brittle as white cast iron. The graphite in cast iron is in the form of flakes, which reduce its strength materially. Slow cooling and the presence of silicon tend to increase the amount of graphite in cast iron, whereas rapid cooling and the presence of manganese and sulfur tend to hold the carbon in the combined state. Cast iron increases in strength as the amount of combined carbon is increased up to about 1.2%. Further increases cause a loss of strength. The hardness and brittleness increase as the amount of combined carbon is increased. In general, cast iron has a high compressive strength and a low tensile strength.

Chilled Castings.   Chilled castings are produced by using molds with certain surfaces made of iron. The iron that comes in contact with these surfaces is therefore cooled suddenly and the carbon in the iron remains in a combined state (Fe_3C) for a certain depth, forming white cast iron, which is very hard. The remainder of the iron cools naturally and forms gray cast iron, which is not as hard but is less brittle. Surfaces of cast iron which are subject to wear are often chilled to increase their life.

Malleable Cast Iron. Castings made of white cast iron may be made malleable and ductile by subjecting them to an annealing process which converts the combined carbon into free carbon in a very finely divided state. The annealing may be accomplished by packing the castings in some inert material, such as sand or clay, and heating to a red heat which is maintained for several days, after which the castings are slowly cooled.

Better results may be secured by packing the castings in an oxidizing material such as iron oxide. The oxide draws the carbon from the castings to a depth of $\frac{1}{16}$ in. or more, forming a skin of soft iron on the castings in addition to converting the combined carbon in the body of the castings into free carbon.

White cast iron is used so that all the free carbon will be in the form of very minute particles evenly distributed throughout the castings and not in the form of large flakes which have such a weakening effect on gray cast iron.

Malleable castings are used for small articles such as builders' hardware. Malleable iron washers are extensively used in timber construction.

Stainless Steel

Stainless steel, sometimes referred to as an *architectural metal,* is used where corrosive conditions prevail and maintenance is costly and where appearance may be a factor. It is ideal for roofing parts that require a metal, such as spires, gutters, facias, downspouts, flashing, expansion joints, and flagpoles. Stainless steel is widely used on interiors such as lobbies, elevators, entrances, stairways, and doorways—nearly any location where a material may be subjected to abuse.

Because it has a greater strength than most other metals that are used for architectural finishes, it is more resistant to wear and tear. Stainless steel has a nonstaining character which permits its use in conjunction with other nonstaining materials without the danger of deterioration through galvanic action with dissimilar metals.

Like most other sheet metals, stainless steel is most frequently designated by gage numbers. As manufactured, it is referred to as *plate* if more than $\frac{3}{16}$ in. thick and in widths greater than 10 in.; *sheet* if less than $\frac{3}{16}$ in. thick and 24 in. or more in width; and *strip* if less than $\frac{3}{16}$ in. in thickness and less than 24 in. in width.

Grades.   There are three basic groups of stainless steel [2]:

"Austenitic grades contain both chromium and nickel and possess the highest corrosion resistance. They are not hardenable by heat treatment and are nonmagnetic in their annealed state, although some types become magnetic with cold working.

Ferritic grades generally contain 14 per cent or more of chromium. They are not hardenable by heat treatment, are hardened only moderately by cold working, and are always magnetic.

Martensitic grades generally contain a maximum of 14% chromium, and are of lower corrosion resistance than austenitic or ferritic grades. They are hardenable by heat treatment and are magnetic. Because of their lower corrosion resistance they find limited application in architectural uses."

Types.   There are more than 40 types of stainless steel; however, only 5 concern building construction. Types have been established according to a numbered system by the American Iron and Steel Institute. The types listed below are of interest to the building industry, especially Type 302, which is probably the most useful. The descriptions below are from reference [2] with slight editorial changes.

AISI Type 302.   This, a United States Steel 18-8 grade stainless steel, is the most commonly used in architecture and is virtually the all-purpose grade in either interior or exterior applications. It is readily formed and welded and may be specified for such exterior uses as column covers, doors and door frames, facias, exposed or concealed flashing, gutters and rain carrying equipment, mullions, panels, and soffits.

AISI Type 304.   This stainless steel is sometimes recommended in lieu of Type 302 and differs from it only in having a lower carbon content. For architectural purposes, it is interchangeable with Type 302 both functionally and economically.

AISI Type 301.   This, a United States Steel 17-7 grade, is a stainless steel which contains slightly smaller amounts of chromium and nickel and has somewhat higher work-hardening characteristics. Otherwise, it retains essentially the same characteristics and properties as Type 302. In strip widths (less than 24 in.) with cold-rolled finishes (not including No. 4 finish), it is slightly less expensive than Type 302. In applications that must perform structurally, the use of Type 301 may be advantageous since high strengths can be achieved by cold rolling and with greater facility than with Type 302.

AISI Type 316. This is a modified United States Steel 18-8 grade to which molybdenum has been added to increase resistance to attack in highly corrosive atmospheres. While this grade is somewhat higher in cost than Type 302, it is the preferred material in marine and harsh industrial-chemical environments, and similar areas where the other chromium-nickel grades might not be adequate.

AISI Type 430. This United States Steel 17 grade stainless steel contains 16–18% chromium and no nickel. It has satisfactory resistance to weathering and atmospheric corrosion in the average urban environment but is not quite as good as Type 302 in resistance to many atmospheres. It costs less than Type 302 and is commonly used for interior applications, trim, and hardware items.

Finishes. Stainless steel is finished and marketed in several standard mill finishes and in special textured finishes. The standard mill finishes are referred to as *unpolished* and *polished finishes*. When stainless steel is specified, the desired finish should be indicated [2].

"Unpolished finishes:

No. 2D Special. A frosty, matte finish produced on either hand sheet mills or continuous mills by cold rolling to the specified thickness, annealing, and descaling. The dull finish may result from the descaling or pickling operation or from a final light temper pass on dull rolls. The comparable strip finish is referred to as 'No. 1 Strip Finish.'

No. 2B. A dense, bright, cold-rolled finish of higher reflectivity. It is produced in the same way as No. 2D finish except that the annealed and descaled sheet receives a final pass on polished rolls. Should polishing be required after fabrication, this finish is the easiest to polish. The comparable strip finish is referred to as 'No. 2 Strip Finish.'

No. 4. A bright, but not highly reflective finish, having a fine-grained linear appearance. It is the most frequently used architectural finish and is achieved by coarse grinding followed by polishing with 120-150 grit abrasives. This finish has a distinct advantage in that weld areas may be easily and inconspicuously blended into it.

No. 6. A soft satin finish having lower reflectivity than No. 4 finish. It is produced by Tampico brushing the No. 4 finish in a medium of abrasive and oil.

No. 7. A highly reflective finish produced by buffing a surface which has first been finely ground with abrasive of approximately 240-280 grit. The grit lines are not removed by the buffing."

Textured finishes are most useful where the stainless steel is to be subjected to abuse or where a particular richness is desired. An *engine turned* finish, which is a pattern of several concentric circles, is particularly attractive for door push plates.

Shapes. Stainless steel can be made into a variety of shapes, including sheet, strip and bars, solid rounds and tubes which are round, rectangular, square, or oval. Stainless steel members such as channels and angles are cold-rolled as well as hot-rolled. (See Figure 16.2.)

(a)   (b)

FIGURE 16.2 Cold-rolled members. (*a*) Angle. (*b*) Channel.

Joining. This material may be joined by welding, brazing, soldering, mechanically formed lap or batten seams, and by mechanical fasteners, such as screws and bolts. Mechanical fasteners should be of the same type of stainless steel.

92. NONFERROUS METALS AND ALLOYS

Concentrating. The first step in the production of most nonferrous metals from their ores consists of increasing the proportion of metal present by removing much of the waste material. This process is called *concentrating*. It includes various mechanical treatments, such as crushing, grinding, and screening, which do not involve heat and cause no chemical changes.

Aluminum

Ores. Aluminum occurs abundantly in nature in combination with oxygen, sodium, fluorine, and silicon. The chief source of aluminum is *bauxite,* which is hydrated oxide of aluminum and iron with some silicon. The most extensive occurrence of aluminum is as the oxide *alumina* (Al_2O_3), which is the principal constituent of clay. No process has yet been devised to obtain aluminum from clay in commercial quantities.

Extraction. The stages in the extraction of aluminum from bauxite are:

(1) *Roasting* bauxite to drive off the water.

(2) *Grinding* roasted bauxite and heating under pressure with a solution of sodium hydrate forming sodium aluminate.

(3) *Precipitating* aluminum hydroxide by heating sodium aluminate solution with aluminum hydroxide or with carbon dioxide.

(4) *Separating* aluminum hydroxide with filtering and dehydrating by heating to form alumina (Al_2O_3).

(5) *Extracting* aluminum from alumina by the electrolytic decomposition of alumina in a molten bath of cryolite which is the fluoride of alumina and sodium.

In the last-named process, the containing vessel is lined with carbon in the form of graphite or coke, which forms the cathode. Carbon rods suspended in the bath form the anode, to which the oxygen goes. The cryolite is melted by the heat generated by the passage of the electric current across a gap which exists between the anodes and the cathodes. When the cryolite bath becomes molten, alumina is thrown on the bath; as it melts, it is broken up into molten aluminum, which settles on the cathode, and oxygen, which goes to the anode with which it combines, forming carbon monoxide which escapes as a gas. The molten aluminum is tapped off as it accumulates. The heat for the process is supplied by the electric current.

Properties and Uses.   Aluminum is formed by forging, casting, drawing, rolling, and extruding. The latter two methods of forming are of particular interest to building construction. In the extrusion process, heated aluminum is squeezed through a die by a hydraulic actuated press. The extruded aluminum is formed into the cross-sectional shape of the die, which can be of almost any design. Thus, aluminum can be economically shaped into most any section desired. This accounts for an abundance of various aluminum mouldings, edgings, window and curtain wall mullions, and sections. Generally, these aluminum members are referred to as *extrusions*.

Aluminum has a low electrical resistance which makes it valuable for transmission lines and other electrical uses. It is a good heat conductor and is noncorrosive. Because of its lightness, it has many special uses; its weight is only about one-third that of iron. It is very malleable, quite ductile, noncorrosive, and strong in proportion to its weight.

Aluminum Alloys.   Because aluminum in its refined state is soft and of low strength, it is of limited use in the building industry. Fortunately, this material forms numerous alloys. Manganese, copper, zinc, silicon, and magnesium are alloyed with aluminum to improve its properties. Most of these alloying elements improve both the strength and the hardness of aluminum.

Wrought alloy types of aluminum are classified by a series code number in a system established by the Aluminum Association. In the 2000 Series, copper is the principal alloying material. The tensile strength is improved, and materials of this alloy are suitable for building structures and bolts. The 3000 Series has manganese as the principal alloy material. The result is a strong workable alloy that is used for corrugated and stamped roofing, siding, and other sheet metal work. Silicon produces an alloy that is numbered in the 4000 Series. This alloy is used for castings and forgings. This material may be finished with a fine anodized finish. Magnesium alloy, the 5000 Series, is used for architectural trim and nails. The magnesium silicon alloys are indicated in the 6000 Series and are used for structural members, nails, railings, windows, doors, and extrusions.

Other symbols that are a part of the series code number indicate temper: annealed or strain hardened.

Copper

Ores.   Copper is found in the native or pure state and in a great variety of ores, of which sulfides are the most common.

Extraction.   The stages in the extraction of copper from concentrated sulfide ores are as follows.

(1) *Roasting* the sulfide ore by heating it in an oxidizing atmosphere where a part of the iron sulfide ore is converted into the oxide by the burning out of the sulfide. The temperature is not high enough to melt the ore. The roasted ore contains chiefly copper sulfide, iron oxide, and siliceous gangue, although some copper oxide may be present.

(2) *Smelting* in a blast furnace with coke as a fuel or in a reverberatory furnace. Copper sulfide and iron sulfide settle to the bottom forming a *matte*, and the slag, containing the *gangue* or earthy materials and iron oxide, forms on top where it may be drawn off.

(3) *Converting* the molten *matte*, containing the sulfides of copper and iron, in a converter similar to that used in the manufacture of Bessemer steel. Air is blown through the molten matte and provides oxygen, which combines with the iron sulfide to form a slag of iron oxide on top of the remaining copper sulfide. By continuing the process, the sulfur is burned out of the sulfides, forming sulfur dioxide

which passes off as a gas and leaving an impure form of copper known as *blister copper.*

(4) *Refining* the blister copper either in a reverberatory furnace or electrolytically. In *furnace* or *fire refining*, the blister copper is melted in a reverberatory furnace; the slag which is formed is skimmed off; and air is blown over the molten bath and oxidizes the impurities which either pass off as gas or form slag on top of the molten metal. Some of the copper is oxidized and must be reduced by adding charcoal and stirring with a green wood pole. Steam formed by the water in the green pole agitates the bath, and the carbon from the charcoal combines with the oxygen of the copper oxide, forming carbon monoxide which passes off as a gas, thus leaving fairly pure copper which is cast into ingots.

In the *electrolytic process,* anodes of blister copper or fire-refined copper and cathodes of pure copper are placed in a copper sulfate bath. An electric current is passed through the bath and causes pure copper to be transferred from the blister copper anode to the pure copper cathode. The impurities, which often include the precious metals, are not soluble in copper sulfate, and being of higher specific gravity, they fall to the bottom of the bath and are removed.

Fire-refined copper is not as pure as that obtained by the electrolytic process. The former is used for wires, tubes, and plates, and for making brasses and bronzes. The latter is used primarily for electrical purposes which require the purer form.

Properties and Uses.   The most important properties of copper are its electrical conductivity and its resistance to corrosion. Its strength and ductility are greatly affected by the mechanical and heat treatment it receives. If heated to a red heat and cooled slowly it is brittle; but if cooled quickly, it is malleable and ductile. It may be cast and welded and may be rolled or drawn when hot or cold. Cold working increases its strength but decreases its ductility. When exposed to moist air, as when used for a roofing material, a thin coating of the green basic carbonate is formed. This protects the copper so that no further treatment is required.

Copper is used primarily for electrical purposes, but it is also used extensively as a constituent of brasses and bronzes and for sheet-metal roofing and shingles, gutters, rain conductors, flashing, and pipe for plumbing. Copper wire is formed by drawing through dies. If it is cold-drawn, it produces what is called *hard-drawn wire*, which is springy.

The alloys of copper, like those of aluminum, are classified by a number series code. The copper that is used mostly for building construction in the form of sheet, roll, and strip is known as alloy copper 110. This alloy is further known as *electrolytic tough pitch copper* and is produced in flat form as cold-rolled and soft-rolled *sheets*, cold-rolled *strip*, and soft-rolled *rolls*. These forms of copper are specified by the number of ounces of the material in a square foot. Sheet is available from 16 oz. to 32 oz., while strip and roll is available only in 16 oz. weights.

This copper alloy may be joined by mechanical bending in seams of various types, or by soldering and brazing. Generally, welding is not recommended.

Copper, a soft metal, takes several mechanical finishes. It may be brushed, burnished, polished, sandblasted, or buffed. However, the natural turquoise patina that it acquires is its most outstanding finish on exterior applications.

Zinc

Ores. The ores of zinc are the sulfide (ZnS), the carbonate (ZNCO₃), the silicates (Zn₂SiO₄H₂O) and (Zn₂SiO₄), and franklinite, which is composed of the oxides of zinc, manganese, and iron. The most common source of zinc is the sulfide zinc blende.

Extraction.   The stages in the extraction of zinc from its ores are as follows.

(1) *Roasting* the ore to drive off the sulfur from sulfides and the carbon dioxide from the carbonates to form zinc oxide.

(2) *Mixing* the oxide with a nearly equal amount of finely ground anthracite coal and placing it in fire-clay retorts.

(3) *Heating* retorts to a white heat so that the coal will form carbon monoxide which will combine with the oxygen of the zinc oxide to form carbon dioxide and zinc vapor.

(4) *Condensing* the zinc vapor at a temperature above the melting point of zinc to form molten zinc, which is poured into molds and allowed to cool, forming *zinc spelter*, the form commonly used. If the temperature at which the condensation is carried on is too low, *zinc dust* is formed. This cannot be melted to form a solid mass and has limited use in the powdered form.

Properties and Uses.   Zinc has a fairly high resistance to corrosion. For this reason, it is used as a roofing material in the form of sheet zinc. Exposure to the weather causes dull-gray zinc carbonate to form on the surface and protect the remainder of the metal.

If zinc is cooled suddenly from the molten state, it is malleable, but if cooled slowly, it is hard and brittle. Commercial zinc is quite brittle, but if heated to about 250°F it becomes malleable and ductile. It may then be drawn into wire or rolled into sheets which are ductile and malleable when cool. Molten zinc shrinks very little in solidifying and casts well.

Zinc is extensively used as a protective coating for steel in the *galvanizing* and *sherardizing* processes. It is also used as one of the constituents in brass, nickel, silver, and other alloys and in making electric batteries.

Lead

Ores.   The only important ore of lead is *galena*, the sulfide (PbS) mixed with gangue.

Extraction.   There are two stages in the extraction of lead from its ores.

(1) *Roasting* by charging crushed ore into a reverberatory furnace where it is heated at a low temperature in an oxidizing atmosphere. Part of the sulfur in the sulfide is oxidized forming lead oxide and sulfur dioxide gas, and some lead sulfate is formed:

$$PbS + 3\,O = PbO + SO_2$$
$$PbS + 4\,O = PbSO_4$$

(2) *Smelting* the lead oxide, the lead sulfate, and the remaining lead sulfide in a blast furnace with limestone, coke, and some other materials to form lead and sulfur dioxide:

$$PbS + 2\,PbO = 3\,Pb + SO_2$$
$$PbS + PbSO_4 = 2\,Pb + 2\,SO_2$$

The molten lead thus formed is drawn off and cast into ingots. This lead must usually be purified before it is ready for the market.

Properties and Uses.   The important physical properties of lead are its resistance to corrosion, its plasticity, and its malleability.

Lead sheets are used to a considerable extent for watertight pans under the floors of shower baths and in similar positions. They are used to a limtied extent for roofing, particularly for curved or irregular surfaces to which lead can be easily fitted by stretching and working. Lead has a high coefficient of expansion and is difficult to hold in place, particularly on pitched roofs.

A roofing known as *hard lead* is composed of lead and antimony. It is stronger and has a lower coefficient of expansion than ordinary lead and can be used on any slope.

Lead pipes are used in plumbing but not as much as they were in former years. The pipe is formed by forming the metal through dies by means of hydraulic presses. The term *plumber* is derived from the Latin word for lead, *plumbum*.

Solder is an alloy of lead and tin.

Tin

Ores. The principal source of tin is the black oxide (SnO_2), known as *cassiterite* or *tinstone*. Tin is the only important metal not found in the U. S.

Extraction. The stages in the extraction of tin from its concentrated ores are as follows.

(1) *Roasting* in a reverberatory furnace to oxidize the sulfur and arsenic which exist as impurities. These oxides pass off as gases.

(2) *Smelting* in reverberatory or blast furnaces. The oxygen of the tin oxide combines with carbon to form carbon dioxide, the carbon being mixed directly with the ore in the furnace, or the oxygen combines with carbon monoxide derived from the partial combustion of coal used as a source of heat. The molten tin accumulates and is drawn off. This crude or raw tin contains copper, iron, arsenic, sulfur, and other impurities.

(3) *Refining* by liquidation and boiling. The ingots of crude tin are placed in a reverberatory furnace, and as the temperature is gradually increased, the tin melts and is removed, leaving the unfused impurities. The molten tin thus obtained is still further purified by boiling while bundles of green twigs are held submerged in the molten tin. The steam given off by the green twigs develops violent boiling which causes the impurities to become oxidized by coming in contact with the air. These oxides form a scum on the surface. This scum is removed and the tin is cast into ingots. Some of the impurities are heavier than the tin and do not pass into the scum but settle toward the bottom. For this reason, the tin that comes from the top of the vessel is purer than the tin that comes from the bottom. The former is called *refined tin* and the latter *common tin*. The common tin is often liquefied and boiled again.

Properties and Uses. Tin is extremely malleable and may be rolled into very thin sheets called *tinfoil*. It is very resistant to corrosion. Its principal use is in coating sheet iron or steel for use as a roof covering.

Tin plate used for roofing does not have a coating of pure tin but of an alloy of 25% tin and 75% lead. This is known as *terne plate* and is much less expensive than *bright tin plate*, which is coated with pure tin.

Tin is used in making bronzes and other alloys.

Alloys

The Brasses.   The *brasses* are alloys of copper and zinc. The most useful brass alloys range in composition from 60% copper and 40% zinc to 90% copper and 10% zinc. Standard brass contains two parts of copper to one part of zinc and is the most commonly used of all the brasses. Those carrying a large amount of copper are copper red in color, and those with a small amount of copper are a silvery white.

Brass may be shaped by casting, hammering, stamping, rolling into sheets, or drawing into wire or tubes.

Muntz metal contains 60% copper and 40% zinc.

The addition of even small amounts of tin to brass greatly increases its resistance to corrosion.

The Bronzes.   The *bronzes* are alloys of copper and tin ranging in composition from 95% copper and 5% tin to 75% copper and 25% tin. The chief effect of tin on copper is to increase its hardness.

Gun metal contains 90% copper and 10% tin. It is the strongest of the bronzes. *Bell metal* contains 80% copper and 20% tin. It is hard and is used for making bells. *Speculum metal* contains two parts copper and one part tin. It is a hard white metal which will take a polish and is used for making mirrors.

Phosphor bronze is a copper-tin alloy to which a small amount of phosphorus has been added as a deoxidizer to eliminate copper oxide. This results in a very marked improvement in the strength and quality of the bronze. It is highly resistant to corrosion.

Manganese bronze is really a brass, for it contains a large amount of zinc and little or no tin. The manganese acts as a deoxidizer and does not appear in the resultant alloy since it has been oxidized and removed in the flux. The addition of manganese greatly improves the strength of the alloy. Manganese bronze is an excellent material for castings and is very resistant to corrosion.

Monel Metal.   An important alloy of nickel and copper is *Monel metal*, which is about two-thirds nickel, somewhat less than one-third copper, and has small amounts of iron, manganese, carbon, and silicon. It is resistant to corrosion, takes an attractive finish, and is easily worked. Monel metal is used for counter tops, sinks, and other purposes in buildings.

REFERENCES AND RECOMMENDED READING

1. Melvin, Nord, *Engineering Materials*, John Wiley and Sons, 1952.
2. *USS Stainless Steel, Architectural Design Details*, United States Steel Corporation, San Francisco, California.
3. *ASTM 1972 Annual Book of Standards, Part 3*, American Society for Testing and Materials, Philadelphia, Pennsylvania.
4. *Contemporary Copper*, Copper Development Association, Inc., New York, New York, 1966.
5. *Aluminum Standards & Data*, 2nd Ed., December 1969, The Aluminum Association, 750 Third Avenue, N.Y., N.Y. 10017.

17

PLASTICS

93. DEFINITIONS AND GENERAL DISCUSSION

Definition. Many materials used in building construction are plastic, that is, they will continue to deform under a constant load, for varying conditions of temperature and stress intensity, and with duration of stress. The large and ever-increasing group of materials known as plastics, however, includes certain common characteristics. A plastic may be defined as follows.

A *plastic* is a material that contains as an essential ingredient an organic substance of high molecular weight which, although solid in the finished state, is soft enough at some stage of its manufacture to be formed into various shapes, usually through the application, either singly or together, of heat and pressure.

Essential Ingredients. All *organic substances* contain carbon. The ones that are involved in the manufacture of plastics are complex substances called *resins*. These may be natural resins, but they are usually synthetic resins built up by chemical processes from such materials as coal, petroleum, natural gas, water, salt, and air. The chemical elements whose atoms form the molecules of plastic resins are principally carbon, hydrogen, oxygen, and nitrogen.

Many kinds of resins are used, with other materials, in the manufacture of plastics to form products with a great variety of physical and chemical properties. The other materials included in plastics are called plasticizers, fillers, and colorants. These are considered in a subsequent paragraph.

Formation of Large Molecules. The large size of the molecules in a plastic resin is an important factor in determining its properties and those of the plastic of which it is the essential ingredient. In forming synthetic resins, long chain-like molecules are built up by linking small molecules together by processes involving various combinations of heat, pressure, or chemical action. In some plastics, the long molecular chains become attached to each other by cross links, when heated, forming net-like structures.

Broad Classification of Plastics.   All plastics include tangled masses of long resin molecules. They are divided into two classes according to their behavior when heated and cooled during manufacture.

Thermoplastics are softened by heat during the manufacturing process and regain their original properties as they solidify during cooling to form the finished products. No chemical changes occur, and the process can be repeated.

The resin molecules of thermoplastics are chain-like with few if any cross links between molecules. They form a tangled mass, but they can slide over each other when the plastic is stressed. If the stress is not too large, they will slowly recover their original relative positions when the stress is removed, and the deformation caused by the stress will disappear.

Thermosetting plastics change chemically when heated during manufacture, solidify while still hot, and assume the form of the finished products. The process cannot be repeated.

Cross links form between the chain-like molecules of the resin and bind them together into a net-like structure which is retained on cooling. The cross links restrict the relative movement of the molecules when the plastic is stressed so it tends to become rigid. Plasticity decreases as the number of cross links increases.

Other Ingredients.   As has been stated, the essential ingredient of all plastics is some type of resin. However, nonresinous ingredients are included in plastics to achieve certain desired properties, facilitate processing, or reduce the cost. These are classified as plasticizers, fillers, and colorants.

Plasticizers are materials which may be included in plastics to make them more plastic or flexible at ordinary temperatures and during the manufacturing process. They are usually organic compounds. Their primary, but not exclusive, use is in thermoplastics. Plasticizers are usually liquids which have high boiling points and act as solvents for the resins.

Fillers are materials included in plastics to provide or improve certain properties and lower their cost. They may be powders or fibrous materials. Powdered quartz or slate may be added for hardness; mica, clay, and asbestos fiber for heat resistance; wood flour for bulk and malleability; and glass and cotton fiber for tensile strength and flexibility. By the addition of appropriate fillers, plastics with widely varying properties can be made from the same kind of resin. A considerable portion of the volume of a plastic may be due to the filler. Fillers are extensively used in thermosetting plastics.

Colorants are responsible for one desirable feature of plastics. This is the great variety of colors in which they are available. These are obtained by the addition of coloring materials which include organic dyes and pigments, and inorganic pigments which are usually metallic oxides.

Solvents.   Plastics are used in liquid form as one of the ingredients of some surface-coating materials. Among such materials are various kinds of paint, varnish, enamel, and lacquers. Most of the resins suitable for use in such products are hard and brittle. To convert them to liquids, as required in the manufacture and application of these products, appropriate organic solvents are used. The resin hardens as the solvent evaporates after a coating is applied. Solvents are also used to liquify the granular or powdered compounds as required in some manufacturing processes.

Shaping or Processing Plastics.   Plastic products can be manufactured in a great variety of forms or shapes by selecting the appropriate kinds of resin and other materials and using suitable equipment and processing procedures. The compounds to be converted into manufactured products are supplied in granular or powdered forms, both commonly called *molding powder,* or in liquid form. The molding powders are heated to form viscous liquids during processing.

The temperatures required to make plastics sufficiently fluid for processing are relatively low compared with those required for most metals. Plastics cannot withstand high temperatures.

In the *molding processes,* the liquid is forced by pressure into molds to produce the desired shape. There it solidifies either by cooling, as with thermoplastics, or by the cross linking of the molecules of thermosetting plastics.

In the usual *extrusion process,* a viscous liquid thermoplastic is forced through an orifice or die shaped to yield a product of the desired cross section. After passing through the die, the extruded plastic is deposited on a moving belt or other device. The extruded material may be cooled in various ways. The processes are usually continuous. Simple dies can be shaped to form solid rods and sheets. Special devices are used to form tubes and to coat wires with plastic. Instead of using long shallow dies or orifices for extruding films or sheets, such forms are more commonly produced by splitting large extruded tubes longitudinally and spreading them out to flatten.

In the *calendering process*, molding powder warmed to a dough-like consistency is fed between the first pair of a series of heated horizontal rolls and emerges from the last pair of heated rolls as a flat plastic film or sheet whose thickness is determined by the gap between the rolls. It is then cooled by a chill roll and may be stored for further processing on a takeoff roll. This process is extensively used for producing thermoplastic film or sheeting.

In the *casting process*, liquid plastics are poured into molds, where they harden slowly. The plastic may be supplied in liquid form or in granular or powdered form which may be liquefied by heat or adding solvents. Both thermoplastics or thermosetting plastics are used. The former will solidify at room or relatively low temperatures, but the latter require heating. No pressure is involved for either.

Laminates are usually formed by impregnating sheets of cloth, paper, asbestos, or woven glass fibers with liquid thermosetting resins; stacking them in layers; and curing by heat and pressure to form solid sheets. If low pressures are used, they are called *low-pressure laminates* or, more commonly, *reinforced laminates*. The latter designation is used especially if the impregnated sheets are woven glass fabric. Reinforced laminates differ from molded plastics in which the fibrous materials are not in layers but are included to increase their flexural strength. If high pressures are used, the products are called *high-pressure laminates*.

Plastic foams may be made by introducing air or some other gas into liquid thermosetting resin and solidifying by heating and curing. The foam may be formed mechanically by whipping air into the liquid resin, by dissolving a gas in the resin under pressure and then lowering the pressure to permit the gas to expand, or by producing gas in the liquid resin by chemical action of materials introduced into the resin. Such foams may be rigid or flexible. They are effective as heat insulation and are very lightweight.

In *sandwich constructions*, two thin, dense, strong, and hard facing layers of plastic, metal, or wood are bonded with adhesives to a core of lightweight material to form a panel. For the panel to function effectively, the core and adhesive must be strong enough to resist the shearing and compressive stresses due to normal loads and impacts. Plastics used for cores are either foamed or honeycombed, the former having superior heat-insulating properties.

Common Properties.   The physical and chemical properties of plastics vary over wide ranges depending upon the type of resin, the kinds and proportions of fillers and plasticizers, the presence of reinforcing fabrics, and the temperatures, pressures, types of equipment, and other factors involved in the manufacturing processes. The following comments refer to many plastics used in building construction:

They are formable into a great variety of shapes and available in a range of colors.

Most plastics have strengths comparable with those of wood or concrete, but those of some laminates and reinforced plastics are very high.

Stiffness is relatively low.

They have relatively light weight as compared with most building materials except wood and lightweight concrete.

They have a high coefficient of expansion, a low thermal conductivity, and the range of temperature resistance, from low to high, is relatively limited.

They will burn but not support combustion.

They are noncorrosive for conditions encountered in buildings.

Deformation due to creep tends to disappear when load is removed.

When considering many other. properties, distinction should be made between thermoplastics and thermosetting plastics.

Thermosetting plastics are superior to thermoplastics in resistance to heat.

Thermoplastics are usually tough but may be soft, flexible, and tough or hard, rigid, and brittle, depending upon composition.

Thermosetting plastics are usually rigid.

Thermosetting plastics are elastic for the stresses to which they are usually subjected.

All plastics are subject to creep under load, but this tendency is much greater for thermoplastics than for thermosetting plastics.

Thermoplastics are soluble in some solvents, but thermosetting plastics are mostly insoluble.

Some thermoplastics have excellent light-transmitting qualities.

Names of Plastics.   A plastic may be designated by the chemical type of resin used as its essential ingredient or by a trade name given by the manufacturer. For a comprehensive list of trade names, their chemical types, and manufacturers, see reference 2.

Specific Properties, Forms, and Uses.   The pertinent properties, forms, and uses of various plastics employed in building construction are summarized in Table 17-1.

Plastic Sheets.   Perhaps the widest application of acrylic sheets is in glazing windows and flat, curved, or domed skylights. This plastic, a

TABLE 17-1
Properties, Forms, and Uses of Plastics

Chemical Type	Properties	Forms	Uses
THERMOPLASTICS			
Acrylics	Transparent, hard, weather-resistant, shatter-resistant, easily scratched	Cast sheets	Window and skylight glazing
Polyethylene	Flexible, tough, translucent, low cost, easily scratched	Film and sheet	Vapor barriers, temporary glazing and building enclosing, protection and materials stored outdoors
Polystyrene	Hard, clear, brittle, water- and chemical-resistant, low cost	Open mesh Tile and sheet	Window screen Wall covering and tile
Vinyls	Tough, wear- and stain-resistant	Tile and sheet	Floor and wall tile, and sheet covering
Polyamides (Nylon)	Tough, hard, wear-resistant, expensive	Coated glass fiber Cast	Window screen Rollers and bearings
THERMOSETTING PLASTICS			
Alkyds	Weather-resistant, tough, good adhesive properties	Liquid and solid	Surface coatings such as paints, enamels, etc., molded products
Melamines	Hard, durable, abrasion-resistant, chemical- and heat-resistant	Sheets	Decorative laminates, high-pressure laminates, counter tops
Polyesters	Weather- and chemical-resistant, stiff, hard	Corrugated and flat translucent laminates, woven glass reinforced	Window glazing and skylights

688

thermoplastic material, is usually ⅛ to ¼ in. thick for glazing, although it may be as thick as 4 in. Sheets are made as large as 10 ft. × 14 ft. Acrylic is somewhat soft when compared to glass and is subject to scratching; however, it is highly resistant to shattering and breaking. This plastic is transparent in several colors, colorless or white in translucent sheets, and in several colors in opaque sheets. The latter are used for facias and as curtain wall panels. Translucent acrylics are used for diffusers in luminous ceilings and light fixtures (luminaires) and as glazing where privacy is required.

Acrylic sheets may be coated with a plastic that produces an abrasion resistance that approaches the scratch resistance of plate glass.

The surfaces of these plastics are usually smooth for transparent glazing, although they may be pebbled or orange-peel surfaces for transculent and opaque sheets. This material is flexible, colorful, durable, and is made into attractive colored mosaic windows. Decorative sheets are also made by embedding materials. Designs vary— burlap, leaves, colored paper, etc. In glazing, these sheets require frames, glazing, or fastenings which allow them to seat upon the frame in a manner which permits thermal expansion and contraction in the frame without rupturing the seal or mechanical fastening.

Acrylic plastic is marketed under several trade names such as *Acrylite* and *Plexiglas*.

The reason that thermoplastic plastics have not been permitted more extensive use in construction is that they are soft and melt or deform under heat.

A polycarbonate plastic sheet, which is used for glazing windows and skylights, has exceptional resistance to impact and breakage. It is colorless or may be of various colors, including a glare-reducing bronze or gray in different intensities. It is known by the proprietary name of Lexan.

Modified acrylic plastic with glass fibered reinforcement is formed into flat, corrugated, or ribbed panels. These translucent panels in several colors measure from 26 in. to 50 in. in width and from 8 ft. to 14 ft. in length. They have exceptional strength, especially the corrugated panels. These sheets are used primarily for siding and roofing of industrial buildings; however, they are used for a variety of other building applications such as fences, visual barriers, etc.

Plastic Films. Polyethylene film is made in several thicknesses measured in mils. The material, sold in rolls with widths of 10 ft. or more, is used as a vapor barrier under floors, in walls, as drop cloths for painting, and temporarily encloses buildings under construction for protection against weather and heat loss. It is colorless or black.

Cellular Plastics. Thermosetting rigid *foam* plastic, mentioned above, is also known as *cellular* plastic. *Polyurethane* and *polystyrene* are cellular plastics used in the building industry. Although a cellular plastic is usually applied in its processed solid form, the ingredients may be purchased and the cellular plastic cast, foamed, or sprayed in place. These plastics, which are snow white, have extraordinary thermal insulating value, are lightweight, and are resistant to moisture and decay. As stated above, they are used as an insulating core in *sandwich* panels that have a structural and finished surface, such as plywood, sheet metal, or cement asbestos. Although it is not a structural material itself, cellular plastic can support limited imposed loads such as thin sections of concrete. They are used with masonry construction as an insulating wythe in a cavity wall or applied to the interior surface with mechanical fasteners or adhesives. Foamed plastics are used on roofs, walls, ceilings, and under floors. This material will not support combustion, although it will sublimate into a gas when exposed to a flame or intense heat. *Styrofoam* is a trade name of a well-known polystyrene.

Epoxy Resins or Epoxies

Introduction. Epoxy resins, often called epoxies, are thermosetting plastics introduced into building construction in about 1954. That epoxies have remarkable properties is known to architects and engineers. They have not yet made full use of them, largely because they are cautious in the use of materials whose value has not been demonstrated by experience, and few workmen are skilled in their use. As time goes on and experience demonstrates the wide field of usefulness of epoxies in building constructions, they seem destined to assume an important role in this field.

The following comments are based on material presented by the American Railway Enginering Association in reference 6 and by the *Engineering News-Record* in reference 7.

Definitions. The term *epoxy* refers to a three-member ring structure containing two hydrogen atoms and one oxygen atom. Materials containing an average of more than one epoxy group per molecule are considered as *epoxy resins* [6].

Epoxy resins alone are chemically stable and may be stored indefinitely, but they are useless. By adding a *curing agent* or *hardener* to the resin in its liquid state, infusible and insoluble solids are formed. The length of time that an epoxy mixture is usable after the curing agent is added is called the *pot life*.

Substances added to epoxy resins during the manufacturing process to alter the properties of the cured resins are called *modifiers*.

Manufacture. The basic resins, together with the modifiers, are cooked under pressure by the manufacturer.

For resins used in construction, the cooking is terminated when the resins are honey-colored liquids having viscosities about the same as those of a 20-to-30-weight motor oil.

The characteristics of the manufactured product depend upon the ingredients, the temperature, the pressure, the cooking time during manufacture, and other factors. Each manufacturer produces a variety of resins suitable for various uses.

Modifiers. The basic resins without modifications are useless. The modifiers added during the manufacturing process to provide the desired properties are classified as flexibilizers, fillers, pigments and dyes, and diluents.

Flexibilizers or *plasticizers* are always required to increase the flexibility. Ordinarily they are mixed with the curing agent by the manufacturer. Among the materials used are a synthetic rubber in liquid form and various coal tar products.

Fillers are added to provide certain desired properties or to reduce the cost. They extend the pot life, lower shrinkage while curing, and reduce the coefficient of expansion. Materials used as fillers are coal tar products, which react chemically with epoxy resins, and inert materials such as clay, asbestos or glass fibers, powdered metals, aluminum oxide, silica, mica, powdered glass, fine sand, or marble dust, which merely occupy space. Inert minerals are the most commonly used fillers for epoxies used in construction. Powdered metals and aluminum oxide are used in coatings and paints.

Pigments and Dyes. The color of many epoxy formulations is satisfactory for most uses but may be modified as desired by using appropriate pigments or dyes.

Diluents are added to improve workability, permit the application of thinner coats, reduce viscosity, or increase penetration. There are few applications except in paints and other coatings.

Curing Agent or Hardener. The curing agent or hardener may be reactive and become a part of the molecules in which case it imparts to the molecules some of its own characteristics. It may also act as a catalyst to facilitate the reaction but not become a part of the molecules.

The pot life may be controlled to a few minutes or a few hours, depending upon the ingredients and the temperature created by the chemical changes which take place or by heat applied externally, both of which are affected by the size of the batch. On construction work, only the former is used, but factory production may utilize the latter for some products. The curing agent completes the chemical changes which were interrupted when the manufacturing process was terminated.

Properties.   Epoxies cannot be classed together as materials with specified properties.

"An epoxy can be as brittle as glass or as resilient as rubber. It can be a high molecular weight solid of great density, or a low molecular weight liquid of such low viscosity that it will leak out of containers capable of holding water. And it can be anywhere between these two extremes" [7].

The properties of cured resins may be varied within wide limits with the proper choice of curing agents, modifiers, and diluents.

Some of the more significant properties that can be achieved are the following [6]:

(1) The liquid resin is convenient to use in many different types of applications.

(2) The curing time may be controlled by selecting the curing agent from the variety of those available.

(3) The properties of cured resins may be varied within wide limits with the appropriate choice of modifiers and curing agents.

(4) The overall strength properties such as tensile, compressive, and flexural strength are excellent. In general, their strength is greater than that of most of the materials they are used on.

(5) The shrinkage during curing is small.

(6) Hardness, toughness, and resistance to abrasion are outstanding.

(7) The resistance to corrosion, salts, acids, petroleum products, solvents, and other chemicals of many kinds is exceptionally good.

(8) Adhesion to the surfaces of most materials is excellent. Their ability to bond similar or dissimilar materials is exceeded by no other organic compounds.

(9) The color is satisfactory for most purposes. Epoxies can be formulated to be almost any color.

(10) Modification of resins to improve one property usually results in a sacrifice in one or more of the other properties.

Application. Cured epoxy compounds may be applied by brush, spraying equipment, trowel, squegee, or other means appropriate for the characteristics of the material being applied and the type of operation.

The curing agent is added to the modified resin immediately before application. The operations must be completed during the pot life of the mixture.

Before applying, the materials must be properly mixed. Careful preparation of the surfaces to be joined, repaired, or protected is required. Any dust, loose material, oil, or grease must be removed from any surfaces involved and rust, surface coatings, and mill scale must be removed from steel.

Epoxies should not be applied when the air temperatures are below 60° or above 95° unless special precautions are taken to protect them from the effects of lower or higher temperatures. Workmen experienced with the use of epoxies are required.

As stated in reference 6,

"Prolonged or frequent skin contact of materials used in epoxy resin systems may cause dermatitis for some individuals. The reactive diluents used in epoxy resins are found to be sources of skin irritation. ... Most persons can work with these materials for some time without taking any precautions to avoid skin irritation. A few persons under these conditions may suddenly break out with skin irritations which will disappear when transferred to other types of work."

Among the precautions suggested in the references, the following are included.

(1) Care in preventing skin contact.

(2) Regular washing of hands, arms and face with warm soapy water. Solvents should not be used.

(3) Use of plastic or rubber gloves and protective clothing which have not been contaminated by prior use.

(4) Complete protection of the body by clothing including a hat when spraying epoxies. Exposed body surfaces should be coated with protective cream, and a respirator should be used.

Uses. As has been stated, epoxy compounds have many remarkable properties and seem destined to assume an important role in building construction.

In general, they are used as bonding materials to join most similar and dissimilar materials, as bonding agents for aggregates, as watertight crack and joint sealers, and as protective coatings including paints. They are expensive, but where their special characteristics are important they can be extremely economical. An epoxy compound must be specially suitable for the purpose for which it is to be used. Some of the uses which have been made or suggested are given in the following paragraphs.

Terazzo floor surfaces are installed with epoxies as the matrix rather than portland cement. The required thickness is only ½ in. instead of the usual thicknesses of 1 or 2 in., with the resultant saving in dead load.

Concrete overlays can be bonded to eroded or worn concrete surfaces.

Corners which have broken off concrete slabs and pieces which have been broken off concrete members can be replaced by coating the broken edges with epoxies or by replacing the broken parts with new concrete sealed with epoxies to the remaining parts.

Bonding the components in composite steel and concrete construction.

"Cold-welding" structural steel components instead of welding, bolting, or riveting under certain conditions.

Repair of cracked or checked timber members.

Adhesive for laminated timber and other timber parts, and for bonding precast or cast-in-place concrete slabs to timber or steel girders.

Waterproofing concrete surfaces by brush or spray coatings.

Corrosion-resistant coatings for steel.

The full uses of epoxies will provide many entirely new methods of construction.

REFERENCES AND RECOMMENDED READING

1. *Modern Plastics Encyclopedia*, Modern Plastics, Berskin Publications, 1960.
2. Herbert R. Simonds, *A Concise Guide to Plastics*, Reinhold, 1957.
3. Edward B. Cooper, "Plastics Used in Building Construction," *Plastics in Building*, Building Research Institute, 1955.
4. *Plastics as Building Materials*, Circular Series D.4.0, Small Homes Council, University of Illinois, 1956.
5. William Dudley Hunt, Jr., *The Contemporary Curtain Wall*, F. W. Dodge Corporation, 1958.

6. Application of Synthetic Resins and Adhesives to Wood Bridges and Trestles, *American Railway Engineering Assocation* Bulletin 562, February 1961, p. 526.
7. Those Wonderful Epoxies, *Engineering News-Record*, July 12, 1962, p. 28.
8. G. H. Dietz, and Marvin E. Goody, ed., *Plastics in Architecture*, Proceedings of a Summer Session, Department of Architecture, Massachusetts Institute of Technology, Cambridge, Massachusetts, 1967.
9. *Sweet's Catalog*, 1971 Edition, F. W. Dodge Corp., New York, New York.

18

INSULATING MATERIALS

94. GENERAL DISCUSSION AND DEFINITIONS

General Discussion. There has been an increasing interest in making the air in occupied buildings more comfortable and healthful and in conditioning the air in manufacturing buildings to suit the manufacturing processes. There has also been interest in reducing the operating costs involved in such improvements. Two important factors in bringing about these conditions are the temperature and the humidity of the air in the buildings. Both factors are closely associated with the construction of the building. In winter, the problem is largely one of providing additional heat and additional humidity, whereas in summer these may be present in excessive amounts. There is a considerable operating cost involved in increasing or decreasing the temperature of the inside air. Usually the savings in this cost which can be brought about by proper design and construction, including insulation, will more than justify the additional expenditure involved. The direct cost of increasing humidity is not large, but humidities of 35 to 40% and more, which are often maintained, may cause condensation in the outside walls and the roof or ceiling of the top story, which can only be avoided by appropriate types of construction. Decreasing the humidity, as may be desirable in many parts of the country during the summer, is not as easily and cheaply accomplished as increasing the humidity, but this does not require any special consideration as far as the construction of buildings is concerned, except where unusually low humidities or temperatures are employed.

Heat Transmission. Heat is transmitted from one side of the walls and roof of a building to the other in the following ways.

(1) By *air infiltration*, or leakage through cracks and other open spaces. The volume of air entering a building is offset by an equal volume leaving the building. Since the temperature of the two vol-

umes would normally be different, there is a transfer of heat with the air change.

(2) By *transmission* through the walls, windows, doors, and roofs.

Air infiltration is minimized by good construction, which reduces or eliminates cracks and other openings through the walls, around cornices, along the sills on foundation walls, and through the roof construction; by the proper design, construction, and installation of doors and window sash and their frames; by calking between frames and surrounding masonry; by covering the sheathing of frame walls with a good quality of building paper with tight joints; and by plaster surfaces on masonry walls. Air infiltration through an unplastered brick wall is many times as great as through the same wall after plastering. The effectiveness of storm sash depends upon how tightly they fit and is relatively high on windows that are not weatherstripped.

Because of the "chimney effect" in tall buildings, the tendency for air infiltration at the lower stories, and for a corresponding outward movement in the upper stories, is large enough to require special provisions to be made to reduce this effect. These provisions consist of closing, with doors or by other means, all openings from floor to floor, such as stairways and elevator shafts. This is also a necessary fire-protection measure. Heat losses by air infiltration increase rapidly as the wind velocity increases and may exceed the losses due to the difference in temperatures inside and outside a building.

Heat is transmitted through walls and roofs in three ways.

(1) By *conduction* from molecule to molecule of the wall material and, to a certain extent, from molecule to molecule of the air in open spaces in the wall.

(2) By *convection* by air currents, which circulate in open spaces within the wall or roof construction and which absorb heat as they pass upward over a warm boundary surface of the air space and release heat as they pass downward over a cold boundary surface of that space. The circulation is produced by the decrease in density of air, which accompanies increase in temperature.

(3) By *radiation*, the process by which energy, called *radiant energy*, is transmitted from one body or surface to another body or surface by electromagnetic waves.

Heat passes through a solid wall entirely by conduction; but if there is an air space inside the wall, it will cross this space by radiation, convection, and conduction. The resistance which a given homogeneous material offers to the passage of heat varies directly with its thickness and inversely with its density, and decreases with increases in the moisture content of the material. The resistance of an air space, such as a stud space in a frame wall or a cell space in a hollow tile wall, to the passage of heat by radiation is independent of the width of the space but is greatly affected by the nature of the boundary surfaces, being low for the surfaces of ordinary building materials but very high for bright metallic surfaces. The resistance by an air space to the passage of heat by convection is practically independent of the width of this space if it exceeds ¾ in., but it decreases very rapidly as the width decreases below ¾ in. More than half of the heat transmitted through an air space over ¾ in. wide bounded by ordinary materials is radiant heat.

Solar Heat.   Solar heat, or radiant heat from the sun's rays, which is transmitted through windows, is an important factor in heating the air in a building. During the winter months, this effect is usually desirable, but in many parts of the country it increases the inside temperatures above the comfort range unless excess heat is removed by air conditioning, which significantly increases the initial and operating costs of a building.

As outlined in Article 84, there are several ways for reducing the solar heat gain through windows. These include shading windows from the sun's rays, double glazing, and the use of heat-absorbing glass as described in Article 83.

Canvas awnings, which were extensively used in the past, and the more modern types made of aluminum or plastics are very effective, especially if they are so designed as to permit the heated air beneath them to escape.

A detailed discussion of window shading is included in reference 11, which has been helpful in preparing these comments and Table 18-1. The table gives the relative values of solar heat transfer for various fenestrations under the conditions stipulated in the table.

The relative values in Table 18-2 will be of interest in comparing results achieved by several methods for reducing heat gain from all sources, including solar heat, through various types of glazing with the heat gain through ordinary glass which is considered 100 for comparative purposes. Values for sunlit surfaces, with and without sunscreens, are given in the first column, and those for surfaces which are not sunlit are given in the second column. It will be noted that the values for sunlit areas with sunscreens are approximately the same as those for corresponding areas which are not sunlit. Some of the areas listed as sunlit would be sunlit if sunscreens were not provided.

TABLE 18-1
Relative Value of Solar Heat Transfer for Various Fenestrations*

Type of Fenestration	Relative Value
Regular window glass	100†
Heat-absorbing glass	69
Double glass—heat-absorbing glass outside, regular plate inside	52
Regular window glass, inside venetian blinds	58
Glass block panel	50

* Arranged from paper by Donald J. Vild in reference 11.
† 100 equals heat transfer of 183 Btu per sq. ft. per hour.
Regular window glass considered as 100.
Conditions: typical clear afternoon in midsummer at 40° North latitude with fenestration facing southwest.

Heat loss by residential buildings during the winter months is often reduced by replacing window screens with storm sash. Heat gain by such buildings during the summer is sometimes reduced by attic fans, which ventilate the attic and replace air heated by solar heat transmitted through the roof with cooler outdoor air. Hot attic air, of course, transmits heat to the rooms beneath. An insect screen is available with the horizontal wires replaced by thin flat metal strips and vertical wires about ½ in. apart to form venetian blinds with miniature slats. Because they are located outside, they act as sunscreens and are quite effective for windows with south exposure but less effective for those with east or west exposure.

TABLE 18-2
Relative Merits of Heat Gain Reducing Methods through Glass*

Type of Area	Sunlit	Not Sunlit
Single glass	100	33
Double glazing	73	22
Heat-absorbing glass	68	24
Double-glazed heat-absorbing glass	51	17
Sunscreen on single glass	35	17
Sunscreen on double glazing	26	12
Sunscreen on heat-absorbing glass	24	12
Sunscreen on double-glazed heat absorbing glass	18	9

* As given by G. R. Munger in reference 12.

Radiant Heat and Comfort. The effect of the temperature of the surrounding air on the comfort of a person is well known. Air temperatures which are considered desirable for comfort depend upon various factors, but normally they are within the range of 70 to 75°F. The effects of the temperatures of the surfaces of the walls, ceilings, and floors which surround a person are not, however, always understood. If any such surfaces are cooler than the desired air temperatures, a person will feel too cool even though the temperature of the intervening air is within the desirable range. Conversely, if any of the surrounding surfaces are warmer than the desirable range in air temperature, he will feel uncomfortably warm. These conditions are due to the loss or gain of heat by the body caused by the radiation of heat between the body and the surrounding surfaces. Therefore, it is important that the temperatures of these surfaces be maintained to a reasonable extent within the comfort range. This factor is significant in selecting insulation.

Condensation. The amount of water vapor that air can contain increases with the temperature of the air. The ratio of the amount actually present to the maximum amount that can be present at that temperature is called the *relative humidity* or simply the *humidity*. It is expressed in percentage. As the temperature of air containing a given amount of water vapor falls, the relative humidity rises until the saturation or *dew-point temperature* is reached and some of the vapor is condensed. This condensation may occur on cold interior wall surfaces and be apparent; but the water vapor tends to pass through the walls and, since the interior of a wall becomes progressively colder toward the outside, a temperature may be reached at which the vapor will condense within the wall. The conditions favorable for this action are an inside humidity of 35 or 40% and higher and a long-continued cold outside temperature. The amount of water that condenses gradually increases and, if the temperature where the water collects is below freezing, ice will form. If porous insulation is present, it will accumulate water as ice and largely lose its effectiveness. The ice will melt when the outside temperature rises sufficiently. Condensation may also occur in insulation over the ceiling of the top story or on the underside of the roof sheathing, particularly around the points of protruding roofing nails. The condensation that forms in the walls, ceiling, or roof construction may come through the finished wall and ceiling surfaces and spoil the decorations, disintegrate the plaster, cause any wood present to swell with resultant cracking, cause paint on exterior wood surfaces to peel off, and damage a building in other ways. These effects are sometimes wrongly attributed to leaking walls and roofs. In wood construction, the wall may be so tightly sealed with sheathing paper that evaporation occurs very slowly, causing the studs and sheathing to decay. Insulation may make conditions worse by causing the outer portions of a wall to be colder, and porous insulation accumulates the water. Some of the higher grades of sheathing paper used on stud walls retard or prevent the escape of water vapor from the walls and increase the condensation effect.

To prevent condensation of this type, avoid high humidities during long periods of low temperatures; place a *vapor barrier* or *seal* under the lath in stud walls and on the bottoms of ceiling joists of the top story and use a sheathing paper which is sufficiently airtight but not an effective vapor barrier; place a vapor seal under the lath on furred masonry walls; place a vapor seal under the insulation on wood or concrete roof-decks, if insulation is used; and ventilate attic spaces. A glossy-surfaced tar paper or a polyethylene film are good vapor seals for use under lath. They must be lapped and tightly fastened at all joints. The longitudinal joints should be over the studs or joists, and the paper should extend from the floor to the ceiling without end joints in the paper. Slaters' felt is considered a satisfactory sheathing paper. On roof decks, a 2-ply seal should be made of saturated felt applied by mopping with hot asphalt or roofing pitch. The vapor seal should always be on the inside.

Some of the rigid insulating materials are coated with bituminous material or encased in bituminous paper to exclude water vapor; many insulating quilts have so-called vaporproof paper coverings; and mineral wood batts are available with vaporproof backs which are placed next to the lath in wood stud walls. These are not regarded as high types of vapor barriers.

95.   TYPES OF INSULATING MATERIALS

Insulating materials may be divided into two general classes according to the way they function. In the first group may be included all the low-density, porous, or fibrous materials with low conductivity or high resistance to the passage of heat, whose effectiveness is due to the minute air spaces of which they are largely composed. This group includes three general types of insulating materials as follows: *rigid* or board and slab insulation, *flexible* or quilt, blanket, and batt insulation, and *fill* insulation. As stated in reference 7, "Still air is the best heat insulator we know. But air is seldom still. The slightest change in temperature will make air expand or contract; the warmer air becomes lighter and floats above the cooler, heavier air. So the job is to keep still air still."

Without going into detail, it may be said that insulating materials confine the air which they contain, and upon which their effectiveness depends, in cells, voids, or other spaces, some or all of whose dimensions are so small that there can be little if any movement of the air within their boundaries.

The second group of insulation materials includes those of the *reflective* type which is used to form boundaries of air spaces and whose insulating value lies in its effectiveness in reflecting radiant energy.

The rigid, flexible, and fill insulators consist of wood, cane, and other vegetable fibers, mineral wool, cork, hair felt, expanded mica, foamed glass or plastics, or light granular materials. They owe their insulating properties to the minute air spaces they contain. The insulating value per inch of thickness is about the same for all these materials. It is approximately equal to the insulating value of 3 in. of wood, 17 in. of glass, 31 in. of brickwork, and 40 in. of concrete.

Reflective insulation consists of some form of sheet metal, metal foil, or a metallic coating which is made very thin because its effectiveness is practically independent of the thickness. The most effective insulation of this type will reflect about 95% of the radiant energy that strikes its surface. Since a large proportion of the heat which passes through an insulated wall is produced by radiant energy, reflective insulation can be very effective.

Rigid Insulation.   Fiberboards, described in Article 67, are used for insulating puropses. To be effective, the fibers are not highly compressed as in the hardboards. Moisture and water vapor are partially excluded from some of the fiberboards if the boards are coated with asphalt or encased in a bituminous waterproof paper. For increased insulating value, fiberboards and gypsum boards are available with aluminum foil coating on one surface to serve as reflective insulation. Another form of rigid insulation is corkboard made of pressed cork. It is available in thicknesses up to 6 in. Some forms of rigid insulation serve the dual purposes of insulation and sheathing or insulation and lath. The common thickness for sheathing is 1 in., but any desired thickness may be obtained.

Flexible Insulation.   This type of insulation may be in the form of *quilts* or *blankets* and *batts*, often spelled *bats*. The quilts and blanket consist of a fibrous material such as treated wood fiber, hair felt, flax fiber, eel grass, or shredded paper stitched between sheets of waterproof paper to form a flexible material available in various thicknesses up to 1 in.

Another form of flexible insulation is made from *mineral* or *rock wool*, formed by blowing molten rock into fibrous form by steam under pressure. This produces a fluffy, noncombustible product weighing about 6 lbs. per cu. ft. It is usually furnished in batts 15 × 24 or 48 in., to fit between studs and ceiling joists spaced 16 in. center to center. The thickness is ordinarily about 4 in. to fill the space completely between 2 × 4 in. studs, but batts 2 in. thick are available. Batts are furnished plain or with a waterproof paper cemented to the back and projecting about 2 in. on each side to provide *nailing flanges*, which lap over the studs or joists to which they are nailed and form a seal. This paper is supposed to serve as a vapor barrier or seal, but if it is to be effective the end joints must be tight. It is better practice to provide an additional vapor barrier, as previously described, over the batt insulation after it is in place, and under the lath. Mineral-wool insulation is also furnished in roll form and as fill insulation, as described in the next paragraph. *Glass wool* is a fibrous glass insulating material similar to mineral wool. It is available in batt, blanket, and fill form.

Fill Insulation.   This material consists of granulated rock wool in the forms of nodules or pellets, granulated cork, expanded mica, and other material which is blown through large tubes or dumped into open spaces in the walls and ceilings of buildings, such as the stud space in frame walls and the ceiling joist space. Fibrous mineral wool and glass wool are also furnished in loose form for packing by hand into open spaces.

Reflective Insulation.   Reflective insulation is placed in air spaces and functions by reflecting a large percentage of the radiant energy which strikes it. The materials used for reflective insulation are very thin tin plate, copper or aluminum sheets, or aluminum foil on the surface of rigid fiberboards or gypsum boards. A common form of reflective insulation is aluminum foil mounted on asphalt-impregnated kraft paper, the strength of which may be increased by the use of jute netting. This material is used for lining air spaces or for curtains to increase the number of air spaces in a given overall space. To be effective, the edges of such curtains must be tightly sealed.

Methods of Installation.   *Rigid insulation* can be used as sheathing on the outside of wood studs (Figure 18.1(a)); as building board without plaster on the inside of wood studs (*b*); and on the lower side of

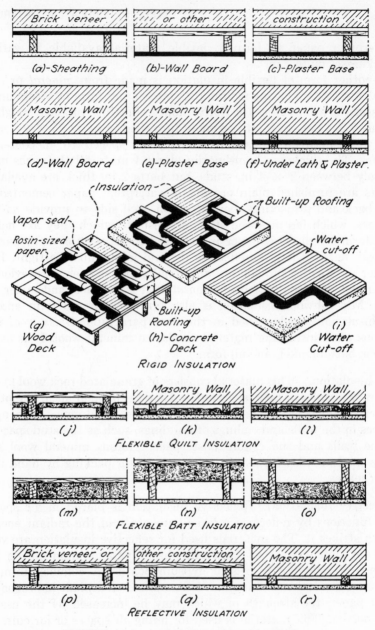

FIGURE 18.1   Installation of heat insulation.

ceiling joists or rafters. It may also be used as lath to receive plaster on the inside of wood studs (c) and on the lower side of ceiling joists. When used with masonry walls, it is nailed to furring strips (d). It may be exposed; it may serve as a plaster base (e); or it may be covered with lath and plaster, with furring strips between the insulation and the lath and on top of the other furring strips or at right angles to them (f).

Rigid insulation may be used on wood and concrete roof decks under built-up roofing (g) and (h). Wood decks are first covered with one or two layers of rosin-sized building paper, lapped at the joints and securely tacked down. This paper keeps asphalt from running through the cracks between the roofing boards. Over this paper or directly on concrete roofs a layer of asphalt-saturated felt is mopped with hot asphalt or roofing pitch. This layer serves as a temporary roofing. If there is to be considerable vapor in the room below, another layer of asphalt-saturated felt is applied by mopping with hot asphalt or roofing pitch to improve the vapor seal. The insulation is applied by coating the surface of the felt with hot asphalt or roofing pitch just before placing the insulation. Each board is bedded in the hot asphalt or roofing pitch. As many layers of insulation as desired are placed in this way. Finally, with staggered joints, a built-up roofing is applied by mopping the first ply to the top of the top layer of insulation. Slate and tile wearing surfaces can be placed on top of the built-up roofing if desired.

In placing the insulation, precautions must be taken to protect the insulation from rain or other moisture by placing, at one time, only such an area as can be covered in a short period and never leaving the insulation unprotected over night. It is also good practice to localize the effect of any leakage that may occur and to provide a place where work can be stopped, by dividing the roof into about 30-ft. squares with *water cutoffs*. These are made by inserting a 16-in. strip of asphalt-saturated felt at joints (Figure 18.1(i)), the strip being tightly cemented to the surface below the insulation and to the insulation itself. Other cutoffs should be located parallel to the parapet walls and about 2 ft. from them. All interruptions in work should occur at waterstops, and the insulation that has been placed should be protected temporarily with at least one ply of the built-up roofing extending to waterstops on all sides. Insulation laid on wood decks sloping more than 3 in. to the foot should be nailed to the deck. Insulation applied in this manner should not be used on concrete decks with such slopes.

Flexible insulation is installed in open spaces in the wall, ceiling, or roof construction where it will not be subjected to loads, for it is easily compressible. Quilts or blankets are installed between the studs and ceiling joists in frame construction or between furring strips on masonry walls (*j*), (*k*), and (*l*). More than one layer of quilt, together or separated, can be used if desired. Quilt insulation divides the space it occupies into two or more air spaces. To be efficient, the insulation should be so fastened at the sides and ends that there will be no air leakage between spaces. Batts are inserted between studs (*o*). If they are provided with vaporproof back and with nailing flanges, they are lapped over the studs, as shown. If the studs, joists, or rafters are deeper than the thickness of the batts, the nailing flanges are nailed to the sides of these members (*n*). Ceilings are insulated (*o*), or the batts may be inserted from below (*n*), before the lath are placed.

Fill insulation is installed by pouring the loose granulated or nodulated form into open spaces between the studs of stud walls and between ceiling joists, or it is installed by blowing this material into place through large flexible tubes using low air pressure. Fibrous mineral wool and glass wool are available in bulk form. They are packed by hand into the space between studs and joists. Great care is necessary to insure the complete filling of stud spaces.

Reflective insulation is applied between studs, joists, or rafters (Figure 18.1(p)). Nailing strips may be used to hold it more tightly. Some forms may be installed by lapping over the studs (*q*), or by inserting between furring strips on masonry walls (*r*). This insulation divides the space into two or more air spaces. For the most effective results the insulation must be fastened at the sides, top, and bottom, and at all laps, so that there will be no circulation of air between the two spaces.

96. ACOUSTIC REQUIREMENTS

Definitions. *Acoustics* may be defined as the science of sound. *Sound* may be defined as the sensation of hearing caused by stimulation of the auditory nerves, usually by vibrations transmitted in air affecting the ear. It may also be defined as the vibrational energy which is the physical cause of that stimulation.

General Comments. Improvements in construction during recent years have often resulted in buildings with poorer acoustical properties owing to the more rigid materials and construction used. On the other hand, radios, television, motion pictures, the increased use of mechanical office equipment, the increase in street noise from the automobile, and many other factors have emphasized the importance of improvement in acoustical conditions.

Largely because of its increasing importance, knowledge in the field of architectural acoustics has been developed so that the necessary acoustical features can be determined when the plans and specifications of a building are being prepared. By this procedure the construction of buildings with good acoustics can be assured, and the embarrassment and inconvenience of remedying defective acoustical conditions while a building is in use, usually at much greater cost, can be avoided.

The various acoustical considerations which enter into the design of buildings are considered briefly in this article so that the problems involved will be appreciated, but no effort is made to present material which can be used in the solution of specific problems. To secure good results, specialized advice is usually required.

Good acoustics of building space contribute to the comfort of its occupants, the efficiency of office and factory workers and their supervisors, the effectiveness of classroom instructors, and understanding and appreciation by the audiences in assembly and concert halls.

Nature of Sound. As stated by Paul E. Sabine [1],

"A vibrating body will impart a portion of its energy to the air surrounding it, i.e. it will generate sound waves. The *frequency of vibration*—the number of complete round trip excursions per second —is expressed as so many cycles per second or, more briefly, *cycles*. Consider one to and fro movement of the surface of a vibrating body. Its forward movement compresses the layer of air adjacent to it, causing an increase in pressure above the normal atmospheric pressure. The returning movement will cause a decrease in pressure below the normal. This fluctuation of the air pressure above and below the normal will be transmitted with a definite velocity in successive layers. The transfer of energy through the air by successive pressure variations in it is called a *sound wave*. Experiment shows that the velocity of sound of all frequencies is approximately 1120 feet per second. . . . It varies slightly with temperature. . . . The *pitch* of a sound is determined by its frequency."

The *intensity* of sound at a given point is the amount of energy transmitted by the fluctuations in air pressure at that point per unit of area perpendicular to the direction of travel.

Measurement of Sound Intensity. In dealing with acoustical problems, it is necessary to make use of a unit of measure for the intensity of a sound. The range of intensities in the sounds of everyday experience is very great. In terms of watts, the unit commonly used for measuring energy,

"The intensity of the faintest sound which a normally acute ear can hear is slightly more than 1. The intensity of the sound of the voice speaking in a confidential tone is of the order of 10,000, ordinary conversation is 100,000 to 1,000,000, street noise 10,000,000 to 1,000,000,000, and a painfully loud sound is from one trillion to ten trillion of these units. As an acoustical device, the ear is unsurpassed in the vast range of intensities to which it will respond without being damaged. Because of the awkwardness of handling numbers of such magnitudes and because of the roughly logarithmic response of the ear, the practice has been adopted of expressing sound intensities on a logarithmic scale" [1].

"On this basis, intensities, or amounts of energy, proportional to 10, 100 and 1000 produce sensations in the ear proportional to 1, 2 and 3 respectively. A slight modification of this logarithmic scale has come into general use to measure sound energy and the amount of noise reduction. It is called the *decibel scale*. The scale merely multiplies the numbers of the logarithmic scale by 10. The unit of the scale, the *decibel*, is a convenient unit, since it is approximately the smallest change in energy that the average ear can detect. For this reason, the unit has frequently been called a *sensation unit*. The decibel scale is suitable for measuring ratios of sound intensity. To measure absolute noise levels the zero value is assigned to a definite level, that is, a level of 20 decibel corresponds to an energy 100 times that corresponding to the zero value" [3].

Loudness.   The relation between loudness and intensity is explained as follows:

'The loudness of sound (sensation) depends upon the intensity, but it also depends upon the frequency of the sound and the characteristics of the human ear. The intensity of sound is a purely physical quantity, whereas, the loudness depends upon the characteristics of the ear. . . .

The relationship between frequency, intensity and loudness is quite involved. We do have, however, a sense of relative loudness in that there is a fair measure of agreement among trained observers in their judgments as to when one sound is one-half, one-third and so on as loud as another. . . . Speaking generally one may say that the quantitative evaluation of the magnitude of sensations is a psychological problem so that the acoustical engineer prefers to deal with those aspects of sound that are subject to physical measurement"[1].

The significance of the decibel scale is illustrated by the comparisons in Table 18-3. By consulting this table it is seen that, according to the decibel scale, the difference in loudness between a very faint sound and one that is deafening is about 120 decibels.

Classification of Sounds. Sounds are usually classified into three types, noise, music, and speech, but this classification is not always clear-cut. In general, *noise* is unwanted sound [4]. If one is listening to music, sounds produced by a person speaking nearby are considered as noise; if one is speaking to this person, the music is considered noise.

Since the physical properties of speech differ considerably from

TABLE 18-3
Noise Levels in Decibels*

Db.	Condition	Noise	Db.	Condition	Noise
		Threshold of feeling	60		
				Noisy home	
120				Average office	
	Thunder of artillery		50		Moderate
	Nearby riveter			Average conversation	
110		Deafening			
	Elevated train			Quiet radio	
	Boiler factory		40		
100				Quiet home	
	Loud street noise			Private office	
	Noisy factory		30		Faint
90		Very loud		Average auditorium	
	Unmuffled truck				
	Police whistle			Quiet conversation	
80			20		
	Noisy office			Rustle of leaves	
	Average street noise			Whisper	
			10		Very faint
70		Loud		Soundproof room	
	Average radio		0		Threshold of audibility
	Average factory				

* Arranged from Reference 1.

those of music, the acoustical properties of speech rooms should differ appreciably from those of music rooms [4].

Requirements of Good Acoustics. "In the design of rooms intended for speaking purposes the prime objective is intelligibility of speech. In the design of music rooms the prime objective is the most favorable enrichment of tonal quality and total blending of the sounds"[4].

Basic Factors in Acoustical Design. Some of the more important factors which must be considered in the planning of buildings to obtain satisfactory acoustical conditions are as follows.

(1) Selection of the building site.

(2) Arrangement, dimensions, and shapes of rooms and other planning features.

(3) Insulation and control of sound transmission within the building.

(4) Installation of absorptive materials within the building.

Each of these factors will be discussed briefly in the following paragraphs. For a more comprehensive and descriptive list see reference 4.

Site Selection. In selecting the site for a building for which acoustical considerations are important, attention should be paid to any effects which present or foreseeable prevailing noises in that area may have on the acoustical conditions in the building. Some of the important sources of objectionable noises are automobile traffic on busy streets, traffic arteries, and highways; railroads; airfields; and industrial establishments.

In considering future developments that may be objectionable, the zoning ordinances and any long-range plans prepared for the city should be consulted. Often the site for a building has already been selected, but even then the factors mentioned are important because they affect the acoustical design of the building.

Building Plans. After the site has been selected and when the plans are being prepared, much can be done to promote good acoustical conditions, often at little or no increase in cost or interference with the efficient functioning of a building. Involved in this phase of the planning are certain factors which affect the acoustics in the various rooms in the building and do not include special acoustical treatment. Some of the desirable planning provisions follow:

(1) Locate rooms that require a quiet environment where they will have the minimum exposures to objectionable outside noises.

(2) Provide horizontal and vertical separation of rooms that require a quiet environment from those that are inherently noisy.

(3) Stagger the doorways on the two sides of a hall or corridor.

(4) Provide doors at the entrances to long corridors, stairways, and elevator foyers.

(5) Provide appropriate shapes for auditoriums, recital halls, music studios, and practice rooms.

(6) Provide appropriate shapes for wall and ceiling surfaces of auditoriums.

(7) Provide room sizes acoustically compatible with their proposed uses.

Sound Transmission, Insulation, and Absorption. When a sound originating in a room under consideration strikes the surface of the wall, floor, ceiling, or other barrier, a part of it is *reflected* from the surface, a part is *transmitted* through the barriers, and a part is *absorbed* by the barriers and its energy dissipated in the form of heat. The reflected part of the sound remains in the room and is represented by the *reflection coefficient*. The part absorbed by the barrier and the part transmitted through it are considered together as being absorbed and are represented by the *absorption coefficient*. When the part of the sound transmitted through a barrier is considered separately, the reduction in sound energy in passing through a barrier is called the *transmission loss*. It is expressed in decibels and is a measure of the effectiveness of a barrier in *insulating* against the transmission of outside sound into a room.

Obviously, a barrier with a low transmission loss is effective in reducing the sound level in a room caused by inside sounds but is ineffective in insulating the room against outside sounds. For example, fiberboard or porous concrete block partitions whose surfaces are not painted or plastered have low transmission losses and are effective in reducing the sound level in a room caused by inside noises but are ineffective in insulating a room against sounds originating in adjoining rooms or from other outside sources. This difference is not always appreciated.

Transmission Loss and Insulation. The following comments are based largely on the National Bureau of Standards Report BMS 17 by W. L. Chrisler entitled *Sound Insulating of Wall and Floor Constructions* [2], on BMS Report 144 with the same title [3], and on two supplements to these reports.

Sounds may enter a building or room in the following ways.

(1) By transmission of *airborne sounds* through openings, such as windows or doors, cracks around doors, windows, water pipes, conduits, ducts of ventilating systems, and in other ways.

(2) By transmission of structural vibrations or *structure-borne sounds* from one part of a building to another. Such sounds, with rare exceptions, finally reach the ear through the air.

(3) By direct transmission through various portions of the structure itself which act as diaphragms and are set in motion by the sound waves striking them.

Transmission loss is determined by measuring, under specified conditions, the difference between the intensity of sound in decibels in the room where the sound originates and the intensity of the sound in a room separated from the first by the wall, ceiling, or floor being tested.

For a given barrier, the transmission loss varies with the frequency or pitch of the sound being lower for low pitched sounds than for those with high pitch. Studies have demonstrated that a frequency of 512 cps is the most satisfactory frequency to use for acoustical design under ordinary conditions.

Airborne sounds are lowered by reducing the openings to a minimum and even eliminating windows entirely. The amount of sound admitted through a closed window of a room may be many times that admitted through the walls, ceiling, and doors, and that admitted by a closed door may be as much as is admitted by the remainder of the enclosing structure of a room except the window. The amount of sound admitted through a window or door only partly opened is many times that admitted by a closed window. These qualitative relationships emphasize the importance of windows in noise control. Most of the noise from ventilating ducts can be eliminated by inserting acoustic filters.

Sounds transmitted by structural vibrations may be reduced by giving special consideration to them when designing the building and by selecting materials which do not transmit vibrations readily.

The weight of a homogeneous wall per unit of area is the most important factor in determining its sound-insulation efficiency. The kind of material and the way it is held in position are of secondary importance. Because of its lightness, fiberboard is not an effective sound insulator. The sound-insulating value of a given material does not increase directly with the thickness or weight per unit of area, but as the logarithm of this weight; therefore a high degree of sound insulation cannot be secured with a homogeneous wall unless it is excessively thick. This relationship is illustrated in Table 18-4.

TABLE 18-4
Relation of Transmission Loss to Weight [13]

Wt.	Loss	Wt.	Loss	Wt.	Loss	Wt.	Loss
1	22.7	10	37	40	45	100	51.3
5	32	20	41	60	48	400	60

Key: Weight in pounds per square foot. Transmission loss in decibels. For frequency of 512 cps.

The insulating value of a wall of a given weight can be increased considerably by dividing the wall into two or more layers. In an ordinary lath and plaster partition, with wood studs to which the lath are fastened, most of the sound is transmitted directly through the studs and only a small portion is transmitted indirectly from one layer of lath and plaster across the air space to the other layer. Stiff studs transmit less sound than flexible studs, but hard strong plaster is a poorer insulator than soft weak plaster, which unfortunately is not sufficiently durable for use. A partition constructed of gypsum lath fastened to the studs with resilient metal clips is a more effective insulator than one in which the lath are nailed directly to the studs.

Staggered studs may be used, with each plaster layer fastened to a different set of alternate studs. This prevents the transmission of sound directly through the studs, but a considerable amount of sound is transmitted indirectly by the studs through the top and bottom plates to which they are attached. The sound-insulating value of a stud partition may be decreased, rather than increased, by using a filling material between the studs. If the filling material is elastic and exerts pressure against the layers of lath and plaster, it may improve the sound-insulating properties of the partition.

A double or cavity masonry wall is more effective in sound insulation than a single wall of the same weight, but fillers placed in the intervening space seem to have little value. Partitions constructed of 3- or 4-in. hollow tile with plaster applied directly to the tile may be too light to give satisfactory sound insulation. If furring strips are fastened to the tile, a waterproof paper is placed over the furring strips to cut off any possible contact of the plaster with the tile, and lath and plaster are then applied to the furring strips, the insulating properties of the partition are increased considerably. Experiments indicate that the method used in fastening the furring strips to the tile is of little importance.

The sound insulation of a concrete floor can be improved by using a *floating floor* of wood and a suspended ceiling. The method of attaching the nailing strips seems to be of little importance, but rigid hangers should not be used for the suspended ceiling. Flexible supports such as springs or wires are satisfactory.

Impact noises caused by walking or moving furniture or by a direct transfer of vibration from machines and musical instruments, such as pianos and radios, form another class of noise which is more difficult to insulate than airborne noise. A machine often sounds as noisy in the room below as in the room where it is located. A so-called floating floor is sometimes built by laying a rough subfloor on wood joists and over this placing a layer of fiberboard which supports a finished wood floor nailed through the fiberboard to the rough floor. Experiments show that the fiberboard, laid in this manner, has no sound-insulating effect. A floor constructed in a similar manner to that just described, but with nailing strips above the fiberboard to receive the full length of the nails holding the finished flooring, is much more effective. The method of fastening the nailing strips is not very important. They can be nailed every 3 or 4 ft. or can be held in position by straps, springs, or small metal chains containing felt. Conversation is not audible through such a floor, but it is not effective in reducing impact noises such as those caused by footsteps.

Floors constructed with separate wood joists for the floor and ceiling below do not give experimental results quite as good as the floating floor. A floating floor added to this construction is very satisfactory as far as airborne noises are concerned, but not as satisfactory in reducing impact noises.

Impacts applied directly to a masonry floor are almost as audible in the room below as in the room where they are applied. A floating floor results in decided improvement, and a suspended ceiling gives still further improvement. The reduction in airborne noise is better than that for impact noise, but the latter noise is much less than that for a concrete slab alone. Concrete construction with a floating floor and a suspended ceiling gives better results than a wood floating floor with floor and ceiling poists separated as described.

Soft and yielding floor coverings act as cushions in reducing the transmission of impact noises produced by walking on the floor or from other causes. Heavy carpet on a pad is very effective, and rubber or cork tile are beneficial but to a much lesser degree.

The noise level in a room caused by outside noises can be reduced by increasing the total absorption units in the room, but this reduction is not large. A much greater reduction can usually be obtained at less cost by increasing the sound insulation of the walls, ceiling, and floors of the room. Absorbent materials are necessary to keep down the noise level resulting from noises originating in the room. Absorbent materials in corridors prevent them from acting as speaking tubes transmitting sound from one room to another when the doors are open.

The *masking effect* due to other noises is important. If a room is located in a quiet area, it may be possible to hear sounds clearly from an adjoining room, but if the room is located where the sound level is high, very little may be heard.

Airborne machinery noises are usually much smaller than those caused by the vibration of the foundation or other support for the machinery. The noise caused by the vibration of the support can be reduced by placing machines on layers of cork, asbestos, rubber, or felt, and sometimes by mounting them on springs.

Values for the transmission losses in decibels through many types of partition, floor, and ceiling construction are given in references 2, 3, and 4. The losses through various common types of construction are given in Table 18-5 for illustrative purposes [3]. Larger losses can be achieved by special types of construction as have been described and more fully considered in reference 3. The addition of acoustical materials has little effect on transmission losses because of the light weight of these materials.

97. ACOUSTIC MATERIALS

Although several materials serve acoustic purposes exclusively, acoustic materials also have other functions. They may be heat insulating materials, or finished ceiling materials. Acoustic problems in most buildings are concerned with noise abatement, so the vast majority of acoustic materials are for this purpose. Actual acoustic problems which address the design of a building to enhance hearing are usually confined to auditoriums and concert halls. This is a science and art which is better left to the experts. This treatise is confined primarily to the acoustic problems concerning the abatement of noise.

Many acoustic problems can be partially alleviated by careful planning of the building. The sources of noise may be classified as those that are brought in from without, or those that are generated within the structure. Most acoustic noise problems from within are generated by mechanical equipment. Careful selection, placement and correct mounting of mechanical equipment on isolation pads can abate sources of noise. Metal duct work, which can transmit noise, should have several expansion joints with flexible materials that will not transmit sounds. Walls and ceilings of toilet rooms should be insulated to isolate objectionable noise caused by flushing toilets and

TABLE 18-5

Transmission Losses in Decibels for Various Constructions [2, 3]

Construction	Weight	Loss
Partitions		
Load-bearing structural clay tile		
⅝ in. sanded gypsum plaster both sides		
1. 8 × 12 by 12 in. 6 cell	48.0	44
2. 6 × 12 by 12 in. 6 cell	39.0	42
3. 4 × 12 by 12 in. 3 cell	29.0	40
4. 3 × 12 by 12 in. 3 cell	28.0	36
Hollow cinder block, ⅝ in. gypsum plaster both sides		
5. 4 × 8 by 16 in.	35.8	44
6. 3 × 8 by 8 in.	32.2	42
7. 2 × 4 in. wood studs 12 in. c. metal lath, ⅞ in. gypsum plaster both sides	20.0	38
8. Same except lime plaster	19.8	44
9. 2 × 4 in. wood studs 16 in. c., ⅜ in. gypsum lath ½ in. gypsum plaster both sides	14.2	42
10. 2 × 4 in. wood studs 16 in. c., ⅝ in. tapered edge gypsum wall board. Joints taped	7.2	37
11. 2 × 4 in. wood studs 16 in. c., ½ in. wood fiberboard both sides, joints filled	5.1	24
12. Same with ½ in. gypsum plaster both sides	13.3	47
13. ¾ in. steel channels 12 in. c. metal lath one side, gypsum plaster both sides, solid 2 in. thick	16.4	34

surging water. Most noise abatement efforts are made by using ceiling materials which absorb the more common sounds. While fiber carpets are used because they have marked acoustic absorptive ability in addition to presenting a comfortable and attractive walking surface, they are now widely used in public buildings because custodial firms usually charge less for cleaning carpeted floors than floors covered with resilient tile.

Acoustic Ceiling Materials. The most widely used acoustic ceiling materials are *acoustic tiles* and metal *pans* or *panels*. These materials are made in several types and sizes. Tiles and panels are of fibered glass, vegetable cane fibers, mineral fibers, wood fibers, asbestos fibers, and kiln fired ceramics. Mineral fiber tile are noncombustible.

Generally, tile faces are usually patterned with a *perforated* design of small holes of various sizes and arrangements, or they have a face

Table 18-5 (*continued*)

14. ¾ in. steel channels 16 in c. perforated gypsum lath one side, gypsum plaster both sides, solid 2 in. thick, ¾ in. channels	19.4	31
15. 12 in. brick, not plastered	121.0	53
16. 8 in. brick, ⅝ in. gypsum plaster both sides	97.0	49
Floors		
1. 2 × 8 in. wood joists 16 in. c. Ceiling-metal lath ⅞ in. plaster Subfloor ¹³⁄₁₆ in., finish floor ¹³⁄₁₆ in. oak	17.1	34
2. 4 in. reinforced concrete slab	53.4	45
3. 6 × 12 by 12 in., 3-cell hollow tile 18 in. c., 6 in. concrete joists between tile, 2 in. slab	83.0	47
4. Same as 3 except 2 in. cinder concrete and 1 in. topping on floor side	109.0	48
Single Sheets		
1. ¼ in. 3 ply plywood	0.78	20
2. ½ in. wood fiberboard	0.75	20
3. ⅛ in. double-strength glass	1.60	27
4. ¼ in. plate glass	3.50	31

Key: Values in decibels for frequency of 512 cps. Weights in pounds per square foot. Expanded metal lath. Sanded plaster.
c.=center-to-center.

pattern that is referred to as *fissured*, which is supposed to resemble natural travertine rock.

The metal pan type of tile are of aluminum, stainless steel, or steel. Some of the metal pans are covered with a plastic film. Most metal panels are perforated with small holes made in various sizes, patterns, and concentrations. These pans are usually 12 in. or 24 in. wide and in several lengths intended to fit a building module. The pans are frequently covered, from the top, with a batt of fibered glass which serves as a heat insulation blanket as well as for its acoustic properties.

Perforations in the panels often serve another purpose. In some ceiling systems, the space between a suspended ceiling and the floor above forms a plenum to conduct air. Thus, the perforations in the tile serve as vents for forced air to circulate into the room. These tile are called *ventilating tile*. Panels are also made integrally with lighting luminaires.

Usually acoustic ceiling panels are installed in frameworks of various light gage metal specially designed to hold the pans or tile. Panels that are set into a Tee-like suspended frame are called *lay-in* ceilings or lay-in systems. The fibered tiles can also be attached to wood splines by stapling, nails, or screws, or adhesive.

Aluminum sheets that are corrugated or deformed with a crimp are perforated with small, closely spaced holes. These sheets are more than 3 ft. wide and up to 5 ft. long. They are suspended by some mechanical method. Loose sound attenuation blankets made of fibered glass bats can be placed on top of these sheets thereby improving their acoustic absorptive ability.

Lead sheets are used for sound barriers in plenums. If the space above a suspended ceiling extends over more than one room, sound from one room may travel into the plenum and reverberate down into another room. This may be effectively deterred by constructing a wall or barrier in the plenum. Lead sheets about $\frac{1}{64}$ in. thick have been used for this purpose. They are suspended from the bottom of the floor above and located immediately over the wall that separates the rooms. The dense lead, serving as an efficient sound barrier, will stop most noise.

Acoustic Properties of Partitions.   The determination of sound transmission for walls is somewhat more rational than that for ceilings. Indices have been established which relate to the ability of a material or assembly of materials in a wall to resist penetration by sound. These index numbers also correspond to various common sounds or noises which are generated in a building. These index numbers, which relate to the performance of a material or assembly, are known as *sound transmission class*. They are usually called simply *STC ratings*.

If a material or assembly has a low ability to resist sound penetration, its STC rating is a low number. Correspondingly, materials and construction methods that have greater ability to resist sound penetration have higher STC ratings. For example, a 4 in. brick partition has an STC rating of 41, whereas a 12 in. brick wall has an STC rating of 54, a 12 in. lightweight concrete block has a rating of 51, and a dense reinforced concrete wall 12 in. thick has an STC rating of 56. Thus, not only is thickness of material a factor in the STC rating, but the density of the material is important.

From a practical point of view, sound isolation criteria have been developed which considers the type of occupancy and the room-to-adjacent-room sound isolation requirements, with a corresponding

STC number. For instance, walls between apartments (separate occupancy) have a higher STC rating than walls within the same apartment (same occupancy). See Ramsey-Sleeper, *Architectural Graphic Standards* [16].

Sound Absorbent Materials.   As stated by Paul E. Sabine [1]:

"Absorption of sound involves the dissipation in the form of heat of the vibrational energy of sound waves. Speaking generally, materials that are absorbent in any considerable degree are either *porous*, inelastically *flexible* or inelastically *compressible*, or they may possess two or more of these properties in varying degrees. In porous absorbent materials, the pores are intercommunicating and penetrate the surface. The alternating pressure in the sound wave forces the air particles into the narrow channels of the pore structure where their vibrational energy is dissipated by the viscosity of the air and the friction against the walls of the channels. Sealing the surface of such a material may decrease in considerable degree the sound absorbing efficiency. Felts, fabrics and fibrous materials of vegetable and mineral fiber absorb sound largely by virtue of their porosity and to a certain extent because of their inelastic flexibility and compressibility. Hard nonyielding absorbents owe their absorbent properties entirely to their porosity. Fibrous wallboards with an impervious surface and plywood owe what absorbent properties they have to their forced, inelastic flexural vibration under the alternating pressure of the sound waves at their surface.

Various means have been found of increasing the absorbent coefficients of commercial absorbents as by slotting, perforating, fissuring or otherwise providing small apertures into the body of the materials. The mechanics of this effect is not completely understood. Depth, diameter and distribution of the holes over the surface of the material have been found to have an important effect on the sound absorbing efficiency. An important property of absorbents of this type is that painting does not, to any measurable degree, decrease the absorbing efficiency so long as the paint does not clog the holes.

It has also been found that covering the surface of a porous material with a thin perforated screen of metal or other hard material produces a negligible effect on its sound absorbing efficiency. The explanation lies in the fact that a thin membrane of this type in which the perforated area may be as small as 10 per cent of the total area transmits practically 100 per cent of the sound energy to the absorbent back of it. At high frequencies, say 2000 cps, the effect of the perforated screen is measurable."

The *absorption coefficient* of a material is equal to the proportion of the sound energy it absorbs to that which strikes it. The absorption coefficient of an open window is 1.00.

Absorption coefficients for various common materials are given in Table 18-6 for illustrative purposes.

Many kinds of special acoustical materials are available with absorption coefficients varying from 0.4 to 0.99 [6]. They include tile made up of various materials such as felted or perforated wood or mineral fiber, thin perforated metal or asbestos-cement board facing backed with a mineral wood sound absorbing pad, and mineral wool with fissured surface. Surface coatings, consisting of mineral fibers with a binder, which can be sprayed on and granular aggregates with

TABLE 18-6
Absorbent Coefficients for Various Common Materials [6]

Item	Coefficient
Open window	1.00
Brick wall, painted	0.017
unpainted	0.03
Plaster, gypsum, or lime, smooth finish on tile or brick masonry	0.025
Same as above on lath	0.03
Plaster, gypsum, or lime, rough finish on lath	0.06
Wood panelling	0.06
Glass	0.027
Marble or glazed tile	0.01
Concrete and terrazzo floors	0.015
Wood floors	0.03
Linoleum, asphalt, rubber, or cork tile on concrete	0.3–0.8
Carpet, unlined	0.20
Carpet, felt-lined	0.37
Fabric curtains	
Light, 10 oz. per sq. yd., hung straight	0.11
Medium, 14 oz. per sq. yd., hung straight	0.13
Heavy, 18 oz. per sq. yd., draped	0.50
Audience seated, units per person depending on character of seats, etc.	3.0–4.3
Chairs, metal or wood, each	0.17
Theater and auditorium chairs, each	
Wood veneer seat and back	0.25
Upholstered in leatherette	1.6
Heavy upholstered in plush or mohair	2.6–3.0

Values for 512 cps.

portland cement, lime, or gypsum binders which can be troweled on, are available.

The thicknesses of most types of tile vary from ½ in. to 1 in., not including an air space and furring if these are provided. Some tile are as thick as 2 or 3 in. without air space or furring. Tile may be square or rectangular. A common size of square tile is 12 × 12 in. and of rectangular tile 12 × 24 in., but larger sizes are also available. Perforated tile have holes from ⅛ in. to ³⁄₁₆ in. in diameter, usually spaced about ½ in. each way, although some have a random spacing.

Tile may be attached in various ways to the surface over which they are applied including cementing, nailing to wood furring strips spaced in accordance with the tile sizes and with or without mineral wool between the strips, or mounting on special metal supports with or without furring strips.

Detailed information about these materials is given in reference 6, and illustrations and brief descriptions in references 13 and 14.

Ordinarily the ceilings of rooms are the most feasible locations for the installation of acoustical materials. If rooms are wide and have low ceilings, side-wall installation locations may be preferable to ceiling locations.

Echoes. When the reflection of a sound is heard as a distinct repetition of the original sound, the reflection is called an *echo.* For an echo to be formed, the time difference between the two sounds must be at least $\frac{1}{20}$ of a second. If the difference is smaller than this, the reflected sound merely reinforces the original sound. Echoes do not occur in small rooms. Successive repetitions of the same sound by reflections from several surfaces are called *multiple echoes.*

Build-Up and Decay of Sound. When a constant source of sound is introduced into a room, each successive wave it produces spreads in all directions. As it strikes the various surfaces in the room, it is partially absorbed and partially reflected from the several surfaces it strikes. The average intensity of the sound builds up to a maximum at which it continues because the rate of introduction of sound energy equals the rate of absorption by the enclosing surfaces and the contents of the room. If the sound source is cut off, the sound in the room does not cease immediately but gradually decays because of absorption. This phenomenon is illustrated in Figure 18.2

Reverberation. Sound which continues in a room after its source has been cut off is called *reverberation.* It may be considered as a series of multiple echoes of decreasing intensity which are so closely

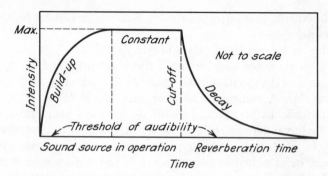

FIGURE 18.2   Build-up, continuation, and decay of sound.

spaced that they produce a continuous sound. Reverberation tends to cause spoken syllables to overlap and become indistinct and individual musical notes to be prolonged. The *reverberation time* is the period included between the cut-off time and the time when the sound becomes inaudible. A quantitative definition for reverberation time and specific procedures for its measurement are used in acoustical studies. Reverberation time is illustrated in Figure 18.2.

The reverberation time for existing rooms can be established by experiment under standardized conditions. W. C. Sabine developed the following formula, based on theoretical studies and many experiments.

$$T = 0.5 \frac{V}{A} \qquad \text{or} \qquad A = 0.5 \frac{V}{T}$$

T is the reverberation time in seconds for 512-cycle sounds, V is the volume of the room in cubic feet, and A is the total of the absorption units in the room and is equal to the sum of the products of each absorbing area in the room and its absorption coefficient. Appropriate allowances are made for occupants and for seats and other furnishings including carpets, window draperies, etc., such as those included in Table 18-6.

This formula is extensively used in acoustical design but has limitations. Other more precise but more complex procedures are also used. From this formula, the value of A which must be achieved by installing special sound absorbing materials to yield the desired reverberation time can be computed for a given value of V.

Auditoriums. The reverberation time of a large room such as an auditorium, assembly hall, concert hall, or theater is of paramount importance. The most desirable or *optimum reverberation time* for a room depends upon the purposes for which it is to be used. It is longer for music than for speech. It also depends upon the volume of the room, which increases with the size of the room.

The acoustical properties of rooms used for speech may be impaired by long reverberation times, but such rooms with very short reverberation times seem dead and unnatural because the ear is accustomed to some reverberation. The acoustical properties of a room used for music may be impaired by reverberation times which are too short or too long. Optimum reverberation times for various conditions have been established by measuring the reverberation times of rooms of various sizes used for various purposes and which are considered by trained observers to have good acoustical properties. One factor which must not be overlooked is the size of the audience for which the time is to be selected because audiences account for much of the sound absorbed.

Reverberation times recommended by the Acoustical Materials Association may be obtained from Figure 18.3 [1]. In this figure, the volumes are plotted on a logarithmic scale and the reverberation times on a linear scale. The shaded area represents acceptable reverberation times for various room sizes and a frequency of 512 cycles per second. For moving picture theaters or auditoriums which have public address

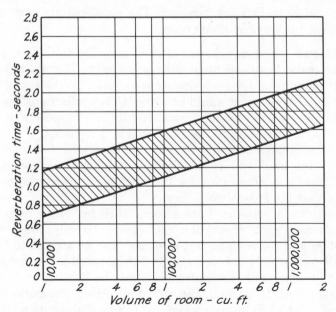

FIGURE 18.3 Optimum reverberation times [1].

systems, the reverberation times should fall near the lower limit of the shaded area and for churches and concert halls, they should fall near the upper limit. It will be noted that for a room with a volume of 10,000 cu. ft., the optimum reverberation times vary from 0.88 to 1.16 seconds and for a room with a volume of one million cubic feet the variation is from 1.5 to 2.0 sec. The seating capacities for rooms with these volumes depend upon their ceiling heights, whether or not there are balconies in the larger rooms, and other factors, but they are of the order of magnitude of 80 to 100 and 3000 to 4000 respectively.

Another factor is the shape of the enclosing surfaces. Concave surfaces such as curved rear walls and domed ceilings tend to focus reflected sound waves to produce concentrations of sound in certain areas which result in annoying echoes and loss of intelligibility of speech. Since convex surfaces reduce the intensity of reflected sound and do not cause echoes, they are not objectionable and may be desirable.

Consideration should be given to reducing the sounds transmitted into the auditorium from outside sources. The size and proportions of a room are important. It has been recommended that the volume of a room in cubic feet should not be greater than two hundred times the number of seats [1]. The presence of a stage or balcony requires special consideration.

Many other factors are involved. The problems are very complex and require the services of an expert.

Work Rooms.   The significance of good acoustics to the occupants of building space has been mentioned previously. The multiple reflections of sound from the enclosing surfaces build up the sound level in a room far above what it would be without such reflections. Increasing the total absorption units in a room results in a proportionate reduction in the contributions to the noise level by reflections.

Except in small rooms, reverberation in untreated rooms may interfere with the intelligibliity of speech, including telephone conversations. When the occupants raise their voices to overcome this effect, they increase the general noise level.

Reductions in noise level are accomplished by means of acoustical materials with high absorption coefficients. Usually these need be applied only to the ceilings, but they may also be applied to the walls of small rooms with low ceilings. It is not possible to compute the amount of acoustical material which must be applied to accomplish a desired result. In general, the use of absorbent materials for office quieting is not likely to be overdone [1]. Sound absorption treatment is usually required in nearly all parts of school buildings and restaurants and in library reading rooms and in rooms used for many other purposes. Corridors often require special consideration, even though other parts of a building may not require treatment.

"Acoustical treatment of offices, banks, schools, etc. often reduces the noise level 6-db. compared with that before treamtent. A 6-db. reduction is equivalent to 30 to 50% noise reduction judged by the human ear" [14].

Acknowledgments. This article is based largely on the reports by V. L. Chrisler prepared for the National Bureau of Standards, references 2 and 3, and a bulletin prepared by Paul E. Sabine for the Acoustical Materials Association [1].

REFERENCES AND RECOMMENDED READING

1. Paul E. Sabine, *Theory and Use of Architectural Acoustical Materials*, Second Edition, Acoustical Materials Association.
2. V. L. Chrisler, *Sound Insulation of Wall and Floor Constructions*, National Bureau of Standards Report BMS 17, 1939, Supplement 1940.
3. *Sound Insulation of Wall and Floor Constructions*, National Bureau of Standards Report 144, 1955.
4. Vern O. Knudsen and Cyril M. Harris, *Acoustical Designing in Architecture*, John Wiley and Sons, 1950.
5. Richard H. Holt and Robert B. Newman, "Architectural Acoustics," included in *Architectural Engineering*, F. W. Dodge Corp., 1955.
6. *Sound Absorption Coefficients of Architectural Acoustical Materials*, Bulletin XII, Acoustical Materials Association, 1950.
7. *Fundamentals of Building Insulation*, Insulation Board Institute, 1950.
8. Tyler Stewart Rogers, *Design of Insulated Buildings for Various Climates*, F. W. Dodge Corp., 1951.
9. *Heating, Ventilating and Air-Conditioning Guide*, American Society of Heating and Ventilating Engineers, published annually.
10. *Sweet's Architectural Catalog File*, F. W. Dodge Corp., published annually.
11. Donald J. Vild, *Principles of Heat Transfer Through Glass Fenestrations*; Alfred L. Jaros, Jr., *Design for Solar Heat Gain and Loss; Windows and Glass in the Exterior of Buildings*, Publication 478, Building Research Institute, National Research Council, 1957.

12. G. R. Munger, "There Are Six Ways to Bring Down the Cost of Air Conditioning", *Refrigerating Engineering and Air Conditioning*, October 1957, p. 46.

13. C. W. Glover, *Practical Acoustics for the Constructor*, Chapman and Hall, Ltd., 1933.

14. G. W. Handy, 'Acoustics,' Section 17, *Building Construction Handbook*, Frederick S. Merritt, Ed., McGraw-Hill, 1958.

15. *Sweet's Catalog*, 1971 Edition, F. H. Dodge Corp., New York, New York.

16. Charles G. Ramsey and Harold R. Sleeper, *Architectural Graphic Standards*, Sixth Edition, John Wiley and Sons, 1970.

INDEX

Page numbers in parentheses indicate tables; page numbers in brackets illustrations.

A A, 4
AA quality glass, 612
Abbreviations, lumber, (241)
Abrasion, 192
Abrasive aggregate, 517
  resistance of rock, 198
Absorbed water, 228
Absorption coefficient, 724, (720)
  of sound, 711
Abutment of arch, 162, 163
Accelerating admixture, 400
Accordion doors, 628, [629]
ACI, 3, 409, 411
  Building Code Requirements for Reinforced Concrete, 3
  318, 3
  318-71, 414, 416
Acid, muriatic, 190, 516
  open-hearth process, 316
  resistance of flooring, 529
Acoustic filters, 712
  materials, installation, 721
  plaster, 555
  tiles, 716
Acoustical Materials Association, 723, 725
Acoustics, 707-725
  domes, 464
Acrylic plastic, corrugated sheets, 583
  sheet, 687, 688
Acrylics, 689, (688)
Acrylite, 689
Actinic glass, 613
Action, pile, 93
Actual size, lumber, 227
Acute arch, 165
Additives, 399; *see also* Admixtures
AD lumber, 227
Adhesive, 535, 536, 686
  dry-use, 287

epoxy, 536
organic, 536
wet-use, 288
Adjustable brackets, 659, [660]
Admixtures, 399, 402
  classified, 401
  functions, 400
  retarding, 407
Adobe, 39
  brick, 173
Adsorbed water, 34
Advancing of caisson, methods, 119
  pneumatic caisson, 123
Aesthetic requirements, joints, 182
Agent, bonding, 543
  liquid, 545
  curing, 691, 692
  reducing, 666, 667, 668, 669
Aggregate, 31, 171, 396, 398-399, 407, 538, 543, 720
  abrasive, 517
  colored, 516
  decorative, 517
  fine, 170
  inert, 518
  particles, 403
  porous, 555
  preplaced, 402, 409
  surfaces, 409
Aging, lime, 540
Agitator, 405
A grade, plywood, 281
Air conditioning, 698
  dried lumber, 227, 237
  entrained cements, 397
  entraining admixtures, 401
  entrainment, 543
  hammer, 85
  infiltration, 696, 697

locks, on open caisson, 121
  of pneumatic caisson, 122
  removal, 123
  in plastic manufacture, 683
  -slaked quicklime, 540
Airborne sounds, transmission, 712
A.I.S.C., 3
  specifications, 334
  structural steel classification, (329)
A.I.S.I., 3
  stainless steel types, 672, 673
A.I.T.C., 4
Alabaster, 540
Alkali resistance of flooring, 529
Alkyds, table, 688
Allowable bearing pressure, 41, 63, 70, 71, (64)
  pressures, code values, 63
  soil bearing value, 66
  pressure, 66, 71, 96
  selection, 63-64
  for underlying stratum, 64
  stresses, 26
Alloys, 681, 682
  aluminum, 675, 676
  copper, 110, 677, 678
Alloy sheet-metal roofing, 578
  steel, 317
  bolts, 335
  high-strength, 474
  zinc, 545
Alluvial soil, 30
Alternate freezing and thawing, 192
Alumina, 396, 667, 674, 675
Aluminum Association, 4, 676
Aluminum, 674-676